W0227726

Cambridge Studies in Biological and Evolutionary Anthropology 37

Patterns of Growth and Development in the Genus *Homo*

It is generally accepted that the earliest human ancestors grew more like apes than like humans today. If they did so, and we are now different, when, how, and why did our modern growth patterns evolve? This book focuses on species within the genus *Homo* to investigate the evolutionary origins of characteristic human patterns and rates of craniofacial and postcranial growth and development, and to explore unique ontogenetic patterns within each fossil species. Experts examine growth patterns found within available Plio-Pleistocene hominid samples, and analyze variation in ontogenetic patterns and rates of development in recent modern humans in order to provide a comparative context for fossil hominid studies. Presenting studies of some of the newest juvenile fossil specimens and information on *Homo antecessor* – the newest species assigned to the genus – this book will provide a rich data source with which anthropologists and evolutionary biologists can address the questions posed above.

JENNIFER L. THOMPSON is a paleoanthropologist and Associate Professor in the Department of Anthropology and Ethnic Studies at the University of Nevada, Las Vegas. Her research interests include paleoanthropology, paleobiology, hominid growth and development, human origins, skeletal biology, odontology, and morphometrics.

GAIL E. KROVITZ is a postdoctoral research associate in the Department of Anthropology at Pennsylvania State University. A paleoanthropologist, she pursues research interests in craniofacial morphology and growth patterns, the origins of modern humans, modern human skeletal variability, and functional morphology of the hominid skeleton.

ANDREW J. NELSON is an Associate Professor in the Department of Anthropology at the University of Western Ontario. He is a paleoanthropologist and bioarchaeologist, and his research is centered in two areas: hominid growth and development, and bioarchaeology on the north coast of Peru, but he also has interests in broad applications of analytical techniques and in the exploration of the interface between science and the media.

Cambridge Studies in Biological and Evolutionary Anthropology

Series Editors

HUMAN ECOLOGY
C. G. Nicholas Mascie-Taylor, University of Cambridge
Michael A. Little, State University of New York, Binghamton
GENETICS
Kenneth M. Weiss, Pennsylvania State University
HUMAN EVOLUTION
Robert A. Foley, University of Cambridge
Nina G. Jablonski, California Academy of Science
PRIMATOLOGY
Karen B. Strier, University of Wisconsin, Madison

Patterns of Growth and Development in the Genus *Homo*

EDITED BY

J. L. THOMPSON
University of Nevada, Las Vegas

G. E. KROVITZ
Pennsylvania State University

A. J. NELSON
University of Western Ontario

CAMBRIDGE
UNIVERSITY PRESS

CAMBRIDGE UNIVERSITY PRESS
Cambridge, New York, Melbourne, Madrid, Cape Town, Singapore,
São Paulo, Delhi, Dubai, Tokyo, Mexico City

Cambridge University Press
The Edinburgh Building, Cambridge CB2 8RU, UK

Published in the United States of America by Cambridge University Press, New York

www.cambridge.org
Information on this title: www.cambridge.org/9780521184229

First published 2003
First paperback edition 2010

A catalogue record for this publication is available from the British Library

ISBN 978-0-521-82272-5 Hardback
ISBN 978-0-521-18422-9 Paperback

Contents

Contributors

Susan C. Antón
Center for Human Evolutionary Studies
Department of Anthropology
Rutgers, The State University of New Jersey
131 George Street
New Brunswick, NJ 08901, USA
scanton@rci.rutgers.edu

José Maria Bermúdez de Castro
Museo Nacional de Ciencias Naturales (CSIC)
C/José Gutiérrez Abascal 2
28006 Madrid, Spain
mcnbc54@mncn.csic.es

Barry Bogin
Behavioral Sciences and Social Sciences
University of Michigan–Dearborn
Dearborn, MI 48128, USA
bbogin@umd.umich.edu

Helen Coqueugniot
UMR 5809
Laboratoire d'Anthropologie des Populations du Passé
Université Bordeaux 1
Avenue des Facultés
33405 Talence Cedex, France

Laurie R. Godfrey
Department of Anthropology
University of Massachusetts
Machmer Hall
240 Hicks Way
Amherst, MA 01003, USA
lgodfrey@anthro.umass.edu

Louise Humphrey
Human Origins Group
Department of Palaeontology
The Natural History Museum
Cromwell Road
London SW7 5BD, UK
lth@nhm.ac.uk

Hajime Ishida
Department of Anatomy
Faculty of Medicine
University of the Ryukyus
Okinawa 903-0215, Japan

Osamu Kondo
Department of Biological Sciences
Division of Anthropology
Graduate School of Science
University of Tokyo
Bunkyo-ku
Tokyo 113-0033, Japan
kondo-o@ss.iij4u.or.jp

Gail E. Krovitz
Department of Anthropology
Pennsylvania State University
University Park, PA 16802, USA
gek10@psu.edu

Kevin L. Kuykendall
School of Anatomical Sciences
University of the Witwatersrand
7 York Road
Parktown
Johannesburg 2193, South Africa
kuykendallkl@anatomy.wits.ac.za

Steve Leigh
Department of Anthropology
University of Illinois
Urbana, IL 61801, USA
s-leigh@staff.uiuc.edu

Dan Lieberman
Department of Anthropology
Harvard University
Peabody Museum
11 Divinity Avenue
Cambridge, MA 02138, USA
danlieb@fas.harvard.edu

Helen Liversidge
Department of Paediatric Dentistry
Royal London Hospital
Turner Street
London E1 2AD, UK
h.m.liversidge@qmul.ac.uk

Tona Majó
UMR 5809
Laboratoire d'Anthropologie des Populations du Passé
Université Bordeaux 1
Avenue des Facultés
33405 Talence Cedex, France

Maria Martinón-Torres
Museo Nacional de Ciencias Naturales (CSIC)
C/José Gutiérrez Abascal 2
28006 Madrid, Spain

Brandeis McBratney-Owen
Harvard University
Harvard Medical School
Department of Cell Biology
240 Longwood Avenue
Boston, MA 02115, USA
mcbratn@fas.harvard.edu

Nancy Minugh-Purvis
Department of Cell and Developmental Biology
University of Pennsylvania School of Medicine
421 Curie Boulevard
Philadelphia, PA 19104, USA
ttt@tradenet.net

Paul O'Higgins
The Hull York Medical School and Department of Biology
University of York
Heslington
York YO10 5DD, UK
paul.ohiggins@hyms.ac.uk

Andrew J. Nelson
Department of Anthropology
University of Western Ontario
London, ON N6A 5C2, Canada
anelson@uwo.ca

Fernando V. Ramírez Rozzi
UPR 2147 – Centre National des Recherches Scientifiques
44 rue de l'Almiral Mouchez
75014 Paris, France
ramrozzi@ivry.cnrs.fr

Antonio Rosas
Museo Nacional de Ciencias Naturales (CSIC)
C/José Gutiérrez Abascal 2
28006 Madrid, Spain

S. Sarmiento Pérez
Museo Nacional de Ciencias Naturales (CSIC)
C/José Gutiérrez Abascal, 2
28006 Madrid, Spain

Una Strand Viðarsdóttir
Evolutionary Anthropology Research Group
Department of Anthropology
University of Durham
43 Old Elvet
Durham DH1 3HN, UK
una.vidarsdottir@durham.ac.uk

Michael R. Sutherland
Statistical Consulting Center
Lederle Graduate Research Tower
University of Massachusetts
Amherst, MA 01003, USA
mikes@math.umass.edu

Jennifer L. Thompson
Department of Anthropology and Ethnic Studies
University of Nevada, Las Vegas
Las Vegas, NV 89154, USA
thompsoj@unlv.edu

Anne-marie Tillier
UMR 5809
Laboratoire d'Anthropologie des Populations du Passé
Université Bordeaux 1
Avenue des Facultés
33405 Talence Cedex, France
am.tillier@anthropologie.u-bordeaux.fr

Frank L'Engle Williams
Department of Anthropology and Geography
Georgia State University
33 Glimer Street
Atlanta, GA 30303, USA
antflw@langate.gsu.edu

Acknowledgments

This volume is based on the 2001 American Association of Physical Anthropologists symposium, "Patterns of growth and development in the genus *Homo*." Thanks to all the original participants in the symposium, including Leslie Aiello who served as a discussant for the session. We sincerely thank our colleagues who contributed to this volume (including Kevin Kuykendall, who did not participate in the original symposium) for their willingness to take part in this project. We also would like to thank Dr Tracey Sanderson for her help in making this project a reality; Laura Petrella for assistance with final assembly of the manuscript; and Jennifer Omstead for the cover art. The cover photo of Dederiyeh 2 appears courtesy of Prof. Akazawa, Dr Kondo, and Dr Ishida; the Le Moustier 1 reconstruction is courtesy of J. L. Thompson; and the photo of La Chapelle aux Saints is courtesy of A. J. Nelson. J. L. Thompson would like to acknowledge the support of the Department of Anthropology at UNLV. G. E. Krovitz would like to thank Dan Lieberman and Alan Walker for their support at different stages of this project. A. J. Nelson would like to acknowledge the Department of Anthropology at The University of Western Ontario, and Chris Nelson.

1 *Introduction*

G. E. KROVITZ
Pennsylvania State University

A. J. NELSON
University of Western Ontario

J. L. THOMPSON
University of Nevada, Las Vegas

Background

The field of paleoanthropology has traditionally concentrated on the adult form of our fossil predecessors. This work has led to detailed insights in the fields of phylogeny, biomechanics, function, and environmental interactions, and has made enormous contributions to our understanding of hominid adaptation and evolution. There are many reasons for concentrating on adult morphology. First, most of the preserved fossils are adult individuals. Second, the adult form is relatively stable over many years of an individual's life, and thus represents a manifestation of the many evolutionary pressures acting on a particular individual and other members of the taxon. Lastly, there has been historical bias against the study of juvenile individuals (Johnston, 1968; Johnston & Schell, 1979; Johnston & Zimmer, 1989).

However, even the most detailed understanding of the adult form still gives an incomplete picture of the adaptation and evolution of our hominid predecessors. There are many reasons that the study of adult individuals alone is not sufficient. First, humans have a life-history pattern that includes a relatively (and absolutely) long juvenile period. Thus, if typical life expectancy among our recent hominid predecessors was only three or four decades (e.g., Bermúdez de Castro & Nicolas, 1997; Trinkaus, 1995; Trinkaus & Tompkins, 1990), less than half of the total life span of a typical individual would have been spent as an adult. As natural selection acts on individuals at every point throughout their lifespan, and not only on adults, it is necessary to understand how juvenile individuals were shaped by natural selection. Additionally, since individuals that die before

Patterns of Growth and Development in the Genus Homo, ed. J. L. Thompson, G. E. Krovitz, and A. J. Nelson. Published by Cambridge University Press. © Cambridge University Press 2003.

1

reproducing do not pass on their genes, natural selection on juvenile individuals can have profound evolutionary consequences (Frisancho, 1970). Second, much of the debate regarding hominid phylogeny centers on the interpretation of morphological characters and their use as taxonomic markers. Determining what ancestral and derived characters are present due to underlying differences in the genetic program and what characters arise through epigenetic interactions or functional and behavioral adaptations is of central importance in this process. The latter set of characters may well be shared across taxa, irrespective of their phylogenetic relationships, while characters with a genetic basis are more useful in taxonomic studies. It is therefore essential to consider the functional and developmental basis of a trait to determine if it would be useful in phylogenetic analysis (Lieberman, 1995, 2000). Additionally, since growth is the process that creates variations in adult form (Howells, 1971; Johnston, 1968; Johnston & Zimmer, 1989), studying growth is necessary for understanding the basis of adult variation. Finally, a third reason why the study of adults alone is not sufficient is that morphological evolution frequently results from modifications of development (e.g., de Beer, 1958; Gould, 1977; Minugh-Purvis & McNamara, 2002; Montagu, 1955; O'Higgins & Cohn, 2000; Raff, 1996). Having a comparative understanding of both growth and development can therefore provide a mechanism for evolutionary change and for understanding morphological differences between species. Since growth and development are strongly influenced by environmental as well as genetic factors (Eveleth & Tanner, 1991; Hoppa, 1992; Johnston & Schell, 1979; Johnston & Zimmer, 1989; Lampl & Johnston, 1996; Saunders, 1992), studying ontogeny not only reveals genetic and evolutionary history, but can illuminate health and environmental factors as well.

Juvenile hominid fossils have always played an important role in paleoanthropology, as the type specimens for *Australopithecus africanus* (Taung 1: Dart, 1925) and *Homo habilis* (OH 7: Leakey *et al.*, 1964), and the first Neandertal ever discovered (Engis 2: see Schwartz & Tattersall, 2002) are all juveniles. However, increased awareness of the wealth of unique information available from ontogenetic studies has recently generated interest in using an ontogenetic perspective in the study of human evolution. Sparked by the conclusions of Mann (1975), the study of growth and development in Plio-Pleistocene early hominids took off in the 1980s, primarily with the examination of enamel microstructures and the application of other new methods to the study of dental development (e.g., Beynon & Dean, 1988; Bromage, 1987; Bromage & Dean, 1985; Conroy & Vannier, 1987; Dean, 1987; Smith, 1986; see review in Kuykendall, this volume). The use of enamel microstructures to assess dental developmental status was also applied to a juvenile Neandertal from Devil's Tower, Gibraltar (Dean *et al.*, 1986; Skinner, 1997; Stringer & Dean,

1997), which fueled further debate on cranial and dental developmental rates in Neandertals (e.g., Stringer *et al.*, 1990; Trinkaus & Tompkins, 1990). However, the lack (or scarcity) of juvenile fossils for most species in the genus *Homo* has precluded developmental research on most species in this genus, with the exception of the Neandertals, which are relatively plentiful and have been well studied (for reasons discussed in Trinkaus, 1990). In fact, the early work of Tillier (e.g., 1979, 1981, 1982, 1983a, 1983b, 1984, 1986a, 1986b, 1987, 1988, 1989, 1992) and Minugh-Purvis (1988) set the tone for later studies of craniofacial and skeletal growth and development on Later Pleistocene *Homo*. Discovery of the relatively complete juvenile *Homo erectus* KNM-WT 15000 (see papers in Walker & Leakey, 1993) made it possible (and necessary) to study cranial, postcranial and dental growth in *H. erectus*. The recent discovery of Lower and Middle Pleistocene material from two sites in the Sierra de Atapuerca, Spain (see papers in Arsuaga *et al.*, 1997; Bermúdez de Castro *et al.*, 1999) provides important information on growth and development in the temporal and evolutionary "gap" between *H. erectus*, Neandertals, and modern humans. And, of course, additional juvenile Neandertal remains continue to increase our knowledge about that important sample (e.g., Akazawa *et al.*, 1995; Golovanova *et al.*, 1999; Maureille, 2002; Rak *et al.*, 1994).

There are several difficulties that arise when studying any juvenile specimen, which are often magnified in the study of juvenile fossil hominid remains. First, some measure of developmental age must be calculated for that specimen, and it is assumed that this developmental age estimate is a proxy for the chronological age of the specimen (although even under the best of circumstances developmental age estimates rarely match actual chronological age exactly; see Lampl & Johnston, 1996). Dental development (primarily assessed through either tooth formation or eruption) is the favored method of age estimation (Johnston & Zimmer, 1989; Lewis & Garn, 1960; Smith, 1991; Ubelaker, 1989). Dental remains are well preserved in the fossil record, they provide an excellent measure of the timing and rate of development, and dental development ties closely in with other measures of life history (e.g., Smith, 1989a, 1989b). Even studies that examine cranial or postcranial remains usually rely on developmental age estimates based on the dentition.

Difficulties with age estimation can arise when different maturational systems (such as dentition and postcranial epiphyses) give different developmental age estimates, suggesting that postcranial growth may be accelerated or delayed relative to dental development, as is the case with the *H. erectus* individual KNM-WT 15000 (documented in Smith, 1993) and several Neandertals (Thompson & Nelson, 2000). The process of age estimation is also complicated when no dental remains are present, as with Mojokerto (see Antón, 1997) and La Ferrassie 6 (see Majó & Tiller, this volume, and Tompkins & Trinkaus,

1987). An additional, and more theoretical question, is whether dental (or skeletal) age estimates for earlier hominids are better determined through using human or ape dental reference standards. Certainly some hominid species are more appropriately modeled using ape standards, and some human standards, but we cannot know a priori which ones to use (see Smith, 1993 for example). Also, through normal biological variation, some individuals will be advanced, and some individuals delayed relative to the average growth patterns of that population or species (Johnston, 1968; Johnston & Zimmer, 1989; Saunders, 1992).

Another difficulty in the study of juvenile fossil hominid remains is the availability and interpretation of appropriate comparative ontogenetic samples. It is first necessary to find comparative samples that span the developmental age range being considered in the study, preferably with large enough numbers to capture the range of variation in the comparative species, and to allow statistical testing. Careful analysis is then necessary to determine if the morphological features under study are truly species specific, and not simply due to the young developmental age of the specimen. It is difficult not to interpret juvenile remains using adult standards of morphology, and to appreciate that younger specimens might have more subtle morphological features (Minugh-Purvis, 1988).

A final difficulty, that is present in any analysis of the fossil or archaeological record, is the possibility of mortality bias. Juvenile individuals from skeletal populations (i.e., individuals who died before reaching adulthood) might not be an adequate representation of healthy individuals from that population (Saunders & Hoppa, 1993; Wood et al., 1992). However, it is likely that error introduced by other factors, such as small sample size and unknown age and sex of juvenile individuals, exceeds that introduced by mortality bias (Saunders & Hoppa, 1993). Ultimately, problems like these plague many aspects of paleoanthropology or bioarchaeology. Careful attention to theoretical context, choice of methodology, and clear attention to maximizing available samples can minimize these confounding issues.

Rationale for (and layout of) this volume

A review of what is known about the patterns of growth and development expressed by different taxa within the genus Homo is timely. Several other excellent edited volumes have recently been published that dealt with the study of growth and development from an evolutionary or archaeological perspective. For example, the recent volume by Hoppa & FitzGerald (1999) assembled and integrated papers dealing with dental and skeletal growth from paleoanthropological and bioarchaeological perspectives. The volume edited by O'Higgins & Cohn (2000) explored the link between development and evolution, and presented methodology suitable for developmental analysis of vertebrate morphology. Finally, the volume edited by Minugh-Purvis & McNamara (2002) dealt primarily with heterochronic theory, the application of heterochronic methodology to the hominid fossil record, and the relationship of developmental change to aspects of hominid life history. The tremendous amount of activity within the field of hominid growth and development in the past 20 years shows the increasing importance of this analytical/theoretical approach in the field of physical anthropology. And yet, even with the large amount of new, and sometimes innovative, research being conducted in this area, there have been few attempts to synthesize what is known about developmental patterns during the later stages of human evolution. This book differs from those edited volumes mentioned above by focusing explicitly on growth and development in the genus *Homo* (or other genera that help put *Homo* in perspective). This book presents a synthesis of what is currently known about growth patterns in the genus *Homo* and explores what is unique about modern human ontogenetic patterns, and when and how those unique features evolved.

One of the key questions in hominid paleontology is *when did anatomically modern humans first appear?* There is widespread public and academic interest in the origins of modern *Homo sapiens*, and examination of this process from an ontogenetic perspective will shed new light on this issue. Thus, the question can be reformulated to ask *when did the modern human pattern of growth and development first appear?* It has been well established that the australopithecines demonstrated an ape-like pattern of growth and development (see review in Kuykendall, this volume). Thus, we must focus on the genus *Homo* in order to explore the origins of human patterns of growth and development.

One of the central goals of this volume is to address the question *when and how did the modern human pattern of growth and development first appear?* Before we can answer this question, we must first ask two other questions. First of all, *what unique aspects (such as elongated subadult period, adolescent growth spurt, etc.) are present in the extant modern human pattern of growth and development?* Then we must ask the related question, *what patterns of growth and development are demonstrated by our closest living hominoid relatives?* Answers to these questions will provide us with the broad evolutionary context necessary to understand how any fossil taxon may or may not conform to the modern human or hominoid patterns.

Next we must examine the hominid fossil record, as this is the only way to directly address the central questions of *when* and in *what fossil taxon* aspects of modern patterns of growth and development first appeared. Ultimately we want to assess *how* aspects of the modern pattern of growth and development first appeared, and what evolutionary mechanisms were behind those changes. Within this framework, we can formulate and test hypotheses derived from our central questions. For example, did the Neandertals demonstrate aspects of the modern human pattern of growth and development? And if so, what are the behavioral and/or evolutionary consequences of these similarities?

The papers in this volume address the questions outlined above. We have invited contributions from many of the leading scholars who are currently active in research on growth and development in the genus *Homo*. These contributors include new researchers as well as leading scholars who have already made substantial contributions to the field of hominid growth and development. We have sought to combine papers that are fundamentally data oriented, in order to provide the basic evidence, with more conceptually oriented papers, which seek to put the basic evidence in a larger context. Additionally, studies presented in this volume consider some of the most recently discovered juvenile fossil specimens, including the Neandertals Dederiyeh 1 and 2, and the Atapuerca material.

The book is organized into three parts, focusing first on studies of modern humans (to define the interpretive context); second, on the earliest evolutionary history of the genus *Homo*; and, third, on the Neandertal and early modern human fossil record. The strength of these papers lies in the fact that they focus on different aspects of the dentition, skull, and postcranial skeleton and represent the state of our knowledge of the patterns of growth in extinct members of the genus *Homo*. Another unique aspect of this book is the summary chapter at the end of each section. These chapters review the most important findings of the papers, and discuss them in the context of previously published research on the evolution of growth and development.

The first part of the book, "Setting the stage: what do we know about human growth and development?" addresses the two questions introduced above: *what patterns of growth and development are demonstrated by our closest hominoid relatives?* and *what is the pattern of growth and development demonstrated by modern humans?* Papers in this part outline the key differences in the pattern of growth and development between modern humans and non-human primates, and provide a comparative analysis of growth and maturation for different parts of the modern human skeleton (i.e., craniofacial, dental, and postcranial). Craniofacial growth is studied by McBratney & Lieberman, and by Strand Viðarsdóttir & O'Higgins, who examine facial positioning in humans and chimpanzees, and variation in facial growth in modern humans, respectively. Variation in modern human dental development is considered by Liversidge, while variation in postcranial growth in modern human archaeological samples is studied by Humphrey. The background presented in these papers provides the basic context for the examination of the fossil hominid species.

The second part of the book, "The first steps: from australopithecines to Middle Pleistocene *Homo*," examines what aspects of modern growth and development were present in each pre-Neandertal hominid fossil taxon. Kuykendall sets the stage for interpreting growth in the genus *Homo* by reviewing what is currently known about growth and development in the australopithecines. Antón & Leigh consider neurocranial growth in *H. erectus*, and the life-history implications of their results. Finally, Bermúdez de Castro and colleagues examine dental development in the hominids from Sierra de Atapuerca, Spain (representing *H. antecessor* and *H. heidelbergensis*). Thus, individual chapters directly examine the available juvenile skeletal and dental material, and document ontogenetic patterns within each Lower and Middle Pleistocene species in the genus *Homo* for which juvenile skeletal or dental material is present.

The third part of the book, "The last steps: the approach to modern humans," considers the Neandertal and early anatomically modern human fossil record. The Neandertals have been extensively studied, as they preserve the most complete ontogenetic sample of any fossil group. Williams and colleagues carry out a heterochronic study of the craniofacial skeleton in *Homo* and *Pan*, and Krovitz examines craniofacial shape differences and growth patterns in Neandertals and modern humans. The ontogenetic patterning and phylogenetic significance of mental foramen number and position is considered by Coqueugniot & Minugh-Purvis. Variation in long-bone dimensions and growth is considered by Kondo & Ishida, and pelvic morphology is considered by Majó & Tillier.

Our summary and concluding chapters integrate the major findings within each section and revisit our central question: *when, and in what mosaic pattern, did the modern human pattern of growth and development first appear?* We also examine when in the ontogenetic process particular taxonomic traits appear, and how these data contribute to the origin of our species.

A review and synthesis of the patterns of growth and development expressed by different taxa within the genus *Homo* is timely. There has been a tremendous amount of activity within the field of hominid growth and development in the last 10–15 years with no real synthesis focused on our genus. This activity has included both theoretical and empirical advances which now permit the detailed examination of such important concepts as neoteny and phylogeny – issues which have formed the focus of debate in our field for more than a century. It is our hope that professionals, students, and the interested lay public alike will find these papers of interest, and that the original and synthetic chapters will help provide direction for future research.

References

Akazawa, T., Muhesen, S., Dodo, Y., Kondo, O., & Mizoguchi, Y. (1995). Neanderthal infant burial. *Nature,* **377**, 585–586.

Antón, S. C. (1997). Developmental age and taxonomic affinity of the Mojokerto Child, Java, Indonesia. *American Journal of Physical Anthropology,* **102**, 497–514.

Arsuaga, J. L., Bermúdez de Castro, J. M., & Carbonell, E. (1997). The Sima de los Huesos hominid site. *Journal of Human Evolution, 33*.

Bermúdez de Castro, J. M., & Nicolas, M. E. (1997). Palaeodemography of the Atapuerca-SH Middle Pleistocene hominid samples. *Journal of Human Evolution, 33*, 333–355.

Bermúdez de Castro, J. M., Carbonell, E., & Arsuaga, J. L. (1999). Gran Dolina Site: TD6 Aurora Stratum (Burgos, Spain). *Journal of Human Evolution, 37*.

Beynon, A. D., & Dean, M. C. (1988). Distinct development patterns in early fossil hominids. *Nature, 335*, 509–514.

Bromage, T. G. (1987). The biological and chronological maturation of early hominids. *Journal of Human Evolution, 16*, 257–272.

Bromage, T. G., & Dean, M. C. (1985). Re-evaluation of the age at death of immature fossil hominids. *Nature, 317*, 525–527.

Conroy, G. C., & Vannier, M. W. (1987). Dental development of the Taung skull from computerized tomography. *Nature, 329*, 625–627.

Dart, R. A. (1925). *Australopithecus africanus*: The man–ape of South Africa. *Nature, 115*, 195–199.

de Beer, G. R. (1958). *Embryos and Ancestors*, 3rd edn. Oxford: Clarendon Press.

Dean, M. C. (1987). The dental developmental status of six East African juvenile fossil hominids. *Journal of Human Evolution, 16*, 197–213.

Dean, M. C., Stringer, C. B., & Bromage, T. G. (1986). Age at death of the Neanderthal child from Devil's Tower, Gibraltar and the implications for studies of general growth and development in Neandertals. *American Journal of Physical Anthropology, 70*, 301–309.

Eveleth, P. B., & Tanner, J. M. (1991). *Worldwide Variation in Human Growth*, 2nd edn. Cambridge: Cambridge University Press.

Frisancho, A. R. (1970). Developmental responses to high altitude hypoxia. *American Journal of Physical Anthropology, 32*, 401–408.

Golovanova, L. V., Hoffecker, J. F., Kharitonov, V. M., & Romanova, G. P. (1999). Mezmaiskaya Cave: A Neanderthal occupation in the Northern Caucasus. *Current Anthropology, 40*, 77–86.

Gould, S. J. (1977). *Ontogeny and Phylogeny*. Cambridge: Harvard University Press.

Hoppa, R. D. (1992). Evaluating human skeletal growth: An Anglo-Saxon example. *International Journal of Osteoarchaeology, 2*, 275–288.

Hoppa, R. D., & FitzGerald, C. M. (eds.) (1999). *Human Growth in the Past: Studies from Bones and Teeth*. Cambridge: Cambridge University Press.

Howells, W. W. (1971). Applications of multivariate analysis to craniofacial growth. In *Craniofacial Growth in Man*, eds. R. E. Moyers & W. M. Krogman, pp. 209–218. Oxford: Pergamon Press.

Johnston, F. E. (1968). Growth in the skeleton of earlier peoples. In *The Skeletal Biology of Earlier Human Populations*. Symposia of the Society for the Study of Human Biology no. 8, ed. D. R. Brothwell, pp. 57–66. Oxford: Pergamon Press.

Johnston, F. E., & Schell, L. M. (1979). Anthropometric variation of Native American children and adults. In *The First Americans: Origins, Affinities, and Adaptations*, eds. W. S. Laughlin & A. B. Harper, pp. 275–291. New York: Gustav Fischer.

Johnston, F. E., & Zimmer, L. O. (1989). Assessment of growth and age in the immature skeleton. In *Reconstructing Life from the Skeleton*, eds. M. Y. Iscan & K. A. R. Kennedy, pp. 11–21. New York: Alan R. Liss.

Lampl, M., & Johnston, F. E. (1996). Problems in the aging of skeletal juveniles: Perspectives from maturation assessments of living children. *American Journal of Physical Anthropology,* **101**, 345–355.

Leakey, L. S. B., Tobias, P. V., & Napier, J. R. (1964). A new species of the genus *Homo* from Olduvai Gorge. *Nature,* **202**, 7–9.

Lewis, A. B., & Garn, S. M. (1960). The relationship between tooth formation and other maturational factors. *Angle Orthodontics,* **30**, 70–77.

Lieberman, D. E. (1995). Testing hypotheses about recent human evolution from skulls: Integrating morphology, function, development, and phylogeny. *Current Anthropology,* **36**, 159–197.

Lieberman, D. E. (2000). Ontogeny, homology, and phylogeny in the hominid craniofacial skeleton: The problem of the browridge. In *Development, Growth and Evolution: Implications for the Study of the Hominid Skeleton*, eds. P. O'Higgins & M. J. Cohn, pp. 85–122. London: Academic Press.

Mann, A. (1975). *Paleodemographic Aspects of the South African Australopithecines*. Philadelphia: University of Pennsylvania Press.

Maureille, B. (2002). A lost Neanderthal neonate found. *Nature,* **419**, 33–34.

Minugh-Purvis, N. (1988). Patterns of craniofacial growth and development in Upper Pleistocene hominids. PhD dissertation, University of Pennsylvania.

Minugh-Purvis, N., & McNamara, K. J. (eds.) (2002). *Human Evolution through Developmental Change*. Baltimore: Johns Hopkins University Press.

Montagu, M. F. A. (1955). Time, morphology, and neoteny in the evolution of Man. *American Anthropologist,* **57**, 13–27.

O'Higgins, P., & Cohn, M. J. (eds.) (2000). *Development, Growth and Evolution: Implications for the Study of the Hominid Skeleton*. London: Academic Press.

Raff, R. A. (1996). *The Shape of Life: Genes, Development, and the Evolution of Animal Form*. Chicago: University of Chicago Press.

Rak, Y., Kimbel, W. H., & Hovers, E. (1994). A Neandertal infant from Amud Cave, Israel. *Journal of Human Evolution,* **26**, 313–324.

Saunders, S. R. (1992). Subadult skeletons and growth related studies. In *Skeletal Biology of Past Peoples: Research Methods*, eds. S. R. Saunders & M. A. Katzenberg, pp. 1–20. New York: Wiley Liss.

Saunders, S. R., & Hoppa, R. D. (1993). Growth deficit in survivors and non-survivors: Biological mortality bias in subadult skeletal samples. *Yearbook of Physical Anthropology,* **36**, 127–151.

Schwartz, J. H., & Tattersall, I. (2002). *The Human Fossil Record*. Vol. 1, *Terminology and Craniodental Morphology of Genus* Homo *(Europe)*. New York: John Wiley & Sons.

Skinner, M. (1997). Age at death of Gibraltar 2. *Journal of Human Evolution,* **32**, 469–470.

Smith, B. H. (1986). Dental development in *Australopithecus* and early *Homo*. *Nature,* **323**, 327–330.

Smith, B. H. (1989a). Dental development as a measure of life history in primates. *Evolution,* **43**, 683–688.

Smith, B. H. (1989b). Growth and development and its significance for early hominid behaviour. *Ossa,* **14**, 63–96.

Smith, B. H. (1991). Standards of human tooth formation and dental age assessment. In *Advances in Dental Anthropology*, eds. M. A. Kelley & C. S. Larsen, pp. 143–168. New York: Wiley-Liss.

Smith, B. H. (1993). The physiological age of KNM-WT 15000. In *The Narioko-tome* Homo erectus *Skeleton*, eds. A. Walker & R. Leakey, pp. 196–220. Cambridge: Harvard University Press.

Stringer, C. B., & Dean, M. C. (1997). Age at death of Gibraltar 2 – a reply. *Journal of Human Evolution,* **32**, 471–472.

Stringer, C. B., Dean, M. C., & Martin, R. D. (1990). A comparative study of cranial and dental development within a recent British sample and among Neandertals. In *Primate Life History and Evolution*, ed. C. J. de Rousseau, pp. 115–152. New York: Wiley-Liss.

Thompson, J. L., & Nelson, A. J. (2000). The place of Neandertals in the evolution of hominid patterns of growth and development. *Journal of Human Evolution,* **38**, 475–495.

Tillier, A. M. (1979). Restes crâniens de l'enfant moustérien *Homo* 4 de Qafzeh (Israël): La mandible et les maxillaires. *Paléorient,* **5**, 67–85.

Tillier, A. M. (1981). Evolution de la région symphysaire chez les *Homo sapiens* juvéniles du Paléolithique moyen: Pech de l'Azé 1, Roc de Marsal, et la Chaise 13. *Comptes Rendus de l'Académie des Sciences de Paris,* **293**, 725–727.

Tillier, A. M. (1982). Les enfants néanderthaliens de Devil's Tower (Gibraltar). *Zeitschrift für Morphologie und Anthropologie,* **73**, 125–148.

Tillier, A. M. (1983a). Le crâne d'enfant d'Engis 2: Un example de distribution des caractères juvéniles, primitifs et néanderthaliens. *Bulletin de la Société royale Belge d'Anthropologie et de Préhistoire,* **94**, 51–75.

Tillier, A. M. (1983b). L'enfant néanderthalien du Roc de Marsal (Campagne du Bugue, Dordogne): Le squelette facial. *Annales de Paléontologie,* **69**, 137–149.

Tillier, A. M. (1984). L'enfant *Homo* 11 de Qafzeh (Israël) et son apport à la compréhension des modalités de la croissance de squelettes Moustériens. *Paléorient,* **10**, 7–48.

Tillier, A. M. (1986a). Ordre d'apparition des caractères Néanderthaliens sur le squelette crânien au cours de la croissance: Problèmes d'analyse phylogénétique. In *Definition et Origines de l'Homme*, ed. M. Sakka, pp. 263–271. Paris: Editions du CNRS.

Tillier, A. M. (1986b). Quelques aspects de l'ontogenèse du squelette cranien des Néandertaliens. *Anthropos (Brno),* **23**, 207–216.

Tillier, A. M. (1987). L'enfant néanderthalien de la Quina H 18 et l'ontogénie des Néanderthaliens. In *Préhistoire de Poitou-Charentes: Problèmes actuels*, ed. B. Vandermeersch, pp. 201–206. Paris: Comité des Travaux Historiques et Scientifiques.

Tillier, A. M. (1988). La place de restes de Devil's Tower (Gibraltar) dans l'ontogenèse des Néanderthaliens. *Bulletins et Mémoires de la Société d'Anthropologie de Paris,* **14**, 257–266.

Tillier, A. M. (1989). The evolution of modern humans: Evidence from young Mousterian individuals. In *The Human Revolution: Behavioral and Biological Perspectives on the Origins of Modern Humans*, eds. P. Mellars & C. Stringer, pp. 286–297. Edinburgh: Edinburgh University Press.

Tillier, A. M. (1992). The origins of modern humans in Southwest Asia: Ontogenetic aspects. In *The Evolution and Dispersal of Modern Humans in Asia*, eds. T. Akazawa, K. Aoki, & T. Kimura, pp. 15–28. Tokyo: Hukusen-sha.

Tompkins, R. L., & Trinkaus, E. (1987). La Ferrassie 6 and the development of Neandertal pubic morphology. *American Journal of Physical Anthropology*, **73**, 233–239.

Trinkaus, E. (1990). Why bother with the Neandertals? *Quaternaria Nova*, **1**, 639–652.

Trinkaus, E. (1995). Neanderthal mortality patterns. *Journal of Archaeological Science*, **22**, 121–142.

Trinkaus, E., & Tompkins, R. L. (1990). The Neandertal life cycle: The possibility, probability, and perceptability of contrasts with recent humans. In *Primate Life History and Evolution*, ed. C. J. de Rousseau, pp. 153–180. New York: Wiley-Liss.

Ubelaker, D. H. (1989). *Human Skeletal Remains*, 2nd edn. Washington, DC: Smithsonian Institution Press.

Walker, A., & Leakey, R. (eds.) (1993). *The Nariokotome* Homo erectus *Skeleton*. Cambridge: Harvard University Press.

Wood, J. W., Milner, G. R., Harpending, H. C., & Weiss, K. M. (1992). The osteological paradox: Problems from inferring prehistoric health from skeletal samples. *Current Anthropology*, **33**, 343–370.

Part I
Setting the stage: What do we know about human growth and development?

2   The human pattern of growth and development in paleontological perspective

B. BOGIN

University of Michigan–Dearborn

Introduction

This volume, *Patterns of Growth and Development in the Genus* Homo, presents an up-to-date review of the evolution of the pattern of human growth. Historically, there are many outstanding research developments in the study of this field, which may be called auxological paleontology. The Greek word *auxein* is used most often to denote a group of acidic organic substances that promote and regulate growth processes in plants. Human growth and development researchers appropriated this word and sometimes call themselves auxologists. In this chapter the phrase auxological paleontology is used to describe research into patterns of growth and development of fossil species. I briefly review a few milestones of research. Following this review, I describe my own research on the evolution of human life history.

Life-history theory is a field of biology concerned with the strategy an organism uses to allocate its energy toward growth, maintenance, reproduction, raising offspring to independence, and avoiding death. For a mammal, it is the strategy of when to be born, when to be weaned, how many and what type of pre-reproductive stages of development to pass through, when to reproduce, and when to die (Bogin, 1999; Smith & Tompkins, 1995).

Human life-history stages are presented in Table 2.1. Note that two of these stages, childhood and adolescence, are present only in living humans, and not in any other living primate species. The childhood stage is essentially the time between weaning (the cessation of breast-feeding) and the eruption of the first permanent molar tooth. In traditional human societies, weaning takes place at a median age of 2.5 years after birth (Detwyller, 1995), but the first permanent

Patterns of Growth and Development in the Genus Homo, ed. J. L. Thompson, G. E. Krovitz, and A. J. Nelson. Published by Cambridge University Press. © Cambridge University Press 2003.

Table 2.1 *Stages in the human life cycle*

Stage	Growth events/duration (approximate or average)
Prenatal life	
First trimester	Fertilization to 12th week: embryogenesis
Second trimester	Fourth through sixth lunar month: rapid growth in length
Third trimester	Seventh lunar month to birth: rapid growth in weight and organ maturation
Birth	
Postnatal life	
Neonatal period	Birth to 28 days: extrauterine adaptation, most rapid rate of postnatal growth and maturation
Infancy	Second month to end of lactation, usually by 36 months: rapid growth velocity, but with steep deceleration in growth rate, feeding by lactation, deciduous tooth eruption, many developmental milestones in physiology, behavior, and cognition
Childhood	Years 3 to 7: Moderate growth rate, dependency on older people for care and feeding, small growth spurt at age 7, eruption of first permanent molar and incisor, cessation of brain growth by end of stage
Juvenile	Years 7–10 for girls, 7–12 for boys: slower growth rate, capable of self-feeding, cognitive transition leading to learning of economic and social skills
Puberty	Occurs at end of juvenile stage and is an event of short duration (days or a few weeks): reactivation of central nervous system of sexual development, dramatic increase in secretion of sex hormones
Adolescence	The stage of development that lasts for 5–10 years after the onset of puberty: growth spurt in height and weight; permanent tooth eruption almost complete; development of secondary sexual characteristics; sociosexual maturation; intensification of interest in and practice of adult social, economic, and sexual activities, integration into adult social and economic networks
Adulthood	
Prime and transition	From 20 years old to end of child-bearing years: homeostasis in physiology, behavior, and cognition; menopause for women by age 50
Old age and senescence	From end of child-bearing years to death: decline in the function of many body tissues or systems

Source: Bogin (2001).

molar does not erupt until a median age of 6.3 years. For almost all other mammals the ages of weaning and first permanent molar eruption are almost coincident, with a correlation coefficient of $r > 0.90$ (Smith *et al.*, 1994). The eruption of some permanent teeth allows the young mammal to move from a diet of mother's milk to the adult diet. In a few mammal species the adults provide some transition foods after the young are weaned but before they are able and responsible to feed themselves on the adult diet (e.g., lions, marmosets, and tamarins). This is especially the case for human beings, as the weaned child up to age six years must be supplied with food and care by older individuals or she/he dies (Bogin, 1999). For human beings, the length of time from weaning to first permanent tooth eruption (which is at least three years) and the total dependency on older individuals during that time are two key features that define the childhood stage. Other features of childhood relating to physical growth, behavior, and cognition are explained below and in Bogin (1999).

Human adolescence is the stage of development following puberty, a biological event that initiates the final process of sexual maturation. One notable feature of human adolescence is the skeletal growth spurt, that is, the acceleration and then deceleration in the rate of growth of most bones in the body (Figure 2.1). Non-human primate species may show postpubertal growth spurts in body mass, but none shows a global spurt in skeletal growth (Bogin, 1999;

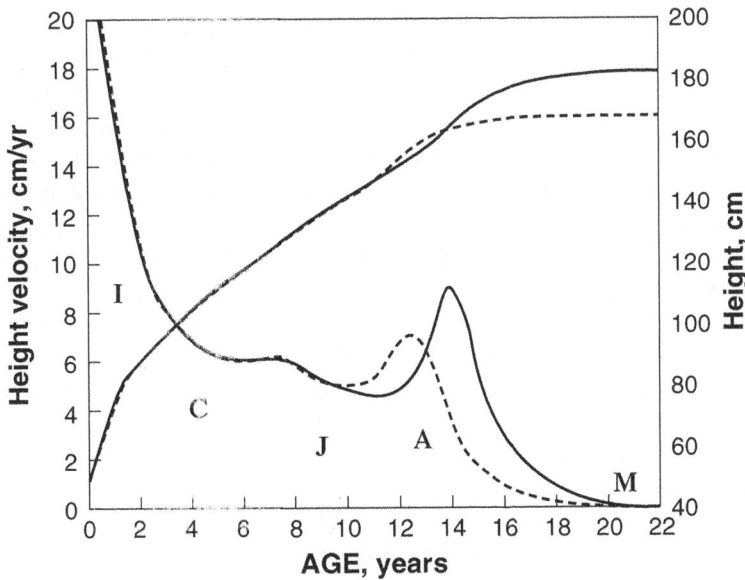

Figure 2.1 Mean distance (amount of growth) and velocity (rate of growth) curves of growth in height for healthy girls (dashed lines) and boys (solid lines). The velocity curves show most clearly the postnatal stages of human life history. Note the spurts in growth rate at late childhood and adolescence for both girls and boys. The stages of postnatal growth are abbreviated as follows: I, infancy; C, childhood; J, juvenile; A, adolescence; M, mature adult. Data used to construct the curves come from Prader (1984) and Buck & Thissen (1980). (From Bogin (1988).)

Hamada & Udono, 2002; Leigh, 1996). Other features of human adolescence are development of muscle, fat, and body hair particular to human men or women (Tanner, 1962) and the development of species-specific human adult behavior patterns and cognitive capacities (Bogin, 1999; Parker, 1996; Schlegel & Barry, 1991). More details of these biological, behavioral, and cognitive changes during adolescence are given later in this chapter. Suffice it to state here that the adolescent skeletal growth spurt and the sexual characteristics of men and women seem to be species-specific sociosexual signals, that influence the behavior of the adolescent and all other members of the adolescents' social group.

A brief history of auxological paleontology

The human life cycle and reproductive behavior stands in sharp contrast to other species of social mammals, even to other primates. Human beings give birth to relatively helpless newborns, provide these babies with a short duration of breast-feeding which is followed by a vastly extended period of offspring dependency. These traits are successfully combined with a mixture of delayed onset of reproduction, but relatively short birth intervals, unusual secondary sexual characteristics, such as the peculiar distribution of both hair and fat in women and men, menopause for women, and about 20 years of often energetic life after menopause. Two other unusual human features are relatively rapid increases in height growth – a smaller spurt which takes place at about age 7 years in many boys and girls, and a larger spurt occurring during adolescence in almost all boys and girls. No other primate or mammalian species is known to have these two postnatal skeletal growth spurts (Bogin, 1999).

Interest in accounting for these traits is ancient (Boyd, 1980; Tanner, 1981), but evolutionary approaches date to the twentieth century. D'Arcy Wentworth Thompson's books, *On Growth and Form* (1917, 1942, 1992), are a *tour de force* combining the classical approaches of natural philosophy and geometry with modern biology and mathematics to understand the growth, form, and evolution of plants and animals. Thompson visualized growth as a movement through time. Scientists from Buffon to Boas had studied the velocity of growth, but Thompson made it clear that growth velocities in stature or weight were only special cases of a more general biological process. That process includes the development of flower parts in plants, the evolution of antler size in mammals, and anything else that grows. Thompson developed the concept and methodology of using transformational grids to illustrate the process of growth during the lifetime of an individual or during the evolutionary history of a species. His application of transformational grids to human and chimpanzee growth is relevant

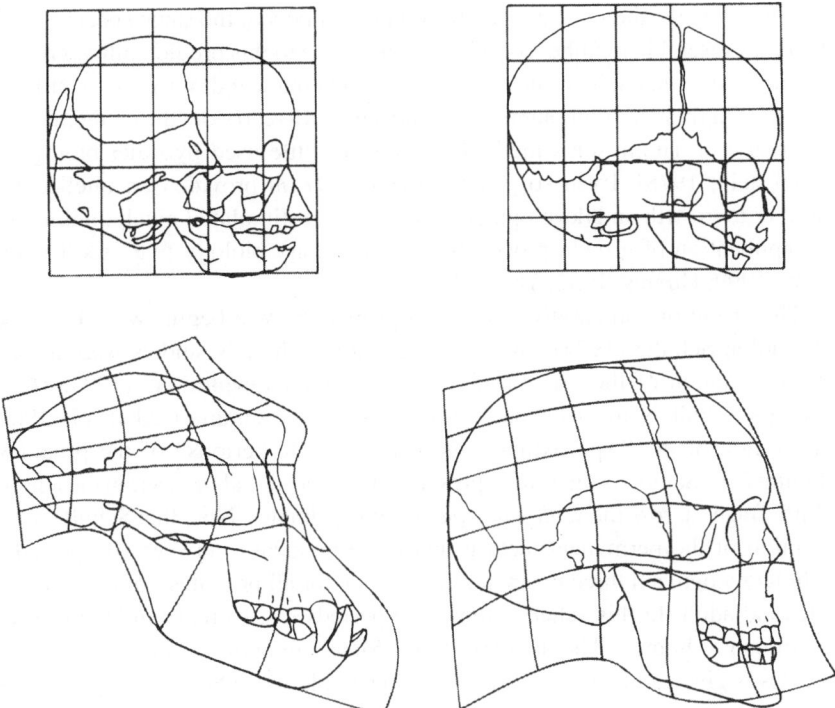

Figure 2.2 Transformation grids for the chimpanzee (left) and human (right) skull during growth. Fetal skull proportions are shown above for each species. The relative amount of distortion of the grid lines overlying the adult skull proportions indicate the amount of growth of different parts of the skull. (Inspired by the transformational grid method of D'Arcy Thompson (1942) and redrawn from Lewin (1993).)

here (Figure 2.2). The anatomical differences between human and chimpanzee are achieved, in part, through alterations in growth rates. Thompson showed in 1917 that the growth of the adult chimpanzee and adult human skull might be derived from a common neonatal form. Although we now know that this is not true, Thompson's work did show that different patterns of growth of the cranial bones, maxilla, and mandible are all that are required to produce the adult differences in skull shape. In a similar manner, some of the differences in the postcranial anatomy between chimpanzee and human being result from unequal rates of growth for common skeletal and muscular elements.

Thompson also described the biological form and growth of many organisms with mathematical functions. The mathematical treatment of growth is made possible by the predictability of biological development. Growth must produce a biological form that meets the ecological requirements of life for the species.

New individuals, then, must resemble other members of the same species more than they resemble members of other species. Due to this predictability, growth and form are amenable to the precision of mathematical description. Until the advent of high-speed computers, which are needed to carry out the mathematical procedure of Thompson's methods, they were little used by other biologists (Bookstein, 1978). Even so, *On Growth and Form* provided an intellectual validity to growth and development research, and stimulated much thinking in the application of growth processes to evolutionary biology (e.g., see Bogin, 1988, 1999; Huxley, 1932; Thom, 1983).

The origin of comparative studies of primate growth begins with the work of Adolph Schultz (1924), and from inception, Schultz's studies were aimed at "the relation of the growth of primates to man's evolution" (1924: 163). Perhaps Schultz's most lasting contributions were summarized in his 1960 illustration of the "Approximate ages of some life periods" of the primates (Figure 2.3). Schultz's diagram represents the proportional increase in the length of life stages across the *scala naturae* of living primates. Note that Schultz used eruption of the permanent teeth to mark the boundary between life periods. Schultz recognized three postnatal life periods for all primates: infantile, juvenile, and adult. In this scheme, these life periods just increase in length from prosimian to human. The data for "Early Man" are entirely speculative as no species is given and very little data were available when Schultz prepared this figure.

Despite the grandeur and vision of Schultz's research on primate growth and primate evolution, very few primate species were actually studied in detail by 1960. In fact, until the 1980s details of skeletal, dental, and somatic growth were known from only three species, the rhesus monkey (*Macaca mulatta*), the chimpanzee (*Pan troglodytes*), and humans. Ana K. Laird used data derived from those three species to continue Schultz's interest in the "Evolution of the human growth curve" – the title of Laird's 1967 paper. In that paper she reviewed studies of the growth of the rhesus monkey, the chimpanzee, and humans. In the style of D'Arcy Thompson, Laird took a mathematical approach to the study of growth. By fitting mathematical functions to the growth data, she hoped to reveal more precisely the stages in the evolutionary development of the human growth curve.

Laird found that growth in weight of the rhesus monkey and the chimpanzee could be explained with two separate growth curves. The first curve described growth up to puberty and the second described growth from puberty to adulthood. Laird found that the velocity curve of human growth required three mathematical functions to model its course. Laird's work was confirmed, independently, by Bock & Thissen (1976), Bogin (1980), and Karlberg (1987). The need for the third function is one aspect of human growth that makes it different

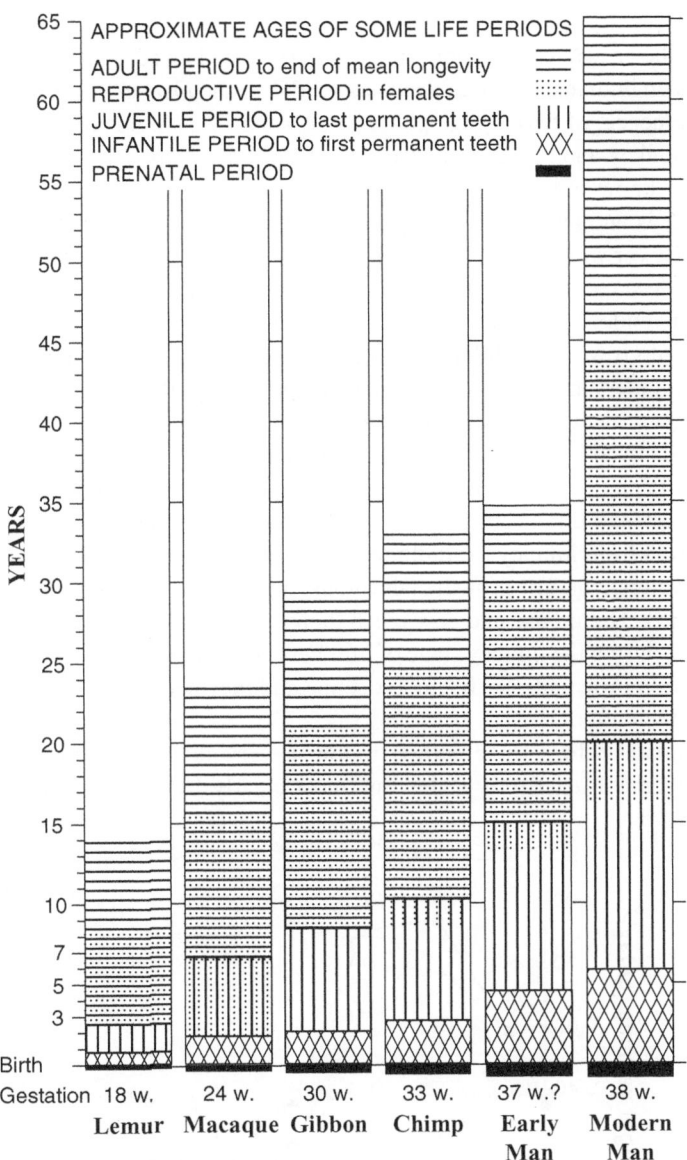

Figure 2.3 Schultz's diagram of the proportional increase in the length of life stages across the *scala naturae* of living primates. Note that Schultz used eruption of the permanent teeth to mark the boundary between life periods. Also Schultz did not recognize the childhood or adolescent stages for modern humans. Indeed, all primate species have the same life stages, which just increase in length from prosimian to human. The estimates for total length of life are based on average expectations rather than theoretical maximums. The data for "Early Man" are entirely speculative as no species is given and very little data were available when Schultz prepared this figure. (From Schultz (1960).)

from the growth of the other primates. Laird stated that the difference in growth between humans and other primates is "due to the *insertion*, between birth and adolescence, of two growth phases, rather than the single phase identifiable in the monkey and the chimpanzee . . ." (1967: 352, emphasis added).

We know today that Laird was not quite correct in her conclusion. Most monkeys and apes have two phases, or life-history stages, between birth and puberty (or what Laird called "adolescence"). These are called the infancy and juvenile stages of growth. Human beings have three stages, infancy, childhood, and juvenile. Nevertheless, Laird was correct that human beings have a new stage of growth, the childhood stage, inserted into our life history.

To understand Laird's mathematical approach to the phases of primate, and human, growth, it is necessary to go back in time to the work of Samuel Brody (1945). Brody discovered a new mammalian growth phase. He called it the juvenile phase of growth, and showed that of all the mammals he studied only primates had this juvenile phase. Brody found that the majority of mammals progress from infancy to adulthood seamlessly, without any intervening stages of development. The growth of the mouse and the cow are examples (Figures 2.4 and 2.5). The mouse and cow reach their maximum rate of growth after birth, that is, during the infancy stage of life. Very soon after weaning the growth rates of both mouse and cow begin to decline and they achieve puberty and fertility soon thereafter.

Human beings, in contrast, follow a very different path from conception to maturity. Brody states this most clearly by writing that his analysis

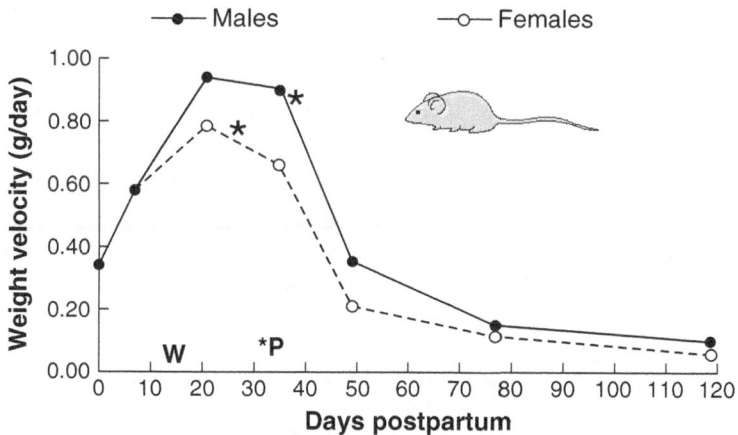

Figure 2.4   Velocity curves for weight growth in the mouse. Weaning (W) takes place between days 15 and 20. In both sexes puberty (P), meaning vaginal opening for females or spermatocytes in testes of males, occurs just after weaning and maximal growth rate. (Redrawn from data reported in Tanner (1962).)

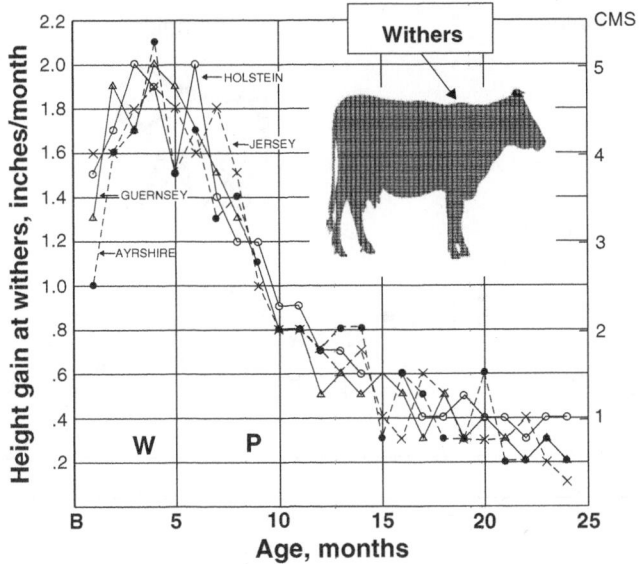

Figure 2.5   Velocity curves based on monthly gains in height at the withers for the several varieties of cows. The figure is from Brody (1945) who reports monthly values for height as the mean value for many animals (sample size varies from 67 to 239 at each age) measured monthly from birth to 24 months.

"demonstrates the close similarity between the age [growth] curves of different animal species. The human age curve, however, differs from the others in having a very long juvenile period, a long interval between weaning and puberty (approximately 3 to 13 years); this period is almost absent in laboratory and farm animals. In these animals, weaning merges into adolescence without the intervention of the juvenile phase found in man." (Brody, 1945: 495)

Brody's terminology is out of date by today's standards, for we know today that between weaning and puberty humans pass through a childhood growth stage and then the juvenile stage (Bogin, 1999). As defined earlier in this chapter, childhood is a period following weaning of three or more years of total dependency on older people. The current definition of the juvenile stage is the period of time between the onset of feeding independence and sexual maturity (Pereira & Altmann, 1985; Pereira & Fairbanks, 1993). Mammals in the juvenile stage typically have slow and slightly decelerating rates of growth. This may be seen for human juveniles in Figure 2.1. The baboon is another species with a juvenile growth stage. As illustrated in Figure 2.6 both male and female baboons have velocity curves for body length that show typical juvenile stages.

Even though Brody's terminology has been superseded, he correctly identified the fact that of all the animals he studied, only humans and the chimpanzee

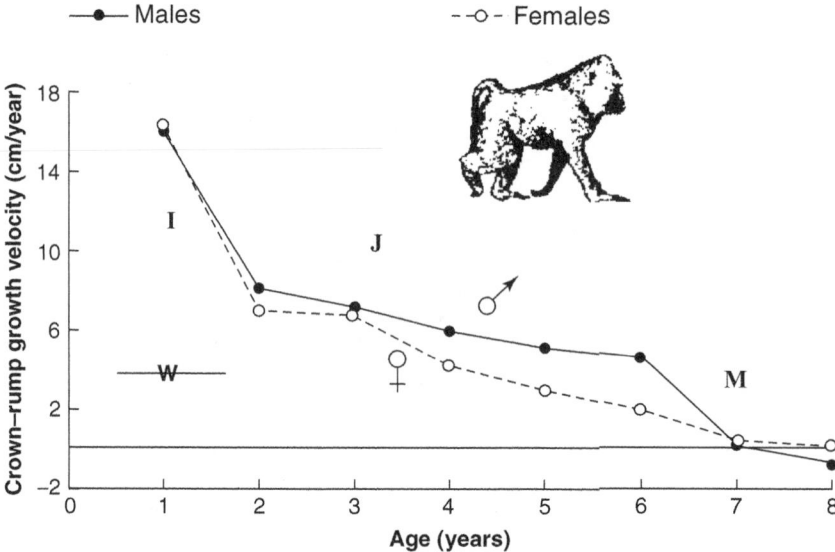

Figure 2.6   Velocity curves for crown–rump length in baboon males and females. The infant (I), juvenile (J), and mature adult (M) stages are labeled. The period of weaning (W) is also indicated. The male and female symbols are placed at the average ages of puberty. (From Coehlo (1985).)

have a juvenile growth period and this is one of his lasting contributions to science. Since Brody's time, juvenile growth stages have been discovered for several other mammalian species. The highly social mammals, such as social carnivores (wolves, lions, hyenas), elephants, many cetaceans (porpoises, whales), and most primates all evolved a new stage of development between infancy and adulthood – the juvenile stage (Bekoff & Byers, 1985; Pereira and Fairbanks, 1993).

We can analyze growth curves of living mammals to identify stages of growth. How to go beyond the discoveries of Brody and Laird for living species to the speculations of Schultz for the growth of fossil species was a mystery until the work of Alan Mann in the 1970s. Mann (1975) analyzed the dental development of *Australopithecus africanus* and *A. robustus* specimens from South Africa. He did so by taking radiographs of the fossils of juvenile australopithecine mandibles. These fossils were formed by replacement of the original mineral elements with minerals from the surrounding soil. This process preserves the details of both surface and internal anatomy of bones and teeth. Mann focused on those fossils with the first or second permanent molar just erupted. The radiographs revealed the state of formation of teeth still encrypted in the mandible. He then compared these with similar radiographs of chimpanzees and humans at the same stage of eruption.

Mann found that the amount of development of the unerupted second or third australopithecine molars was, in every case, like that of living humans. According to Mann, *Australopithecus* fossils did not have the rapid molar development of chimpanzees and were not "halfway" between the chimpanzee and the human pattern. Mann argued that a human-like delay in dental development for australopithecines indicates a similar delay in overall maturation of the body. Given this, Mann characterized the genus *Australopithecus* as human-like in its pattern of growth, with a human-like dependence on a long developmental period for the learning of human-like technological, social, and cultural skills.

Subsequent fossil discoveries and new interpretations disputed Mann's claims. Bromage & Dean (1985) developed a new method of estimating the age at death of hominoids that is supposed to be applicable to all species, living and extinct. The method, based on the formation of microstructural features of tooth crowns (see Bermúdez de Castro (2002) for a clear description of this method) indicated that *Australopithecus* and early *Homo* might have grown at a rate more similar to apes than to humans. Smith (1986, 1991) developed an alternative method of dental aging, based on pattern profiles of the development of tooth crowns and roots. She found that no species of early hominin (*A. afarensis, A. africanus, A. robustus, A. boisei, Homo habilis*, or early *H. erectus*) conforms precisely to either the pongid or the human pattern profiles. Additional studies by Beynon & Wood (1987), Conroy & Vannier (1991), and Smith & Tompkins (1995) reached similar conclusions.

Dean *et al.* (2001) confirmed these earlier studies via a careful analysis of 13 hominins, including fossils belonging to the taxa studied by Smith. Dean and colleagues concluded that "It therefore seems likely that truly modern dental development emerged relatively late in human evolution." By "relatively late," they mean sometime after 1.5 Ma (million years ago). Bermúdez de Castro and colleagues (Bermúdez de Castro, 2002; Bermúdez de Castro *et al.*, 1999, and this volume) present evidence that the species they call *Homo antecessor*, dating from 800 000 BP (before present) was the first hominin to show the human-like pattern of dental development.

Mann's pioneering studies led the way in this research and gave us the methodological basis for much of the work that is presented in this volume. Mann showed that it is both possible and essential to look to the fossils for evidence of the evolution for the pattern of human growth.

A synthesis of ideas

My own research builds on, and synthesizes, the contributions of Thompson, Schultz, Laird, Brody, and Mann, and the others described above (Bogin, 1988,

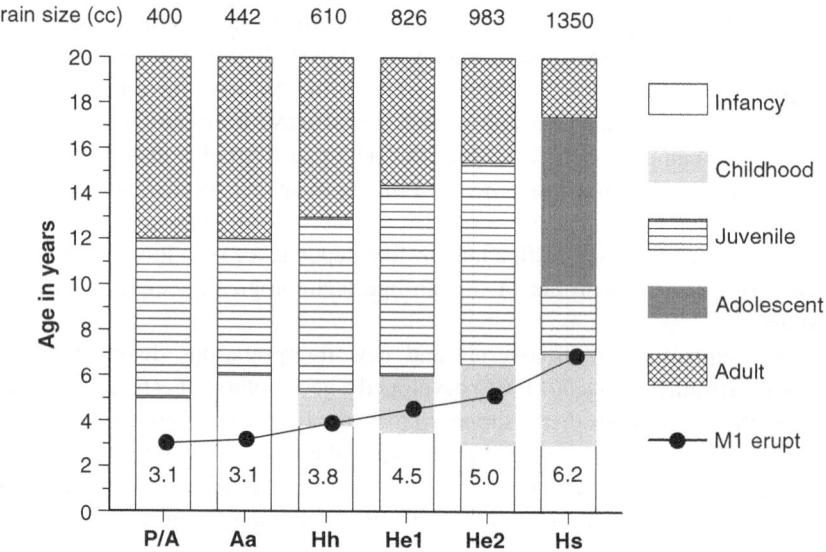

Figure 2.7   The evolution of hominin life history during the first 20 years of life.
Mean brain sizes are given at the top of each histogram. Mean age at eruption of the
first permanent molar (M1) is given near the bottom of each histogram and is graphed
across the histograms. Abbreviated nomenclature as follows: P/A, *Pan* and
Australopithecus afarensis; Aa, *Australopithecus africanus*; Hh, *Homo habilis*; He1,
early *Homo erectus*; He2, late *Homo erectus*; Hs, *Homo sapiens*. (From Bogin (1999).)

1999). I find that human life history consists of five stages of post-natal growth.
These are infancy, childhood, juvenile, adolescent, and adult (Table 2.1;
Figure 2.1). Among the living primates, the childhood and the adolescent
stages (as defined above) are unique to the human species. Based on my review
of the evidence available to date, the childhood stage may have evolved during
the time of *H. habilis*, and the adolescent stage seems to be a feature only
of anatomically modern *H. sapiens* (Bogin, 1999) and, possibly, Neandertals
(Nelson & Thompson, 2002).

Figure 2.7 is my Schultz-inspired summary of the evolution of human life
history. This figure must be considered as "a work in progress" for two reasons.
The first is that only the data for the first and last species (*Pan* and *Homo sapiens*)
are known with some certainty. The second is that new information will alter
what is known about patterns of growth for fossil hominins.

Known ages for eruption of the first permanent molar (M1) are given for
Pan and *H. sapiens*. Smith & Tompkins (1995) calculated estimated ages for
M1 eruption for the fossil species. Age of eruption of M1 is an important life-
history event that correlates very highly with other life-history events. Known
or estimated adult brain sizes are given at the top of each bar; the estimates are

averages based on reports in several textbooks of human evolution. Brain size is another major influence on life-history evolution (Martin, 1983, 1990).

Australopithecus afarensis appears in the fossil record about 3.9 Ma. *Australopithecus afarensis* shares many anatomical features with non-hominin pongid (ape) species including an adult brain size of about 400 cm^3 (Simons, 1989) and a pattern of dental development indistinguishable from that of extant apes (Bromage & Dean, 1985; Conroy & Vannier, 1991; Smith, 1991). Therefore, the chimpanzee and *A. afarensis* are depicted in Figure 2.7 as sharing the typical tripartite stages of postnatal growth of social mammals – infant, juvenile, adult. Following the definitions used throughout this chapter, infancy represents the period of feeding by lactation, the juvenile stage represents a period of feeding independence prior to the onset of sexual maturation, and the adult stage begins following puberty and sexual maturation. The duration of each stage and the age at which each stage ends are based on empirical data for the chimpanzee. A probable descendent of *A. afarensis* is the fossil species *A. africanus*, dating from about 3.0 Ma. To achieve the larger adult brain size of *A. africanus* (average of 442 cm^3) may have required an addition to the length of the fetal and/or infancy periods, as these are the life stages when most mammalian brain growth takes place (Martin, 1983). Figure 2.7 indicates an extension to infancy of one year for *A. africanus*.

The first permanent molar (M1) of the chimpanzee erupts at 3.1 years, but chimpanzees remain in infancy and continue to be nursed to about age five years. Until that age the young chimpanzee is dependent on its mother, and will not survive if the mother dies or is otherwise not able to provide care and feeding (Goodall, 1983; Nishida *et al.*, 1990). After erupting M1 the infant chimpanzee nurses less often and begins to eat adult-type foods. But, the infant must learn how to find and process these foods before becoming independent of its mother. Learning to successfully open fruits that are protected by hard shells and to extract insects from nests (such as ants and termites) requires 12 to 18 months of observation and imitation by the infant of the mother (Goodall, 1983). For these reasons, chimpanzees extend infancy for about two years past the eruption of M1. Based on brain size and details of dental anatomy the mean age of M1 eruption for *A. afarensis* and *A. africanus* is estimated to be very similar to that of the chimpanzee (Dean *et al.*, 2001). It is likely that these early hominins followed a pattern of life history stages identical to chimpanzees.

About 2.2 Ma fossils with several more human-like traits including larger cranial capacities and greater manual dexterity appear. Also dated to about this time are stone tools of the Oldowan tradition. Given the biological and cultural developments associated with these fossils they are considered by most paleontologists to be members of the genus *Homo* (designated as *H. habilis, H. rudolfensis*, or early *H. erectus* – referred to collectively here as *H. habilis*).

The rapid expansion of adult brain size during the time of *H. habilis* (650 to 800 cm^3) might have been achieved with further expansion of both the fetal and infancy periods. Martin (1983) demonstrated that brain sizes of up to 850 cm^3 may be achieved by extending the chimpanzee pattern of fetal and infant growth. However, the insertion of a brief childhood stage into hominin life history may have occurred as *H. habilis* shows evidence of a growth pattern in the femur that is distinct from that of the australopithecines, but consistent with that of later hominins (Tardieu, 1998).

Further brain size increase occurred during early *H. erectus* times (He1 in Figure 2.7), which begin about 1.6 Ma. The earliest adult specimens have mean brain sizes of 826 cm^3, but many individual adults had brain sizes between 850 to 900 cm^3. This places *H. erectus* at or above Martin's 850 cm^3 limit for an ape-like growth pattern and seems to justify insertion and/or expansion of the childhood period to provide the biological time needed for the rapid, human-like, pattern of brain growth. Late *H. erectus* (He2 – dated after 1.0 Ma), with average adult brain sizes of 983 cm^3, are depicted with further expansion of the childhood stage. In addition to bigger brains (some individuals had brains as large as 1100 cm^3), the archaeological record for late *H. erectus* shows increased complexity of technology (tools, fire, and shelter) and social organization (Klein, 1989). These techno-social advances, and the increased reliance on learning that occur with these advances, may well be correlates of changes in biology and behavior associated with further development of the childhood stage of life (Bogin & Smith, 1996). The evolutionary transition to archaic, and finally, modern *H. sapiens* expands the childhood stage to its current dimension.

The *H. sapiens* grade of evolution also sees the addition of an adolescent stage to postnatal development. There is no strong evidence for the evolution of an adolescent life stage in any hominins prior to *H. sapiens* or *H. neanderthalensis* (Bogin, 1999; Nelson & Thompson, 2002; Smith & Tompkins, 1995; Tardieu, 1998).

Why did childhood and adolescence evolve?

The basic pattern of human growth is shared by all living people and is the outcome of the 4+ million-year evolutionary history of the hominins. The pattern of human growth evolved in the context of the biological and social ecology of our ancestors. The term "ecology" is used here to refer to the relationship that an individual organism, or group of individuals of a species, has with its physical, biological, and social environment. At the core of any ecological system are two sets of behaviors; the first is directed toward how an organism acquires food and the second is directed toward how the organism reproduces. All organisms are alike in that they share behaviors related to what may be called simply "food and sex."

Social mammals, including most primates, satisfy their needs for food and reproduction through a complex ecology of biological and social relationships with their conspecifics (i.e., members of the same species) and their environment. The environment includes both other biological species as well as the physical surroundings. Human beings also share in this biosocial ecology, and add to it a significant cultural component. Human beings are cultural animals, meaning that we possess all the potentials and limitations of any living creature, to which we add a cultural trilogy of: (1) dependence on technology, (2) codified social institutions, such as kinship and marriage, and (3) ideology. In its anthropological sense, ideology refers to a set of symbolic meanings and representations particular to any society, through which its members view and interpret nature. Elements of the human capacity for culture may be found in many other species of animals, such as tool use and some aspects of language, but only in the human species do all three aspects of the cultural trio become so intensified, elaborated, and universal. Because of this, the evolution of human growth may be best understood by using a biocultural perspective.

Why childhood?

To understand the place of childhood in human evolution, please consider the data shown in Figure 2.8. Depicted in this figure are several hominoid developmental landmarks. Compared with living apes, human beings in traditional societies experience developmental delays in the eruption of the first permanent molar, age at menarche, and age at first birth. However, humans have a shorter infancy period and a shorter birth interval. Note from Figure 2.8 that in both apes and traditional human societies the infancy stage and the interval between successful births almost coincide.

I discussed earlier that dental development is an excellent marker for life history in the primates. There is a very strong correlation between age at eruption of the first molar (M1 eruption) and cessation of brain growth (Smith, 1991), and, in general, primates wean infants about the time M1 erupts. This timing makes sense, because the mother must nurse her current infant until it can process and consume an adult diet, which requires at least some of the permanent dentition.

The human species is a striking exception to this relationship between permanent tooth eruption and birth interval. Women in traditional societies wait, on average, just three years between births, not the six years expected on the basis of M1 eruption (Blurton Jones *et al.*, 1992; Bogin, 2001; Howell, 1979). The short birth interval gives humans a distinct advantage over other apes, because

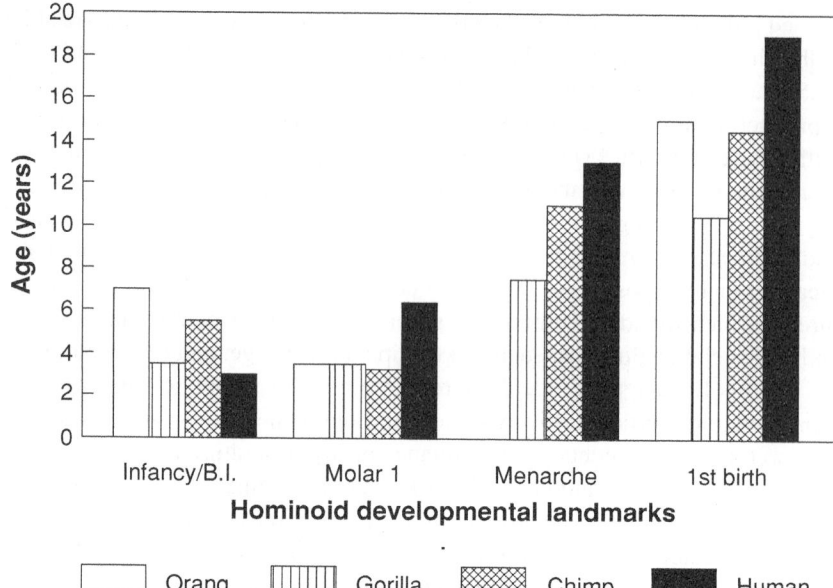

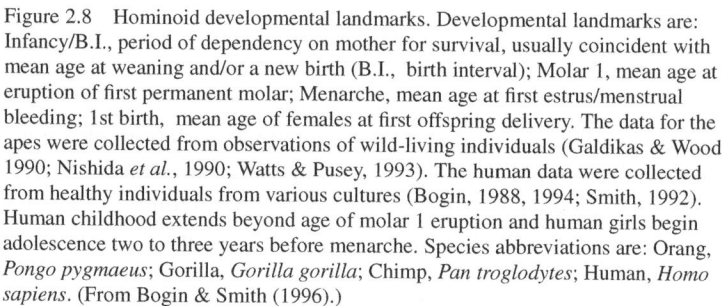

Figure 2.8   Hominoid developmental landmarks. Developmental landmarks are: Infancy/B.I., period of dependency on mother for survival, usually coincident with mean age at weaning and/or a new birth (B.I., birth interval); Molar 1, mean age at eruption of first permanent molar; Menarche, mean age at first estrus/menstrual bleeding; 1st birth, mean age of females at first offspring delivery. The data for the apes were collected from observations of wild-living individuals (Galdikas & Wood, 1990; Nishida *et al.*, 1990; Watts & Pusey, 1993). The human data were collected from healthy individuals from various cultures (Bogin, 1988, 1994; Smith, 1992). Human childhood extends beyond age of molar 1 eruption and human girls begin adolescence two to three years before menarche. Species abbreviations are: Orang, *Pongo pygmaeus*; Gorilla, *Gorilla gorilla*; Chimp, *Pan troglodytes*; Human, *Homo sapiens*. (From Bogin & Smith (1996).)

we can produce and rear two offspring through infancy in the time it takes chimpanzees or orangutans to produce and rear one offspring. By reducing the length of the infancy stage of life (i.e., the period of lactation) and by developing the special features (see below) of the human childhood stage, humans have the potential for greater lifetime fertility than any ape. Shorter lactation translates into greater fertility because in traditional societies, with high energy expenditure and marginal to adequate diets, the physiological demands of milk production and nursing interfere with ovulation and pregnancy (Ellison, 2001).

Selection for increased reproductive success is the force that drives much of biological evolution. The evolution of the human childhood stage gave our species this reproductive advantage, because children no longer are fed by nursing. Children are still dependent on older individuals for feeding and protection.

The child must be given foods that are specially chosen and prepared, but the mother does not have to provide 100% of offspring nutrition and care directly. Traditional societies solved the problem of childcare by spreading the responsibility among many individuals, including juveniles, adolescents, or adults (e.g. Lancaster & Lancaster, 1983; Turnbull, 1983). In Hadza society (African hunters and gatherers), grandmothers and great-aunts supply a significant amount of food and care to children (Blurton Jones, 1993; Hawkes *et al.*, 1997). In Agta society (Philippine hunter–gatherers), women hunt large game animals but still retain primary responsibility for childcare. They accomplish this dual task by living in extended-family groups – two or three brothers and sisters, their spouses, children, and parents – and sharing the childcare (Estioko-Griffin, 1986). Among the Maya of Guatemala (horticulturists and agriculturists), many people live together in extended-family compounds. Women of all ages work together in food preparation, clothing manufacture, and childcare (Bogin, ethnographic fieldwork observations). In some societies, fathers provide significant childcare, including the Agta and the Aka pygmies, hunter–gatherers of central Africa (Estioko-Griffin, 1986; Hewlett, 1992). Summarizing the data from many human societies, Lancaster & Lancaster (1983) call this kind of shared childcare and feeding "the hominid adaptation," because no other primate or mammal does all of this.

Childhood also may be viewed as a mechanism that allows for more precise "tracking" of ecological conditions by allowing more time for developmental plasticity. The fitness of a given phenotype (i.e., the physical features and behavior of an individual) varies across the range of variation of an environment. When phenotypes are fixed early in development, such as in mammals that mature sexually soon after weaning (e.g., rodents), environmental change and high mortality are positively correlated. Social mammals (carnivores, elephants, primates) prolong the developmental period by adding a juvenile stage between infancy and adulthood. Adult phenotypes develop more slowly in these mammals because the juvenile stage lasts for years. These social mammals experience a wider range of environmental variation, such as seasonal variation in temperature and rainfall. They also experience years of food abundance and food shortage as well as changes in the number of predators and in types of diseases. The result on the phenotype is a better conformation between the individual and the environment (Bogin, 1999).

Fitness is increased because more offspring survive to reproductive age than in mammalian species without a juvenile stage. For example, ~4% of infant Norway rats (which have no juvenile stage) born in the wild survive to adulthood versus 14–16% of lions, which have a juvenile stage (Lancaster & Lancaster, 1983). Monkeys and apes have juvenile periods that last as long as or longer than those of social carnivores. Consequently, these primates rear between 12% and

36% of their offspring to adulthood (Lancaster & Lancaster, 1983). The human childhood stage adds an additional four years of relatively slow physical growth and allows for behavioral experience that further enhances developmental plasticity. The combined result is that humans in traditional societies (hunting and gathering or horticultural groups) rear at least 50% of their offspring to adulthood (Bogin, 2001). In the technologically advanced nations today, survival to adulthood is appreciably higher. For the United States, in the year 1999, it was estimated that 98.6% of live-born infants would survive to age 20 years (Anderson & DeTurk, 2002).

The bottom line, in a biological sense, is that the evolution of human childhood, as a new life-history stage added between the infancy and juvenile stages, decreases the interbirth interval and increases reproductive fitness.

Why adolescence and the adolescent growth?

One often cited reason for the additional years of development provided by human adolescence is the "extra time" required to learn and practice technology, social organization, language, and other aspects of culture (Kaplan *et al.*, 2000; Watts, 1985, 1990). According to this hypothesis, human culture is so complex that not even the infant, child, and juvenile stages of growth provide enough time to learn what is need to be a successful adult. Furthermore, this "extra time" hypothesis explains the adolescent spurt in skeletal growth as a consequence of all this delayed maturation. The basic argument is that by the end of adolescence our ancestors were left with proportionately less time for procreation than most other mammals, and therefore needed to attain adult size and sexual maturity quickly.

Empirical field research with human foraging societies is showing that the "extra time" hypothesis is not sufficient to explain adolescence. Ethnography and experimental studies with the Hadza show no difference in essential hunting and gathering skills between adolescents who had lived all their lives in the bush and those who spent many years away at boarding schools (Blurton Jones & Marlowe, 2002). Working with the Merian, a society of Melanesian islanders, Bird & Bird (2002a, 2002b) find that youngsters aged 5 to 15 years are just as efficient as older people in fishing but are less efficient at foraging for food on the reef. However, the youngsters' lower efficiency of reef foraging is due to the effects of smaller body size and less attention to work (playing), and not due to learning or practice. Since it does not take 20 years to learn to forage something else must be the reason for human adolescence.

The adolescent growth spurt is not a consequence of delayed maturation. Consider first that there is no need to experience an adolescent growth spurt to reach adult height or fertility. Historical sources describe the castrati, male opera singers of the seventeenth and eighteenth centuries who were castrated as boys to preserve their soprano voices, as being unusually tall for men (Barbier, 1996; Peschel & Peschel, 1987). Also, children who are born without gonads or have them removed surgically prior to puberty (due to diseases such as cancer) do not experience an adolescent growth spurt, but do reach their normal expected adult height (Prader, 1984). Of course, castrati, whether or not opera singers, do not become reproductively successful. There are, however, gonadally intact individuals, for the most part very late maturing boys and girls, who have virtually no growth spurt. Nevertheless, these late maturing individuals do grow to be normal sized adults, and they become fertile by their early 20s – not significantly later than individuals with a spurt (Bogin, 1999).

Another problem with the "extra time" argument for the adolescent growth spurt is that it does not explain the timing of the spurt. Girls experience the growth spurt before becoming fertile, but for boys the reverse is true. There are other sex-based differences in development that take place during adolescence. Why do these differences exist? The order in which several pubertal events occur in girls and boys is illustrated in Figure 2.9 in terms of time before and after peak height velocity (PHV) of the adolescent growth spurt. In both girls and boys puberty begins with changes in the activity of the hypothalamus and other parts of the central nervous system (Plant, 1994). These changes are labeled as "CNS puberty" in the figure. Note that the central nervous system (CNS) events begin at the same relative age in both girls and boys, that is, three years before PHV. This is also the time when growth rate changes from decelerating to accelerating.

In girls (Figure 2.9 upper panel), the first outward sign of puberty is the development of the breast bud (B2 stage) and wisps of pubic hair (PH2 stage). Tanner (1962) developed a system of staging the breast and pubic hair development of girls, and genital and pubic hair development in boys. The Tanner system has five stages. Stage 1 is the absence of a trait and stage 5 is the adult form of a trait. The first appearance of breast and pubic hair development is followed, in order, by (1) a rise in serum levels of estradiol which leads to the laying down of fat on the hips, buttocks, and thighs; (2) the adolescent growth spurt; (3) further growth of the breast and body hair (B3 and PH3); (4) menarche; (5) completion of breast and body hair development (B5 and PH5); and (6) attainment of adult levels of ovulation frequency.

The path of pubertal development in boys (Figure 2.9 lower panel) starts with a rise in serum levels of luteinizing hormone (LH) and the enlargement of the testes and then penis (G2). Genital maturation in boys begins, on average, only a few months after that of girls. However, the timing and order of other secondary sexual characteristics is unlike that of girls. About a year after CNS puberty, there is: (1) a rise in serum testosterone levels (T) which is followed by

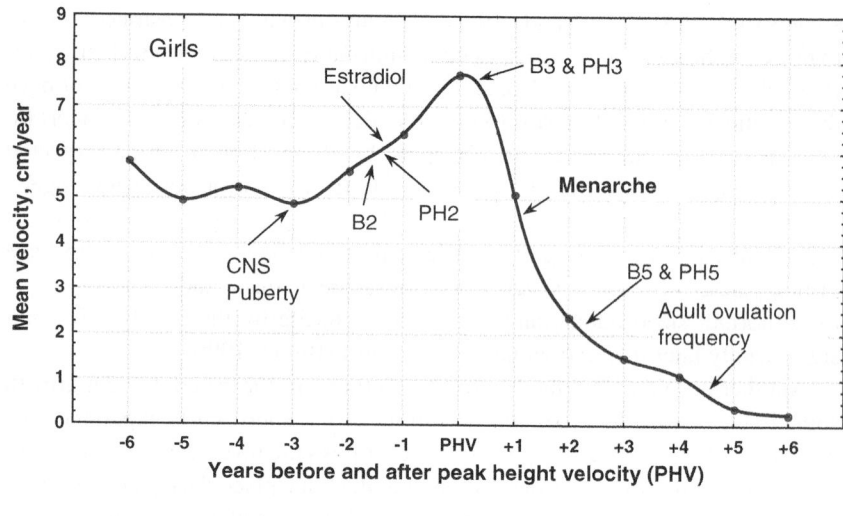

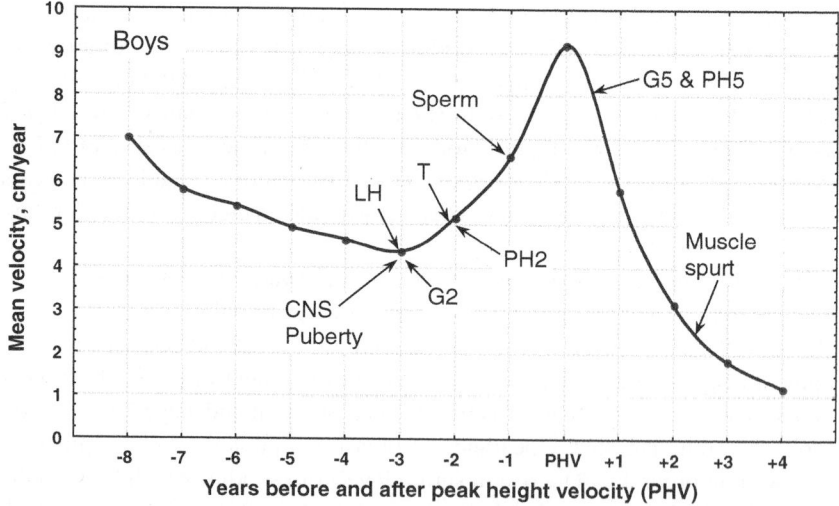

Figure 2.9   The ordering of several sexual maturation events for girls (top panel) and boys (bottom panel) during the adolescent growth spurt. The velocity curves were calculated using data derived from a sample of healthy, well-nourished girls and boys living in Guatemala. See text for an explanation of each labeled event. (From Bogin (1999).)

the appearance of pubic hair (PH2); (2) about a year later motile spermatozoa may be detected in urine; (3) PHV follows after about another year, along with deepening of the voice, and continued growth of facial and body hair; (4) the adult stages of genital and pubic hair development follow the growth spurt (G5 and PH5); and (6) near the end of adolescence boys undergo a spurt in muscular development.

The sex-specific order of pubertal events tends not to vary between early and late maturers. Neither does it vary between well-nourished girls and boys and those who suffered from severe malnutrition in early life, between rural and urban dwellers, or between European, African, and Native American ethnic groups (Cameron *et al.*, 1988, 1990, 1993; Bogin *et al.*, 1992).

Why do girls have adolescence?

For girls an adolescent stage of human growth (defined in Table 2.1) may have evolved to increase their reproductive fitness (Bogin, 1993, 1999). The evolution of childhood afforded adult hominin females the opportunity to give birth at shorter intervals, but producing offspring is only a small part of reproductive fitness. Rearing the young to their own reproductive maturity is necessary for reproductive success.

Studies of yellow baboons, toque macaques, and chimpanzees show that between 50% and 60% of first-born offspring die in infancy (Altmann, 1980). By contrast, in hunter–gatherer human societies, 39% of Hadza first born (Blurton Jones *et al.*, 1992) and 44% of !Kung first-born offspring (Howell, 1979) die in infancy. Studies of wild baboons by Altmann (1980) show that whereas the infant mortality rate for the first-born is 50%, mortality for second-born drops to 38%, and for third- and fourth-born reaches only 25%. The difference in infant survival with each subsequent birth is due, in part, to experience and knowledge gained by the mother.

Human females internalize much maternal information and build important networks of social support during their juvenile and adolescent stages, giving the adult women a reproductive edge. The initial human advantage may seem small, but it means that up to 21 more people than baboons or chimpanzees survive out of every 100 first-born infants – more than enough over the vast course of evolutionary time to make the evolution of human adolescence an overwhelmingly beneficial adaptation.

In human societies, juvenile girls often are expected to provide significant amounts of childcare for their younger siblings (Weisner, 1987), whereas in most other social mammal groups, the juveniles are often segregated from adults and infants (Perieira & Fairbanks, 1993). Thus, human girls enter adolescence with considerable knowledge of the needs of children. Adolescent girls gain knowledge of sexuality and reproduction because they look mature sexually, and are treated as such, several years before they actually become fertile. The adolescent growth spurt and the changes in physical appearance associated with the spurt serve as signals of sexual maturation (Figure 2.9). About a year after peak height velocity, girls experience menarche, an unambiguous external signal of internal reproductive system development. However, most girls experience one to three years of anovulatory menstrual cycles after menarche (Worthman, 1993). The dramatic changes of adolescence make it appear that the girls are sexually mature and this facilitates their participation in adult social, sexual, and economic behavior. The years of adolescent involvement in adult socio-sexual behavior helps the girls build reliable networks of support that are essential to successful reproduction (Hawkes *et al.*, 1997; Lancaster & Lancaster, 1983).

It is noteworthy that female chimpanzees and bonobos, like human girls, also experience about two years of postmenarchial infertility, that is, they have cycles of estrus swelling without ovulation (Pussey, 2001). So, this time of life may be a shared hominoid trait. Like human adolescents, these subadult female chimpanzees and bonobos participate in a great deal of adult social and sexual behavior, which may help new mothers acquire needed resources for themselves and their infants. In terms of reproductive fitness, chimpanzees rear about 36% of their infants to adulthood. After human beings, this is the second best percentage of all the primates (Lancaster & Lancaster, 1983).

Although ape and human females may share a year or more of postmenarche sterility, apes reach adulthood sooner than humans. Full reproductive maturation in human women is not achieved until about 5 years after menarche (Bogin, 1999; Worthman, 1993). The average age at menarche in the United States (where girls are generally healthy and well nourished) is 12.4 years, which means that the average age at full sexual maturation occurs between the ages of 17 and 18 years. One reason for the longer period of human adolescent sterility is that human female fertility is correlated with relatively slow growth of the pelvis. Worthman (1993) and Ellison (2001) find that the crucial variable for successful first birth is size of the pelvic inlet, the bony opening of the birth canal. Moerman (1982) measured pelvic X-rays from a sample of healthy, well-nourished American girls who achieved menarche between 12 and 13 years. These girls did not attain adult pelvic inlet size until 17–18 years of age. Quite unexpectedly, the adolescent growth spurt, which occurs before menarche, does not influence the size of the pelvis in the same way as the rest of the skeleton. Rather, the female pelvis has its own slow pattern of growth, which continues for several years after adult stature is achieved. Ellison (2001: 159) states that the attainment of adult shape of the pelvis is under the control "of the same estrogen hormones that bring growth of the long bones to a halt." Ellison concludes that pelvic maturity and fertility are designed to occur only after adult skeletal size is achieved. This makes sense in that a young woman who is still growing needs her available energy and nutrients for her own growth. Indeed, teenage mothers and their infants are at risk because of the reproductive immaturity of the mother. Risks include a low-birth-weight infant, premature birth, and high blood pressure in the mother (Garn & Petzold, 1983; Scholl & Hediger, 1993). The likelihood of these risks declines and the chance of successful pregnancy and birth increases markedly after age 18.

Cross-cultural studies of reproductive behavior show that human societies acknowledge (consciously or not) this special pattern of human female pelvic growth. For example, the age at first childbirth clusters around 19 years for women from such diverse cultures as the Kikuyu of Kenya, Mayans of Guatemala, Copper Eskimos of Canada, and both the colonial and contemporary United States (reviewed in Bogin, 1994). Despite waiting nearly two decades to begin reproduction, human women have the capacity to eventually produce more offspring and successfully rear them to adulthood than any other primate. The addition of the childhood stage to life history frees the mother from several years of lactation, which shortens the interval between births. The adolescent stage provides the social, economic, and political resources that mothers need to support new infants and their own older children.

Why do boys have adolescence?

The adolescent development of boys is quite different from that of girls. Boys become fertile well before they assume the size and the physical characteristics of men. Analysis of urine samples from boys 11–16 years old shows that they begin producing sperm at a median age of 13.4 years (Muller *et al.*, 1989). Yet cross-cultural evidence indicates that few boys successfully father children until they are into their third decade of life (e.g., Kikuyu men in Kenya: Worthman (1993), Ache men in Paraguay: Hill & Kaplan (1988), and Canadian Inuit: Condon (1990)). In the United States, the National Center for Health Statistics (1999) reports that in the year 1999 adolescents between 15–19 years old fathered infants at a rate of 21 per 1000 young men in that age group. Men aged 20–24 years fathered four times as many babies (84 per 1000), and men aged 25–29 fathered 115 per 1000. Calculated another way, teenage fathers in the United States account for less than 4.0% of all fathers.

The explanation for the lag between sperm production and fatherhood is not likely to be a simple one of sperm performance, such as not having the endurance to swim to an egg cell in the woman's fallopian tubes. More likely is the fact that the average boy of 13.4 years is only beginning his adolescent growth spurt (Figure 2.1) and phenotypic maturation (Figure 2.9). In terms of physical appearance, physiological status, psychosocial development, and economic productivity the 13-year-old boy is still more a juvenile than an adult (Tanner, 1962). Anthropologists working in many diverse cultural settings report that few women (and more important from a cross-cultural perspective, few prospective in-laws) view the teenage boy as a biologically, economically, and socially viable husband and father (Schlegel & Barry, 1991).

The delay between sperm production and reproductive maturity is not wasted time in either a biological or social sense. The obvious and the subtle psychophysiological effects of testosterone and other androgen hormones that are released after gonadal maturation may "prime" boys to be receptive to their future roles as men (Weisfeld, 1999). Alternatively, it is possible that physical changes provoked by the endocrine hormones provide a social stimulus toward adult behaviors (Halpern *et al.*, 1993). Whatever the case, early in adolescence, sociosexual feelings of guilt, shame, anxiety, pleasure, and pride intensify. At the same time, adolescent boys become more interested in adult activities, adjust their attitude to parental figures, and think and act more independently. In short, they begin to behave like men (Weisfeld, 1999).

However – and this is where the survival advantage may lie – they still look like boys. One might say that a healthy, well-nourished 13.5-year-old human male, at a median height of 160 cm (62 inches) "appears" to be more child-like than he really is. Because their adolescent growth spurt occurs late in sexual development, and their development of muscle mass occurs even later, teenage males can begin to integrate into adult networks and practice behaving like adults before they are actually perceived as adults. The sociosexual antics of adolescent boys are often considered to be more humorous than serious. Yet, they provide the experience to fine-tune their sexual and social roles and to involve themselves in the economic and political networks of adults before the lives of the adolescents, or the lives of their offspring depend on them. For example, competition between men for women or limited food resources favors the older, more experienced man. Because such competition may be fatal, the child-like appearance of the immature but hormonally and socially primed adolescent male may be life-saving as well as educational.

Girls and boys – two paths through adolescence

Adolescence became part of human life history because it conferred significant reproductive advantages to our species, in part, by allowing the adolescent to socially integrate into the economic, sexual, and political world of adults. This allows adolescents to learn and practice adult economic, social, and sexual behaviors before reproducing. It also permits adolescents to begin building social networks that are essential to the cooperative style of infant and childcare found in all human societies.

Girls and boys follow two different developmental paths through adolescence. This is because girls best integrate into and learn about adult social roles while they are infertile but perceived by adults as mature, whereas boys best learn their adult social roles while they are fertile but not yet perceived as sexually mature by adults. Without the adolescent growth spurt, and the sex-specific timing of maturation events around the spurt, this unique style of biocultural development could not occur.

The shape of things to come

This chapter highlights only a few of the outstanding contributions that provide the foundation for this volume. Many other researchers helped to build this foundation, but there is not space enough to include them here (see Bogin, 1999 for a more detailed review). Our current knowledge of auxological paleontology –the evolution of human ontogeny – is an outcome of nearly a century of research, with technical inputs from many disciplines. We have advanced considerably from the innovative, but speculative, ideas of D'Arcy Thompson and Adolph Schultz and the early path-breaking empirical studies of Samuel Brody, Ana Laird, and Alan Mann. Life-history theory is now a cornerstone of evolutionary biology (e.g., Stearns, 1992). As this volume demonstrates, the evolution of human life history is an active and exciting area of investigation. The existing research tells us what we know about the evolution of growth and development in the genus *Homo*, and what we would like to know.

References

Altmann, J. (1980). *Baboon Mothers and Infants.* Cambridge: Harvard University Press.

Anderson, R. N., & DeTurk, P. B. (2002). United States life tables, 1999. *National Vital Statistics Report,* **50** (6), March 21, 1–40.

Barbier, P. (1996). *The World of the Castrati: The History of an Extraordinary Operatic Phenomenon.* London: Souvenir Press.

Bekoff, M., & Byers, J. A. (1985). The development of behavior from evolutionary and ecological perspectives in mammals and birds. *Evolutionary Biology,* **19**, 215–286.

Benyon, A. D., & Wood, B. A. (1987). Patterns and rates of enamel growth in the molar teeth of hominids. *Nature,* **335**, 509–514.

Bermúdez de Castro, J. M. (2002). *El Chico de la Gran Dolina.* Barcelona: Editorial Crítica.

Bermúdez de Castro, J. M., Rosas, A., Carbonell, E., Nicolás, M. E., Rodríguez, J., & Arsuaga, J. L. (1999). A modern human pattern of dental development in Lower Pleistocene hominids from Atapuerca-TD6 (Spain). *Proceedings of the National Academy of Sciences of the USA,* **96**, 4210–4213.

Bird, R. B., & Bird, D. W. (2002a). Constraints of knowing or constraints of growing? *Human Nature*, **13**, 239–267.

Bird, D. W., & Bird, R. B. (2002b). Children on the reef: Slow learning or strategic foraging? *Human Nature*, **13**, 269–297.

Blurton Jones, N. G. (1993). The lives of hunter–gather children: Effects of parental behavior and parental reproductive strategy. In *Juvenile Primates: Life History, Development, and Behavior*, eds. M. E. Pereira & L. A. Fairbanks, pp. 309–326. Oxford: Oxford University Press.

Blurton Jones, N., & Marlowe, F. W. (2002). Selection for delayed maturity: Does it take 20 years to learn to hunt and gather? *Human Nature*, **13**, 199–238.

Blurton Jones, N. G., Smith, L. C., O'Connell, J. F., Hawkes, K., & Kamazura, C. (1992). Demography of the Hadza, an increasing and high density population of savanna foragers. *American Journal of Physical Anthropology*, **89**, 159–181.

Bock, R. D., & Thissen, D. M. (1976). Fitting multi-component models for growth in stature. *Proceedings of the Ninth International Biometric Conference*, **1**, 431–442.

Bock, R. D., & Thissen, D. (1980). Statistical problems of fitting individual growth curves. In *Human Physical Growth and Maturation, Methodologies and Factors*, eds. F. E. Johnston, A. F. Roche, & C. Susanne, pp. 265–290. New York: Plenum Press.

Bogin, B. (1980). Catastrophe theory model for the regulation of human growth. *Human Biology*, **52**, 215–227.

Bogin, B. (1988). *Patterns of Human Growth*. Cambridge: Cambridge University Press.

Bogin, B. (1993). Why must I be a teenager at all? *New Scientist*, **137** (6 Mar), 34–38.

Bogin, B. (1994). Adolescence in evolutionary perspective. *Acta Paediatrica*, **406** (Suppl.), 29–35.

Bogin, B. (1999). *Patterns of Human Growth*, 2nd edn. Cambridge: Cambridge University Press.

Bogin, B. (2001). *The Growth of Humanity*. New York: Wiley-Liss.

Bogin, B., & Smith, B. H. (1996). Evolution of the human life cycle. *American Journal of Human Biology*, **8**, 703–716.

Bogin, B., Wall, M., & MacVean, R. B. (1992). Longitudinal analysis of adolescent growth of ladino and Mayan school children in Guatemala: Effects of environment and sex. *American Journal of Physical Anthropology*, **89**, 447–457.

Bookstein, F. C. (1978). *The Measurement of Biological Shape and Shape Change*. New York: Spinger-Verlag.

Boyd, E. (1980). *Origins of the Study of Human Growth*, eds. B. S. Savara & J. F. Schilke, based on unfinished work left by Richard E. Scannon. Eugene: University of Oregon Press.

Brody, S. (1945). *Bioenergetics and Growth*. New York: Reinhold Publishing Co.

Bromage, T. G., & Dean, M. C. (1985). Re-evaluation of the age at death of immature fossil hominids. *Nature*, **317**, 523–527.

Cameron, N., Mitchell, J., Meyer, D., Moodie, A., Bowie, M. D., Mann, M. D., & Hansen, J. D. L. (1988). Secondary sexual development of "Cape Coloured" girls following kwashiorkor. *Annals of Human Biology*, **15**, 65–76.

Cameron, N., Mitchell, J., Meyer, D., Moodie, A., Bowie, M. D., Mann, M. D., & Hansen, J. D. L. (1990). Secondary sexual development of "Cape Coloured" boys following kwashiorkor. *Annals of Human Biology*, **17**, 217–228.

Cameron, N., Grieve, C. A., Kruger, A., & Leschner, K. F. (1993). Secondary sexual development in rural and urban South African black children. *Annals of Human Biology*, **20**, 583–593.

Coehlo, A. M. Jr. (1985). Baboon dimorphism: Growth in weight, length and adiposity from birth to 8 years of age. In *Nonhuman Primate Models for Human Growth*, ed. E. S. Watts, pp. 125–159. New York: Alan R. Liss.

Condon, R. G. (1990). The rise of adolescence: Social change and life stage dilemmas in the Central Canadian Arctic. *Human Organization*, **49**, 266–79.

Conroy, G. C., & Vannier, M. W. (1991). Dental development in South African australopithecines. Part I: problems of pattern and chronology. *American Journal of Physical Anthropology*, **86**, 121–136.

Dean, C., Leakey, M. G., Reid, D., Schrenk, F., Schwartz, G. T., Stringer, C., & Walker, A. (2001). Growth processes in teeth distinguish modern humans from *Homo erectus* and earlier hominins. *Nature*, **414**, 628–631.

Detwyller, K. A. (1995). A time to wean: The hominid blueprint for the natural age of weaning in modern human populations. In *Breastfeeding: Biocultural Perspectives*, eds. P. Stuart-Macadam & K. A. Detwyller, pp. 39–74. New York: Aldine de Gruyter.

Ellison, P. T. (2001). *On Fertile Ground*. Cambridge: Harvard University Press.

Estioko-Griffin, A. A. (1986). Daughters of the Forest. *Natural History*, **95** (May), 36–43.

Galdikas, B. M., & Wood, J. W. (1990). Birth spacing patterns in humans and apes. *American Journal of Physical Anthropology*, **83**, 185–191.

Garn, S. M., & Petzold, A. S. (1983). Characteristics of the mother and child in teenage pregnancy. *American Journal of the Diseases of Childhood*, **137**, 365–68.

Goodall, J. (1983). Population dynamics during a 15-year period in one community of free-living chimpanzees in the Gombe National Park, Tanzania. *Zeitschrift für Tierpsychologie*, **61**, 1–60.

Halpern, C. T., Udry, R. J., Campbell, B., & Suchinddran, C. (1993). Testosterone and pubertal development as predictors of sexual activity: A panel analysis of adolescent males. *Psychosomatic Medicine*, **55**, 436–447.

Hamada, Y., & Udono, T. (2002). Longitudinal analysis of length growth in the chimpanzee (*Pan troglodytes*). *American Journal of Physical Anthropology*, **118**, 268–284.

Hawkes, K., O'Connell, J. F., & Blurton Jones, N. G. (1997). Hadza women's time allocation, offspring provisioning, and the evolution of post-menopausal lifespans. *Current Anthropology*, **38**, 551–578.

Hewlett, B. S. (1992). *Intimate Fathers: The Nature and Context of Aka Pygmy Paternal Infant Care*. Ann Arbor: University of Michigan Press.

Hill, K., & Kaplan, H. (1988). Tradeoffs in male and female reproductive strategies among the Ache. Parts I and II. In *Human Reproductive Behavior: A Darwinian Perspective*, eds. L. Betzig, M. Borgerhoff-Mulder, & P. Turke, pp. 277–289 and 291–305. Cambridge: Cambridge University Press.

Howell, N. (1979). *Demography of the Dobe !Kung*. New York: Academic Press.

Huxley, J. S. (1932). *Problems of Relative Growth*, 2nd edn. London: Methuen. Reprint 1972, New York: Dover.

Kaplan, H. S., Lancaster, J. B., Hill, K., & Hurtado, A. M. (2000). A theory of human life history evolution: Diet, intelligence, and longevity. *Evolutionary Anthropology*, **9**, 156–183.

Karlberg, J. (1987). On the modelling of human growth. *Statistics in Medicine*, **6**, 185–192.

Klein, R. G. (1989). *The Human Career: Human Biological and Cultural Origins*. Chicago: University of Chicago Press.

Laird, A. K. (1967). Evolution of the human growth curve. *Growth*, **31**, 345–355.

Lancaster, J. B., & Lancaster, C. S. (1983). Parental investment: The hominid adaptation. In *How Humans Adapt*, ed. D. J. Ortner, pp. 33–65. Washington, DC: Smithsonian Institution Press.

Leigh, S. R. (1996). Evolution of human growth spurts. *American Journal of Physical Anthropology*, **101**, 455–474.

Lewin, R. (1993). *Human Evolution: An Illustrated Introduction*. Oxford: Blackwell Scientific Publications.

Mann, A. (1975). *Some Paleodemographic Aspects of South African Australopithecines*, University of Pennsylvania Publications in Anthropology no.1. Philadelphia: University of Pennsylvania Press.

Martin, R. D. (1983). *Human Brain Evolution in an Ecological Context*, 52 James Arthur Lecture. New York: American Museum of Natural History.

Martin, R. D. (1990). *Primate Origins and Evolution: A Phylogenetic Reconstruction*. Princeton: Princeton University Press.

Moerman, M. L. (1982). Growth of the birth canal in adolescent girls. *American Journal of Obstetrics and Gynecology*, **143**, 528–532.

Muller, J., Nielsen, C. T., & Skakkebaek, N. E. (1989). Testicular maturation and pubertal growth and development in normal boys. In *The Physiology of Human Growth*, eds. J. M. Tanner & M. A. Preece, pp. 201–207. Cambridge: Cambridge University Press.

National Center for Health Statistics (1999). http://www.cdc.gov/nchs/data/statab/t991x23.pdf

Nelson, A. J., & Thompson, J. L. (2002). Neanderthal adolescent postcranial growth. In *Human Evolution through Developmental Change*, eds. N. Minugh-Purvis & K. McNamara, pp. 442–463. Baltimore: Johns Hopkins University Press.

Nishida, T., Takasaki, H., & Takahata, Y. (1990). Demography and reproductive profiles. In *The Chimpanzees of the Mahale Mountains: Sexual and Life History Strategies*, ed. T. Nishida, pp. 63–97. Tokyo: University of Tokyo Press.

Parker, S. T. (1996). Using cladistic analysis of comparative data to reconstruct the evolution of cognitive development in hominids. In *Phylogenies and the Comparative Method in Animal Behavior*, ed. E. Martins, pp. 433–448. Oxford: Oxford University Press.

Pereira, M. E., & Altmann, J. (1985). Development of social behavior in free-living non-human primates. In *Nonhuman Primate Models for Human Growth and Development*, ed. E. S. Watts, pp. 217–309. New York: Alan R. Liss.

Perieira, M. E., & Fairbanks, L. A. (eds.) (1993). *Juvenile Primates: Life History, Development, and Behavior*. New York: Oxford University Press.

Peschel, R. E., & Peschel, E. R. (1987). Medical insights into the *castrati* of opera. *American Scientist*, **75**, 578–583.

Plant, T. M. (1994). Puberty in primates. In *The Physiology of Reproduction*, 2nd edn, eds. E. Knobil & J. D. Neill, pp. 453–85. New York: Raven Press.

Prader, A. (1984). Biomedical and endocrinological aspects of normal growth and development. In *Human Growth and Development*, eds. J. Borms, R. R. Hauspie, A. Sand, C. Susanne, & M. Hebbelinck, pp. 1–22. New York: Plenum Press.

Pussey, A. (2001). Of genes and apes: Chimpanzee social organization and reproduction. In *Tree of Origin: What Primate Behavior Can Tell Us about Human Social Evolution*, ed. F. B. M. De Waal, pp. 9–38. Cambridge: Harvard University Press.

Schlegel, A., & Barry, H. (1991). *Adolescence: An Anthropological Inquiry*. New York: Free Press.

Scholl, T. O., & Hediger, M. L. (1993). A review of the epidemiology of nutrition and adolescent pregnancy: Maternal growth during pregnancy and its effects of the fetus. *Journal of the American College of Nutrition*, **12**, 101–107.

Schultz, A. H. (1924). Growth studies on primates bearing upon man's evolution. *American Journal of Physical Anthropology*, **7**, 149–164.

Schultz, A. H. (1960). Age changes in primates and their modification in Man. In *Human Growth,* ed. J. M. Tanner, pp. 1–20. Oxford: Pergamon Press.

Simons, E. L. (1989). Human origins. *Science*, **245**, 1343–1350.

Smith, B. H. (1986). Dental development in *Australopithecus* and early *Homo. Nature*, **323**, 327–330.

Smith, B. H. (1991). Dental development and the evolution of life history in Hominidae. *American Journal of Physical Anthropology*, **86**, 157–174.

Smith, B. H. (1992). Life history and the evolution of human maturation. *Evolutionary Anthropology*, **1**, 134–142.

Smith, B. H., & Tompkins, R. L. (1995). Toward a life history of the Hominidae. *Annual Review of Anthropology*, **25**, 257–279.

Smith, B. H., Crummett, T. L., & Brandt, K. L. (1994). Ages of eruption of primate teeth: A compendium for aging individuals and comparing life histories. *Yearbook of Physical Anthropology*, **37**, 177–231.

Stearns, S. C. (1992). *The Evolution of Life Histories*. Oxford: Oxford University Press.

Tanner, J. M. (1962). *Growth at Adolescence*, 2nd edn. Oxford: Blackwell Scientific Publications.

Tanner, J. M. (1981). *A History of the Study of Human Growth*. Cambridge: Cambridge University Press.

Tardieu, C. (1998). Short adolescence in early hominids: Infantile and adolescent growth of the human femur. *American Journal of Physical Anthropology*, **197**, 163–178.

Thom, R. (1983). *Mathematical Models of Morphogenesis*, translated by W. M. Brookes & D. Rand. New York: Halsted Press/John Wiley & Sons.

Thompson, D'Arcy W. (1917). *On Growth and Form*. Cambridge: Cambridge University Press.

Thompson, D'Arcy W. (1942). *On Growth and Form*, revised edn. Cambridge: Cambridge University Press.

Thompson, D'Arcy W. (1992). *On Growth and Form*, reissue of the 1942 revised edn. Cambridge: Cambridge University Press.

Turnbull, C. M. (1983). *The Human Cycle*. New York: Simon & Schuster.

Watts, D. P., & Pusey, A. E. (1993). Behavior of juvenile and adolescent great apes. In *Juvenile Primates: Life History, Development, and Behavior*, eds. M. E. Pereira & L. A. Fairbanks, pp. 148–170. Oxford: Oxford University Press.

Watts, E. S. (1985). Adolescent growth and development of monkeys, apes and humans. In *Nonhuman Primate Models for Human Growth and Development*, ed. E. S. Watts, pp. 41–65. New York: Alan R. Liss.

Watts, E. S. (1990). Evolutionary trends in primate growth and development. In *Primate Life History and Evolution*, ed. C. J. DeRousseau, pp. 89–104. New York: Wiley-Liss.

Weisfeld, G. (1999). *Evolutionary Principles of Human Adolescence.* New York: Basic Books.

Weisner, T. S. (1987). Socialization for parenthood in sibling caretaking societies. In *Parenting across the Life Span: Biosocial Dimensions*, eds. J. B. Lancaster, J. Altmann, A. S. Rossi, & L. R. Sherrod, pp. 237–270. New York: Aldine de Gruyter.

Worthman, C. M. (1993). Biocultural interactions in human development. In *Juvenile Primates: Life History, Development, and Behavior*, eds. M. E. Perieira & L. A. Fairbanks, pp. 339–357. New York: Oxford University Press.

3 *Postnatal ontogeny of facial position in* Homo sapiens *and* Pan troglodytes

B. McBRATNEY-OWEN
Harvard University

D. E. LIEBERMAN
Harvard University

Introduction

Morphological differences between adult taxa of closely related species, including ancestors and descendants, must arise from differences in development. It follows that to understand how and why such differences arose – and how they are manifested in patterns of integrated morphology – we need to understand the shifts in the developmental processes that generate them. For example, are changes in brain shape responsible for most of the differences in overall cranial shape between modern and archaic humans (Weidenreich, 1941), or do the differences in cranial shape between these taxa reflect multiple pathways of selection on particular aspects of the face and neurocranium that may be adaptations to climate, mastication, speech, and other such factors? Recent advances in evolutionary developmental biology (for reviews, see Hall, 1999; Carroll *et al.*, 2001), as well as studies of craniofacial integration and development (e.g., Ackermann & Krovitz, 2002; Krovitz, 2000; Ponce de León & Zollikofer, 2001) indicate that our null hypothesis should probably be the latter. Apparently most (but not necessarily all) evolutionary changes occur through shifts early in development that make use of pre-existing developmental pathways to generate novel but highly integrated morphologies.

Unraveling the complex relationship between patterns of growth and development and their underlying processes is especially interesting but challenging for studies of the origin of our own species. Unlike all other mammals, including closely related species of "archaic" *Homo* spp. (AH) such as the Neanderthals, "anatomically modern" *Homo sapiens* (AMHS) alone possesses an anteroposteriorly short and superoinferiorly vertical face that is tucked almost entirely

Patterns of Growth and Development in the Genus Homo, ed. J. L. Thompson, G. E. Krovitz and A. J. Nelson. Published by Cambridge University Press. © Cambridge University Press 2003.

underneath the anterior cranial fossa (Lieberman *et al.*, 2002). Indeed, facial retraction[1] appears to support a distinct phylogenetic species designation for AMHS. A key question, then, is what causes facial retraction in *Homo sapiens*? Unfortunately, we cannot answer this question in any definitive way because we do not yet understand enough about the many hierarchical levels and interrelated processes of skull development to identify the shifts responsible for such a morphological change. In addition, we are limited (justifiably) by the inability to do experiments on human embryos, and we lack a sufficiently good fossil record of development in any fossil hominid species.

Analyses of ontogeny, however, are a sensible place to start. Ontogenetic analyses are by definition studies of pattern rather than process. However, by studying patterns of craniofacial growth during ontogeny, we can identify morphological regions that appear to be integrated over space and time to identify key units of change. In addition, ontogenetic studies can help pinpoint when major shifts occur, allowing us to focus our studies on development at particular times of interest using all the data and tools at our disposal including experimental manipulations of laboratory animal models, analyses of craniofacial syndromes, quantitative trait locus analysis, etc. Here, we specifically explore the ontogeny of the cranial base and its effects on facial retraction as a means of generating hypotheses about the developmental bases for the origin of human cranial form.

Since the cranial base is the major zone of interaction and integration between the brain superiorly and face inferiorly, it follows that variables such as facial length, cranial base and fossae lengths, and the angle of the cranial base all have effects on facial position (Figure 3.1) (Lieberman, 1998 ; Lieberman *et al.*, 2000, 2002 ; Spoor *et al.*, 1999). Lieberman *et al.* (2002) tested this hypothesis using geometric morphometric analyses of three-dimensional coordinate data from available adult AMHS and AH cranial computed tomography (CT) scans and found that interactions between structural variables, such as cranial base angle, cranial base and fossa lengths and widths, and facial length, do indeed appear to contribute to variation in facial position among fossil and extant humans (Lieberman *et al.*, 2002). Lieberman *et al.* (2002) also did a preliminary analysis of the ontogeny of these structural variables (excluding fossae width) using geometric morphometric analyses of two-dimensional coordinate data from AMHS and chimpanzee (*Pan troglodytes*) radiographs and found that the

---

[1] Facial projection (termed neuro-orbital disjunction by Weidenreich (1941)) is defined here as the anteroposterior position of the face (viscerocranium) relative to the anterior cranial fossa ; prognathism is defined as the anteroposterior projection of the lower face relative to the upper face. Facial retraction (the absence of facial projection) thus differs from orthognathism (the absence of prognathism) (Bilsborough & Wood, 1988). Although AMHS exhibit a lack of prognathism along with reduced facial projection, these two configurations are independent. For example, *Paranthropus boisei* lacks prognathism yet still has a high degree of facial projection (Bilsborough & Wood, 1988).

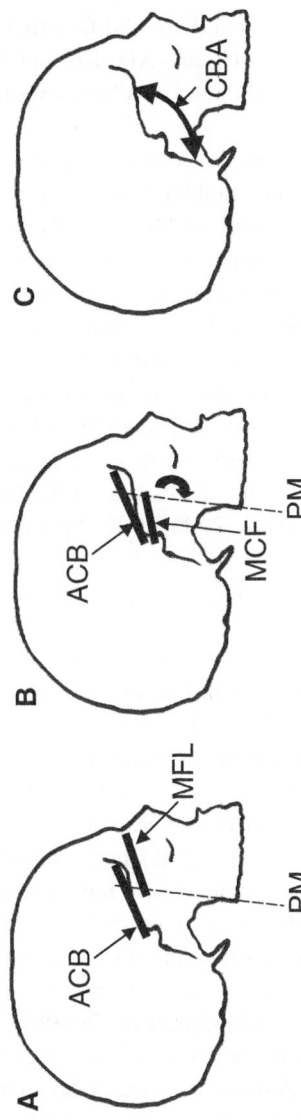

Figure 3.1 Schematic representation of spatial interactions predicted to have an affect on facial position. A, anterior cranial base length relative to midfacial length; B, anterior cranial base length relative to middle cranial fossa length; C, cranial base angle. ACB = anterior cranial base; MFL = midfacial length; MCF = middle cranial fossa length; PM plane = posterior maxillary plane. See text for descriptions of how lengths, angle, and PM plane are defined.

same structural variables of the cranial base and face that contribute to differences in facial position between AMHS and AH also contribute to similar facial positions throughout postnatal growth and development of the extant hominids (Lieberman *et al.*, 2002).

Many questions, however, remain about the facial retraction hypothesis. When in ontogeny is the developmental basis for facial retraction set ? And to what extent is facial retraction a result of changes in the cranial base, the brain, or the face ? Thus, as a prelude to further studies that focus on the developmental bases of particular ontogenetic patterns, we analyze here in depth the hypothesis that AMHS facial retraction derives in part from ontogenetic alterations in structural variables of the cranial base and face (Lieberman *et al.*, 2002). We use Euclidean distance matrix analysis to compare two-dimensional coordinate data from longitudinal and cross-sectional samples of lateral radiographs of AMHS and *P. troglodytes* to examine the ontogeny of specific spatial interactions that are predicted to have an effect on facial position in AMHS and *P. troglodytes* (Lieberman, 1998 ; Lieberman *et al.*, 2000, 2002 ; Spoor *et al.*, 1999).

Anatomical model

Three spatial interactions, illustrated in Figure 3.1, are predicted to influence facial projection. First, because the roof of the face is the floor of the anterior cranial base (mostly the orbital and cribriform plates), the anteroposterior length of the midface relative to the anteroposterior length of the anterior cranial base must influence the position of the anterior aspect of the face relative to the anterior margin of the neurocranium. Put simply, an anteroposteriorly short midface relative to a long anterior cranial base must result in facial retraction (Figure 3.1A). Midfacial length can be measured in more than one way but is defined here as the projected midline anteroposterior distance between the posterior maxillary (PM) points and nasion (see Table 3.1 for landmark descriptions and Figure 3.2 for location of landmarks). The PM points, which are the superior termini of the PM plane (for discussion see McCarthy & Lieberman, 2001), are useful landmarks for defining the back of the midface because they lie on the projected midline boundary between the cranial base (the anterior most points on the middle cranial fossae) and the back of the ethmomaxillary complex (Enlow, 1990). Following other recent studies (Lieberman & McCarthy, 1999 ; Lieberman *et al.*, 2000, 2002 ; Ross & Ravosa, 1993 ; Spoor *et al.*, 1999), anterior cranial base length is defined as the midsagittal distance between sella and foramen caecum.

Another influence on facial projection is the length of the middle cranial fossa relative to the length of the anterior cranial base. Because the ethmomaxillary

Table 3.1 *Landmarks in this study*

Landmark	Abbreviation	Definition
Basion	BAS	Midline point on the anterior margin of the foramen magnum
Anterior nasal spine	ANS	Most anterior midline point on the maxillary body at the level of the nasal floor[a]
Nasion	NAS	Midline point where the two nasal bones and the frontal intersect
Sella	SEL	Center of the sella turcica[a]
Posterior maxillary plane point	PMP	Midline intersection of posterior maxillary plane with planum of sphenoid[b]
Foramen caecum	FOC	Pit on the cribriform plate between the crista galli and the endocranial wall of the frontal bone[a]
Posterior nasal spine	PNS	Most posterior midline point on the maxillary body at the level of the nasal floor at the articulation of the hard and soft palates[a]

Source: All definitions from White & Folkens (1991) except for [a]Lieberman and McCarthy (1999) and [b]defined here.

complex of the face grows forward through displacement under the anterior cranial fossa from its boundary with the middle cranial fossa (see Enlow, 1990), it was previously suggested that a proportionately longer middle cranial fossa may cause the face to project more anteriorly beyond the anterior cranial base (Lieberman, 1998). This hypothesis was falsified (Spoor *et al.*, 1999); however, Lieberman *et al.* (2002) recently showed that the length and width of the middle cranial fossa (which houses the temporal lobes) is significantly greater in AMHS than in AH.

The length of the middle cranial fossa can have two important effects on facial position. First, the middle cranial fossa makes up a significant proportion of the length of the anterior cranial fossa, so that a longer middle cranial fossa can contribute to a longer anterior cranial base relative to facial length (see above). Second, expansion of the middle cranial fossa may contribute to cranial base angulation via its effects on the PM plane (Figure 3.1B) (Lieberman, 2000; Lieberman *et al.*, 2002; McCarthy & Lieberman, 2001). The PM plane is a projected midline chord between two termini (the maxillary tuberosities, inferiorly; and the most anterior points on the greater wings of the sphenoid, superiorly) that defines the back of the face along its junction with the middle cranial fossa (Enlow, 1990; McCarthy & Lieberman, 2001). Since the PM plane is constrained to intersect the neutral horizontal axis of the orbits at 90° in all primates at all postnatal ontogenetic stages, and because the top of the orbits is the floor of the anterior cranial base, the rotation of the PM plane

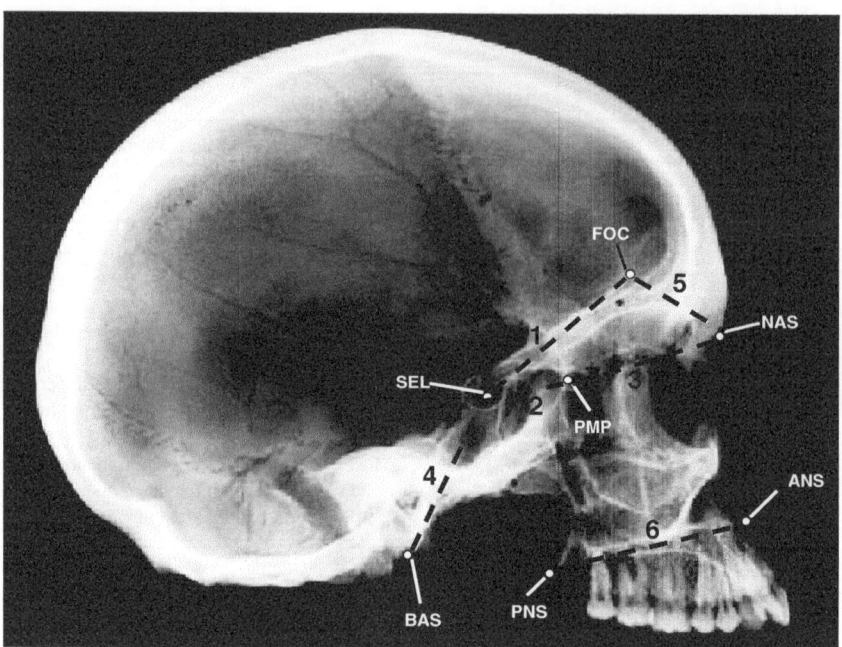

Figure 3.2   Seven sagittal and parasagittal landmarks on lateral radiograph of adult
male AMHS. ANS, anterior nasal spine ; BAS, basion ; FOC, foramen caecum ; NAS,
nasion ; PMP, posterior maxillary plane point ; PNS, posterior nasal spine ; SEL, sella.
See Table 3.1 for descriptions of precise landmark placement. Black dashed lines refer
to linear distances tested in Hypotheses 1–6 (see text).

necessitates a ventral rotation of the face (McCarthy & Lieberman, 2001),
tucking the face under the anterior cranial fossa (Figure 3.1B). The length of
the middle cranial fossa can be difficult to measure but is best defined as the
projected midline distance between sella, which marks the division between
the anterior (prechordal) and posterior (postchordal) portions of the cranial
base (for review, see Lieberman *et al.*, 2000) and the projected midline average
of the PM points (see above). Although Lieberman (1998) originally termed
this distance sphenoid length, it is actually the projected midline length of the
anterior portion of the middle cranial fossae (Lieberman, 2000 ; McCarthy &
Lieberman, 2001 ; Spoor *et al.*, 1999).

As noted above, a third and related interaction that is hypothesized to influence facial projection is the midline angle of the cranial base, most commonly
measured as the angle between two projected midline chords : (1) a postchordal
line from basion to sella and (2) a prechordal line from sella to the foramen caecum (Figure 3.1C) (for discussion, see Lieberman & McCarthy, 1999). The cranial base angle has long been recognized to influence facial projection through

its effects on facial orientation (Biegert, 1963 ; Lieberman, 2000 ; Lieberman *et al.*, 2000 ; May & Sheffer, 1999 ; Moss & Young, 1960 ; Ravosa, 1988, 1991a, 1991b ; Ravosa *et al.*, 2000a, 2000b ; Ross, 1995a, 1995b ; Ross & Ravosa, 1993 ; Shea, 1985a, 1985b, 1986, 1988 ; Weidenriech, 1941). In particular, recent studies (Lieberman *et al.*, 2000, 2002 ; Spoor *et al.*, 1999) suggest that the primary cause for reduced facial projection in AMHS is increased cranial base flexion as the cranial base in adult AMHS is approximately 15° more flexed than in AH (Lieberman *et al.*, 2000, 2002). As noted above, the floor of the anterior cranial base is the roof of the face. In addition, the orientation of the roof of the face and the posterior margin of the face (the PM plane) is constrained to be near 90° in anthropoids in which the midline of the cranial base approximates the orientation of the orbits (McCarthy & Lieberman, 2001). As a result of this constraint, a more flexed cranial base may cause more of the face to be rotated ventrally under the anterior cranial fossa, and a more extended cranial base may cause the face to rotate dorsally in front of the anterior cranial fossa (Figure 3.1C).

While it would be ideal to compare the ontogeny of these variables in AMHS and AH (e.g., Neanderthals) such a comparison is not possible at this time because there are no juvenile AH with sufficiently complete cranial bases. We can, however, test the above model by examining how anterior cranial base length, midfacial length, and middle cranial fossa length affect facial position in ontogenetic samples of AMHS and our most closely related relative, the chimpanzee. We do so in several ways. We first identify when during ontogeny specific combinations of these structural variables (anterior cranial base length, midfacial length, and middle cranial fossa length) contribute to different facial positions between humans and chimpanzees. We then test if different growth patterns contribute to differences in facial projection between *P. troglodytes* and AMHS. In particular, we predict that in a comparison of AMHS and *P. troglodytes* cranial ontogeny, (1) the human midface will remain relatively anteroposteriorly short in relation to the anterior cranial base and (2) the human middle cranial fossa length will be elongated. Finally, we also analyze upper and lower facial length and posterior cranial base length and discuss the effects of cranial base angulation on facial position.

We stress that this study is not a direct test of the developmental bases for facial retraction in AMHS. Such a study will require both a better fossil record and a more complete understanding of many processes of craniofacial development and integration. However, because the general processes of cranial growth are similar in all primates (Moore & Lavelle, 1974) and adult chimpanzees have very projecting faces, the effort to identify structural variables and growth patterns that contribute to facial projection in extant taxa such as humans and chimpanzees may provide a reasonable predictive tool for studying the effects of hypothesized differences in cranial growth between AMHS and AH that affect the ontogeny of the facial retraction. The data should also help clarify how facial projection in AH differs structurally and developmentally from facial retraction in AMHS.

Materials and methods

Samples

Radiographs were used to compare cranial ontogeny in AMHS and *P. troglodytes*. The AMHS sample was from the Denver Growth Study, a longitudinal radiographic study of skull growth and development spanning from 1 month of age to adulthood in Americans of European descent (Maresh, 1948; Maresh & Washburn, 1938; McCammon, 1970; Lieberman & Mc-Carthy, 1999). Lateral radiographs of six males and six females were used at each of four age groups: Stage I = 1.75 years, Stage II = 5.75 years, Stage III = 9.75 years, and Stage IV = 17.75 years. These known ages roughly correlate with dental, skeletal, and neural growth trajectories (Krogman, 1931; Schultz, 1962; Smith, 1989). Stage I was prior to the eruption of dm^2. Stage II was just prior to eruption of M^1, when neural growth was 95% complete (Krogman, 1931; Schultz, 1962; Smith, 1989). Stage III was prior to the eruption of M^2. Stage IV was after M^2 eruption and near the completion of facial growth (Krogman, 1931; Schultz, 1962; Smith, 1989). Nine individuals had longitudinal data through all four stages, but, due to incomplete longitudinal samples for other individuals in the study, the remainder of the sample was cross-sectional (Table 3.2).

Data for *P. troglodytes* were from a cross-sectional sample of cranial radiographs taken of specimens housed at the Peabody Museum, Harvard University; the Museum of Comparative Zoology, Harvard University; the American Museum of Natural History, New York; and the Cleveland Museum of Natural History (see Lieberman & McCarthy (1999) for details). Specimens were allocated to Stages I–IV matched developmentally to the human stages discussed above based on the following criteria of dental eruption and craniofacial development: Stage I, prior to eruption of dm^2; Stage II, after the eruption of dm^2, prior to eruption of M^1, when neural growth is 95% complete; Stage III, after the eruption M^1 and prior to the eruption of M^2; Stage IV, after M^2 eruption and near the completion of facial growth (Krogman, 1931; Schultz, 1962; Smith, 1989). Six chimpanzees of unknown sex from each developmental stage were examined with the exception of Stage I which was limited to only three individuals (Table 3.2).

Table 3.2 *Modern human and chimpanzee samples used in this paper.*
Stage I = approximately half way through the neural growth trajectory;
Stage II = near the end of the neural growth phase; Stage III = half way
between Stages II and IV; Stage IV = completion of the facial growth
trajectory. Homo sapiens *were assigned to each stage based on known*
corresponding ages of each individual. Pan troglodytes *were assigned to*
each stage based on dental eruption phases that correspond to each
growth trajectory. The sex of each chimpanzee is unknown

		Stage I	Stage II	Stage III	Stage IV
AMHS					
Females	#012	████████████████████████████████			
	#064	████			
	#106	██████████████████████████			
	#107				████
	#110	██████████████████████████			
	#114	██████████████████████████████			
	#115	████████████████████████████████			
	#121		█████████		
Males	#505	████			
	#515				████
	#533	██████████████████████████████			
	#570	███████████████████			
	#582	████████████████████████████████			
	#585		███████████████████████		
	#609	██████████████████████████████			
	#616	████████████████████████████████			
Pan troglodytes					
	#51385	████			
	#89427	████			
	#180088	████			
	#410		████		
	#753		████		
	#1395		████		
	#1848		████		
	#51210		████		
	#51387		████		
	#1174			████	
	#1177			████	
	#1183			████	
	#51405			████	
	#77817			████	
	#202412			████	
	#240				████
	#829				████
	#887				████
	#51201				████
	#51386				████
	#54349				████

Measurements

Seven landmarks (mostly midsagittal) were identified on each radiograph (defined in Table 3.1 and shown in Figure 3.2). In order to test hypotheses about variation in facial projection that derive from differences in cranial base and facial dimensions, most of these landmarks were from the face and cranial base (Lieberman, 1998; Spoor *et al.*, 1999). Landmarks for each individual were recorded on tracing paper and the two-dimensional coordinates for each landmark were then measured with a MicroScribe 3DX digitizer (Immersion Corp, San Jose, CA).

In order to test the effects of landmark identification error on measurement variance, we calculated the coefficient of variation (CV) of all 21 possible distance measurements by collecting landmark coordinate data for all seven landmarks once each on five different days (thus five different sets of coordinates per landmark) from the same radiograph of an adult human male. Our null hypothesis, that replicate measurements from the same individual were the same ($P \leq 0.05$), was not rejected for 20 distances but was rejected for BAS–PNS (0.08). This distance was between a pair of landmarks that were very close together thus any variance in landmark identification greatly affected the relatively small distance calculated between the landmarks. The results for this particular distance were subsequently disregarded.

Geometric morphometric analyses

Euclidean distance matrix analysis (EDMA) was used to test the general hypotheses outlined above and defined more specifically below. EDMA is a coordinate-free, quantitative method that compares linear distances between all landmarks in a sample (Lele, 1991, 1993; Lele & Cole, 1996; Lele & Richtsmeicr, 1991, 1995; Richtsmeier & Lele, 1993; Richtsmeier *et al.*, 1993). EDMA was particularly useful in our analyses as we were able to compare shape differences at specific linear distances between AMHS and *P. troglodytes* and then determine if differences in growth patterns in these linear distances contributed to any potential differences in shape. Due to different biological sizes among our samples, and our specific desire to understand differences in shape, we reduced the effects of size by scaling our samples by the geometric mean of all landmarks calculated separately for each sample (Jungers *et al.*, 1995). WinEDMA (Cole, 2002) was used to analyze shape and growth differences between samples.

For the shape difference matrix (SDM) analyses, size-corrected samples of AMHS and *P. troglodytes* from the same stages were compared: AMHS Stage I/*P. troglodytes* Stage I; AMHS Stage II/*P. troglodytes* Stage II; AMHS

Stage III/*P. troglodytes* Stage III ; AMHS Stage IV/*P. troglodytes* Stage IV. The AMHS sample was always in the numerator. Elements of the SDM with a ratio of 1.00 indicated that those particular linear distances were the same between the samples. SDM elements greater than 1.00 indicated that the numerator sample (AMHS) was relatively larger in those linear distances at that particular stage, while SDM elements less than 1.00 indicated that the denominator sample (*P. troglodytes*) was relatively larger in those linear distances at that particular stage.

For the growth difference matrix (GDM) analyses, size corrected samples of AMHS and *P. troglodytes* from successive stages were compared : AMHS from Stage I to II/*P. troglodytes* from Stage I to II ; AMHS from Stage II to III/*P. troglodytes* from Stage II to III ; AMHS from Stage III to IV/*P. troglodytes* from Stage III to IV. The AMHS samples were always in the numerator for the formation of the GDM. GDM elements greater than 1.00 indicated that the numerator sample (AMHS) grew more in those linear distances between two particular stages, while GDM elements less than 1.00 indicated that the denominator sample (*P. troglodytes*) grew more in those linear distances between two particular stages.

In order to determine which linear distances were significantly different between samples or if growth patterns of linear distances were significantly different between samples, non-parametric bootstrapping of 1000 re-samples was used to determine confidence intervals of 0.90 ($\alpha = 0.10$) for each size corrected linear distance (Lele & Richtsmeier, 1995). Due to larger sample sizes ($n = 12$ per stage), the reference sample for the bootstrapping procedure was always the numerator (AMHS). Linear distances with confidence intervals that did not span 1.00 indicated a significant difference in length or growth between samples in that linear distance.

Hypotheses

In order to test the general hypothesis that the AMHS face was retracted relative to the anterior cranial fossa, as the result of a long anterior cranial base relative to facial length and/or a relatively longer middle cranial fossa with a more flexed cranial base, SDM analysis was used to test a series of individual linear distance hypotheses in humans and chimpanzees. Table 3.3 summarizes the structural variables and the corresponding landmark-defined linear distances tested by each hypothesis.

HI_{shape} : *Relative anterior cranial base length (as measured by SEL to FOC) is significantly longer in AMHS than* P. troglodytes *at each ontogenetic stage.* If the ratios for anterior cranial base length between AMHS and *P. troglodytes*

Table 3.3 *Corresponding hypotheses, linear distances, and structural variables tested by SDM and GDM analyses*

Hypothesis	Linear distance[a]	Structural variable
H1	SEL–FOC	Anterior cranial base
H2	SEL–PMP	Middle cranial fossa
H3	NAS–PMP	Midface
H4	SEL–BAS	Posterior cranial base
H5	NAS–FOC	Upper face
H6	ANS–PNS	Lower face

[a] Refer to Table 3.1 for landmark descriptions.

at each ontogenetic stage are greater than 1.00, and the confidence intervals do not span 1.00, then reject H_01_{shape} : Relative anterior cranial base length is not significantly longer in AMHS than *P. troglodytes* at each ontogenetic stage.

$H2_{shape}$: *Relative middle cranial fossa length (as measured by SEL to PMP) is significantly longer in AMHS than in* P. troglodytes *at each ontogenetic stage.* If the ratios for middle cranial fossa length between AMHS and *P. troglodytes* at each ontogenetic stage are greater than 1.00, and the confidence intervals do not span 1.00, then reject H_02_{shape} : Relative middle cranial fossa length is not significantly longer in AMHS than in *P. troglodytes* at each ontogenetic stage.

$H3_{shape}$: *Relative midfacial length (as measured by NAS to PMP) is significantly shorter in AMHS than* P. troglodytes *at each ontogenetic stage.* If the ratios for midfacial length between AMHS and *P. troglodytes* at each ontogenetic stage are less than 1.00, and the confidence intervals do not span 1.00, then reject H_03_{shape} : Relative midfacial length is not significantly shorter in AMHS than *P. troglodytes* at each ontogenetic stage.

$H4_{shape}$: *Relative posterior cranial base length (as measured by SEL to BAS) is significantly longer in AMHS than* P. troglodytes *at each ontogenetic stage.* If the ratios for posterior cranial base length between AMHS and *P. troglodytes* at each ontogenetic stage are greater than 1.00, and the confidence intervals do not span 1.00, then reject H_04_{shape} : Relative posterior cranial base length is not significantly longer in AMHS than *P. troglodytes* at each ontogenetic stage.

$H5_{shape}$: *Relative upper facial length (as measured by NAS to FOC) is significantly shorter in AMHS than* P. troglodytes *at each ontogenetic stage.* If the ratios for upper facial length between AMHS and *P. troglodytes* at each developmental stage are less than 1.00, and the confidence intervals do not span 1.00, then reject H_05_{shape} : Relative upper facial length is not significantly shorter in AMHS than *P. troglodytes* at each ontogenetic stage.

$H6_{shape}$ *: Relative lower facial length (as measured by ANS to PNS) is significantly shorter in AMHS than* P. troglodytes *at each ontogenetic stage.* If the ratios for lower facial length between AMHS and *P. troglodytes* at each ontogenetic stage are less than 1.00, and the confidence intervals do not span 1.00, then reject H_06_{shape} : Relative lower facial length is not significantly shorter in AMHS than *P. troglodytes* at each ontogenetic stage.

If the SDM analyses reveal significant relative differences in individual linear distances between AMHS and *P. troglodytes* at particular ontogenetic stages, then we predict that the growth patterns of those linear distances will also differ between the species. GDM analyses will test the following growth hypotheses, which correspond with the linear distances tested above:

$H1–6_{growth}$ *: Linear distances that differ between AMHS and* P. troglodytes *will also exhibit differences in growth pattern at some or all ontogenetic stages.* If the ratios for linear distance growth between AMHS and *P. troglodytes* are not equal to 1.00, and the confidence intervals do not span 1.00, then reject $H_01–6_{growth}$: Growth patterns do not differ significantly between AMHS and *P. troglodytes.*

Results

SDM analyses

The SDM analyses results were summarized in Figure 3.3 and listed in Table 3.4 (see Table 3.5 for linear distance ratios and confidence intervals).

$H1_{shape}$ *:* Not rejected. Relative anterior cranial base length was significantly longer in AMHS than *P. troglodytes* at each ontogenetic stage.

Table 3.4 *Summary of results of SDM analyses on AMHS and* P. troglodytes. *H1–6 = Hypotheses 1–6*

Stage I	Stage II	Stage III	Stage IV
Larger in AMHS			
Anterior cranial base (H1)	Anterior cranial base (H1)	Anterior cranial base (H1)	Anterior cranial base (H1)
Posterior cranial base (H4)	Middle cranial fossa (H2)	Posterior cranial base (H4)	Middle cranial fossa (H2)
Lower face (H6)	Posterior cranial base (H4)	Lower face (H6)	Posterior cranial base (H4)
Larger in Pan troglodytes			
Upper face (H5)	Midface (H3)	Midface (H3)	Midface (H3)
	Upper face (H5)	Upper face (H5)	Upper face (H5)

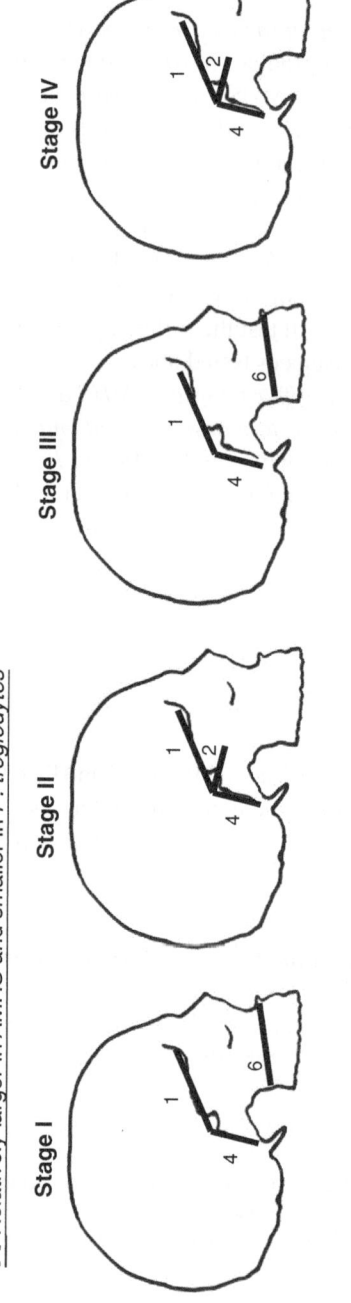
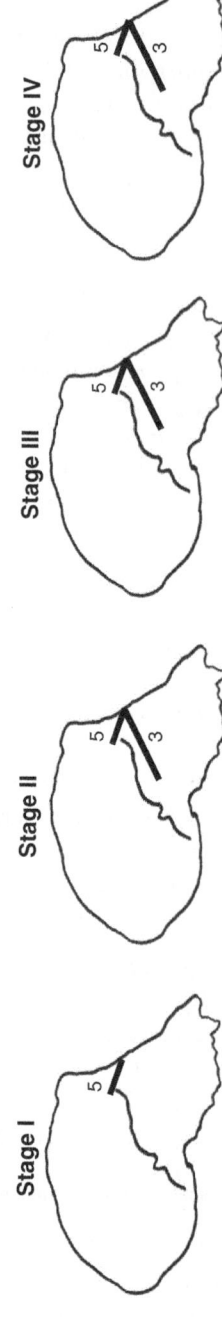

Figure 3.3 Shape difference matrix summaries for comparisons between AMHS and *Pan troglodytes* Stages I, II, III, and IV. The linear distances illustrated in this figure are listed in Table 3.3. Numbers correspond to Hypotheses 1–6 (see text).

Table 3.5 *Form difference matrix (FDM) and growth differences matrix (GDM) results of Euclidean distance matrix analyses (EDMA) of linear distances*

Analysis	Linear distance[a] (Hypothesis)	Results[b]		
		Low	Estimate	High
Form difference matrix	**SEL to FOC** (1)	**1.037**	**1.083**	**1.129**
AMHS Stage I/*P. troglodytes* Stage I	SEL to PMB (2)	0.930	0.993	1.055
	NAS to PMP (3)	0.974	1.023	1.074
	SEL to BAS (4)	**1.010**	**1.064**	**1.116**
	NAS to FOC (5)	**0.568**	**0.666**	**0.788**
	ANS to PNS (6)	**1.029**	**1.071**	**1.113**
Form difference matrix	**SEL to FOC** (1)	**1.126**	**1.174**	**1.221**
AMHS Stage II/*P. troglodytes* Stage II	**SEL to PMP** (2)	**1.075**	**1.153**	**1.235**
	NAS to PMP (3)	**0.877**	**0.930**	**0.985**
	SEL to BAS (4)	**1.032**	**1.097**	**1.165**
	NAS to FOC (5)	**0.406**	**0.488**	**0.599**
	ANS to PNS (6)	0.939	1.063	1.164
Form difference matrix	**SEL to FOC** (1)	**1.033**	**1.087**	**1.140**
AMHS Stage III/*P. troglodytes* Stage III	SEL to PMP (2)	0.973	1.096	1.249
	NAS to PMP (3)	**0.886**	**0.929**	**0.973**
	SEL to BAS (4)	**1.099**	**1.151**	**1.209**
	NAS to FOC (5)	**0.561**	**0.664**	**0.809**
	ANS to PNS (6)	**1.029**	**1.082**	**1.143**
Form difference matrix	**SEL to FOC** (1)	**1.024**	**1.103**	**1.190**
AMHS Stage IV/*P. troglodytes* Stage IV	**SEL to PMP** (2)	**1.153**	**1.269**	**1.444**
	NAS to PMP (3)	**0.843**	**0.895**	**0.946**

Table 3.5 (*cont.*)

Analysis	Linear distance[a] (Hypothesis)	Results[b]		
		Low	Estimate	High
	SEL to BAS **(4)**	**1.049**	**1.157**	**1.281**
	NAS to FOC **(5)**	**0.655**	**0.743**	**0.838**
	ANS to PNS (6)	0.994	1.046	1.103
Growth difference matrix	**SEL to FOC** **(1)**	**1.025**	**1.084**	**1.154**
AMHS from Stage I to II/*P. troglodytes* from Stage I to II	**SEL to PMP** **(2)**	**1.060**	**1.161**	**1.276**
	NAS to PMP **(3)**	**0.847**	**0.909**	**0.984**
	SEL to BAS (4)	0.954	1.031	1.112
	NAS to FOC **(5)**	**0.564**	**0.732**	**0.953**
	ANS to PNS (6)	0.881	0.993	1.102
Growth difference matrix	**NAS to FOC** **(1)**	**1.045**	**1.361**	**1.757**
AMHS from Stage II to III/ *P. troglodytes* from Stage II to III	SEL to PMP (2)	0.832	0.951	1.098
	NAS to PMP (3)	0.927	0.999	1.076
	SEL to BAS (4)	0.980	1.049	1.128
	SEL to FOC **(5)**	**0.865**	**0.926**	**0.992**
	ANS to PNS (6)	0.920	1.017	1.153
Growth difference matrix	NAS to FOC (1)	0.875	1.119	1.375
AMHS from Stage III to IV/ *P. troglodytes* from Stage III to IV	SEL to PMP (2)	0.988	1.158	1.355
	NAS to PMP (3)	0.891	0.963	1.038
	SEL to BAS (4)	0.900	1.005	1.124
	SEL to FOC (5)	0.937	1.014	1.106
	ANS to PNS (6)	0.898	0.967	1.044

[a] See Table 3.1 for landmark abbreviations and Table 3.3 for corresponding hypotheses, linear distances, and structural variables tested by SDM and GDM analyses.

[b] Significant ratio results and confidence intervals are in bold ($P = 0.10$).

H2$_{shape}$: Not rejected. Relative middle cranial fossa length was significantly longer in AMHS than in *P. troglodytes* at Stages II and IV. However, there was no significant difference between the species at Stage I and III.

H3$_{shape}$: Not rejected. Relative midfacial length was significantly shorter in AMHS than *P. troglodytes* at Stages II, III, and IV. There was no significant difference in relative midfacial length between AMHS and *P. troglodytes* at Stage I.

H4$_{shape}$: Not rejected. Relative posterior cranial base length was significantly longer in AMHS than *P. troglodytes* at each ontogenetic stage.

H5$_{shape}$: Not rejected. Relative upper facial length was significantly shorter in AMHS than *P. troglodytes* at each ontogenetic stage.

H6$_{shape}$: Rejected. Relative lower facial length was not significantly different between AMHS and *P. troglodytes* at Stages II and IV and was significantly greater in AMHS at Stages I and III.

GDM analyses

The GDM analyses results were summarized in Figure 3.4 and listed in Table 3.6 (see Table 3.5 for linear distance ratios and confidence intervals).

H1$_{growth}$: Not rejected. Growth patterns for anterior cranial base length were significantly different between AMHS and *P. troglodytes* from Stage I to II and from Stage II to III.

H2$_{growth}$: Not rejected. Growth patterns for middle cranial fossa length were significantly different between AMHS and *P. troglodytes* from Stage I to II.

H3$_{growth}$: Not rejected. Growth patterns for midfacial length were significantly different between AMHS and *P. troglodytes* from Stage I to II.

H4$_{growth}$: Rejected. Growth patterns for posterior cranial base length were not significantly different between AMHS and *P. troglodytes* between any ontogenetic stages.

H5$_{growth}$: Not rejected. Growth patterns for upper facial length were significantly different between AMHS and *P. troglodytes* from Stage I to II and from Stage II to III.

H6$_{growth}$: Rejected. Growth patterns for lower facial length were not significantly different between AMHS and *P. troglodytes* between any ontogenetic stages.

Summary of results

H1 : Anterior cranial base length
As the anterior cranial base length is significantly longer in AMHS than *P. troglodytes* at all four ontogenetic stages in this analysis, the growth pattern

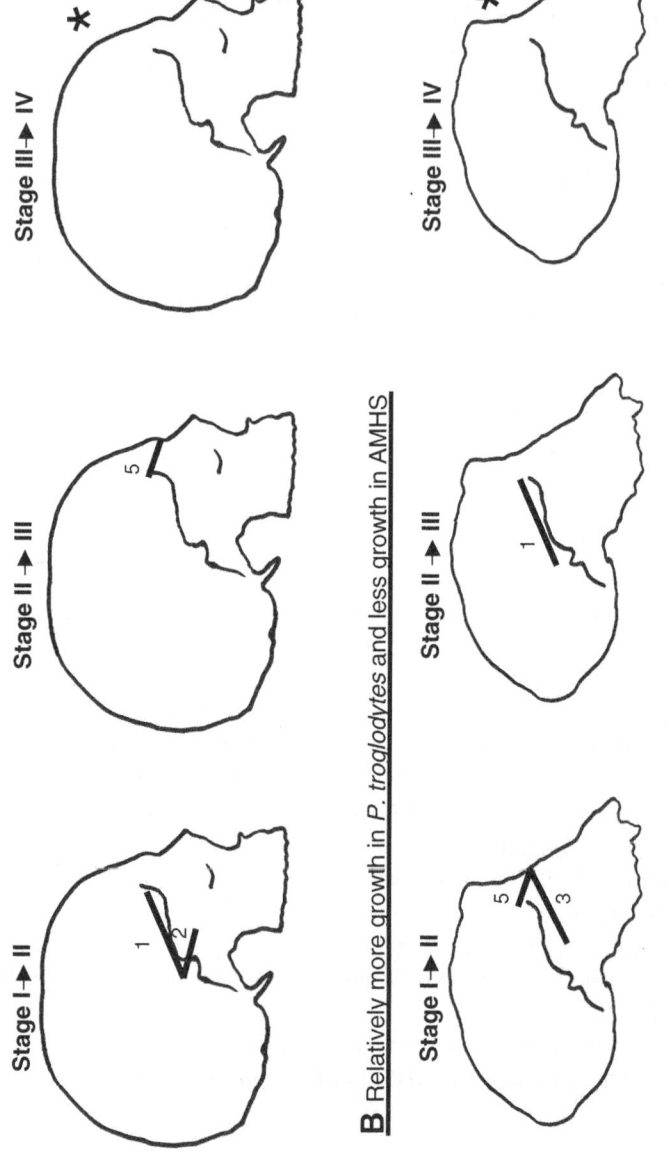

Figure 3.4 Growth difference matrix summaries for AMHS and *P. troglodytes* samples from Stages I to II, II to III, and III to IV. Linear distances illustrated in this figure are listed in Table 3.3. Numbers correspond to Hypotheses 1–6 (see text). *, no significant growth differences from Stage III to IV.

Table 3.6 *Summary of results of GDM analyses on AMHS and P. troglodytes. H1–6 = Hypotheses 1–6*

Stage I to II	Stage II to III	Stage III to IV
More growth in AMHS		
Anterior cranial base (H1)	Upper face (H5)	n.s.
Middle cranial fossa (H2)		
More growth in Pan troglodytes		
Midface (H3)	Anterior cranial base (H1)	n.s.
Upper face (H5)		

n.s., no significant difference.

that first establishes this shape pattern at Stage I must occur prior to 1.75 years in humans. Between Stage I and II, the anterior cranial base grows relatively more in AMHS; however, this growth pattern reverses between Stage II and III when *P. troglodytes* grows relatively more in anterior cranial base length. Despite the fact that *P. troglodytes* experiences relatively more growth between those stages, the anterior cranial base still remains relatively longer in AMHS. Thus, the shape pattern that is established by Stage I is not perturbed by different growth patterns between the two species. These results support the prediction that anterior cranial base length remains relatively elongated in AMHS thus positioning the frontal lobes more anteriorly relative to the short midface.

H2 : Middle cranial fossa length

Relative middle cranial fossa length is only significantly larger in AMHS than in *P. troglodytes* at Stages II and IV. The growth patterns that may underlie these changes in shape pattern are only partially revealed in the growth analyses. There is significantly more relative growth in AMHS middle cranial fossa length between Stage I and II, perhaps explaining how middle cranial fossa length becomes significantly greater in AMHS by Stage II. However, AMHS and *P. troglodytes* do not differ significantly in their growth patterns in the later stages although the significant shape patterns alter. It should be noted, however, that middle cranial fossa length is greater in AMHS than *P. troglodytes* at Stage III, but not significantly. Therefore, we suggest that greater growth in AMHS between Stage I and II increases the relative length of the middle cranial fossa, and it remains larger as a result of similar growth patterns between the species in the later developmental stages.

The relatively larger middle cranial fossa length in humans may be a result of greater volume of the temporal lobes, which sit partially within the middle cranial fossa. Compared to other extant hominoids, modern humans have slightly expanded temporal lobes relative to brain size (Semendeferi & Damasio, 2000). Since the PM plane is constrained to intersect the orbital axis at 90° in primates and the anterior border of the middle cranial fossa is the PM plane, an increase in middle cranial fossa length in AMHS may necessitate a rotation of the PM plane along with the face (the "facial block" as defined by McCarthy and Lieberman, 2001), which would rotate the face ventrally beneath the frontal lobes (see below).

H3 : Midfacial length
As relative midfacial length is significantly shorter in AMHS than *P. troglodytes* at Stages II, III, and IV, the growth pattern that establishes this shape pattern at Stage II must occur prior to 5.75 years in humans. Our results support this as relative midfacial length grows significantly more in *P. troglodytes* than AMHS from Stage I to II. After 5.75 years of age in AMHS, midfacial length remains relatively short in AMHS, perhaps keeping the face in a retracted position.

H4 : Posterior cranial base length
As relative posterior cranial base length is significantly longer in AMHS than *P. troglodytes* at all four ontogenetic stages, the growth pattern that first establishes this shape pattern at Stage I must occur prior to 1.75 years in humans. Although the shape pattern differs significantly at each ontogenetic stage, the growth patterns for posterior cranial base length are not significantly different between AMHS and *P. troglodytes*. Perhaps the shape pattern is established early in ontogeny and subsequently does not change due to similar growth patterns occurring for the remainder of postnatal ontogeny.

H5 : Upper facial length
Similar to results for other linear distances above, relative upper facial length is significantly shorter in AMHS than *P. troglodytes* at all four ontogenetic stages ; thus whatever growth processes first establish this shape pattern at Stage I must occur prior to 1.75 years in humans. *Pan troglodytes* has relatively more growth in upper facial length from Stage I to II, further establishing the shape pattern. However, AMHS has relatively more growth in upper facial length from Stage I to II, although the relative growth is apparently not enough to disturb the shape pattern. May & Scheffer (1999) also found that the upper face grows in early development of AMHS, making the face more projecting but, overall, African apes have more projecting faces. However, since they did not scale their measurements, their results are perhaps a reflection of total facial size being larger in the African apes and are hard to compare with our scaled results.

H6 : Lower facial length

The results for the lower face are the most unexpected in this study. Not only is relative lower facial length not significantly different between AMHS and *P. troglodytes* at Stages II and IV, it is actually significantly relatively greater in AMHS at Stages I and III. Furthermore, the growth patterns for lower facial length are not significantly different between AMHS and *P. troglodytes* between any ontogenetic stages, leaving the changes in shape pattern unaccounted for by any significantly different growth patterns. While one might hypothesize that the face grows considerably more during later ontogeny in *P. troglodytes* than AMHS, the results presented here for lower facial length suggest that although parts of the *P. troglodytes* face may grow more, growth of lower facial length is not relatively greater than in AMHS. Perhaps the lower face appears to elongate anteroposteriorly due to the continued extension of the cranial base in *P. troglodytes* (Lieberman & McCarthy, 1999) as this may cause the entire facial block to rotate dorsally from under the frontal lobes, while the relative dimensions of the lower face remain unchanged.

Discussion

Facial retraction is one of the few autapomorphies that appears reliably to distinguish AMHS from AH crania (Lieberman *et al.*, 2002), but the craniofacial architecture that underlies variation in facial position is not well understood. Three spatial interactions have been predicted to have an effect on facial retraction (Lieberman, 1998 ; Lieberman *et al.*, 2000, 2002) : (1) the human midface remains relatively anteroposteriorly short in relation to the anterior cranial base ; (2) the human middle cranial fossa length is elongated ; and (3) the cranial base remains more flexed. This study examined the bases for this model by testing whether the length variables in these predictions contribute to variation in facial position. As we cannot fully study the shape and growth patterns of these predicted differences in archaic humans, we tested the model by comparing the ontogeny of facial position in AMHS versus *P. troglodytes* with these variables. If any of the length variables contributed to shape patterns that consistently differed between the two extant species, then the same shape patterns might be relevant in predicting variation in, and ontogeny of, Pleistocene hominid facial positions, although not necessarily in the same proportions.

Did any of the predictions within cranial base and facial lengths contribute to facial retraction in AMHS ? Compared to *P. troglodytes*, the AMHS anterior cranial base was relatively longer at all four stages, but relative midfacial length was not significantly shorter in AMHS until Stage II. The growth pattern analyses suggested that this may due to *P. troglodytes* experiencing significantly more relative growth in midfacial length between Stages I and II. Therefore, after at least 5.75 years of age, AMHS had a relatively longer anterior cranial base and relatively shorter midface than *P. troglodytes*. This spatial relationship in AMHS apparently positions the frontal lobes more anteriorly relative to the short midface, reflecting the larger brain in humans both absolutely and relative to facial size. In other words, a short face relative to a long anterior cranial base was a factor in facial retraction in AMHS after neural growth was 95% complete (Krogman, 1931 ; Schultz, 1962 ; Smith, 1989).

The prediction that the human middle cranial fossa length is relatively elongated was also only partially supported by our results. At Stage I, the middle cranial fossa was relatively shorter (but not significantly so) in AMHS than *P. troglodytes* (see Table 3.5), and significantly more relative growth occurred in this linear distance between Stage I to II. From Stage II to IV, the middle cranial fossa was relatively larger in AMHS than in *P. troglodytes*, but only significantly so at Stages II and IV and not at Stage III. Although some shape patterns that appear to affect facial position were established by 1.75 years, middle cranial fossa length did not become significantly relatively larger in AMHS than *P. troglodytes* until sometime between stage I and II (1.75 to 5.75 years). Therefore a relatively long middle cranial fossa does not contribute to establishing facial retraction early in development in AMHS, but may keep it retracted later in postnatal development relative to patterns observed in *P. troglodytes*.

Although we did not test the ontogeny of cranial base angulation in this study, the results of a previously published study of the same data set (Lieberman & McCarthy, 1999) are relevant here with respect to the third variable, the cranial base angle, predicted to have an effect on facial retraction. The cranial base of modern humans flexes rapidly after birth but then remains stable after about two years of age, while the *P. troglodytes* cranial base gradually extends throughout postnatal ontogeny (Lieberman & McCarthy, 1999). The lack of cranial base extension after two years of age in AMHS may keep the face tucked under the elongated anterior cranial base. Interestingly, AH cranial bases are more extended (about 15°) than AMHS (Lieberman *et al.*, 2000, 2002). Although it is unlikely that AH cranial bases extended in an identical ontogenetic pattern as in *P. troglodytes*, it would be interesting to know when this extension occurs in AH ontogeny and what processes drive it.

In short, the general prediction that shape patterns of length variables contribute to different patterns of facial position and that these linear distances exhibit different growth patterns was not rejected. Our results specifically identified a relatively long anterior cranial base and relatively short upper face as contributing structurally to the unique facial retraction of modern humans prior to Stage I. By Stage II, increased relative middle cranial fossa length and decreased relative midfacial length may contribute to keeping the modern human face retracted beneath the frontal lobes.

One of the most important results of this study was that two key shape patterns were established by Stage I in both chimpanzees and humans and do not change significantly at any subsequent point in ontogeny. Furthermore, growth patterns for the linear distances studied here differed more between the early ontogenetic stages and had no significant differences between the later stages. Our ontogenetic analyses suggested that the unique facial retraction and attendant shape patterns of two linear distances appeared by 1.75 years in AMHS. These results are similar to the ontogenetic patterns observed in Krovitz's (2000) and Ponce de León & Zollikofer's (2001) comparisons of juvenile Neanderthal and AMHS crania, both of which found that cranial traits that distinguish the two taxa were already present by 2.2 years (age of the youngest Neanderthal in their samples). Thus, the AMHS morphological pattern likely occurred via developmental shifts that took place prenatally or very early in postnatal ontogeny. Our results did not support the accelerated endochondral growth model (Green, 1990; Green & Smith, 1991). If relative anterior and posterior cranial base lengths are longer in AMHS by Stage I and the relatively smaller distances in *P. troglodytes* are proxies for what to expect in AH juveniles then the cranial base of AH does not undergo accelerated growth compared to AMHS.

As noted, the above analyses were inspired by an effort to understand the developmental bases for differences in AH and AMHS ontogeny but, because of deficiencies in the fossil record, a complete ontogenetic comparison between AH and AMHS was not possible. The results of Krovitz (2000) and Ponce de León & Zollikofer (2001) highlight how comparative analyses among extant taxa such as chimpanzees and humans are nonetheless useful for testing relevant models of craniofacial growth and development, as well as for helping to pinpoint when during ontogeny we should expect to see manifestations of key differences in growth patterns. Although overall growth and development of the skull was obviously more similar between AMHS and AH than between AMHS and *P. troglodytes*, the degree of facial projection was more similar between AH and *P. troglodytes* in terms of morphology. In particular, the above results, in combination with comparisons of adult AH and AMHS (Lieberman *et al.*, 2002), suggest that the structural and ontogenetic bases for facial projection in *P. troglodytes* and AH are comparable (a long face relative to anterior cranial base length, a shorter middle cranial fossa, and more extended cranial base). However, future work is needed to compare more completely early postnatal and prenatal craniofacial development between AMHS and *P. troglodytes* in order to clarify where and when differences in facial position first begin to appear prior to two years of age in AMHS.

Further research on the developmental bases of facial retraction in AMHS is important for several reasons. From a phylogenetic perspective, a better understanding of the variables that contribute to facial retraction will help to identify better, more independent, autapomorphies of AMHS. Like most cranial variables, facial retraction is an integrated feature that derives from underlying structural relationships among many cranial components (in this case a longer anterior cranial base, a shorter face, a longer middle cranial fossa, and a more flexed cranial base). Thus, future studies should focus on what developmental factors influence the growth and interactions among these variables.

Even more interesting are the clues that ontogenetic analyses provide for constructing and ultimately testing hypotheses about natural selection. Ontogenetic integration of complex phenotypes, such as the position of the face, occurs on multiple levels of development and it is quite possible that the developmental mechanisms that drive facial position and cranial base architecture have their basis not only in the cranial base itself, but rather in key processes that determine brain shape. Cranial base form is clearly correlated to brain shape, but how much neural development influences cranial base morphology, or vice versa, is not known. The above results indicate that the anterior cranial base is significantly longer relative to face size at all four developmental stages, and that middle cranial fossa length is longer from Stage II to IV in AMHS than in *P. troglodytes*. One interesting but as yet untested hypothesis is that neural growth, especially relative enlargement of the temporal lobes in AMHS (Semendeferi & Damasio, 2000), may have driven the above described shifts in cranial base morphology that, in turn, led to changes in facial retraction. In other words, facial retraction may be a consequence of temporal lobe expansion in combination with a reduction in facial size.

Conclusions

This study has focused solely on ontogenetic patterns. Yet, ultimately, we need to understand what developmental and genetic processes underlie these differences. Theoretically, the most significant sources for change in complex morphological structures, such as the cranial base, are expected to arise from alterations in embryonic development. The above data indicate that we should focus future research efforts on two parts of the skull. First, developmental alterations in condensations that eventually give rise to the cartilaginous anlage (which influence the general shape of mature ossified bones) are hypothesized to be important in generating evolutionary change in morphology (Atchley & Hall, 1991). Since the cranial base arises from such condensations, it may be pertinent to understand how changes in condensation developmental mechanisms may affect adult cranial base morphology. Our current research examines the development of cranial base condensations to test hypothesized effects of mechanistic alterations in their growth on adult osseous morphology, in ways predicted to affect AMHS facial position and overall craniofacial morphology.

Second, the brain interacts with the cranial base in many ways, and it is possible that shifts in brain growth have important epigenetic influences on cranial base growth, and hence overall cranial shape. Thus, developmental interactions between the brain and cranial base should be investigated to test if embryonic neural shape is a driving force in determining cranial base form by altering condensation formation or the later occurring processes of chondrogenesis and ossification. Understanding the developmental mechanisms that contribute to the cranial base and facial architecture that underlie facial retraction in AMHS is necessary to help identify the developmental context of the origin and evolution of AMHS cranial form. However, ontogenetic studies as presented here are also an important component of hypothesis building and hypothesis testing in the study of the evolution of complex morphological structures.

Acknowledgments

We thank the editors of this volume for inviting us to contribute this chapter and for their patience. We also thank the following for their assistance: R. McCarthy, K. Mowbray, A. Moore (Denver Growth Study), A. W. Crompton and M. Rutzmoser (Museum of Comparative Zoology, Harvard), B. Latimer and L. Jellema (Cleveland Museum of Natural History), R. McPhee, B. Mader, and the late W. Fuchs (American Museum of Natural History, New York), and D. Pilbeam (Peabody Museum, Harvard). And, finally, we thank G. Krovitz for assistance with EDMA and for many enjoyable discussions.

References

Ackermann, R., & Krovitz, G. E. (2002). Common patterns of facial ontogeny in the hominid lineage. *Anatomical Record*, **269**, 142–147.

Atchley, W. R., & Hall, B. K. (1991). A model for development and evolution of complex morphological structures. *Biological Reviews of the Cambridge Philosophical Society*, **66**, 101–157.

Biegert, J. (1963). The evaluation of characteristics of the skull, hands and feet for primate taxonomy. In *Classification and Human Evolution,* ed. S. L. Washburn, pp. 116–145. Chicago: Aldine Press.

Bilsborough, A., & Wood, B. A. (1988). Cranial morphometry of early hominids: Facial region. *American Journal of Physical Anthropology*, **76**, 61–86.

Carroll, S. B., Grenier, J. K., & Watherbee, S. D. (2001). *From DNA to Diversity.* Oxford: Blackwell Science.

Cole, T. M., III. (2002). *WinEDMA: Software for Euclidean Distance Matrix Analysis.* Kansas City: University of Missouri, Kansas City School of Medicine.

Enlow, D. H. (1990). *Facial Growth*, 3rd edn. Philadelphia: W. B. Saunders.

Green, M. D. (1990). Neanderthal craniofacial growth: An ontogenetic model. MA thesis, University of Tennessee, Knoxville.

Green, M., & Smith, F. H. (1991). Neanderthal craniofacial growth. *American Journal of Physical Anthropology*, **12** (Suppl.), 164.

Hall, B. K. (1999). *Evolutionary Developmental Biology*, 2nd edn. Dordrecht: Kluwer Academic Publishers.

Jungers, W. L., Falsetti, A. B., & Wall, C. E. (1995). Shape, relative size, and size-adjustments in morphometrics. *Yearbook of Physical Anthropology*, **38**, 137–161.

Krogman, W. M. (1931). Studies in growth changes in the skull and face of anthropoids. IV: Growth changes in the skull and face of chimpanzee. *American Journal of Anatomy*, **47**, 325–342.

Krovitz, G. E. (2000). Three-dimensional comparisons of craniofacial morphology and growth patterns in Neandertals and modern humans. PhD dissertation, Johns Hopkins University.

Lele, S. (1991). Some comments of coordinate-free and scale-invariant methods in morphometrics. *American Journal of Physical Anthropology*, **85**, 407–417.

Lele, S. (1993). Euclidean distance matrix analysis (EDMA): Estimation of mean form and mean form differences. *Mathematics and Geology*, **25**, 573–602.

Lele, S., & Cole, T. M. III (1996). A new test for shape differences when variance-covariance matrices are unequal. *Journal of Human Evolution*, **31**, 193–212.

Lele, S., & Richtsmeier, J. (1991). Euclidean distance matrix analysis: A coordinate-free approach for comparing biological shapes using landmark data. *American Journal of Physical Anthropology*, **86**, 415–427.

Lele, S., & Richtsmeier, J. (1995). Euclidean distance matrix analysis: Confidence intervals for form and growth differences. *American Journal of Physical Anthropology*, **98**, 73–86.

Lieberman, D. E. (1998). Sphenoid shortening and the evolution of modern human cranial shape. *Nature*, **393**, 158–162.

Lieberman, D. E. (2000). Ontogeny, homology, and phylogeny in the hominid craniofacial skeleton: The problem of the browridge. In *Development, Growth and Evolution: Implications for the Study of Hominid Skeletal Evolution*, eds. P. O'Higgins & M. Cohn, pp. 85–122. London: Academic Press.

Lieberman, D. E., & McCarthy, R. C. (1999). The ontogeny of cranial base angulation in humans and chimpanzees and its implications for reconstructing pharyngeal dimensions. *Journal of Human Evolution*, **36**, 487–517.

Lieberman, D. E., Ross, C. F., & Ravosa, M. J. (2000). The primate cranial base: Ontogeny, function, and integration. *Yearbook of Physical Anthropology*, **43**, 117–169.

Lieberman, D. E., McBratney, B. M., & Krovitz, G. (2002). The evolution and development of cranial form in *Homo sapiens*. *Proceedings of the National Academy of Sciences of the USA*, **99**, 1134–1139.

Maresh, M. M. (1948). Growth of the heart related to bodily growth during childhood and adolescence. *Pediatrics*, **2**, 382–402.

Maresh, M. M., & Washburn, A. H. (1938). Size of the heart in healthy children. *American Journal of Disabled Children*, **56**, 33–60.

May, R., & Sheffer, D. B. (1999). Growth changes in measurements of upper facial positioning. *American Journal of Physical Anthropology*, **108**, 269–280.

McCammon, R. (1970). *Human Growth and Development*. Springfield: C. C. Thomas.

McCarthy, R. C., & Lieberman, D. E. (2001). The posterior maxillary (PM) plane and anterior cranial architecture in primates. *Anatomical Record*, **264**, 247–260.

Moore, W. J., & Lavelle, C. L. B. (1974). *Growth of the Facial Skeleton in the Hominoidea*. London: Academic Press.

Moss, M. L., & Young, R. W. (1960). A functional approach to craniology. *American Journal of Physical Anthropology*, **18**, 281–292.

Ponce de León, M. S., & Zollikofer, C. P. (2001). Neanderthal cranial ontogeny and its implications for late hominid diversity. *Nature*, **412**, 534–538.

Ravosa, M. J. (1988). Browridge development in Cercopithecidae: A test of two models. *American Journal of Physical Anthropology*, **76**, 535–555.

Ravosa, M. J. (1991a). Ontogenetic perspectives on mechanical and nonmechanical models of primate circumorbital morphology. *American Journal of Physical Anthropology*, **85**, 95–112.

Ravosa, M. J. (1991b). Interspecific perspective on mechanical and nonmechanical models of primate circumorbital morphology. *American Journal of Physical Anthropology*, **86**, 369–396.

Ravosa, M. J., Noble, V. E., Hylander, W. L., Johnson, K. R., & Kowalski, E. M. (2000a). Masticatory stress, orbital orientation and the evolution of the primate postorbital bar. *Journal of Human Evolution*, **38**, 667–693.

Ravosa, M. J., Vinyard, C. J., & Hylander, W. L. (2000b). Stressed out: Masticatory forces and primate circumorbital form. *Anatomical Record*, **261**, 173–175.

Richtsmeier, J. T., & Lele, S. (1993). A coordinate-free approach to the analysis of growth patterns: Models and theoretical considerations. *Biological Reviews*, **68**, 381–411.

Richtsmeier, J. T., Cheverud, J. M., Danahey, S. E., Corner, B. D., & Lele, S. (1993). Sexual dimorphism of ontogeny in the crab-eating macaque (*Macaca fasicularis*). *Journal of Human Evolution*, **25**, 1–30.

Ross, C. F. (1995a). Allometric and functional influences on primate orbit orientation and the origins of the Anthropoidea. *Journal of Human Evolution*, **29**, 201–227.

Ross, C. F. (1995b). Muscular and osseous anatomy of the primate anterior temporal fossa and the functions of the postorbital septum. *American Journal of Physical Anthropology*, **98**, 275–306.

Ross, C. F., & Ravosa, M. J. (1993). Basicranial flexion, relative brain size and facial kyphosis in nonhuman primates. *American Journal of Physical Anthropology*, **91**, 305–324.

Schultz, A. H. (1962). Metric age changes and sex differences in primate skulls. *Yearbook of Physical Anthropology*, **10**, 129–154.

Semendeferi, K., & Damasio, H. (2000). The brain and its main anatomical subdivisions in living hominoids using magnetic resonance imaging. *Journal of Human Evolution*, **38**, 317–332.

Shea, B. T. (1985a). On aspects of skull form in African apes and orangutans, with implications for hominoid evolution. *American Journal of Physical Anthropology*, **69**, 329–342.

Shea, B. T. (1985b). Ontogenetic allometry and scaling: A discussion based on the growth and form of the skull in African apes. In *Size and Scaling in Primate Biology*, ed. W. L. Jungers, pp. 175–205. New York: Plenum Press.

Shea, B. T. (1986). On skull form and the supraorbital torus in primates. *Current Anthropology*, **27**, 257–259.

Shea, B. T. (1988). Phylogeny and skull form in the hominoid primates. In *The Biology of the Orangutans*, ed. J. H. Schwartz, pp. 233–246. New York: Oxford University Press.

Smith, B. H. (1989). Dental development as a measure of life history in primates. *Evolution*, **43**, 683–688.

Spoor, C. F., O'Higgins, P., Dean, M. C., & Lieberman, D. E. (1999). Anterior sphenoid shortening in modern humans. *Nature*, **397**, 572.

Weidenreich, F. (1941). The brain and its role in the phylogenetic transformation of the human skull. *Transactions of the American Philosophical Society*, **31**, 321–442.

White, T. D., & Folkens, P. A. (1991). *Human Osteology*. San Diego: Academic Press.

4   *Variation in modern human dental development*

H. LIVERSIDGE

Royal London Hospital

Introduction

Dental development includes the formation of enamel and dentine and the eruption of both dentitions. Every healthy, normal child reaches a state of dental maturity when all the permanent teeth complete root formation and are functional. Tooth development can be assessed in many ways and as dental growth data are used by many disciplines, reference data are available in several ways to accommodate these needs. Orthodontists and pediatric dentists may need to assess the best age for treatment and require different information than anthropologists who might wish to compare growth or growth patterns between population or species. Tooth formation and eruption occur throughout the growth period from the middle trimester up to adulthood and this long maturation period provides a useful tool in assessing growth and maturation. Eruption of teeth is useful between 6 months and about 3 years for the deciduous teeth and about 5 to 16 years for the permanent teeth excluding the third molars. Assessing tooth formation and eruption from radiographs allows many more stages to be assessed; both crown and root formation stages as well as position of the developing tooth relative to the alveolar bone or occlusal levels. In addition developing teeth are less influenced by environmental factors than other growth systems (see Demirjian, 1986). This chapter is an extensive review of published studies in different human populations that document the timing of eruption of deciduous and permanent teeth into the mouth, eruption stages and tooth formation (excluding the third molar). It brings together details from as many published sources as possible, detailing sample size, age range and methods of analyses, and builds on previous review chapters and books (Demirjian, 1986; Eveleth & Tanner, 1976, 1990; Smith, 1991).

Patterns of Growth and Development in the Genus Homo, ed. J. L. Thompson, G. E.Krovitz and A. J. Nelson. Published by Cambridge University Press. ©Cambridge University Press 2003.

Methods

Recording dental formation

Dental formation can be quantified in several ways, outlined by Smith (1991) and data presentation can be divided into three groups. Firstly, the timing of specific stages for individual teeth can be documented. Examples are eruption into the mouth, position of an erupting tooth within the jaw, and crown and root fractions. These types of results make up the main part of this chapter. The timing of events can also be presented not in chronological age, but in relation to a defined reference stage of one particular tooth, for example, describing growth stages of M_2 in a group of children with M_1 all at a specific reference stage such as crown complete. The second group is to present results as the average tooth stage at specific ages; several populations have been described in this way: young children from Finland (Nyström *et al.*, 1977) and a Swedish and French group (Herbert, 1977; Nielsen & Ravn, 1976). Thirdly, dental maturation as a whole can be quantified as dental age. The most widely known method is that of Demirjian *et al.* (1973), but others such as Nolla (1960) exist.

Growth statistics

Investigation of discrete growth events is best done using the status quo method. Calculating the mean age-of-attainment of a discrete growth event is based on the principle of a cumulative frequency curve (Healy, 1986). For example, the eruption age of the first permanent molar is investigated in a group of girls. One question is asked of each girl: is the first permanent molar visible in the mouth or not? In a large group of sufficient age range, the curve will be S-shaped recording 0% of children in the youngest age group having the tooth present in the mouth, up to 100% of the oldest age group having erupted the tooth. The percentages of girls with the tooth present at successive years are plotted against age. This way ensures that a sufficient age range of children are examined to include early, average, and late developers (see Smith, 1991). Several statistical methods are appropriate to derive the mean age and range at which 50% of the children are most likely to have a tooth emerged (equivalent to the median or 50th percentile). These include probit analysis, Kärber's method, life tables and maximum likelihood (Dahlberg & Menegaz-Bock, 1958; Eveleth & Tanner, 1990; Finney, 1952; Healy, 1986; Holman & Jones, 1991; Taranger, 1976). This review includes studies using these methods mentioned above.

If a cumulative frequency distribution method is not used, the mean age of any growth event reflects only the age distribution of the sampled children and not the biological event. For example, if the age range of a sample is 9 to 12 years of age, the non-cumulative mean age of eruption of M_1 is of little meaning if 100% of children at age 9 have the tooth erupted.

A few growth studies are longitudinal, where each child is examined or X-rayed at intervals, and data of this type give information about the timing of events in each individual child as well as individual rate of growth. A tooth can be observed in the mouth only after it erupts and not before, so the timing of eruption must be corrected by subtracting half the interval between examinations (Dahlberg & Menegaz-Bock, 1958). This retrospective nature of observing a growth event also has a bearing on the use of age-of-attainment data to predict age. Predicting age using age-of-attainment data will tend to overestimate age and one solution is to subtract half the interval between the stage under question and the preceding one (Scheuer & Black, 2000; Smith, 1991).

The way age is recorded and grouped is also of importance and presents a problem if details are not given. Exact decimal age can be calculated from Eveleth & Tanner (1990), some studies group quarter, half or annual cohorts. For example, the category "All 5-year-old children" includes children on their fifth birthday up to children the day before their sixth birthday. Analysis of this type of data should make use of the midpoint of the age interval and not use the minimum age of that group.

Stages of dental development

Teeth form in layers that appear first as cusp tip(s) that join together and develop into the crown and then the root. This sequence of successive stages can be arbitrarily divided into stages of crown and root formation as well as eruption stages of the tooth within the alveolar bone into the mouth. During maturation, each child reaches a stage, is in that stage for some time, and then enters the next stage. Difficulties arise when teeth are observed between stages; this occurs when stages are either too numerous or not well described. Some stages are more easily defined than others, such as initial cusp tip mineralization, root furcation, or apex completion, and the eruption stages such as cusp tip(s) at the alveolar bone level or at the occlusal level. Some early studies provided line drawings of stages with scant descriptions, some of these include fractions or proportion of crown/root stage that assume some knowledge of crown height/root length for that individual or group. Another approach has been to make use of relative size, i.e., root length less than or greater than crown height (Demirjian *et al.*, 1973). This particular stage assessment has the advantage of written criteria in addition to a radiographic example and line drawing for each stage.

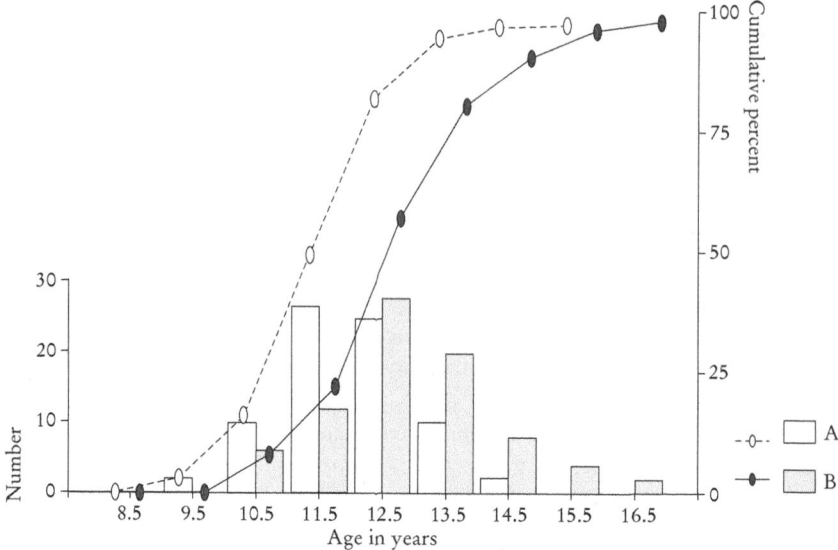

Figure 4.1   Representation of histogram and cumulative frequency distribution of two growth stages. (After Taranger (1976).)

The duration of stages and the rate of change between stages can only be assessed from longitudinal studies. Other studies are cross-sectional where children are examined once, and the appropriate statistical method should be employed for the type of study.

An example illustrates some of the principles described. Figure 4.1 shows a graphic representation of a histogram and a cumulative frequency curve. A group of children is examined for two defined growth stages. The youngest child in stage A is 9 years old while all children had reached this stage by 14 years. The youngest child in stage B is 10 years old while all had reached this stage by 16 years.

Results

Timing of eruption

The first part of this section concerns the timing of eruption; then stages of eruption and lastly crown and root formation. Geographic populations are after Eveleth & Tanner (1976). Where bilateral data are given, only the left is reported. Timing and variation of the mean age between populations in the figures are mostly data for girls; the exceptions are when data are combined in the original source.

Studies of eruption are detailed in Tables 4.1 and 4.2. Data for several populations have been calculated using probit analysis from original papers and are included in Tables 4.3 and 4.4. Probit analysis involves transforming the cumulative percentages to standard deviation units (15.9% is −1 SD, 84.1% is +1 SD) and fitting these to a straight line (the slope of which indicates the standard deviation of the mean). In this type of regression, values from the tails of the distribution have greater weighting than those in the middle, dictated by the slope of the cumulative curve. For example, the normal deviate difference between 2% and 3% is 0.17 SD units, whereas between 44% and 45% is 0.03 SD units (after Taranger, 1976). Timing of eruption of deciduous teeth (Table 4.3) was calculated using probit for children in Korea (Yun, 1957) and the Czech Republic, although no sample size breakdown is given in that paper (Holnerová, 1975) so the age of 50th percentile was interpolated by eye from cumulative curves of the raw data. Eruption timing of permanent teeth including children from Nigeria (Ajmani & Jain, 1984), China (Diao, 1981; Yang, 1988), Italy (Adorni-Braccesi, 1965; Caltabiano *et al.*, 1979), and Russia (Fedorov *et al.*, 1984) are presented in Table 4.4. These reports give proportion and number of each age group but no details of age groupings. One study gives no details of how age was recorded. If one assumes that the age group "5 years" indicates the age at last birthday, and includes all children aged between five and six years (midpoint 5.5), the mean age of eruption of M_1 in Nigerian children (Ajmani & Jain, 1984) calculated using probit is 5.70 years (±1.07). However, if the age group "5 years" is assumed to indicate that 5 is the midpoint, the mean age of eruption of M_1 in Nigerian children is 5.41 (±1.27). This paper uses the age-at-last-birthday approach, and these recalculated results should be interpreted with this in mind. As with the Czech Republic study, the study of children near Lake Baikal in Siberia (Fedorov *et al.*, 1984) gives no sample size breakdown for each age group and 50th percentile was interpolated by eye from cumulative curves. Data for children from Africa are reproduced in Table 4.5; the mean age from two studies have been moved forward six months so that age is reported as the age at last birthday.

Eruption is a process involving movement of the tooth from within the alveolar bone through the gingivae into the oral cavity and clinical eruption is only one part of this progression (Demirjian, 1986). Emergence is usually defined as the first evidence of the tooth through the soft tissue, although a few studies document one cusp and/or entire occlusal surface for molars. A few reports score clinical stages of eruption of deciduous (Hulland *et al.*, 2000) and permanent teeth (Pahkala *et al.*, 1991; Sato, 1990), though none uses cumulative frequency analyses.

Table 4.1 *Summary of studies of eruption of deciduous teeth*

Country	Place	Reference	Sample size[a] Boys	Sample size[a] Girls	Study type[b]	Age (months/years)	Analysis
Europeans in Europe							
Croatia	Zagreb	Rajic et al., 1999	(1288)		C	0–42 m	Percentile
Czech Republic	Prague, Liberec	Holnerová, 1975[c]	566	571	C	5–3 m	Percentage
Finland	Helsinki	Nyström et al., 1985	200	180	L	0–3 yrs	Percentile
Hungary	Debrecen	Tegzes, 1961	582	543	C	5–28 m	Probit
Iceland	Reykjavik	Magnusson, 1982	498	429	C	0–83 m	Probit
Spain	Madrid	Ramirez et al., 1994	62	52	L	0–3 yrs	Mean
Sweden	Umea	Lysell et al., 1962	96	75	L	0–41 m	Mean
Sweden	Regional	Hägg & Taranger, 1986	122	90	L	0–18 yrs	Probit
United Kingdom	London	Leighton, 1968	39	45	L	0–30 m	Mean
European descendants in Australia, Africa, and the Americas							
Australia	Perth	Hitchock et al., 1984	81	83	L	0–4 yrs	Centile
Canada	Montreal	Lauzier & Demirjian, 1981	202	165	L	0–3 yrs	Percentile
Africans in Africa and of African ancestry							
Senegal	Dakar	Yam et al., 2001	275	298	L	0–30 m	Mean
Asiatics in Asia and the Americas							
China	Hong Kong	Low et al., 1973	1509	1504	C	4–45 m	Probit
Guatemala	El Progresso	Delgado et al., 1975	(273)		L	0–3 yrs	Percentile
Guatemala	El Progresso	Delgado et al., 1975[d]	(273)		L	0–3 yrs	Life table

Country	Location	Reference	(n)[a]	n	[b]	Age	Method
Indonesia	Java	Holman & Jones, 1991	(468)		L	0–2.5 yrs	Life table
Japan	Tokyo	Kitamura, 1942[d]	(114)		L	0–3 yrs	Life table
Korea	Chulla-Book-Do	Yun, 1957[c]	(1838)		C	3–36 m	Percentage
	Gwangju	Choi & Yang, 2001	567	503	C	4–36 m	Percentile
Indo-Mediterraneans in the Near East, North Africa, and India							
Bangladesh	Meheran	Holman & Jones, 1998	(397)		L	0–3 yrs	Life table
Egypt	El-Manial, Shibeen El Kanater	El-Beheri & Hussein, 1987	1887	1860	C	4–36 m	Percentile
India	Gulbargo	Reddy, 1981	591	533	C	?0–36 m	Percentile
	Delhi	Kaur & Singh, 1992	–	1643	C	0–20 yrs	Probit
	Ludhiana, Punjab	Kaul et al., 1992	165	147	C	4–31 m	Probit
Iraq	Baghdad	Baghdady & Ghose, 1981	510	507	C	1–40 m	Kärber
Pakistan	Lahore	Saleemi et al., 1994	245	198	L	0–5 yrs	Probit
Saudi Arabia	Tabuk	Lawoyin et al. 1996	(167)		L	0–2 yrs	Mean
Tunisia	Chott El Djerid	Bouterline & Tesi, 1972	(1450)		C	3–36 m	Probit
Australian Aborigines and Pacific Islanders							
Papua New Guinea	Bougainville	Friedlander & Bailit, 1969	121	118	C	0–3 yrs	Probit
	Gulf Province	Ulijaszek, 1996	(135)		C	0–30 m	Probit

[a] Sample size in parentheses (*n*) represents combined boys and girls.
[b] C, cross-sectional; L, longitudinal.
[c] 50th percentile calculated from data for present review.
[d] Data reanalyzed by Holman and Jones (1998).

Table 4.2 Summary of studies of eruption of permanent teeth

Country	Place	Reference	Sample size[a] Boys	Sample size[a] Girls	Study type[b]	Age (years)	Analysis
Europeans in Europe							
Croatia	Zagreb	Rajic et al., 2000	1398	1370	C	4–16	Percentile
Denmark	Copenhagen	Helm & Seidler, 1974	3983	3946	C	3–17	Kärber
	Regional	Parner et al., 2001	(850000)		L	0–16	Maximum likelihood
Finland	Regional	Virtanen et al., 1994	456	455	L	3–21	Mean
	Juuka	Eskeli et al., 1999	525	483	C	5–15	Probit
	Vimpeli	Eskeli et al., 1999	265	304	C	5–15	Probit
	Helsinki	Nyström et al., 2001	86	101	L	0.5–16	Percentile
France	Lille	Rousset et al., 2001	280	294	C	5.5–15	Cumulative frequency
Germany	Jena	Kromeyer & Wurschi, 1996	98	109	L	4–11	Probit
Iceland	Reykjavik	Magnusson, 1976	791	850	C	5–18	Kärber
Italy	Catania	Caltabiano et al., 1979[c]	–	1000	C	6–13	Percentage
	Florence	Adorni Braccesi 1965[c]	2639	2494	C	6–13	Percentage
Spain	Madrid	Mesa, 1988	(778)		C	5–14	Probit
Sweden	Regional	Hägg & Taranger, 1986	122	90	L	0–18	Probit
United Kingdom	Birmingham	Clements et al., 1953	1427	1365	C	5–13	Probit
	Manchester	Miller et al., 1965	1000	1000	C	3–15	Kärber
	Regional	Lavelle, 1976	(4000)		C	5–14	Probit
	London	Smith et al., 2000	1500	1423	C	5–15	Kärber
	Belfast	Richardson et al., 1975	139	130	C	5–9	Probit
	Belfast	Kochhar & Richardson, 1998	146	130	L	5–15	Mean
European descendants in Australia, Africa, and the Americas							
Argentina	Regional	Muniz, 1988	2366	2369	C	3–14	Probit
Australia	Fremantle	Halikis, 1961	435	427	C	2–15	Kärber

Country	Location	Reference	(n)	n	L/C	Age range	Method
Brazil	Rio de Janeiro	Eveleth, 1966	(198)		L	5.5–15.5	Probit
	Sao Paulo	Souza Freitas et al., 1970	1968	1720	C	4.3–14.2	Kärber
	Sao Paulo	Marques et al., 1978	878	863	C	4–14	Percentile
Canada	Montreal	Perreault et al., 1974	797	738	L	3–14	Probit
	Montreal	Demirjian & Levesque, 1980	2732	2705	C + L	2.5–19	Percentile
	N. Quebec	Masson, 1980	(175)		C	6–17	Probit
South Africa	Johannesburg	Monk, 1974	1231	1227	C	?4–20	Probit
United States of America	Regional (ten states)	Garn et al., 1973	(5788)		C	4.5–16.5	Cumulative frequency
	Portland	Savara & Steen, 1978	124	163	L	4–15	Percentile

Africans in Africa and of African ancestry

Country	Location	Reference	(n)	n	L/C	Age range	Method
Gambia	Mandinka	Billewicz & McGregor, 1975	(635)		L	4.5–14	Maximum likelihood
Ghana	Sunyani	Houpt et al., 1967[d]	455	260	C	6–19	Probit
Ivory Coast	Abidjan	Bakayoko et al., 1989	451	418	C	6–15	Percentile
	Abidjan	Bakayoko et al., 1994	(823)		C	4–8.5	Percentile
Kenya	Nairobi	Hassanali & Odhiambo, 1981	881	802	C	4–14	Kärber
	Nairobi (Asians)	Hassanali & Odhiambo, 1981	582	582	C	4–14	Kärber
	Nairobi	Manji and Mwaniki, 1985	3490	3424	C	5–14	Probit
Nigeria	Jos	Ajmani & Jain, 1984[c]	1022	646	C	6–23	Percentage
South Africa	Soweto	Blankenstein et al., 1990a	505	519	C	3.9–9.6	Probit
South Africa	Lenasia (Asians)	Blankenstein et al., 1990b	507	529	C	3.9–10	Probit
Togo	Lome	Richardson et al., 1975	200	98	C	5–9	Probit
Uganda	Regional	Krumholt et al., 1971[d]	311	311	C	2–15	Kärber
United States of America	Regional	Garn et al., 1973	(3868)		C	4.5–16.5	Cumulative frequency
Zambia	Choma	Gillett, 1997	256	287	C	5.5–14	Probit

(cont.)

Table 4.2 (cont.)

Country	Place	Reference	Study type[b]	Sample size[a] Boys	Girls	Age (years)	Analysis
Asiatics in Asia and the Americas							
Brazil	São Paolo (Japanese)	Eveleth & de Souza Freitas, 1969	C	509	481	4.5–14.5	Probit
Canada	N. Ontario (Cree, Ojibway)	Mayhall et al., 1977	C	(599)		4–20	Probit
	Sioux Lookout Zone	Titley, 1984	C	(1191)		5.5–17.4	Probit
	Foxe Basin (Inuit)	Mayhall et al., 1978	C + L	(368)		?0–22	Probit
Caribbean	Curacao	Debrot, 1972	C	2155	2175	6–12	50th percentile
China	Hong Kong	Lee et al., 1965	C	3024	3308	6–16	Probit
	Rizeke, Tibet	Diao, 1981[c]	C	320	312	6–17	Percentage
	Nanjing (Sichuan?)	Quoted in Diao, 1981[c]	C	(27606)		6–17	Percentage
	Lhasa	Yang, 1988[c]	C	1581	1439	4–22	Percentage
Greenland	Angmagssalik, Julianehaab	Boesen et al., 1976	C	(1027)		6–17	Kärber
Russia	Chita	Fedorov et al., 1984[c]	C	1125	1255	5–14	Percentage
Japan	Kagamigahara	Höffding et al., 1984	C	964	855	6–15	Kärber
Taiwan	Taipei	Lan, 1971	C	2794	2786	4–16	Kärber
Thailand	Bang Chan	Kamalanathan et al., 1960	C	125	115	7–14	Percentile
United States of America	Arizona (Pima)	Dahlberg & Menegaz-Bock, 1958	C	470	487	3–15	Probit

Indo-Mediterraneans in the Near East, North Africa, and India

Egypt	Komombo, Nubia	El-Nofely, 1971	570	361	C	5.5–13	Percentile
India	Punjab	Sharma & Mittal, 2001	251	232	C	6–13	Probit
	Punjab	Kaul et al., 1975	564	573	C	6–14	Probit
	Calcutta	Banerjee et al., 1985	125	392	C	5–15	Percentile
	Himachal Pradesh	Singh, 1980	515	–	C	6–13	Probit
	Meghalaya	Jaswal, 1983	615	648	C	5–15	Probit
	Andra Pradesh	Kumar & Sridhar, 1990	533	475	C	5–14	Probit
	Haryana	Kaur & Singh, 1992	–	1643	C	0–20	Probit
	Manipur	Gaur & Singh, 1994	238	202	C	5–14	Probit
Iraq	Baghdad	Baghdady & Ghose, 1981	1387	1456	C	4–15	Kärber
Israel	Regional	Koyoumdjisky et al., 1977, 1981	1076	1040	C + L	3–14	Percentile
	Ashkelon	Barker & Zusman, 1990	644	683	C	4–16	Kärber
Sri Lanka	Jaffna (Tamil)	Pathmanathan et al., 1985	377	534	C	5–20	Probit

Australian Aborigines and Pacific Islanders

Australia	Yuendumu	Brown, 1978	74	51	L	5–19+	Mean
Papua New Guinea	Bougainville	Friedlander & Bailit, 1969	342	316	C	4–13	Probit

[a] Sample size in parentheses (*n*) represents combined boys and girls.

[b] C, cross-sectional; L, longitudinal.

[c] Data reanalyzed by probit for present review, age given as age at last birthday.

[d] Age given as age at last birthday.

Table 4.3 *Eruption of deciduous teeth (mean ± SD)*

Country	Sex	Jaw	i1	i2	c	m1	m2
Czech Republic	Girls	Upper	0.83	0.99	1.54	1.25	2.21
	Boys		0.77	0.99	1.54	1.23	2.08
	Girls	Lower	0.71	1.12	1.58	1.13	2.13
	Boys		0.62	1.10	1.58	1.29	2.06
Korea		Upper	0.80 ± 0.25	0.97 ± 0.21	1.38 ± 0.40	1.51 ± 0.42	1.94 ± 0.66
		Lower	0.68 ± 0.27	0.99 ± 0.27	1.39 ± 0.40	1.52 ± 0.44	1.93 ± 0.62

Source: Czech Republic: Holnerová (1975); Korea: Yun (1957).

Results from this review show that the sequence of deciduous eruption usually follows the sequence of formation (i1, i2, m1, c, m2) with no clear pattern in differences between jaws or between boys and girls. Permanent teeth erupt into the mouth in two phases with the first molars, central incisors, and then the lateral incisors erupting during the first phase. This is followed by a quiescent period before premolars, canines, and second molars emerge; the sequence of premolars and canine shows some sex and population variation. This gap between the two phases ranges between one and three years and is greater in maxillary compared to mandibular teeth. The eruption of successional teeth is influenced by local factors such as disease or early loss, although some studies fail to take this into account (Clements *et al.*, 1953; Miller *et al.*, 1965). Mandibular teeth tend to precede maxillary teeth except for premolars. Sex differences are small in the deciduous dentition and girls are generally ahead of boys in eruption timing of permanent teeth; the permanent mandibular canine shows the largest difference in timing between girls and boys. Small population differences have been reported in the deciduous dentition (reviewed by Holman & Jones, 1998).

The mean ages from Tables 4.1 to 4.4 show that some outliers are apparent when comparing mean values between populations, but few patterns or trends are evident, although populations from Asia show more variation. Compared to other European populations, children from Croatia tend to show later eruption for some teeth; while children in India and Tunisia also show early eruption of one or two tooth types. Variation in the mean eruption timing of m_2 is shown as a boxplot in Figure 4.2.

Population differences in the eruption of the permanent dentition are more evident, although they are greater than those in formation of crown and root stages and less than other non-dental growth systems. Some ethnic groups show earlier clinical eruption than other populations, although no pattern or trend

Table 4.4 *Eruption of permanent teeth (mean ± SD); for all populations age was assumed to be that of last birthday and analyzed after adding 6 months to this age*

Country	Sex	Jaw	I1	I2	C	P1	P2	M1	M2
Italy 1	Girls	Upper	7.12 ± 0.92	8.02 ± 0.78	10.53 ± 0.81	10.20 ± 0.90	10.78 ± 0.87	6.14 ± 1.16	11.54 ± 0.79
		Lower	6.15 ± 1.20	7.59 ± 0.68	9.90 ± 0.90	10.31 ± 0.95	10.76 ± 0.79	–	11.22 ± 0.71
Italy 2	Girls	Upper	7.29 ± 0.94	8.45 ± 0.95	11.39 ± 1.30	10.31 ± 1.22	11.06 ± 1.29	–	12.30 ± 1.03
	Boys		7.65 ± 0.99	8.79 ± 0.91	11.81 ± 1.13	10.63 ± 1.53	11.55 ± 1.69	–	–
	Girls	Lower	–	7.64 ± 0.91	10.15 ± 1.05	10.68 ± 1.28	11.32 ± 1.47	–	11.81 ± 1.16
	Boys		–	7.90 ± 0.96	11.02 ± 1.17	11.06 ± 1.36	11.88 ± 1.53	–	12.30 ± 1.28
Russia[a]	Girls	Upper	6.37	7.15	9.80	8.60	9.60	5.80	10.90
	Boys		6.65	7.50	10.4	9.20	10.00	6.00	11.40
	Girls	Lower	5.57	6.55	8.75	8.83	9.55	5.45	10.25
	Boys		5.75	6.78	9.52	9.25	10.15	5.70	10.75
Nigeria	Girls	Upper	–	7.52 ± 1.21	10.27 ± 1.51	10.51 ± 1.47	11.62 ± 1.68	6.77 ± 0.64	11.16 ± 2.07
	Boys		–	8.19 ± 1.12	10.82 ± 1.43	10.74 ± 1.48	11.38 ± 1.75	6.55 ± 0.57	11.65 ± 2.15
	Girls	Lower	7.19 ± 0.78	7.31 ± 0.90	10.16 ± 1.67	10.00 ± 1.20	10.46 ± 1.83	6.63 ± 0.64	10.68 ± 2.04
	Boys		7.03 ± 0.66	7.77 ± 0.67	10.31 ± 1.50	10.29 ± 1.24	11.08 ± 1.61	–	10.81 ± 1.90
China	Girls	Upper	7.50 ± 1.33	8.47 ± 1.28	10.98 ± 1.75	10.70 ± 1.72	11.86 ± 1.73	–	12.50 ± 1.38
	Boys		7.53 ± 1.24	8.71 ± 1.37	11.74 ± 1.52	11.66 ± 1.76	12.49 ± 1.65	–	13.23 ± 1.02
Tibet	Girls	Lower	–	7.73 ± 1.51	10.68 ± 1.32	10.84 ± 1.58	11.92 ± 1.69	–	12.28 ± 1.24
	Boys		–	7.47 ± 1.63	11.61 ± 1.50	11.75 ± 1.77	12.90 ± 1.83	–	12.96 ± 1.08
Nanjing		Upper	7.69 ± 1.40	9.00 ± 1.27	11.47 ± 1.64	10.85 ± 1.72	11.86 ± 1.72	–	12.98 ± 1.51
		Lower	–	7.61 ± 1.53	10.88 ± 1.63	11.27 ± 1.66	12.03 ± 1.83	–	12.37 ± 1.54
Lhasa		Upper	7.85 ± 1.14	9.06 ± 1.18	11.51 ± 1.51	10.41 ± 1.77	11.42 ± 1.78	6.95 ± 0.83	12.77 ± 1.16
		Lower	6.92 ± 0.77	7.85 ± 1.04	10.66 ± 1.30	10.59 ± 1.53	11.66 ± 1.81	7.05 ± 1.31	12.30 ± 1.38

[a] Mean age for Russian interpolated by eye because the proportion but not sample number given in original paper.

Sources: Italy: Caltabiano et al. (1979), Adorni-Braccesi (1965); Russia: Fedorov et al. (1984); Nigeria: Ajmani & Jain (1984); China: Diao (1981); Lhasa: Yang (1988).

Table 4.5 *Time of eruption of permanent teeth (left side if both are given): Africa*

Group	Sex	I^1	I^2	C'	P^1	P^1	M^1	M^2	I_1	I_2	C	P_1	P_2	M_1	M_2
Gambia	Girls	7.13	8.10	10.56	9.78	10.60	5.80	11.19	6.08	7.07	9.64	9.96	10.69	5.46	10.91
	Boys	7.38	8.60	11.29	10.26	11.23	6.00	11.96	6.22	7.46	10.55	10.70	11.44	5.67	11.56
Ghana[a]	Girls	6.5	7.8	10.0	9.5	10.5	5.5	11.4	5.6	7.0	9.4	9.7	10.8	5.0	11.0
	Boys	6.8	8.0	10.9	10.0	11.0	5.5	11.4	5.8	6.6	10.5	10.5	11.1	5.4	11.3
Ivory Coast[b]	Girls	6.42	7.42	11.00	10.00	11.25	6.00	11.67	5.50	6.50	10.00	10.08	11.00	5.33	11.50
Kenya 1	Girls	6.55	7.71	10.26	9.40	10.15	6.13	11.40	5.62	6.56	9.20	9.62	10.23	5.70	11.07
	Boys	6.91	7.99	10.93	9.87	10.74	6.32	11.54	5.83	6.86	9.96	10.05	10.90	6.03	11.39
Kenya 2	Girls	6.95	7.97	10.60	9.74	10.69	6.27	11.54	6.34	7.19	9.66	9.84	10.66	6.08	11.13
	Boys	7.24	8.36	11.24	9.97	11.10	6.67	12.20	6.57	7.52	10.58	10.58	11.37	6.47	11.90
Kenya 3	Girls	6.53	7.59	10.23	9.69	10.42	5.35	11.04	5.44	6.61	9.30	9.79	10.37	4.74	10.81
	Boys	6.78	7.99	10.85	10.11	10.92	5.74	11.70	5.23	6.94	10.13	10.35	10.98	5.15	11.43
Uganda[a]	Girls	6.6	7.4	9.7	9.3	10.1	6.0	10.3	5.8	6.4	8.5	9.5	10.3	5.8	10.1
	Boys	6.6	7.8	10.5	9.6	11.0	5.6	11.0	6.0	6.7	10.1	10.1	11.0	5.8	10.6
USA	Girls	6.75	7.64	10.66	10.06	10.73	5.95	11.61	5.87	6.55	9.81	10.09	10.75	5.67	11.21
	Boys	6.96	7.97	10.97	10.45	11.22	6.25	12.32	6.11	6.98	10.38	10.40	11.18	6.10	11.96
Zambia	Girls	6.47	7.32	9.81	9.30	10.45	5.06	11.18	5.31	6.55	9.51	8.87	10.59	5.35	10.74
	Boys	6.63	7.89	10.06	10.42	10.94	5.77	11.46	5.79	6.62	9.95	9.90	11.23	5.19	11.30

[a] Six months have been added to data for Ghana and Uganda as age was reported as age of last birthday.
[b] Data combined from Bakayoko *et al.* (1989, 1994).

Sources: As in Table 4.2. Data for Nigerian children are in Table 4.4.

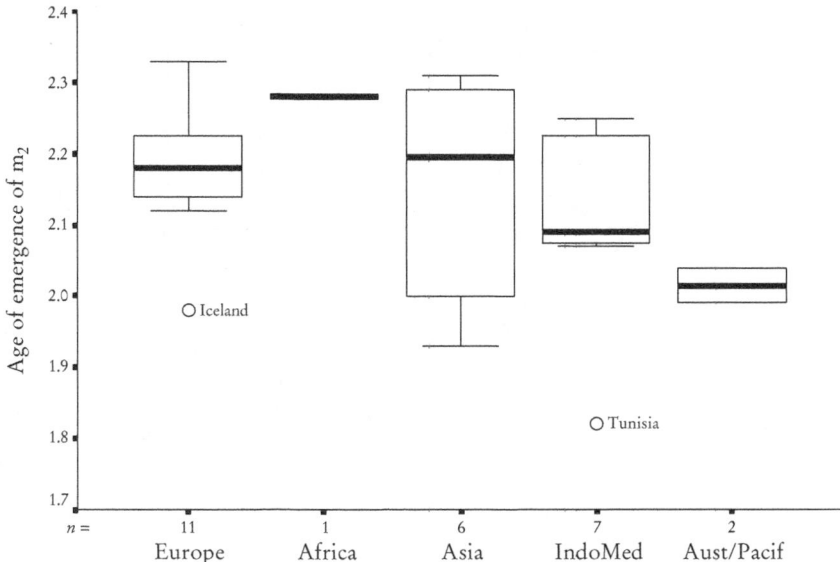

Figure 4.2   Boxplot of mean eruption time of m_2 for populations. The box represents the interquartile range, whiskers extend to the highest and lowest values, excluding outliers. A line across the box indicates the median. (See Tables 4.1 and 4.3 for sources.)

is clear and thus data for only one deciduous and one permanent tooth are depicted.

Variation is greatest for the populations from Asia especially for later emerging teeth. Mean eruption times of permanent mandibular teeth in European, African, and Asian girls are shown in Figure 4.3. Variation in the mean timing of M_1 eruption between populations is shown in Figure 4.4; children from Croatia, Nigeria, China and Thailand are late outliers, while children from Zambia and Argentina are early outliers from the rest of their groups. Other teeth show slightly different outliers.

Comparing eruption directly is more useful in the determination of differences between groups. Debrot (1972) found that canine and premolars emerged earlier in "Negro" children in Curaçao, Antilles compared to children in Maryland, USA, although it is not clear if cumulative analyses were used. Garn *et al.* (1973) found that "Negro" children emerged all teeth earlier than "White" children in the Ten State study. The average advancement was around three months and was more apparent in early emerging teeth. Children in Togo emerged first molars and first incisors about three months earlier than children in Belfast (Richardson *et al.*, 1975). African children were also up to six months advanced in eruption compared to Asian children in Kenya with the

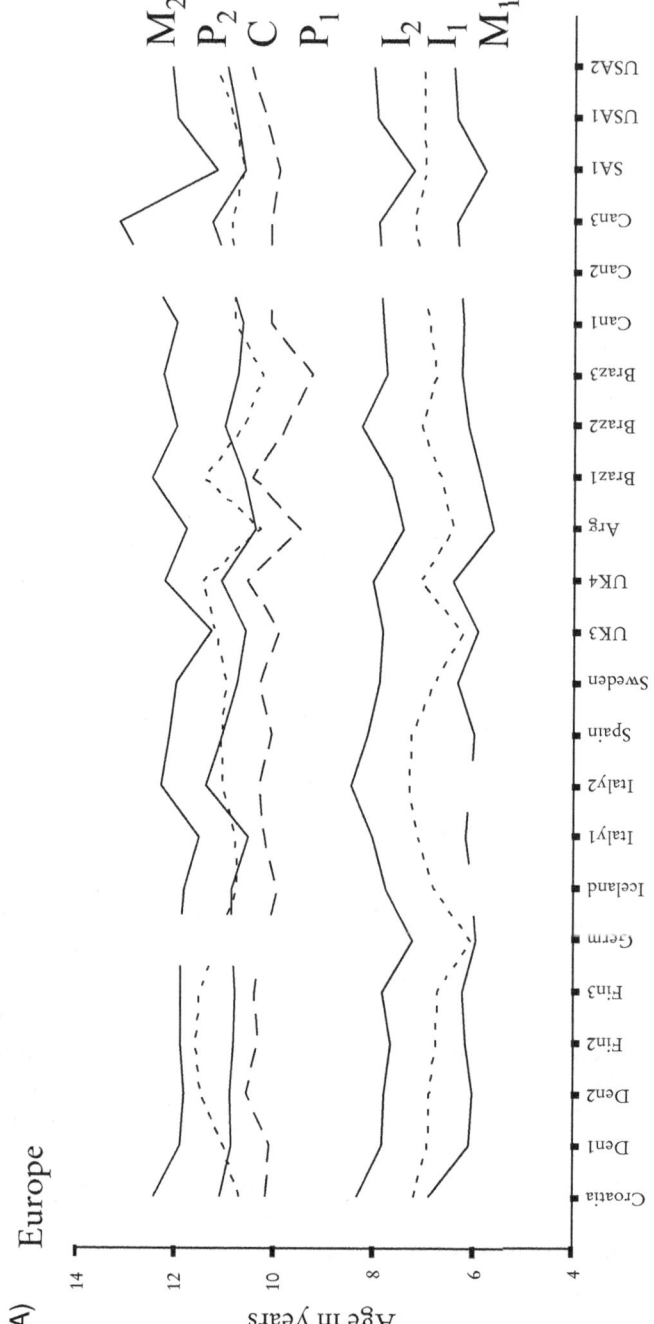

Figure 4.3 Diagram of permanent tooth eruption for populations of (A) European origin, (B) Africa, and (C) Asia. (See Table 4.2 for sources.)

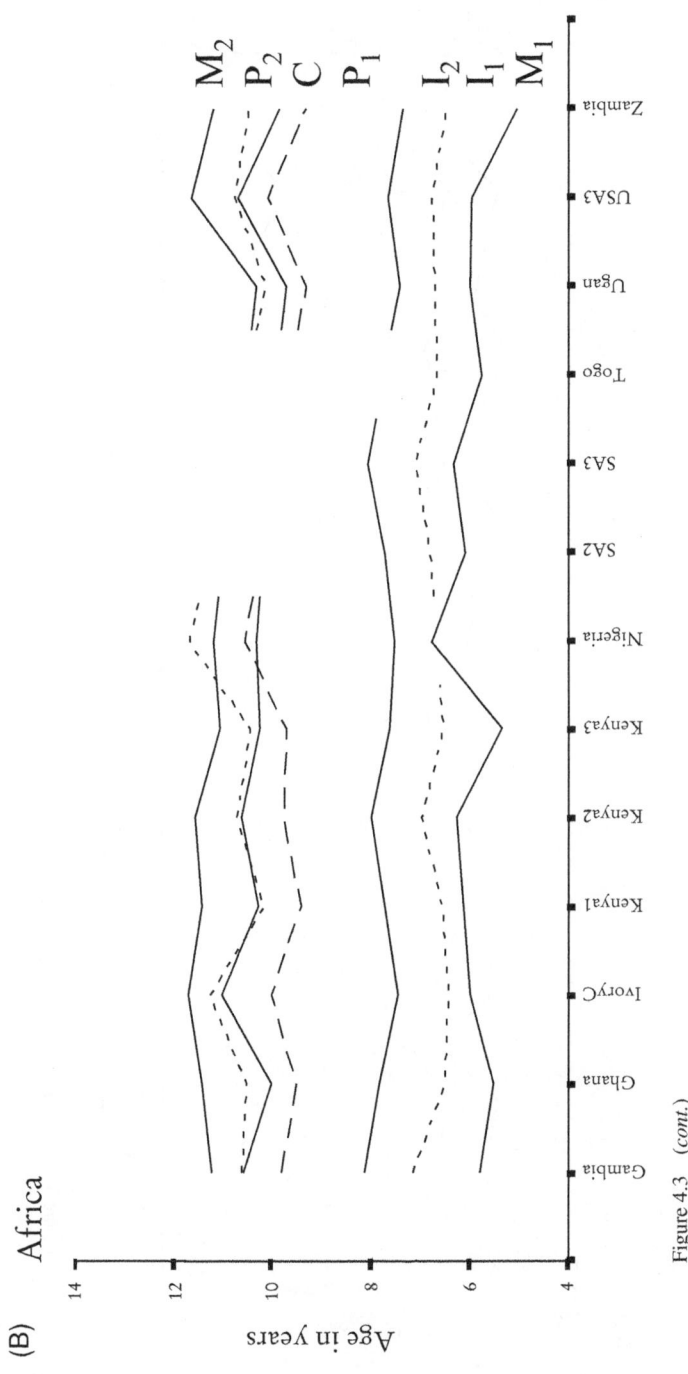

(B)

Africa

Age in years

Gambia · Ghana · IvoryC · Kenya1 · Kenya2 · Kenya3 · Nigeria · SA2 · SA3 · Togo · Ugan · USA3 · Zambia

M_2 P_2 C P_1 I_2 I_1 M_1

Figure 4.3 (cont.)

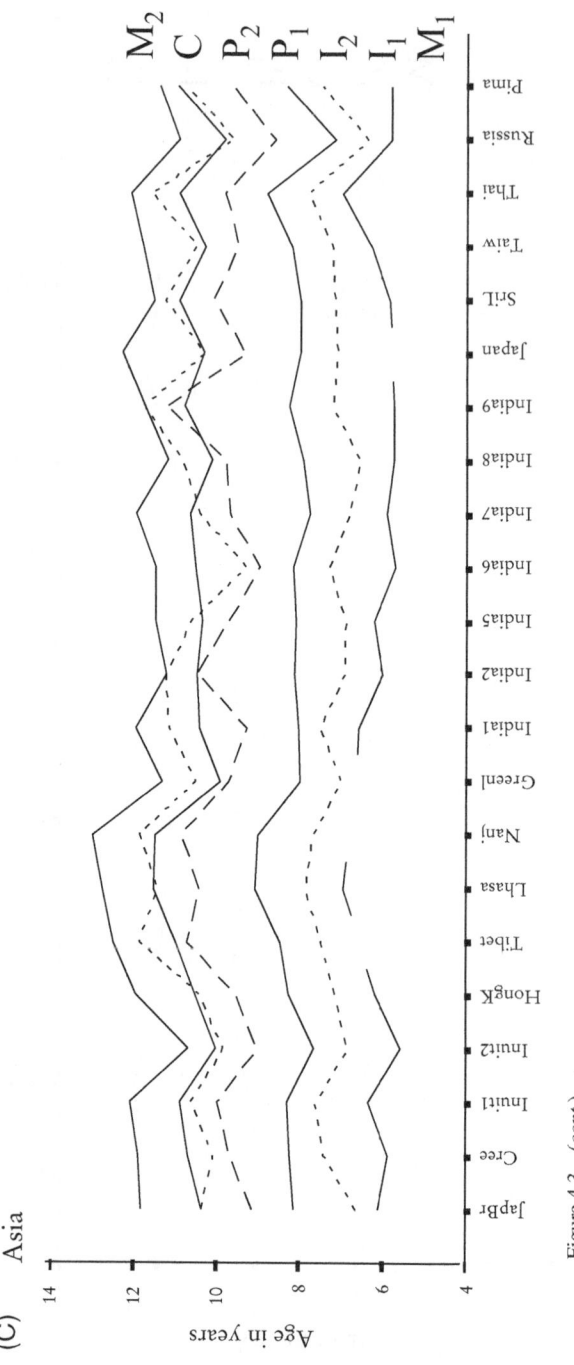

(C) Asia

Age in years

M_2
C
P_2
P_1
I_2
I_1
M_1

JapBr Cree Inuit1 Inuit2 HongK Tibet Lhasa Nanj Greenl India1 India2 India5 India6 India7 India8 India9 Japan SriL Taiw Thai Russia Pima

Figure 4.3 (cont.)

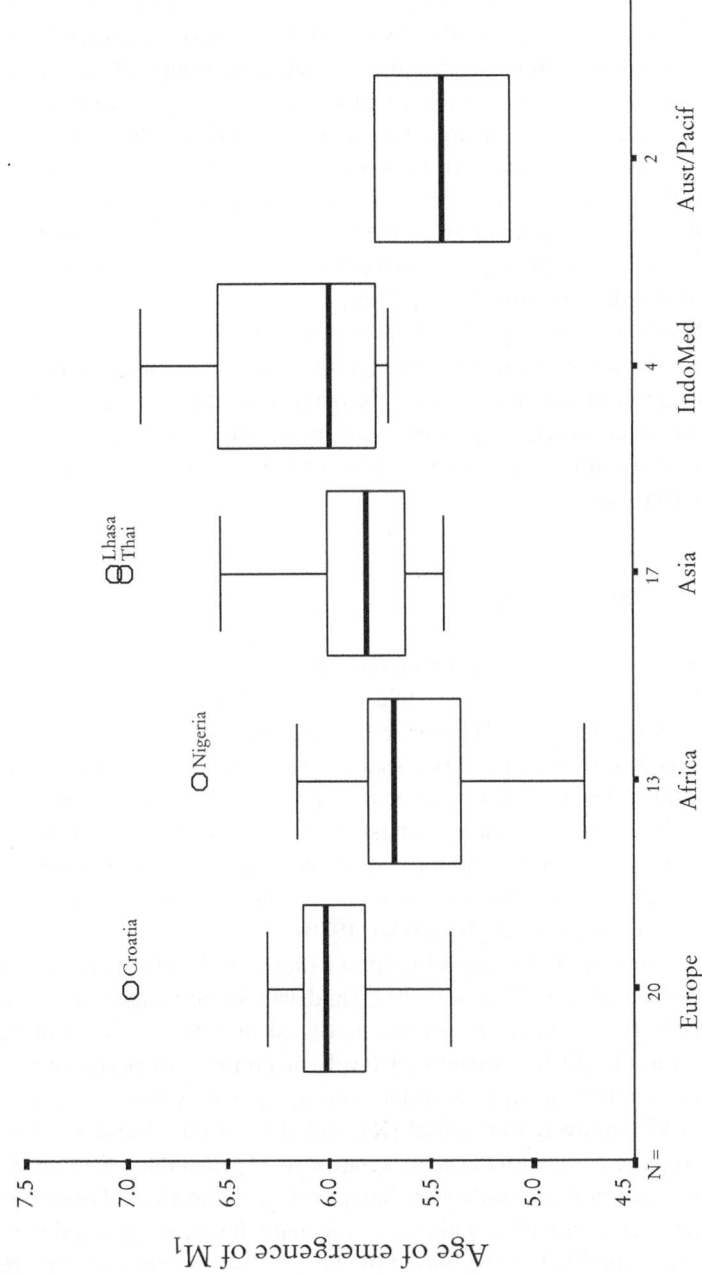

Figure 4.4   Boxplot of mean eruption time of M_1 for different populations. (See Tables 4.2, 4.4, and 4.5.)

mandibular teeth showing a bigger difference than maxillary teeth (Hassanali & Odhiambo, 1981). A study of seven ethnic groups of children living in Israel found eruption timing to be uniform between groups (Koyoumdjisky-Kaye *et al.*, 1981), although some differences of early emerging teeth were apparent. A large random study in Argentina show no significant differences in timing of eruption of permanent teeth between Amerindian and Caucasian populations as well as rural and town children, although there was a tendency for earlier eruption in metropolitan areas (Muniz, 1988). South African black children emerged the first phase of permanent teeth 0.3 to 0.6 year earlier than Indian children (Blankenstein *et al.*, 1990b).

Results from this review show that although the mean age of eruption between populations in geographic regions shows some variation, no clear pattern is evident; different populations are outliers for different teeth. The few studies that make direct comparisons show that population differences are between three to six months and more apparent in the mandibular teeth compared to maxillary teeth.

Stages of eruption

Radiographically, tooth eruption has been quantified in relation to the inferior border of the mandible, the occlusal surface, the inferior alveolar canal, and to M_1 (Carlson, 1944; Darling & Levers, 1975; Feasby, 1981; Schopf, 1970; Shumaker & El Hadary, 1960; Tsai, 2000). Others use the alveolar bone level or occlusal levels and a number of reports describe these stages as well as the midpoint between them (Bengston, 1935; Schopf, 1970). How these stages correspond to clinical eruption has not been documented although one report documents the time interval between alveolar and clinical eruption, which is illustrated in Figure 4.5 (Haavikko, 1970).

A couple of studies report eruption stages of deciduous teeth (Liversidge, 2001; Liversidge & Molleson, 2003) and data for permanent tooth eruption are available from several investigations (listed in Table 4.6, data in Table 4.7). Bengston (1935) first quantified levels of eruption of permanent teeth with stages including the occlusal surface or cusp tips below bone level, at the alveolar level (AE), midway to occlusal (X), and at the occlusal levels. Another study assessed some mandibular teeth relative to M_1 including AE and X (Schopf, 1970). Both of these studies are hampered by the method chosen for analysis; neither uses cumulative statistics to calculate the mean ages and the early and late stages are likely to be biased by the age distribution. They are included in this review despite this as so few studies report quantitative data on the stages of eruption.

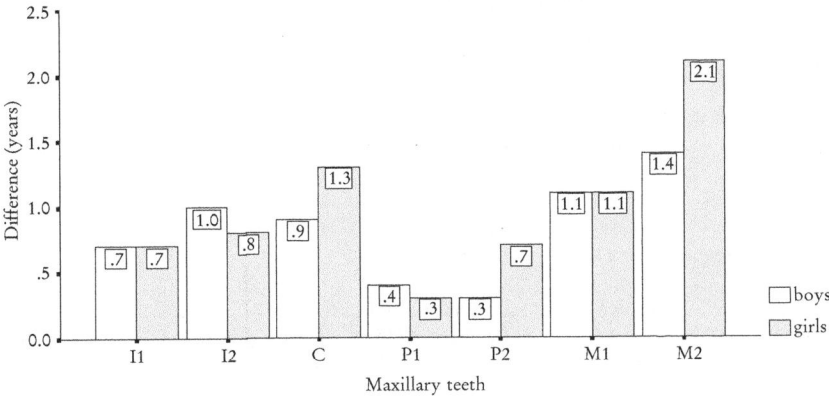

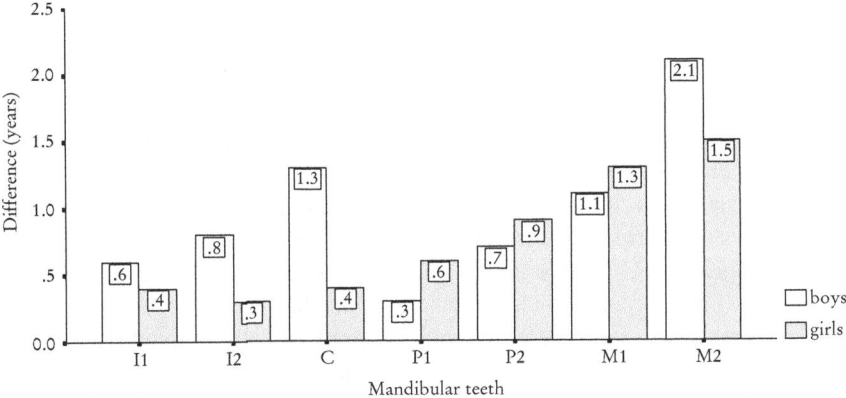

Figure 4.5 Bar chart of time between alveolar and clinical eruption of permanent teeth. (From Haavikko (1970).)

Studies that report age of alveolar eruption and use cumulative statistics include one cross-sectional (Haavikko, 1970) and two longitudinal (Ando *et al.*, 1965; Garn *et al.*, 1958), although the Japanese study (Ando *et al.*, 1965) defines the position of the occlusal surface of the tooth as "slightly higher than the alveolar crest." Mean ages from Ando *et al.* (1965) have been calculated using probit regression for this review and are found in Table 4.8. The mean age of alveolar eruption varies considerably between the studies in Tables 4.7 and 4.8, although data from both longitudinal studies are similar. The timing for M_1 alveolar eruption varies between around 4 and almost 6 years; the pioneer longitudinal study by Gleiser & Hunt (1955) reports the age as 5.4 years. The timing for M_2 alveolar eruption is between 9 and 13 years. A larger proportion

Table 4.6 *Summary of studies of permanent tooth eruption stages*

| Country | Place | Reference | Sample size[a] | | Study type[b] | Age range (years) |
			n Boys	n Girls		
Canada	Ontario	Feasby, 1981	54	49	L	3–14
Finland	Helsinki	Haavikko, 1970	615	547	C	2–21
Germany	Mainz	Schopf, 1970	(350)		C	6–13
Japan	Takada	Ando et al., 1965[c]	94	82	L	5.2–11.2
United Kingdom	Bristol	Darling & Levers, 1975	(362)		C	2–22
	London	Liversidge, 2001	(186)		C	0–14
United States of America	Chicago	Bengston, 1935	(563)		C	5.5–13.5
	Yellow Springs	Garn et al., 1958	(255)		L	0–16

[a] Sample size in parentheses (*n*) represents combined boys and girls.
[b] C, cross-sectional; L, longitudinal.
[c] Data reanalyzed by probit; see Table 4.8.

of white children from the Midsouth showed earlier eruption stages in M^2 and M_2 compared to children in Midwest America suggesting regional differences (Mappes *et al.*, 1992).

Root stage development during eruption has also been studied. Eruption begins during early root formation and the tooth reaches occlusion before root completion. Alveolar eruption of deciduous teeth occurs during early root growth, when root length is less than crown height. Once root length $\geq$ crown height, deciduous teeth are at the midway mark or fully erupted (Liversidge, 2001). Permanent teeth that form early erupt with less root than later forming teeth. Root stage of M_1 at alveolar eruption is reported as $R^1/_4$ from one of the earliest longitudinal radiographic studies (Gleiser & Hunt, 1955). This root stage is also reported for alveolar eruption of first molars and upper incisors (Haavikko, 1970). Clinical emergence of I_1 and M_1 occurs when the root is about half final length; other permanent teeth emerge when the roots are around three-quarter length (Grøn, 1962). Demirjian & Levesque (1980) found that clinical emergence of permanent mandibular teeth coincided with, or was very closely followed by, stage G (root length complete with open apex).

A number of longitudinal studies report the rate of eruption of permanent teeth. Between 2 and 4 mm of root growth occurs prior to any occlusal movement (Carlson, 1944); after 6 mm of root length of M_2, the eruption rate accelerates, reaching a peak during the latter half of root formation (Kuno, 1980). This rate is greatest during the year before full occlusion and has been noted for other permanent teeth (Feasby, 1981).

Table 4.7 *Mean timing (age in years) of eruption stages for permanent teeth*

Study	Level[a]	Sex	I^1	I^2	C'	P^1	P^1	M^1	M^1	I_1	I_2	$C,$	P_1	P_2	M_1	M_2
Garm et al., 1958	AE	Girls											9.7	10.3	5.7	10.7
		Boys											10.1	11.1	5.8	11.2
	Occ	Girls											10.3	11.3	6.9	11.8
		Boys											10.9	12.2	6.9	12.7
Haavikko, 1970	AE	Girls	6.1	7.0	9.3	9.0	9.5	5.3	10.3	6.7	7.8	10.6	9.6	10.2	6.4	12.4
		Boys	6.2	7.3	11.2	9.8	11.1	5.3	11.4	6.9	8.3	12.1	10.2	11.4	6.4	12.8
Liversidge, 2001	AE														4.20	9.29
	X														5.87	10.82
	Occ														6.85	12.25

[a] AE, cusp tips at alveolar level; X, cusp tips half way to occlusal level; Occ, cusp tips at occlusal level.

Table 4.8 *Calculated mean timing in years of eruption stage "slightly higher than the alveolar crest" of Japanese children*

Tooth	Sex	n	Mean	SD	Tooth	Sex	n	Mean	SD
I^1	Girls	82	7.14	1.04	I_1	Girls	82	6.30	0.59
	Boys	94	7.01	0.82		Boys	94	6.28	0.59
I^2	Girls	82	8.11	0.84	I_2	Girls	82	7.13	0.72
	Boys	94	8.16	0.73		Boys	93	7.14	0.69
C'	Girls	57	10.28	1.16	C	Girls	60	9.24	1.02
	Boys	35	10.37	1.03		Boys	33	9.54	1.17
P^1	Girls	81	9.22	1.44	P_1	Girls	82	9.59	1.31
	Boys	94	9.07	1.40		Boys	93	9.61	1.37
P^2	Girls	82	10.26	1.60	P_2	Girls	81	10.46	1.24
	Boys	94	10.29	1.53		Boys	94	10.54	1.43
M^1	Girls	82	5.61	0.85	M_1	Girls	82	–	–
	Boys	94	5.77	0.67		Boys	94	5.27	0.92
M^2	Girls	82	11.23	0.71	M_2	Girls	82	10.86	0.62
	Boys	94	11.36	0.76		Boys	94	10.98	0.10

Source: Ando *et al.* (1965).

It seems clear that these studies of eruption stages only begin to describe the eruption process in humans and further research is needed in this area.

Tooth formation

A considerable amount of information about pre- and perinatal dental growth is known from anatomical and autopsy material (see Scheuer & Black, 2000), while there are a few radiographic studies of the deciduous dentition that report data for individual teeth. One longitudinal investigation from birth documents the formation of the mandibular deciduous canine and molars (Moorrees *et al.*, 1963a). Gilster *et al.* (1964) document the formation of m_2 and another cross-section investigation combines recent archaeological and radiographic investigation of both maxillary and mandibular deciduous tooth formation and eruption (Liversidge & Molleson, 2003). Permanent tooth formation is far better described. Studies with data for crown and root stages of individual permanent teeth are listed in Table 4.9. The only longitudinal study from birth documents the permanent mandibular canine, premolars and molars (Moorrees *et al.*, 1963b). Several other publications are based on this study (Fanning, 1961; Fanning & Brown, 1971). The timing of growth stages from Moorrees *et al.* (1963b) is considerably earlier than other studies. This is clear from a report of tooth formation in Sardinian children, where the age range for the Moorrees

Table 4.9 *Summary of radiographic studies giving data for individual permanent tooth formation*

Country	Place	Reference	Sample size		Study type[a]	Age range (years)
			n Boys	n Girls		
Canada	Montreal	Demirjian & Levesque, 1980	2705	2732	C + L	2–19
	Toronto	Anderson *et al.*, 1976	121	111	L	3.5–18
China	Chengdu	Zhao *et al.*, 1990	465	438	C	3–16
Finland	Helsinki	Haavikko, 1970	615	547	C	2–21
United Kingdom	London	Liversidge & Speechly, 2001	263	258	C	4–9
	London	B. Poolsanguan, unpublished data	279	288	C	5–15
United States of America	Boston, Yellow Springs	Moorrees *et al.*, 1963a[b]	184	161	L	0–10
	Cleveland	Simpson & Kunos, 1998	152	151	C + L	2.2–18

[a] C, cross-sectional; L, longitudinal.
[b] See updated data in Smith (1991).

et al. (1963b) stages is plotted against stage for each tooth (Diaz *et al.*, 1993). Although the method of statistical analysis used in Diaz *et al.* (1993) was inappropriate, the plots show that the age range in Sardinian children for each stage is later than the range provided by Moorrees *et al.* (1963b). The timing of tooth formation from Moorrees *et al.* (1963b) is noticeably earlier than other studies, possibly due to the large number of crown and root stages; the precision of repeated observation is compromised when successive stages are only marginally different. Despite this, it remains the most widely cited radiographic study of tooth formation and will remain important because of the ethical issues of using X-rays in growth studies.

Timing of Demirjian's stages for individual teeth are documented in Table 4.10 and 4.11 for Canadian (Demirjian & Levesque, 1980) and Chinese children (Zhao *et al.*, 1990). The Chinese children attain most stages between 0.5 and 1.5 years earlier than the Canadian children. Data of living children in London are similar to the Chinese except for P_2 which attains stages D, E and F about a year earlier (Liversidge & Speechly, 2001). Timing of root and apex maturation (stages G and H) of I_1 is about 1 year later in the London group compared to the sample of Demirjian & Levesque (1980) but I_2 timing was similar (B. Poolsanguan, unpublished data).

Table 4.10 *Median age in years of permanent tooth formation in Canadian children*

Tooth	Sex	Stage							
		A	B	C	D	E	F	G	H
I₁	Girls					3.5	5.3	6.5	8.1
	Boys					3.9	5.7	6.8	8.5
I₂	Girls					3.7	6.1	7.3	9.2
	Boys					4.4	6.5	7.7	9.6
C	Girls				2.9	4.9	7.6	9.6	12.2
	Boys				3.3	5.4	8.5	10.6	13.4
P₁	Girls			3.5	4.2	6.0	8.6	10.1	12.7
	Boys			3.6	4.5	6.5	9.1	10.8	13.4
P₂	Girls	3.8	4.1	4.7	5.6	7.1	9.3	11.1	13.6
	Boys	3.8	4.1	4.7	5.9	7.6	9.6	11.6	14.2
M₁	Girls					3.7	5.2	6.3	9.5
	Boys					4.1	5.4	6.7	10.2
M₂	Girls	3.5	4.0	4.6	5.9	7.9	9.9	11.5	14.9
	Boys	3.5	4.0	4.9	6.3	8.5	10.4	12.0	15.3

Source: Demirjian & Levesque (1980).

Table 4.11 *Median age in years of permanent tooth formation in Chinese children*

Tooth	Sex	Stage							
		A	B	C	D	E	F	G	H
I₁	Girls				3.0	4.0	5.7	6.7	8.5
	Boys				3.2	4.2	5.8	7.2	8.8
I₁	Girls				3.1	4.6	6.6	7.6	9.9
	Boys				3.3	4.9	6.8	8.0	10.4
C	Girls			3.3	4.4	6.1	8.1	10.4	13.1
	Boys			3.4	4.8	6.5	9.0	11.6	14.0
P₁	Girls			4.0	5.8	7.0	8.9	11.0	13.4
	Boys			4.1	6.2	7.4	9.4	11.6	13.9
P₂	Girls	3.3	4.0	5.0	6.3	7.6	9.9	12.2	14.2
	Boys	3.3	4.0	5.5	6.6	8.0	10.3	12.6	14.4
M₁	Girls				3.0	4.1	5.6	7.1	10.0
	Boys				3.1	4.3	5.8	7.2	10.9
M₂	Girls	3.1	4.1	5.5	7.1	8.9	10.5	12.3	14.3
	Boys	3.1	4.1	5.6	7.3	9.5	10.8	13.0	14.7

Source: Zhao *et al.* (1990).

Comparing tooth formation in two or more populations directly is an effective way to quantify regional and ethnic differences in tooth formation; however, not all reports of such differences use cumulative analyses and reported differences may reflect sampling (Harris & McKee, 1990; Loevy, 1983). Two studies do use appropriate analyses; no significant differences in mandibular tooth formation were observed between Bangladeshi and white London children aged 4 to 9 years (Liversidge & Speechly, 2001). In contrast 12 to $13^{1}/_{2}$ year olds in the Midsouth of the United States were considerably delayed compared to Midwest American children (Mappes *et al.*, 1992). This interesting result is hampered by lack of details of sample size; in addition the age interval extends over a very short time of growth and the stages considered are difficult to assess ($R^{3}/_{4}$, Rc, $A^{1}/_{2}$). Several investigations have demonstrated differences in relative timing of anterior and posterior teeth or early- and late-forming teeth within or between populations (Fanning & Moorrees, 1969; Tompkins, 1996; Watt & Lunt, 1999) and these patterns are easily explored if data for individual teeth are available. Few differences in one tooth relative to another were apparent before age 9 between Bangladeshi and white London children (Liversidge & Speechly, 2001). Medieval Scottish children were advanced in permanent tooth formation relative to M_1 root stages in North American children; however, this was for later root stages only (Watt & Lunt, 1999). Australian Aborigines and North American children show few differences in permanent molar formation prior to $R^{1}/_{4}$ root length of M_2, thereafter crown formation of M_3 was advanced in the Aboriginal children (Fanning & Moorrees, 1969). African children from Botswana and prehistoric American Indians were ahead in both M_2 and M_3 formation, relative to M_1 from the root cleft stage compared to Canadians (Tompkins, 1996). Although sampling and sample size in these studies are not ideal, these findings suggest this shift is related to age and possibly available space in the jaw.

Formation of permanent teeth can also be expressed as dental age using a maturity scale, the most widely used is that described initially for Canadian children by Demirjian *et al.* (1973), updated in 1976 (Demirjian & Goldstein, 1976). Since then numerous studies around the world have shown an advancement in dental maturation compared to the Canadian standard, particularly between the ages of 5 and 8 (Table 4.12). These are detailed for French, Italian, Chinese, German, Australian, and Iranian children (Frucht *et al.*, 2000; Malagola *et al.*, 1989; McKenna *et al.*, 2002; Proy *et al.*, 1981; Zhao *et al.*, 1990; S. M. Fatemi, pers. comm.). A few studies compare maturity directly between populations but using a single maturity score fails to clarify when or how these differences occur and are possibly inherent in the method, as the timing of transitions between stages and greatest standard deviation occurs around six years of age

Table 4.12 *Studies using dental maturity scale*

Country	Place	Reference	Sample size Boys	Girls	Study type[a]	Age (years)
Australia	Adelaide	McKenna et al., 2002	288	327	C	4.9–16.9
Belgium	Leuven	Willems et al., 2001	1029	1087	C	1.8–18
China	Chengdu	Zhao et al., 1990	465	438	C	3–16
	Hong Kong	Davis & Hägg, 1994	101	103	C	5–7
Finland	Helsinki	Nyström et al., 1986	349	389	C + L	2.5–16.5
	Helsinki	Nyström et al., 1988	50	40	C	4–15
	Kuhmo	Nyström et al., 1988	181	214	C	4–15
	Helsinki, Turku	Kataja et al., 1989	506	556	C	2.5–17.2
France	Venissieux	Proy et al., 1981	98	108	C	3.5–14.5
	Venissieux	Proy & Gautier, 1985	(1610)		C	3.5–?16
Germany	Freiburg	Frucht et al., 2000	489	514	C	2–20
Holland	Nymegen	Prahl-Andersen et al., 1979	232	254	C + L	4–14
India	Manipal	Koshy & Tandon, 1998	93	91	C	5–15
Italy	Rome	Malagola et al., 1989	(157)		C	5–14
Korea	Kwangju	Tievens & Mörnstad, 2001a	173	137	C	3–17.2
Norway	Oslo	Nykänen et al., 1998	128	133	L	5.4–12.7
Sweden	Regional	Teivens & Mörnstad, 2001b	243	242	C	2.6–17.2
United Kingdom	London	Liversidge et al., 1999	263	258	C	4–9
	Sheffield	Davidson & Rodd, 2001	42	39	C	2.6–15.8
	Sheffield (Somali)	Davidson & Rodd, 2001	42	39	C	2.6–15.8

[a] C, cross-sectional; L, longitudinal.
Source: Demirjian et al. (1973) and later versions.

when so many teeth contribute to the total score (see Nyström *et al.*, 2001). Few substantial population differences are apparent and this is made more difficult when the method is adapted. Bangladeshi and white children living in London, England were not significantly different in dental maturation (Liversidge *et al.*, 1999). A small group of Somali children were about nine months ahead of white children in dental maturity in Sheffield, England (Davidson & Rodd, 2001). A larger group of Korean children were a few months later in dental maturation compared to Swedish children (Teivens & Mörnstad, 2001a). Adapting this maturity scale to other populations has been approached in several ways. The self-weighted score for each stage has been recalculated for French children (Proy & Gautier, 1985, 1986). This is most useful if the score values are presented in real age rather than percentages (Willems *et al.*, 2001) and these scores

Table 4.13 *Age scores for stage expressed directly in years Belgian children*

Tooth	Sex	Stage							
		A	B	C	D	E	F	G	H
I_1	Girls			1.83	2.19	2.34	2.82	3.19	3.14
	Boys			1.68	1.49	1.50	1.86	2.07	2.19
I_2	Girls			0.00	0.29	0.32	0.49	0.79	0.70
	Boys			0.55	0.63	0.74	1.08	1.32	1.64
C	Girls			0.60	0.54	0.62	1.08	1.72	2.00
	Boys			0.00	0.04	0.31	0.47	1.09	1.90
P_1	Girls	−0.95	−0.15	0.16	0.41	0.60	1.27	1.58	2.19
	Boys	0.15	0.56	0.75	1.11	1.48	2.03	2.43	2.83
P_2	Girls	−0.19	0.01	0.27	0.17	0.35	0.35	0.55	1.51
	Boys	0.08	0.05	0.12	0.27	0.33	0.45	0.40	1.15
M_1	Girls				0.62	0.90	1.56	1.82	2.21
	Boys				0.69	1.14	1.60	1.95	2.15
M_2	Girls	0.14	0.11	0.21	0.32	0.66	1.28	2.09	4.04
	Boys	0.18	0.48	0.71	0.80	1.31	2.00	2.48	4.17

Source: Willems *et al.* (2001).

are reproduced in Table 4.13; dental age is calculated by adding the scores for each tooth. Other ways to adapt the method of dental maturity are to adjust the total score conversion tables (Kataja *et al.*, 1989; Nyström *et al.*, 1986, 1988) or recalculate by formulae (Frucht *et al.*, 2000; Teivens & Mörnstad, 2001b) or in ways that are not entirely clear (Koshy & Tandon, 1998). Some of these methods do not clarify the differences between populations and make comparisons more difficult. Teivens & Mörnstad (2001b) use a polynomial regression, where the children below the median do not reach full maturity. Koshy & Tandon (1998) assume that the dental maturation curve is advanced as a whole, in addition to estimating dental age up to almost 19 years. Frucht *et al.* (2000) report a significant delay of canine and first premolar formation, yet the median age of canine apex closure (girls 12 years, boys 13.7 years) is not dissimilar to that given by Demirjian & Levesque (1980).

These findings suggest if population differences in cross-sectional studies do exist, they are small. However, a cross-sectional study fails to take different rates of growth into account. Two investigations have shown that only about a quarter of children stay within their percentile throughout growth; three-quarters begin below the median and shift above it, either before age six, between six and eight years or after age nine (Loevy & Goldberg, 1996, 1999).

Several studies have used other scoring methods (Carvalho *et al.*, 1990; Nolla, 1960). The principle of a continuum of dental formation is appropriate and the developmental "curves" of timing of stages from these studies may be of interest, although, results are of questionable validity if nominal data are treated as parametric. Tooth formation in Croatian children (Stefanac, 1987) has been quantified in this way; some of these children were found to be delayed compared to Syrian children at stage $R^2/_3$ (Stefanac-Papic *et al.*, 1998). This root stage is Nolla's stage 8, whereas stage 9 is "root almost complete, apex open" and teeth are scored subjectively. For instance M_1 is scored as 8.98 in Zagreb boys and 8.61 in boys from Damascus. It is difficult to interpret this finding as no intra-observer error is mentioned and subjective root length stages are notoriously difficult to assess. Tooth formation using this method has also been reported for children in Brazil (Ferreira *et al.*, 1993) and Berlin (Holtgrave *et al.*, 1997).

Another score is described by Simpson & Kunos (1998) where tooth formation is expressed as a percentage of crown or root completion. Data are given as bivariate plots and age of attainment of crown completion for some permanent teeth are given.

Results from this review of crown and root formation suggest that population differences are small. In summary, longitudinal data from Moorrees *et al.* (1963b) are considerably earlier than data from other studies. The only substantial population difference in dental maturation is between the original reference sample Demirjian *et al.* (1973) and more recent studies. This method of calculating dental age tends to overestimate maturation, has been updated by Willems *et al.* (2001) and needs validating. Relative formation of the later forming permanent teeth is another area of research opportunity.

Conclusions

Several factors need to be remembered when comparing tooth formation data between studies, populations, tooth types, or even taxa. First, the timing of clinical eruption of teeth varies more than the timing of crown and root growth stages. Second, and more importantly, a time-related gradient exists for both eruption and formation of teeth. The mean age ± 2 standard deviations of eruption of deciduous and early permanent teeth is considerably smaller than later emerging teeth; similarly deciduous tooth formation varies less than permanent tooth formation. This becomes clear if one compares the *y*-axis scale for age in Figures 4.2 and 4.4; variation of the deciduous tooth is considerably smaller than for the permanent tooth.

Figure 4.6 illustrates the mean age ± 2 standard deviations of some formation stages of m_2, M_1, and M_3 (data for girls from Moorrees *et al.*, 1963a, 1963b) showing that early formation stages vary less than later stages and early-forming teeth vary less than later-forming teeth. This gradient exists within the

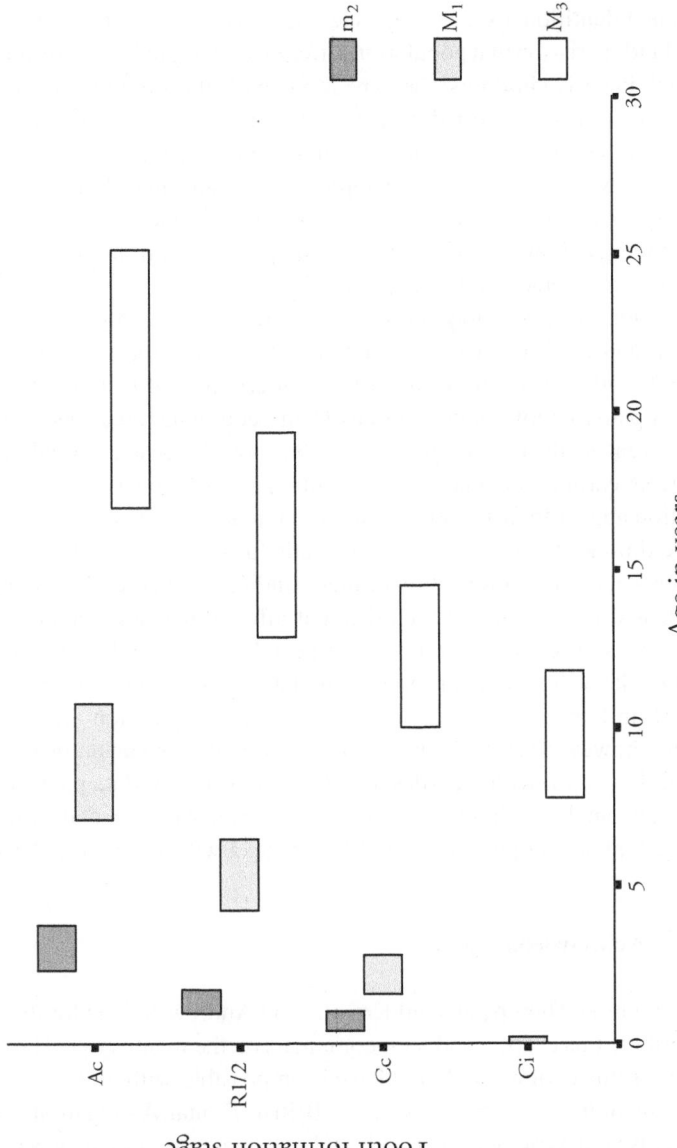

Figure 4.6   Mean age ± 2SD for some tooth formation stages of m_2, M_1 and M_3 in girls (data from Moorrees *et al.*, 1963a, 1963b). Stages of formation: Ci, cusp initiation; Cc, crown completion; R1/2, root half length; Ac, apex closed.

permanent dentition also; M_1 varies less than M_3 which is the most variable of all. It stands to reason that population differences in tooth formation and eruption are small in early childhood, become more evident during the mixed dentition, and are probably greatest during root formation of the later-forming teeth. Meaningful comparisons are not possible if the age range and sample, or the methodology and analyses, are inappropriate or unsound. Further information from careful histological studies of enamel and dentine also add to knowledge of chronology; however, the laborious preparation and analyses required for histology are impractical for large samples.

Although there is a suggestion of a difference between populations in both eruption and tooth formation in humans, this is less than one standard deviation. The differences in average timing of eruption of teeth do not appear to follow a pattern between populations from geographic areas and outliers vary for different teeth. For instance, some but not all African populations erupt some teeth earlier than others in different regions of the world. Crown and root formation appear to differ little between populations; however, variation in the relative timing of later-forming teeth needs further study, as does the study of eruption stages. The implication of these findings is that tooth formation is not complicated by the same difficulties that affect other non-dental growth systems such as different environments, especially nutrition. Furthermore, if data are unavailable for the appropriate population, predicting age from developing teeth in humans is not subject to substantial population differences. This does not, however, refute the need for more population standards from various parts of the world, with well-described methods, useful data presentation, and appropriate analyses. In this way the specific patterns of dental formation and development adaptations between different taxa will be better understood.

Acknowledgments

I thank Jennifer Thompson, Gail Krovitz, and Andrew Nelson for the invitation to contribute this chapter and their comments on the manuscript. The collection of data for this review would not have been possible without the help of many people, including the librarians of the British Dental Association and British Library, as well as numerous friends and colleagues around the world who sent me reprints and unpublished data.

References

Adorni-Braccesi, M. (1965). Variations in the chronology and the order of eruption of the permanent teeth. *Orthodontie Française*, **36**, 429–442. (In French.)

Ajmani, M. L., & Jain, S. P. (1984). Eruption age of teeth in Nigeria. *Anatomischer Anzeiger*, **157**, 245–252.

Anderson, D. L., Thompson, G. W., & Popovich, F. (1976). Age attainment of mineralization stages of the dentition. *Journal of Forensic Science*, **21**, 191–200.

Ando, S., Aizawa, K., Nakashima, T., Shinbo, K., Sanka, Y., Kiyokawa, K., & Oshima, S. (1965). Studies on the consecutive survey of succedaneous and permanent dentition in the Japanese children. I: Eruption processes of permanent teeth. *Journal of the Nihon University School of Dentistry*, **7**, 141–181.

Baghdady, V. S., & Ghose, L. J. (1981). Eruption time of primary teeth in Iraqi children. *Community Dental and Oral Epidemiology*, **9**, 245–246.

Bakayoko, R. L., Djaha, K., Adiko, E. F., Assi, K. D., & Egnankou, J. K. (1989). The dental age of the appearance of teeth amongst Negroid children in Ivory Coast. *Odonto-Stomatologie Tropicale*, **3**, 97–101. (In French.)

Bakayoko, R. L., Roux, H., Eboi-Agneroh, G., Mansila, E., Kone, D., & Pass, C. (1994). Incisors and first permanent molars: What eruption age for the Ivoirian child? *Odonto-Stomatologie Tropicale*, **66**, 22–24. (In French.)

Banerjee, A. R., Manerjee, P. & Dutta, K. (1985). Eruption of permanent teeth among the Bengalees. In *Dental Anthropology: Applications and Methods*, ed. V. Rami-Reddy, pp. 27–34. New Delhi: Inter India Publications.

Barker, D. K., & Zusman, S. P. (1990). Study of the eruption times of the permanent dentition of the children in Israel. *Refuat Hashinayim*, **8**, 8–13.

Bengston, R. G. (1935). A study of the time of eruption and root development of the permanent teeth between six and thirteen years. *Northwest University Bulletin*, **35**, 3–9.

Billewicz, W. Z., & McGregor, I. A. (1975). Eruption of permanent teeth in West African (Gambian) children in relation to age, sex and physique. *Annals of Human Biology*, **2**, 117–128.

Blankenstein, R., Cleaton-Jones, P. E., Luk, K. M., & Fatti, L. P. (1990a). The onset of eruption of permanent teeth amongst South African black children. *Archives of Oral Biology*, **35**, 228–228.

Blankenstein, R., Cleaton-Jones, P. E., Maistry, P. K., Luk, K. M., & Fatti, L. P. (1990b). The onset of eruption of permanent teeth amongst South African Indian children. *Annals of Human Biology*, **17**, 515–521.

Boesen, P., Eriksen, J. H., & Helm, S. (1976). Timing of permanent tooth emergence in two Greenland Eskimo populations. *Community Dental and Oral Epidemiol*, **4**, 244–247.

Brown, T. (1978). Tooth emergence in Australian Aboriginals. *Annals of Human Biology*, **5**, 41–54.

Bouterline, E., & Tesi, G. (1972). Deciduous tooth eruption in a region of southern Tunisia. *Human Biology*, **44**, 433–442.

Caltabiano, M., Cicciu, D., & Russo, S. (1979). Eruption timing in permanent teeth in 1000 students of 6 to 13 years old in the school population of Catania. *Rivista Italiana di Stomatologia*, **48**, 15–26. (In Italian.)

Carlson, H. (1944). Studies on the rate and amount of eruption of certain human teeth. *American Journal of Orthodontic and Oral Surgery*, **30**, 575–588.

Carvalho, A. A. F., de Carvalho, A., & dos Santos Pinto, M. C. (1990). Radiographic study of the development of the permanent dentition of Brazilian children with a chronological age of 84 and 131 months. *Revista de Odontologia da Universidade Estadual Paulista*, **19**, 31–39. (In Portuguese.)

Choi, N. K., & Yang, K. H. (2001). A study on the eruption timing of primary teeth in Korean children. *Journal of Dentistry for Children*, **68**, 244–249.

Clements, E. M. B., Davies-Thomas, E., & Pickett, K. G. (1953). Time of eruption of permanent teeth in British children in 1947–48. *British Medical Journal*, **i**, 1421–1424.

Dahlberg, A. A., & Menegaz-Bock, R. M. (1958). Emergence of the permanent teeth in Pima Indian children. *Journal of Dental Research*, **37**, 1123–1140.

Darling, A. I., & Levers, B. G. (1975). The pattern of eruption of some human teeth. *Archives of Oral Biology*, **20**, 89–96.

Davidson, L. E., & Rodd, H. D. (2001). Interrelationship between dental age and chronological age in Somali children. *Community Dental Health*, **18**, 27–30.

Davis, P. J., & Hägg, U. (1994). The accuracy and precision of the "Demirjian system" when used for age determination in Chinese children. *Swedish Dental Journal*, **18**, 113–116.

Debrot, A. (1972). A variable influencing tooth eruption age differences between groups. *Journal of Dental Research*, **51**, 12–14.

Delgado, H., Habicht, J. P., Yarbrough, C., Lechtig, A., Martorell, R., Malina, R. M., & Klein, R. E. (1975). Nutritional status and the timing of deciduous tooth eruption. *American Journal of Clinical Nutrition*, **28**, 216–224.

Demirjian, A. (1986). Dentition. In *Human Growth: A Comprehensive Treatise,* vol. 2, *Postnatal Growth and Neurobiology*, 2nd edn, eds. F. Falkner & J. M. Tanner, pp. 269–298. New York: Plenum Press.

Demirjian, A., & Goldstein, H. (1976). New systems for dental maturity based on seven and four teeth. *Annals of Human Biology*, **3**, 411–421.

Demirjian, A., & Levesque, G.–Y. (1980). Sexual differences in dental development and prediction of emergence. *Journal of Dental Research*, **59**, 1110–1122.

Demirjian, A., Goldstein, H., & Tanner, J. M. (1973). A new system of dental age assessment. *Human Biology*, **45**, 211–227.

Diao, C. X. (1981). Survey of permanent teeth eruption of Tibetan (youth) in Rikeze, Tibet. *Chinese Journal of Stomatology*, **16**, 55–59. (In Chinese.)

Diaz, G., Maccioni, P., Zedda, P., Cabitza, F., & Cortis, I. M. (1993). Dental development in Sardinian children. *Journal of Craniofacial Genetics and Developmental Biology*, **13**, 109–116.

El-Beheri, S., & Hussein, M. H. (1987). Sequence and age of emergence for deciduous teeth among a group of children in urban and rural areas of Egypt. *Egyptian Dental Journal*, **33**, 13–30.

El-Nofely, A. (1971). Eruption of permanent dentition and growth in an Egyptian group. *Egyptian Dental Journal*, **17**, 271–280.

Eskeli, R., Laine-Alava, M. T., Hausen, H., & Pahkala, R. (1999). Standards for permanent tooth emergence in Finnish children. *Angle Orthodontist*, **69**, 529–533.

Eveleth, P. B. (1966). Eruption of permanent dentition and menarche of American children living in the tropics. *Human Biology*, **38**, 60–70.

Eveleth, P. B., & de Souza Freitas, J. A. (1969). Tooth eruption and menarche of Brazilian children of Japanese ancestry. *Human Biology*, **41**, 176–184.

Eveleth, P. B., & Tanner, J. M. (1976). *Worldwide Variation in Human Growth*. Cambridge: Cambridge University Press.

Eveleth, P. B., & Tanner, J. M. (1990). *Worldwide Variation in Human Growth*, 2nd edn. Cambridge: Cambridge University Press.

Fanning, E. A. (1961). A longitudinal study of tooth formation and root resorption. *New Zealand Dental Journal*, **57**, 202–217.

Fanning, E. A., & Brown, T. (1971). Primary and permanent tooth development. *Australian Dental Journal*, **16**, 41–43.

Fanning, E. A., & Moorrees, C. F. (1969). A comparison of permanent mandibular molar formation in Australian aborigines and Caucasoids. *Archives of Oral Biology*, **14**, 999–1006.

Feasby, W. H. (1981). A radiographic study of dental eruption. *American Journal of Orthodontics*, **80**, 554–560.

Fedorov, S. D., Bobrovskikh, L. P., & Ivanova, N. S. (1984). Eruption of the permanent teeth in children living in the Lake Baikal area. *Stomatologiia*, **63**, 15–17. (In Russian.)

Ferreira, E. R. Jr., Santos-Pinto, L. A. M., & Santos-Pinto, R. (1993). Stage of tooth mineralization: Comparative analysis according to sex. *Revista da Faculdade de Odontología da Universidade de São Paulo*, **22**, 303–313. (In Portuguese.)

Finney, D. J. (1952). *Probit Analysis.* Cambridge: Cambridge University Press.

Friedlander, J. S., & Bailit, H. L. (1969). Eruption times of the deciduous and permanent teeth of natives on Bougainville Island, Territory of New Guinea. *Human Biology*, **41**, 51–65.

Frucht, S., Schnegelsberg, C., Schulte-Monting, J., Rose, E., & Jonas, I. (2000). Dental age in southwest Germany: A radiographic study. *Journal of Orofacial Orthopedics*, **61**, 318–329.

Garn, S. M., Lewis, A. B., Koski, K., & Polacheck, D. L. (1958). The sex difference in tooth calcification. *Journal of Dental Research*, **37**, 561–567.

Garn, S. M, Sandusky, S. T., Nagy, J. M., & Trowbridge, F. L. (1973). Negro-Caucasoid differences in permanent tooth emergence at a constant income level. *Archives of Oral Biology*, **18**, 609–615.

Gaur, R., & Singh, N. Y. (1994). Emergence of permanent teeth among the Meiteis of Manipur, India. *American Journal of Human Biology*, **6**, 321–325.

Gleiser, I., & Hunt, E. E. (1955). The permanent mandibular first molar: Its calcification, eruption and decay. *American Journal of Physical Anthropology*, **13**, 253–283.

Gillett, R. M. (1997). Dental emergence among urban Zambian school children: An assessment of the accuracy of three methods in assigning ages. *American Journal of Physical Anthropology*, **102**, 447–454.

Gilster, J. E., Smith, F. H., & Wallace, G. K. (1964), Calcification of mandibular second primary molars in relation to age. *Journal of Dentistry for Children*, **31**, 284–288.

Grøn, A. (1962). Prediction of tooth emergence. *Journal of Dental Research*, **41**, 573–585.

Haaviko, K. (1970). The formation and alveolar and clinical eruption of the permanent teeth, an orthopantomograph study. *Proceedings of the Finnish Dental Society*, **66**, 104–170.

108   *H. Liversidge*

Hägg, U., & Taranger, J. (1986). Timing of tooth emergence. A prospective longitudinal study of Swedish urban children from birth to 18 years. *Swedish Dental Journal*, **10**, 195–206.

Halikis, S. E. (1961). The variability of eruption of permanent teeth and loss of deciduous teeth in Western Australian children. I: Times of eruption of the permanent teeth. *Australian Dental Journal*, **6**, 137–140.

Harris, E. F., & McKee, J. H. (1990). Tooth mineralization standards for blacks and whites from the middle southern United States. *Journal of Forensic Science*, **35**, 859–872.

Hassanali, J., & Odhiambo, J. W. (1981). Ages of eruption of the permanent teeth in Kenyan African and Asian children. *Annals of Human Biology*, **8**, 425–434.

Healy, M. J. R. (1986). Statistics of growth standards. In *Human Growth: A Comprehensive Treatise,* 2nd edn, vol.3, eds. F. Falkner & J. M. Tanner, pp. 47–58. New York: Plenum Press.

Helm, S., & Seidler, B. (1974). Timing of permanent tooth emergence in Danish children. *Community Dental and Oral Epidemiology*, **2**, 122–129.

Herbert, H. (1977). Some inquiries into the norms of the dentition between 6 and 12 years. *Pedodontie Française*, **11**, 183–199. (In French.)

Hitchcock, N. E., Gilmour, A. I., Gracey, M., & Kailis, D. G. (1984). Australian longitudinal study of time and order of eruption of primary teeth. *Community Dental and Oral Epidemiology*, **12**, 260–263.

Höffding, J., Maeda, M., Yamaguchi, K., Tsuji, H., Kuwabara, S., Nohara, Y., & Yoshida, S. (1984). Emergence of permanent teeth and onset of dental stages in Japanese children. *Community Dental and Oral Epidemiology*, **12**, 55–58.

Holman, D. J., & Jones, R. E. (1991). Longitudinal analysis of deciduous tooth emergence in Indonesian children. I: Life table methodology. *American Journal of Human Biology*, **3**, 389–403.

Holman, D. J., & Jones, R. E. (1998). Longitudinal analysis of deciduous teeth eruption. II: Parametric survival analysis in Bangladeshi, Guatemalan, Japanese and Javanese children. *American Journal of Physical Anthropology*, **105**, 209–230.

Holnerová, E. (1975). Deciduous teeth and dental age in Czech children. *Scripta Medica (Brno)* **48**, 153–160.

Holtgrave, E. A., Kretschmer, R., & Muller, R. (1997). Acceleration in dental development: Fact or fiction? *European Journal of Orthodontics*, **19**, 703–710.

Houpt, M. I., Adu-Aryee, S., & Grainger, R. M. (1967). Eruption times of permanent teeth in the Brong Ahafo region of Ghana. *American Journal of Orthodontics*, **53**, 95–99.

Hulland, S. A., Lucas, J. O., Wake, M. A., & Hesketh, K. D. (2000). Eruption of the primary dentition in human infants: A prospective descriptive study. *Pediatric Dentistry*, **22**, 415–421.

Jaswal, S. (1983). Age and sequence of permanent-tooth emergence among Khasis. *American Journal of Physical Anthropology*, **62**, 177–186.

Kamalanathan, G. V., Hauck, H. M., & Kittiveja, C. (1960). Dental development of children in a Siamese village, Bang Chan, 1953. *Journal of Dental Research*, **39**, 455–461.

Kataja, M., Nyström, M., & Aine, L. (1989). Dental maturity standards in southern Finland. *Proceedings of the Finnish Dental Society*, **85**, 187–197.

Kaul, S. S., Pathak, R. K., & Santosh, S. (1992). Emergence of deciduous teeth in Punjabi children, north India. *Zeitschrift für Morphologie und Anthropologie*, **79**, 25–34.

Kaul, S., Saini, S., Saxena, B., & Kaul, S. (1975). Emergence of permanent teeth in school-children in Chandigarh, India. *Archives of Oral Biology*, **20**, 587–593.

Kaur, B., & Singh, R. (1992). Physical growth and age at eruption of deciduous and permanent teeth in well-nourished Indian girls from birth to 20 years. *American Journal of Human Biology*, **6**, 757–766.

Kochhar, R., & Richardson, A. (1998). The chronology and sequence of eruption of human permanent teeth in Northern Ireland. *International Journal of Paediatric Dentistry*, **8**, 243–252.

Koshy, S., & Tandon, S. (1998). Dental age assessment: The applicability of Demirjian's method in south Indian children. *Forensic Science International*, **94**, 73–85.

Koyoumdjisky-Kaye, E., Baras, M., & Grover, N. B. (1977). Stages in the emergence of the dentition: An improved classification and its application to Israeli children. *Growth*, **41**, 285–296.

Koyoumdjisky-Kaye, E., Baras, M., & Grover, N. B. (1981). Emergence of the permanent dentition: Ethnic variability among Israeli children. *Zeitschrift für Morphologie und Anthropologie*, **72**, 267–282.

Kromeyer, K., & Wurschi, F. (1996). Tooth eruption in Jena children in the first phase of mixed dentition. *Anthropologischer Anzeiger*, **54**, 57–70. (In German.)

Krumholt, L., Roe-Petersen, B., & Pindborg, J. J. (1971). Eruption times of the permanent teeth in 622 Ugandan children. *Archives of Oral Biology*, **16**, 1281–1288.

Kumar, C. L., & Sridhar, M. S. (1990). Estimation of the age of an individual based on times of eruption of permanent teeth. *Forensic Science International*, **48**, 1–7.

Kuno, T. (1980). On the eruptive process of the mandibular second molars, with particular reference to 45 degrees oblique cephalometric analysis. *Journal of the Nihon University School of Dentistry*, **22**, 108–114.

Lan, W. H. (1971). The Chinese Dentition. IV: The eruption time of permanent teeth in Chinese. *Journal of the Formosan Medical Association*, **70**, 159–165.

Lauzier, C., & Demirjian, A. (1981). Emergence of primary teeth in French-Canadian children. *Union Médical du Canada*, **110**, 1061–1064. (In French.)

Lavelle, C. L. (1976). The timing of tooth emergence in four population samples. *Journal of Dentistry*, **4**, 231–236.

Lawoyin, T. O., Lawoyin, D. O., & Lawoyin, J. O. (1996). Epidemiological study of some factors related to deciduous tooth eruption. *African Dental Journal*, **10**, 19–23.

Lee, M. M., Low, W. D., & Chang, K. S. (1965). Eruption of the permanent dentition of Southern Chinese children in Hong Kong. *Archives of Oral Biology*, **10**, 849–861.

Leighton, B. C. (1968). Eruption of deciduous teeth. *Practitioner*, **200**, 836–842.

Liversidge, H. M. (2001). A radiographic study of dental eruption. In *Dental Morphology 2001, 12th International Symposium of Dental Morphology*, ed. A. Brook, pp. 49–58. Sheffield: Sheffield Academic Press.

Liversidge, H. M., & Molleson, T. (2003). Variation in crown and root formation and eruption of human deciduous teeth. *American Journal of Physical Anthropology* (in press).

Liversidge, H. M., & Speechly, T. (2001). Growth of permanent mandibular teeth of British children aged 4 to 9 years. *Annals of Human Biology*, **28**, 256–262.

110   *H. Liversidge*

Liversidge, H. M., Speechly, T., & Hector, M. P. (1999). Dental maturation in British children: Are Demirjian's standards applicable? *International Journal of Paediatric Dentistry*, **9**, 263–269.

Loevy, H. T. (1983). Maturation of permanent teeth in black and latino children. *Acta Odontologica Pediatrica*, **4**, 59–62.

Loevy, H. T., & Goldberg, A. F. (1996). Shifts in tooth maturation patterns in non-French Canadian girls. In *Biological Mechanisms of Tooth Movement and Craniofacial Adaptation*, eds. Z. Davidovitch & L.A. Norton, pp. 491–94. Boston: Harvard Society for the Advancement of Orthodontics.

Loevy, H. T., & Goldberg, A. F. (1999). Shifts in tooth maturation patterns in non-French Canadian boys. *International Journal of Paediatric Dentistry*, **9**, 105–110.

Low, W. D., Ng, C. K., Chen, D., & Fung, S. H. (1973). Eruption of the deciduous dentition in Chinese children in Hong Kong. *Zeitschrift für Morphologie und Anthropologie*, **65**, 129–142.

Lysell, L., Magnusson, B., & Thilander, B. (1962). Time and order of eruption of the primary teeth. A longitudinal study. *Odontologisk Revy*, **13**, 217–234.

Malagola, C., Caligiuri, F. M., & Barbato, E. (1989). Evaluation of dental age using qualitative radiographic analysis II. *Mondo Ortodontica*, **14**, 471–475. (In Italian.)

Magnusson, T. E. (1976). Emergence of permanent teeth and onset of dental stages in the population of Iceland. *Community Dental and Oral Epidemiology*, **4**, 30–37.

Magnusson, T. E. (1982). Emergence of primary teeth and onset of dental stages in Icelandic children. *Community Dental and Oral Epidemiology*, **10**, 91–97.

Manji, F., & Mwaniki, D. (1985). Estimation of median age of eruption of permanent teeth in Kenyan African children. *East African Medical Journal*, **62**, 252–259.

Mappes, M. S., Harris, E. F., & Behrents, R. G. (1992). An example of regional variation in the tempos of tooth mineralization and hand–wrist ossification. *American Journal of Orthodontic and Dentofacial Orthopedics*, **101**, 145–151.

Marques, G. D., Guedes–Pinto, A. C., & Abramowicz, M. (1978). The eruption sequence of permanent teeth in children from the city of Sao Paolo. *Revista da Faculdade de Odontología da Universidade de São Paulo*, **16**, 187–194. (In Portuguese.)

Masson, J. P. (1980). Permanent tooth emergence timing of Northern Quebec Caucasoid children. *Journal of Canadian Dental Association*, **46**, 643–645.

Mayhall, J. T., Belier, P. L., & Mayhall, M. F. (1977). Permanent tooth emergence timing of Northern Ontario Indians. *Journal of the Ontario Dental Association*, **54**, 8–10.

Mayhall, J. T., Belier, P. L., & Mayhall, M. F. (1978). Canadian Eskimo permanent tooth emergence timing. *American Journal of Physical Anthropology*, **49**, 211–216.

McKenna, C. J., James, H., Taylor, J. A. & Townsend, G. C. (2002). Tooth development standards for South Australia. *Australian Dental Journal*, **47**, 223–227.

Mesa, M. S. (1988). Permanent tooth eruption of Spanish children. *Collegium Antropologicum*, **12**, 141–146.

Miller, J., Hobson, P., & Gaskell, T. J. (1965). A serial study of the chronology of exfoliation of deciduous teeth and eruption of permanent teeth. *Archives of Oral Biology*, **10**, 805–818.

Monk, M. (1974). Eruption of the permanent teeth of Caucasoid children and adolescents in the low income group in Johannesburg. *Journal of the Dental Association of South Africa*, **29**, 525–529.

Moorrees, C. F. A., Fanning, E. A., & Hunt, E. E. (1963a). Formation and resorption of three deciduous teeth in children. *American Journal of Physical Anthropology*, **21**, 205–213.

Moorrees, C. F. A., Fanning, E. A., & Hunt, E. E. (1963b). Age variation of formation stages for ten permanent teeth. *Journal of Dental Research*, **42**, 1490–1502.

Muniz, B. (1988). Chronology of permanent tooth eruption in Argentinean children. *Revista de Associatión Odontologica Argentina*, **76**, 222–228. (In Spanish.)

Nielsen, H. G., & Ravn, J. J. (1976). A radiographic study of mineralization of permanent teeth in a group of children aged 3–7 years. *Scandinavian Journal of Dental Research*, **84**, 109–118.

Nolla, C. M. (1960). The development of the permanent teeth. *Journal of Dentistry for Children*, **27**, 254–266.

Nykanen, R., Espeland, L., Kvaal, S. I., & Krogstad, O. (1998). Validity of the Demirjian method for dental age estimation when applied to Norwegian children. *Acta Odontologica Scandinavica*, **56**, 238–244.

Nyström, M., Kilpinen, E., & Kleemola-Kujala, E. (1977). A radiographic study of the formation of some teeth from 0.5 to 3.0 years of age. *Proceedings of the Finnish Dentistry Society*, **73**, 167–172.

Nyström, M., Kleemola-Kujala, E., Kaartinen, A., Nuottamo, L., Laine, P., & Evalahti, M. (1985). Presence of primary teeth in a series of Finnish children. *Proceedings of the Finnish Dentistry Society*, **81**, 77–81.

Nyström, M., Haataja, J., Kataja, M., Evalahti, M., Peck, L., & Kleemola–Kujala, E. (1986). Dental maturity in Finnish children, estimated from the development of seven permanent mandibular teeth. *Acta Odontologica Scandinavica*, **44**, 193–198.

Nyström, M., Ranta, R., Kataja, M., & Silvola, H. (1988). Comparisons of dental maturity between the rural community of Kuhmo in northeastern Finland and the city of Helsinki. *Community Dental and Oral Epidemiology*, **16**, 215–217.

Nyström, M., Kleemola-Kujala, E., Evalahti, M., Peck, L., & Kataja, M. (2001). Emergence of permanent teeth and dental age in a series of Finns. *Acta Odontologica Scandinavica*, **59**, 49–56.

Pahkala, R., Pahkala, A., & Laine, T. (1991). Eruption pattern of permanent teeth in a rural community in northeastern Finland. *Acta Odontologica Scandinavica*, **49**, 341–349.

Parner, E. T., Heidmann, J. M., Vaeth, M., & Poulsen, S. (2001). A longitudinal study of time trends in the eruption of permanent teeth in Danish children. *Archives of Oral Biology*, **46**, 425–431.

Pathmanathan, G., Prakash, S., & Paul, V. (1985). Permanent tooth emergence in Tamils of Sri Lanka. In *Dental Anthropology: Applications and Methods*, ed. V. Rami-Reddy, pp. 37–47. New Dehli: Inter India Publications.

Perreault, J. G., Demirjian, A., & Jenicek, M. (1974). Eruption of permanent teeth in French-Canadian children. *Journal of the Canadian Dental Association*, **40**, 306–313. (In French.)

Prahl-Andersen, B., Kowalski, C. J., & Heydendael, P. (1979). *A Mixed-Longitudinal, Interdisciplinary Study of Growth and Development*. San Francisco: Academic Press.

Proy, E., & Gautier, N. (1985). Dental maturation: construction of tables. *Revue d'Orthopedie Dentofaciale*, **19**, 523–534. (In French.)

Proy, E., & Gautier, N. (1986). Dental maturation in French children and adolescents. *Revue d'Orthopédie Dentofaciale*, **20**, 107–121. (In French.)

Proy, E., Sempé, M., & Ajacques, J. C. (1981). A comparative study of dental and skeletal maturation of French children and adolescents. *Rev Orthop Dento Faciale*, **15**, 309–320. (In French.)

Rajic, Z., Mestrovic, S., & Vukusic, N. (1999). Chronology, dynamics and period of primary tooth eruption in children from Zagreb, Croatia. *Collegium Antropologicum*, **23**, 659–663.

Rajic, Z., Rajic-Mestrovic, S. & Verzak, Z. (2000). Chronology, dynamics and period of permanent tooth eruption in Zagreb children (Part II). *Collegium Antropologicum*, **24**, 137–143.

Ramirez, O., Planells, P., & Barberia, E. (1994). Age and order of eruption of primary teeth in Spanish children. *Community Dental and Oral Epidemiology*, **22**, 56–59.

Reddy, V. R. (1981). Eruption of deciduous teeth among the children of Gulbarga, south India. *Indian Journal of Medical Research*, **73**, 772–781.

Richardson, A., Akpata, S., Ana, J., & Franklin, R. (1975). A comparison of tooth eruption ages in European and West African children. *Transactions of the European Orthodontics Society*, 161–167.

Rousset, M. M., Boualam, N., & Delfosse, C. (2001). Occlusion and rhythm of eruption. *Bulletin de la Groupement International pour la Recherche Scientifique en Stomatologie et Odontologie*, **43**, 53–61.

Saleemi, M. A., Hägg, U., Jalil, F., & Zaman, S. (1994). Timing of emergence of individual primary teeth: A prospective longitudinal study of Pakistani children. *Swedish Dental Journal*, **18**, 107–112.

Sato, S. (1990). *Eruption Of Permanent Teeth: A Color Atlas*. St Louis: Ishiyaku EuroAmerica.

Savara, B. S., & Steen, J. C. (1978). Timing and sequence of eruption of permanent teeth in a longitudinal sample of children from Oregon. *Journal of the American Dental Association*, **97**, 209–214.

Scheuer, L., & Black, S. (2000). *Developmental Juvenile Osteology*. San Diego: Academic Press.

Schopf, P. M. (1970). Root calcification and tooth eruption in the mixed dentition: A study in panoramic X-rays. *Fortschritte der Kieferorthopädie*, **31**, 39–56. (In German.)

Sharma, K., & Mittal, S. (2001). Permanent tooth emergence in Gujjars of Punjab, India. *Anthropologischer Anzeizer*, **59**, 165–178.

Shumaker, D. B., & El Hadary, M. S. (1960). Roentgenographic study of eruption. *Journal of the American Dental Association*, **61**, 535–541.

Simpson, S. W., & Kunos, C. A. (1998). A radiographic study of the development of the human mandibular dentition. *Journal of Human Evolution*, **35**, 479–505.

Singh, S. P. (1980). Eruption of permanent teeth in Gaddi Rajut males of Dhaula Dhar Range of the Himalayas. *Zeitschrift für Morphologie und Anthropologie*, **70**, 295–301.

Smith, B. H. (1991). Standards of human tooth formation and dental age assessment. In *Advances in Dental Anthropology*, eds. M. Kelley & C.S. Larsen, pp. 143–168. New York: Alan R. Liss.

Smith, J. M., Smith, R. N., Brook, A. H., & Elcock, C. (2000). Timing of permanent tooth eruption in London school children. In *Dental Morphology '98*, eds. J. T. Mayhall & T. Heikkinen, pp. 187–191. Oulu: Oulu University Press.

Souza Freitas, J. A. de, Alvares, L. C., & Lopes, E. S. (1970). Eruption of the permanent dentition of third generation Brazilian white children. *Estomatologia e Cultura*, **2**, 201–208. (In Portuguese.)

Stefanac, J. (1987). Dental maturity norms. *Acta Stomatologica Croatica*, **21**, 35–47. (In Serbo-Croatian.)

Stefanac-Papic, J., Alkadri, K. Z., Legovic, M., & Galic, N. (1998). Comparison of dental maturity between two ethnic groups. *Collegium Antropologicum*, **22** (Suppl.), 123–126.

Taranger, J. (1976). Evaluation of biological maturation by means of maturity criteria. *Acta Paediatrica Scandinavica*, **258** (Suppl.), 77–82.

Tegzes, E. (1961). The time of eruption of the milk teeth. *Acta Paediatrica Academiae Scientiarum Hungaricae*, **1**, 289–300. (In German.)

Teivens, A., & Mörnstad, H. (2001a). A modification of the Demirjian method for age estimation in children. *Journal of Forensic Odontostomatology*, **19**, 26–30.

Teivens, A., & Mörnstad, H. (2001b). A comparison between dental maturity rate in the Swedish and Korean populations using a modified Demirjian method. *Journal of Forensic Odontostomatology*, **19**, 31–35.

Titley, K. C. (1984). A comparative investigation of permanent tooth emergence timing of northern Ontario Indians. *Journal of the Canadian Dental Association*, **50**, 775–778.

Tompkins, R. L. (1996). Human population variability in relative dental development. *American Journal of Physical Anthropology*, **99**, 79–102.

Tsai, H. H. (2000). Eruption process of the second molar. *Journal of Dentistry for Children*, **67**, 275–281.

Ulijaszek, S. J. (1996). Age of eruption of deciduous dentition of Anga children, Papua New Guinea. *Annals of Human Biology*, **23**, 495–499.

Virtanen, J. I., Bloigu, R. S., & Larmas, M. A. (1994). Timing of eruption of permanent teeth: Standard Finnish patient documents. *Community Dental and Oral Epidemiology*, **22**, 286–288.

Watt, M. E., & Lunt, D. A. (1999). Stages of tooth development relative to the first permanent molar in a Mediaeval population from the south west of Scotland. In *Dental Morphology '98*, eds. J. T. Mayhall & T. Heikkinen, pp. 120–127. Oulu: Oulu University Press.

Willems, G., Van Olmen, A., Spiessens, B., & Carels, C. (2001). Dental age estimation in Belgian children: Demirjian's technique revisited. *Journal of Forensic Science*, **46**, 893–895.

Yam, A. A., Cisse, D., Tamba, A., Diop, F., Diagne, F., Diop, K., & Ba, I. (2001). Chronology and date of eruption of primary teeth in Senegal. *Odonto-Stomatologie Tropicale*, **24**, 34–38. (In French.)

Yang, Z. L. (1988). Permanent tooth eruption in 3020 children in Lasa County. *Chinese Journal of Stomatology*, **23**, 49–51. (In Chinese.)

Yun, D. J. (1957). Eruption of primary teeth in Korean rural children. *American Journal of Physical Anthropology*, **15**, 261–268.

Zhao, J., Ding, L., & Li, R. (1990). Study of dental maturity in children aged 3–16 years in Chengdu. *Journal of West China University of Medical Sciences*, **21**, 242–246. (In Chinese.)

5 Developmental variation in the facial skeleton of anatomically modern Homo sapiens

U. STRAND VIÐARSDÓTTIR
University of Durham

P. O'HIGGINS
University of York

Introduction

An in-depth understanding of how our facial skeletal morphology arises during ontogeny is likely to lead to better understanding of the evolutionary, functional, and environmental influences that underpin variation. Thus, without an understanding of comparative postnatal growth we cannot hope to explain fully variation in the adult facial skeletons of modern and fossil hominins. Further, such understanding opens up new possibilities for practical applications such as the forensic classification by geographic subgroup of the faces of infants and juveniles as well as those of adults. In this chapter we will briefly review what is known of geographic variation and postnatal ontogeny in human crania before describing some of the results of our ongoing studies into interpopulation (geographic) differences and their ontogenetic basis. In so doing, we explore the possibility of classifying subadult material in the same way as is presently routinely carried out for adults.

Our recent studies have highlighted a great deal of variation within modern human ontogeny, as well as an early onset of morphologically distinct, population specific morphologies in modern humans (O'Higgins & Strand Viðarsdóttir, 1999; Strand Viðarsdóttir *et al.*, 2002). These differences in ontogeny between modern human groups can sometimes be of equivalent scale to those documented between species of non-human primates (e.g., O'Higgins *et al.*, 2001). This ontogenetic variability needs to be kept in mind when comparing the development of fossil hominin species to that of a single human

Patterns of Growth and Development in the Genus Homo, ed. J. L. Thompson, G. E. Krovitz and A. J. Nelson. Published by Cambridge University Press. © Cambridge University Press 2003.

population, or indeed a conglomerate sample of mixed populations, since such comparisons could lead to overgeneralized or, at worst, erroneous conclusions.

In this chapter we will discuss the variability of modern human ontogeny in the context of population-specific differences in the facial skeleton. To illustrate this we provide an example study to illustrate the differences in facial shape and allometry that exist between modern human populations.

Morphological differences in the form of the craniofacial skeleton in adult modern humans: An overview

Many modern human populations show statistically significant differences in craniofacial form. Early quantitative analyses of differences in the craniofacial skeleton were largely focused on the possibilities of racial identification (e.g., Giles & Elliot, 1962). More recently larger-scale studies have used form variations in the skull as an indicator not only of population affinity, but also of interpopulation relationships, in particular in relation to their evolutionary histories and origins (Hanihara, 1992, 1996; Lahr, 1995). Over the last three decades much of the work on modern human craniofacial diversity has been based on large comparative data sets, such as those collected by Howells (1973, 1989, 1995), and more recently the larger, more comprehensive, data set collected and analyzed by Hanihara (1992, 1996). In addition, studies have been carried out on variation within specific geographic regions, and based on specific morphological or dental traits (Hennessy & Stringer, 2002; Hernandez *et al.*, 1997; Howells, 1986; Lahr, 1995; Stringer *et al.*, 1999; Turner, 1992). Many of these analyses have revealed the same general trends in craniofacial variation in modern humans. Overall, adult cranial variation between populations seems to be relatively limited. Relethford (1994), using Howell's data set, found that only ~10% of morphometric variation in the modern human craniofacial skeleton was expressed as interregional variation amongst major geographic regions. Craniometric variation within regions was greater than average within Africa, and smaller than average within Europe (Relethford & Harpending, 1994).

Some documented patterns of craniofacial variation between modern human groups are repeated in different studies. The most notable is the similarity between Australian and African populations in craniofacial form. Thus, Australian populations align more closely with Africans than they do with geographically closer populations, such as the Melanesians (Howells, 1973, 1989). There are two main hypotheses offered to explain the reason behind the Australian–African similarities. The first emphasizes the possibility of convergent evolution (e.g., due to similarities in climate: Guglielmino-Matessi *et al.*, 1977; Howells, 1989), while the second emphasizes the similarities as being the result of shared, relatively primitive features, retained from the first appearance of modern humans (e.g., Nei & Roychoudhury, 1993; Stringer *et al.*, 1999). Another example of craniofacial similarities that may not have been expected from geographic distribution is the similarity between Tierra del Fuegans and North American Eskimo (Hernandez *et al.*, 1997). However, since both these populations are cold adapted, this similarity may reflect morphological adaptation to cold climates. Both populations have also been reported to use teeth as tools. Thus a large part of their morphological similarities, in particular increased cranial robustness and craniofacial width, could be due to shared adaptations to masticatory stress (see Lahr, 1995).

All in all, there is little evidence in any study of craniofacial form for a straightforward link between geographical distribution and morphological diversity. Thus, if geographical distribution is a good reflection of the diversification and evolution of modern human subgroups, it seems that while human craniofacial form may be population specific it is not a reliable indicator of evolutionary history. The exception to this is Froment's (1992) study based on 536 modern human groups. He found that the scores of the groups on the two major axes of variation indicated correspondence between the major morphological affinities within the sample and a map of the Old World. His finding indicated an increase in facial breadth going from west to east through the Old World and a decrease in nasal breadth from south to north. However, a close scrutiny of Froment's (1992) results reveals the position of the Africans on the Old World map to be midway between the African and Australian continents, reflecting their close morphological similarity, and not at all a good fit to geography.

Thus, morphological diversity need not necessarily mirror geographical diversity. In contrast, patterns of molecular variation accord to a much greater degree with geography than does morphological diversity (Batzer *et al.*, 1994; Cann *et al.*, 1987; Cavalli-Sforza *et al.*, 1994; Nei & Roychoudhury, 1993). Therefore if geographical dispersal reflects evolutionary history, molecular data appear to be a better indicator of this dispersal than does cranial morphology.

This pattern is likely because the translation of genetic variation into phenotypic variation involves multiple, interacting, and complex ontogenetic mechanisms. Thus, during ontogeny genetic information is translated into the phenotype through the processes of development (changes in shape with age), growth (changes in size with age), and allometry (changes in shape with size). These are subject to genetically and epigenetically mediated environmental influences (such as the climatic similarities between Africa and Australia), and to epigenetic interactions between developing tissues (Scheuer & Black, 2000). In consequence, the correspondence between genetic and phenetic variation is not direct, with phenetic variation reflecting both genetic and epigenetic influences on growth.

General principles of growth in the craniofacial skeleton

Although variation in the form of the adult craniofacial skeleton has been extensively studied little is known about the underlying ontogenetic basis of this variation in morphology. When do differences in form develop, and what are the processes involved?

The facial skeleton is made up of numerous independent bones, each of which grows and develops under the influence of various specific and systemic factors. In addition many of the individual bones can be divided into subunits such as the maxillary alveolus, or the orbital part of the maxilla, which grow to some degree independently from each other. Despite this, the facial skeleton has to remain a functional whole throughout the course of development. This is achieved through constant remodeling of individual bones in order to adapt to changes in size and/or shape of other bones.

Although some workers (e.g., Nur & Hasson, 1984) have claimed that there is a predetermined genetic blueprint for craniofacial shape, most would agree that a combination of genetic and epigenetic influences is responsible for adult facial morphology (Hunt, 1998; Lieberman, 1996). Several ways in which these influences could interact have been postulated (e.g., Sperber, 1989). It could be argued that, being insulated from the world to some degree, the prenatal form of the craniofacial skeleton is largely genetically determined and that it is only after birth that epigenetic influences have major impact. However, the human fetus is not suspended in a mechanically inert medium, but is subject to the forces operating within the womb. It is also capable of frequent and strong movements, the force of which may subject the skeleton to considerable stress. It is therefore likely that prenatally the skeleton adapts to epigenetic influences.

The regulation of growth of the cranium is frequently described in terms of functional matrices that regulate bone modeling and remodeling. The theory of functional matrices was first postulated by Moss (1964, 1968; Moss & Saletijn, 1969a, 1969b). In its simplest form the theory postulates that although the sizes, shapes, and positions of skeletal tissues in the craniofacial skeleton are genetically influenced at the initiation of ossification, any further genetic control acts not directly on the bone itself, but on its associated functional matrices. The functional matrix can be any set of tissues or spaces that guide the size, shape, and position of the supporting skeletal tissue (Ranly, 1980). In the case of the facial skeleton these would include the brain, teeth, eyeballs and muscles, the nasal and oral cavities, and the associated respiratory tract. The theory of functional matrices does not allow for direct genetic control of bone formation at facial or cranial sutures. However clinical evidence does suggest that both sutures and cartilage can regulate their own activity to some extent, e.g., in cases of premature fusion of cranial sutures (Burrows *et al.*, 1999; Ranly, 1980;

Van Limborg, 1970; Wong *et al.*, 1991). Thus, the functional matrix theory can be used to explain many aspects of genetic and biomechanical control of growth and development of the craniofacial skeleton, although its limits and restrictions need to be studied in greater detail (Moss, 1997a, 1997b, 1997c, 1997d).

During growth a large amount of bone deposition takes place at the sutural margins of the craniofacial skeleton. This is said to be because the displacement of bone due to the expansion of the functional matrices causes tension, which stimulates the deposition of new bone (McLachlan, 1994). As the growth of the functional matrices influences the form of their associated skeletal units, the bone is constantly modeled and remodeled through the processes of bone formation and resorption, the ultimate goal of which is to maintain the mineralized bone matrix and the lifelong mechanical integrity of the skeleton (Canalis, 1993; Lanyon, 1993). Some workers choose to make a distinction between modeling and remodeling (Bromage, 1986). Bone modeling is associated with the growth of bones in childhood, changing the form of the growing bone to adapt to changes in size and shape of related elements. Unlike remodeling, it is a continuous process and covers a large surface (Eriksen *et al.*, 1993). Remodeling, on the other hand, is cyclical and usually covers a small area, in response to either unusual intermittent strains on load-bearing bones (Fehling *et al.*, 1995; Heinonen *et al.*, 1995; Lanyon & Rubin, 1984), or the body's need to maintain steady levels of serum calcium (Lanyon, 1993; Parfitt *et al.*, 1983). In the context of craniofacial growth, periosteal remodeling maintains the shape and proportion of bones as well as adjusting them to changes in the shape and location of adjoining elements so that the skull remains a functional whole (for a relatively recent summary of modeling and remodeling see Martin *et al.*, 1998).

Comparative growth of the hard
tissues – geometric morphometrics

Much of our recent work has focused on the ways in which ontogenetic processes bring about differences in facial form (e.g., O'Higgins & Strand Viðarsdóttir, 1999; Strand Viðarsdóttir & O'Higgins, 2001; Strand Viðarsdóttir *et al.*, 2002). Recent advances in morphometrics have allowed us to more readily to compare ontogenetic changes in facial shape, without the confounding influence of size (O'Higgins, 2000; Rohlf, 1998; Strand Viðarsdóttir *et al.*, 2002). The methods we employ are those of geometric morphometrics and, in particular, approaches based on the shape space of Kendall (1984). These methods allow us to partition size from shape, while preserving full three-dimensional information on the geometry of the objects under study at all stages of analysis (Dryden & Mardia, 1998). Consequently, the ontogenetic shape changes of a wide range of organisms have now been studied using these techniques (e.g., Collard & O'Higgins, 2001; Djorovic & Kalezic, 2000; Fink & Zelditch, 1995; Monteiro *et al.*, 1997; O'Higgins *et al.*, 2001; Reilly, 1990; Strand Viðarsdóttir & O'Higgins, 2001).

The partitioning of size from shape is particularly relevant in the investigation of allometry and ontogenetic scaling. Thus, many studies have indicated that, in general, where sexual dimorphism exists in the primate face, it arises principally through ontogenetic scaling such that male and female adult morphologies represent different endpoints on a single ontogenetic trajectory, with female adults usually being smaller than males (e.g., Cheverud & Richtsmeier, 1986; Corner & Richtsmeier, 1991, 1992, 1993; Leigh & Cheverud, 1991; Richtsmeier *et al.*, 1993a, 1993b; Shea, 1983, 1986). A similar mechanism may at least contribute to the ontogeny of facial differences amongst modern human subgroups.

Thus, our own work has revealed that intraspecific differences in modern human facial form come about through three closely interrelated aspects of ontogeny, working either singly or in conjunction: (1) an early (possibly prenatal) onset of population specific facial morphologies; (2) a divergence in the ontogenetic shape scaling trajectories of the facial skeleton; and (3) ontogenetic scaling, where populations have different end (adult) points on the same ontogenetic shape scaling trajectory (referred to as "ontogenetic shape trajectory" elsewhere in this chapter for the sake of brevity) (O'Higgins & Strand Viðarsdottir, 1999; Strand Viðarsdóttir & O'Higgins, 2001; Strand Viðarsdóttir *et al.*, 2002). In contrast to earlier studies of primates cited above, it appears that in some human populations sexually dimorphic features develop through a combination of displacement of the ontogenetic shape trajectory and ontogenetic scaling (Strand Viðarsdóttir, 1999). Further, more recent studies by one of us (P O'H) indicate that in many primate species, such as *Cebus apella* and the papionins, sexual dimorphism develops through a late divergence of the growth vector as well as a relative extension of the male ontogenetic shape trajectory (O'Higgins & Collard, 2002; O'Higgins & Jones, 1998; O'Higgins *et al.*, 2001). These findings further elaborate on the role of ontogenetic scaling as expounded by the workers cited above. There has been no in-depth study of non-sexual intraspecific variation in primates other than humans but research on interspecific variation between non-human primate species (O'Higgins & Collard, 1999; O'Higgins *et al.*, 2001) shows the same processes at work (although to different extent) as those which produce interpopulation variation in modern humans.

We present in this chapter a study examining the ontogeny of adult facial morphology in three populations: French/British Caucasians, African Americans, and Arikara Plains Native Americans. It illustrates the influence of early-established form differences, ontogenetic scaling, and divergence of ontogenetic shape trajectories in generating differences between modern human populations.

Background to the study

In this study we concentrate on the facial skeleton excluding the mandible because facial and neurocranial growth are subject to different influences (see earlier discussion on functional matrices). Thus by focusing on the face alone we simplify our analysis and its subsequent interpretation.

Facial shape differences between Caucasians, African Americans, and Native Americans are probably the best-documented example of variation in human craniofacial form (e.g., Giles & Elliot, 1962; Gill & Rhine, 1990; Snow *et al.*, 1979). For many years, workers in the United States of America have been aware of the necessity of distinguishing between skeletal remains from these three population groups, not only to aid forensic identification, but also to identify the remains of individuals for potential repatriation and reburial. Consequently, a considerable number of forensic identification techniques have been developed specifically to distinguish between individuals from these three groups (Bass, 1986; Brues, 1990; Giles & Elliot, 1962; Rhine, 1990). There is therefore an additional advantage in focusing on the facial skeleton, in that it has previously been shown to be the most diagnostic craniofacial region of these groups in adults (Bass, 1986).

Although the above-mentioned studies have proved relatively successful at distinguishing between adult human remains, there are few techniques available to identify subadult remains. This lack of suitable techniques is mainly due to the confounding influence of the large-scale ontogenetic allometric changes on recognizing potential population-specific morphologies. These problems can potentially be overcome with geometric morphometric techniques, in which size is separated from shape, and allometry (i.e., scaling of shape) can be analyzed independently of other aspects of shape variability if necessary. Additionally, this study will assess the possibility of using geometric morphometric techniques to classify subadult specimens to the correct population group on the basis of facial morphology. In order to assess the differences in ontogenetic processes between the three groups, we will test three null hypotheses. The methods section details how these hypotheses will be tested.

$H_0 1$: *There is no difference in the shape of the facial skeleton between the three populations.* Testing of the first hypothesis allows us to examine whether there are statistically different population specific facial shapes for each population, irrespective of the age of the individuals.

Table 5.1 *Description of the samples used in the present study*

	African American	Arikara	Caucasian
Number adults[a]	12 (7M; 5F)	10 (5M; 5F)	10 (5M; 5F)
Number subadults	22	49	49
Total	34	59	59
Minimum age (years) subadults[b]	0.75	2.5	0
Maximum age (years) subadults	16	18	19
Collection[c]	CMNH	UTK	NHM, RCS, MH

[a] M, males; F, females.
[b] Biological ages are given in years, and are estimated on the basis of tooth eruption, as in Ubelaker (1989).
[c] NHM, Natural History Museum London; CMNH, Cleveland Museum of Natural History, Ohio; UTK, Department of Anthropology, University of Tennessee, Knoxville; RCS, Royal College of Surgeons, London; MH, Musée de l'Homme, Paris.

$H_0 2$: *The differences between populations are inadequate to allow subadult individuals of unknown origin to be correctly assigned to population groups on the basis of facial morphology.* The second hypothesis is dependent on the first hypothesis being rejected. The attempt at falsification of the second hypothesis allows us to examine the possibility of correctly identifying subadult individuals to population groups.

$H_0 3$: *There is no difference between the ontogenetic shape scaling trajectories of the three populations.* The attempt at falsification of the third hypothesis examines the evidence for significant divergence in the ontogenetic shape scaling trajectories between groups. It also allows us to consider the extent to which ontogenetic scaling contributes to differences in facial morphology between populations.

Materials

The study includes 152 individuals, ranging from infancy to adulthood, from three geographically distinct populations: Arikara Plains Native Americans (ARIK), African Americans (AFR), and French/British Caucasians (CAUC). The composition and the origins of the populations (i.e., samples) are given in Table 5.1. Great care was taken in specimen selection to avoid bias in sample composition in terms of age and sex, although the availability of material in museum collections did not allow us to gather data from age and sex matched samples. Each subadult individual is assigned a biological/developmental age estimate based on tooth eruption according to the dental standard of Schour & Massler (1941), as adapted for use on non-white populations by Ubelaker

(1989). This estimate of maturation is used simply for the purposes of graphing data and not for subsequent statistical analysis. No attempt has been made to sex the subadult sample. All adults are assigned the arbitrary age of 21 years. In this study, individuals are classified as adults if the third permanent molar has fully erupted and the spheno-occipital synchondrosis has fused. Care was taken to include only young adult specimens, as determined by degree of dental wear, stage of suture closure (Meindl & Lovejoy, 1985), and postcranial parameters (such as pubic symphysis aging and auricular surface aging), where possible (Brooks & Suchey, 1990; Lovejoy *et al.*, 1985).

The facial skeleton of each individual is represented by 26 unilateral homologous landmarks in three dimensions. These were collected using a Polhemus 3-Space Isotrak II electromagnetic digitizer. Unilateral data were used in preference over bilateral data, in order to reduce the number of variables by taking advantage of symmetry (Strand Viðarsdóttir & O'Higgins, 2001). This also allows us to increase the size of the available sample by allowing inclusion of specimens in which one side of the face is damaged. The landmarks are listed in Table 5.2. In order to aid visual interpretation of results, each face is approximated by a three-dimensional surface, obtained by triangulations of landmarks. This surface only approximates the facial skeletal surface and is used solely for visualization purposes; analysis is based only on the three-dimensional coordinates of landmarks.

Methods

The three-dimensional coordinates of landmarks are analyzed using techniques from geometric morphometrics. These techniques preserve complete information about the relative spatial configuration of landmarks throughout an analysis and utilize the properties of Kendall's shape space (see below) (Slice *et al.*, 1996). The shape spaces and associated statistics of these methods are well understood (Dryden & Mardia, 1998) and yield highly visual and readily interpretable results.

The landmarks are registered (superimposed) using generalized Procrustes analysis (GPA) minimizing the sum of squared distances between homologous landmarks by translating, rotating, reflecting, and scaling them to best fit. This registration method does not introduce bias into the distribution of specimens where landmarks vary independently and according to random error (Rohlf, 1999). Scaling is according to centroid size, which is the square root of the sum of squared Euclidean distances from each landmark to the centroid (the mean of landmark coordinates). Centroid size is used in this study as an expression of the overall scale of the landmark configuration, and thus of the face, and to examine allometry and growth. All analyses of shape are carried

Table 5.2 *Landmarks used in the present study*

Number	Landmark definition[a]
1	Bregma
2	Frontomalare orbitale
3	Frontomalare temporale
4	Nasion
5	Glabella
6	Stephanion
7	Frontotemporale
8	Superior rim of the orbit
9	Supraorbital torus
10	Dacryon
11	Zygotemporale superior
12	Zygotemporale inferior
13	Maxillofrontale
14	Zygoorbitale
15	Zygomaxillare
16	Jugale
17	Orbitale
18	Alveolare
19	Nasospinale
20	Alare
21	External alveolus at second incisor
22	External alveolus at canine
23	External alveolus at most posterior tooth
24	Palatine–maxillary suture
25	Infraorbital foramen
26	Staphylion

[a] For further explanation of the anatomical location of individual landmarks, see O'Higgins & Strand Viðarsdóttir (1999).

out on landmark configurations from which centroid size has been partitioned through the scaling parameters outlined above. Information about the centroid size of the individual specimens prior to GPA is retained for the purpose of studying size–shape relationships, i.e., allometry.

The registering of landmark coordinates through GPA results in each specimen being represented as a single point in a non-Euclidean shape space of $km - m - m(\frac{m-1}{2}) - 1$ dimensions, known as Kendall's shape space (Kendall, 1984), where k is equivalent to the number of landmarks, and m denotes the dimensionality of those landmarks. To aid statistical analysis, the points are projected into a linear tangent space (Dryden & Mardia, 1992), and statistical analyses are carried out within that space using standard multivariate methods. This approach is satisfactory when variations are small, i.e., where the data lie within full Procrustes distance of about $d_f = 0.2$, of the mean as in these data (see O'Higgins, 2000; Dryden & Mardia, 1998).

To explore relative relationships between individual shapes, as in the testing of null hypotheses 1 and 2, principal components analysis (PCA) is used to calculate principal axes of variation in the tangent space. Visualization of shape differences along the principal components (PCs) is obtained by warping the triangulated surface of the mean shape to represent shapes at any position within the plot, using the loadings of original landmark coordinates on these PCs (O'Higgins & Jones, 1998; O'Higgins & Strand Viðarsdóttir, 1999). Cartesian transformation grids, calculated using the method of thin plate splines (Bookstein, 1989), are used to further visualize and interpret shape differences.

To test null hypotheses 1 and 2 we explore the significance of differences between populations. This is achieved by computing Mahalanobis' distances between them, and assessing the significance of these differences with Hotelling's T^2. Further examination of differences between populations is carried out by discriminant analysis. All of these are computed between different populations with PC scores from GPA/PCA using (1) the adults alone, and (2) the combined adults and subadults. Discriminant function analysis can be used to predict group affiliation of unknown specimens. In the context of the present study two methodological issues arise. First, in assessing how well discriminant functions perform, they can be expected to better classify data used to calculate them (calibration data), than data not used in the initial computation (test data). Therefore to assess our discriminant functions, cross-validation was carried out repeatedly, so that each time, one arbitrary individual was excluded from the calculation of the classification function, and then assigned to a group using it. Thus in turn, all the individuals were treated as unknowns. Second, in computing the discriminant function the inclusion of "noisy" data that does not differentiate between groups simply adds to dimensionality often at the cost of discriminatory power. To optimize discrimination in the current analyses we computed functions with increasing numbers of PCs. In doing this we sequentially included lower to higher order PCs until discriminatory power of the functions fell off. This is because higher order PCs tend to "noise," i.e., they contribute to total variance but not to between-group variance. In the case of the adults-only discriminant function, optimal discrimination was achieved with 13 PCs, and in the case of the combined adults and subadult discriminant function, 23 PCs.

In order to investigate commonality of ontogenetic shape scaling trajectories and so to test null hypothesis 3, each population is subjected to a separate GPA and PCA and the relationships between variations in shape (principal component scores) and centroid size are assessed using correlation analysis (Sokal & Rohlf, 1995). In each case only the first principal component showed a large or significant correlation with centroid size (and incidentally with maturation), and thus well represents the shape scaling or allometric component of the ontogenetic trajectory (the whole of which includes the relationship between shape, size, and age).

The significance of the angle between ontogenetic shape trajectories (represented by PC1 in each of the populations studied following a combined GPA) of pairs of populations is assessed in relation to the distribution of angles between 1000 random resamplings (Good, 1993). These resamplings are such that two groups of the same sample sizes as the true groups are randomly created from the original data and the angle between them recorded. The proportion of times the permuted angle exceeds the true angle approximates the *P*-value for the significance of the true angle.

Since PC1 represents ontogenetic shape changes in all populations studied, "mean adult" and "mean infant" facial shapes are created for each population by warping the mean shape to the extremes of the ontogenetic shape trajectory (for a detailed explanation of this procedure, see O'Higgins, 2000). The resultant coordinates of these estimates of mean shape are subsequently submitted to a separate GPA/PCA to allow ready visualization of differences in ontogenetic shape trajectories between populations.

Where pairs of populations show no significant angle between their ontogenetic shape scaling trajectories (as represented by the PC1s from the analyses of each population) the possible presence of ontogenetic scaling differences between population pairs is assessed by the computation of the mean adult centroid size for each population and relating this to ontogenetic shape trajectory divergence and shape differences between adults. The significance of any difference in facial centroid size is evaluated using Student's *t*-test paying regard to any differences in variance if necessary.

Results

The results are divided into two main sections. The first examines the differences in morphology between populations and relates to the first and second null hypotheses specified earlier. The second compares the actual ontogenetic shape trajectories between populations, and describes the test of the third null hypothesis.

Differences in facial shape irrespective of maturation

Statistical significance of interpopulation differences in *adult* facial shape is examined through computation of Mahalanobis' distances from the specimen

Table 5.3 *Results of the discriminant analyses between adult populations. The upper value in each table element represents the Mahalanobis' distance between the populations, and the lower the corresponding* P-value from Hotelling's T²

Caucasian			
Arikara	6.0582		
	0.0001		
African American	5.2797	4.4101	
	0.0005	0.0003	
	Caucasian	Arikara	African American

Table 5.4 *Results of the cross-validation test of the adult populations, based on PCs 1–13. The table lists the percentage of individuals from the known groups (listed in the first column), assigned to each group (listed in the last row) in the cross-validation test. It should be read horizontally for correct interpretation*

Caucasian	88.89	0	11.11
Arikara	0	100	0
African American	10	0	90
	Caucasian	Arikara	African American

scores on PCs 1–13 from a GPA/PCA of the adults alone (Table 5.3). These PCs were found to provide the greatest discrimination between the groups (see methods), and account for >85% of the total variance within the overall sample. A discriminant analysis with cross-validation (Table 5.4) shows that between 89% and 100% (mean 93%) of the adult individuals can be correctly classified to population based on these PCs. There is no significant correlation (see definition of correlation analysis above) between the numbers of individuals in each population and the percentage of correct assignations.

Table 5.5 lists the Mahalanobis' distances between the populations with adult and subadult samples combined based on PCs 1–23 (90% of total variance) that optimize discrimination for these data. Again, all the populations are significantly separated on the basis of facial shape irrespective of the variability in the age of the individuals included in the analysis. This finding is supported by a crossvalidated discriminant analysis (based on PCs 1–23) which correctly assigns 88% to 97% (mean 91.98%) of specimens to the correct group (Table 5.6). The individuals incorrectly assigned spanned all developmental ranges (0.75 years to adult) and were not clustered in particular age groups.

Table 5.5 *Results of the discriminant analyses between full populations (adults and subadults). The upper value in each table element represents the Mahalanobis' distance between the populations, and the lower the corresponding P-value from Hotelling's* T^2

Caucasian			
Arikara	3.5093		
	0.0001		
African American	4.5901	4.1586	
	0.0001	0.0001	
	Caucasian	Arikara	African American

Table 5.6 *Results of the cross-validation test of the full populations (adults and subadults), based on PCs1–23. The table lists the percentage of individuals from the known groups (listed in the first column), assigned to each group (listed in the last row) in the cross-validation test. It should be read horizontally for correct interpretation*

Caucasian	88.14	8.47	3.39
Arikara	3.39	96.61	0
African American	10	0	91.18
	Caucasian	Arikara	African American

Differences in ontogenetic shape trajectories

In order to investigate the possibility of divergent shape scaling trajectories in the three populations as postulated in the third null hypothesis, the data from all three populations are submitted to joint GPA and separate PCAs are carried out on each population. PC1 is found to correlate significantly with centroid size (and also with biological age) in all populations (Table 5.7). It is the only PC to do so. Therefore, we interpret it to represent ontogenetic allometric shape changes in each population. The relationships between age, size, and shape in the Arikara are plotted in Figure 5.1. Figure 5.1A represents growth (age vs. size), Figure 5.1B, development (shape vs. age), and Figure 5.1C, allometry (shape vs. size). The loadings of shape coordinates on PC1 are presented as visualizations of shape change along this axis with overlain Cartesian transformation grids in Figure 5.2.

Table 5.7 *Correlations between PC1 and facial centroid size,
PC1 and biological age, and facial centroid size and
biological age, for each of the samples studied*

	PC1 vs. size	PC1 vs. age	Size vs. age
African American	$r = -0.900$	$r = -0.896$	$r = 0.896$
	$P = 4.18*10^{-13}$	$P = 5.27*10^{-11}$	$P = 7.57*10^{-13}$
	$r = 0.900$	$r = 0.890$	$r = 0.90$
Arikara	$P = 1.39*10^{-21}$	$P = 8.93*10^{-21}$	$P = 2.01*10^{-22}$
	$r = -0.930$	$r = -0.901$	$r = 0.842$
Caucasian	$P = 5.11*10^{-11}$	$P = 1.25*10^{-18}$	$P = 3.56*10^{-14}$

Table 5.8 *Pair-wise comparisons of the angle between PC1s.
The upper value denotes the angle, in degrees, between the PC1s of
the populations being compared, and the lower the corresponding
P-value assessed by a permutation test*

African American			
	21.4		
Arikara	$P = 0.0669$		
	23.6	24.1	
Caucasian	$P = 0.0049$	$P < 0.0009$	
	African American	Arikara	Caucasian

Viewed in isolation the relationships between size, PC1, and maturation look similar across all populations (Figures 5.1, 5.3, and 5.4). However, the Cartesian transformation grids reveal individual variations in craniofacial allometry, superimposed on the shared ontogenetic allometric shape changes. Thus the Caucasians show a marked contraction in the supraorbital area with increasing size (Figure 5.5, point 1), not noted in the other populations. This implies that while all populations follow ontogenetic allometric trends with shared features, they also appear to show some distinct aspects of ontogenetic allometry. To assess the possibility of divergence in ontogenetic shape scaling trajectories between populations a pairwise comparison of angles between PC1s is carried out. The results are given in Table 5.8. A permutation test indicates that two out of the three comparisons reveal a statistically significant difference in the direction of the PC1 (ontogenetic shape trajectory) between the two populations compared. Thus, ontogenetic shape changes in the Caucasian sample are

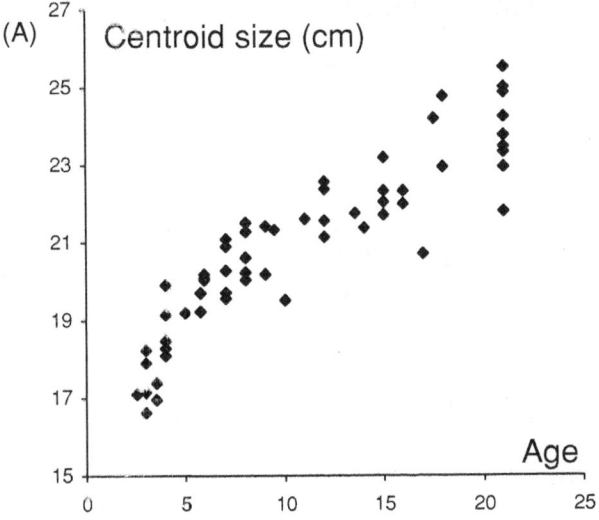

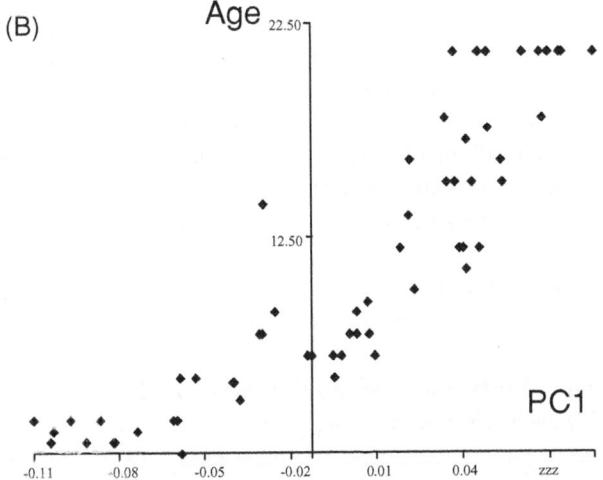

Figure 5.1 Arikara Plains Native Americans: (A) Biological age vs. centroid size; (B) biological age vs. PC1; (C) centroid size vs. PC1.

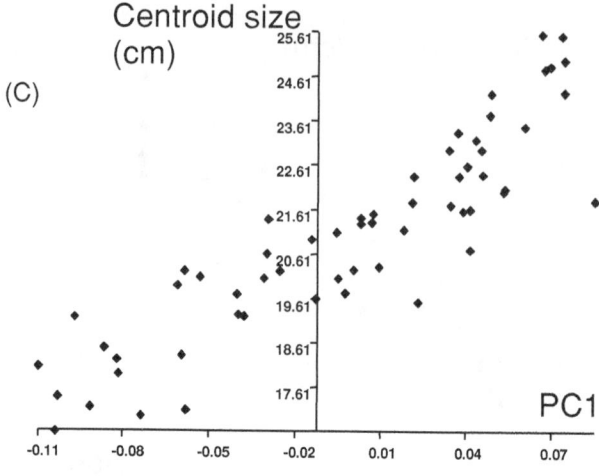

Figure 5.1   (*cont.*).

significantly different from those of both the Arikara and the African American samples. Ontogenetic shape trajectories of the Arikara and African American samples are not significantly different ($P = 0.07$) from one another, although that comparison only just exceeds the 95% significance threshold, thus a slightly larger sample might have given a significant result. These results indicate that at least between the Caucasians and the other two populations (and possibly between all three), differences in adult face shape arise partly through differences in the ontogenetic allometric trajectories, the shape component of which is here represented by PC1. Thus, differences in ontogenetic allometric trajectories actively contribute to differences in adult facial shape.

A comparison of adult facial sizes (Table 5.9) reveals that there are no significant differences in facial centroid size between the adults of the three populations, thus precluding the notion of allometric scaling.

Table 5.9 *Pair-wise Student's* t-*test of the difference in centroid size between adults of the three populations*

Caucasian			
Arikara	$t = 1.941$		
	$P = 0.069$		
African American	$t = 2.23$	$t = 0.077$	
	$P = 0.056$	$P = 0.9400$	
	Caucasian	Arikara	African American

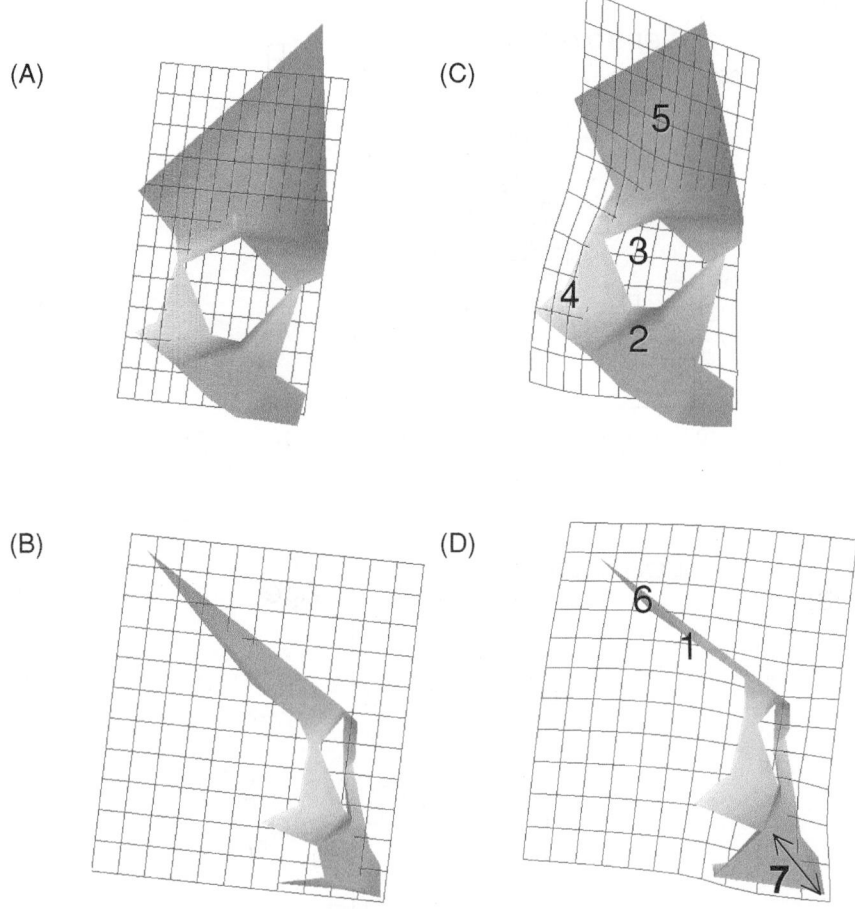

Figure 5.2   Arikara Plains Native Americans: Morphological representations and transformation grids showing the variation in facial shape represented by PC1, from the negative (left, reference: PC1 −0.10) to the positive extremes (right, target; PC1 0.09), frontal (top) and lateral (bottom) views. See text for explanations of numbered labels.

Discussion

The results presented in the previous section have falsified the first and second null hypotheses. Thus, H_01, that there is no difference in the shape of the facial skeleton between the three populations irrespective of age, is falsified by the discriminant analyses. These indicate that differences in facial shape

(A)

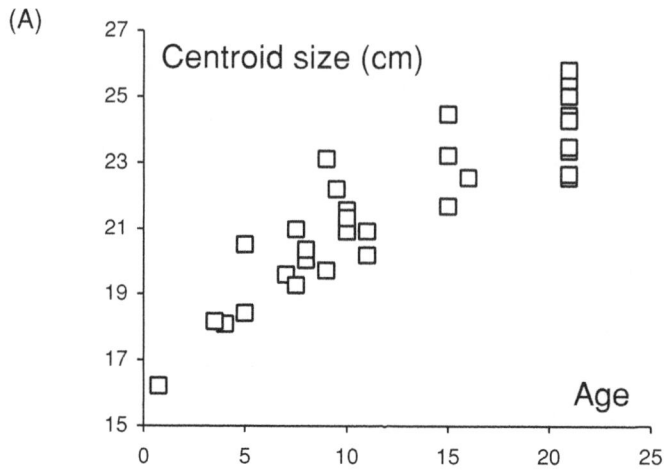

(B)

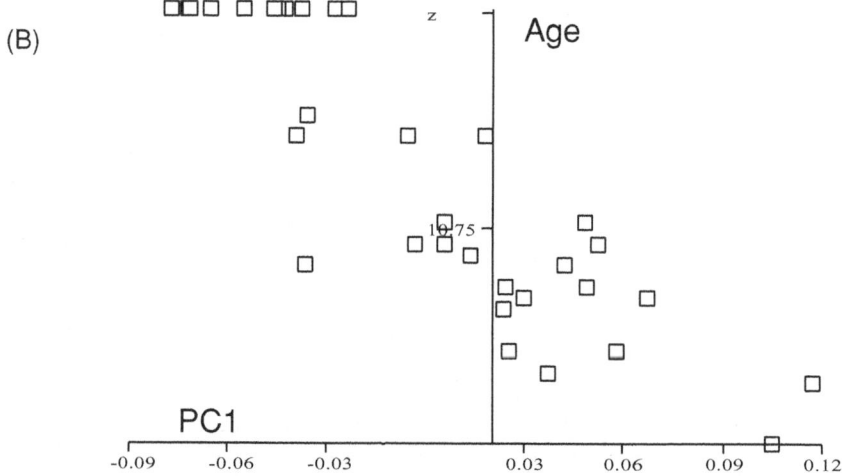

Figure 5.3   African Americans: (A) Biological age vs. centroid size; (B) biological age vs. PC1; (C) centroid size vs. PC1.

between the populations are present even in the smallest/youngest individuals, arising at least in early infancy or even prenatally. These differences are sufficient to allow "unknown" subadults to be correctly assigned to population groups. These findings also falsify $H_0 2$. The significant differences in angles between ontogenetic shape scaling (allometric) vectors falsify the third null hypothesis, $H_0 3$.

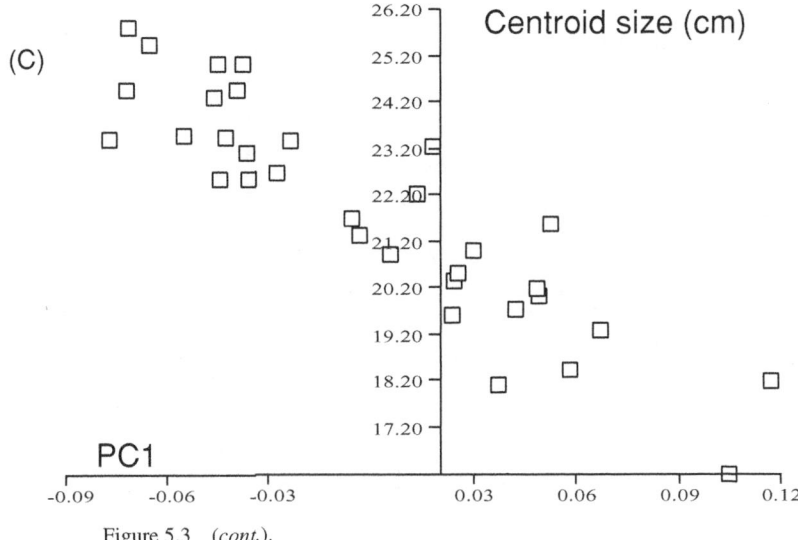

Figure 5.3   (*cont.*).

These findings show that with a sample of three populations we can demonstrate at least two ways by which differences in adult facial shape arise: through an early onset of population specific morphologies, and divergent ontogenetic allometries. Larger-scale studies of a greater number of populations (Strand Viðarsdóttir *et al.*, 2002) have also revealed that for certain population comparisons there is evidence of ontogenetic scaling of non-divergent allometric trajectories.

The analyses of Figures 5.1 and 5.2 examine ontogenetic changes in size and shape in one population, the Arikara. It is interesting to note that in both the plot of growth (Figure 5.1A) and that of development (Figure 5.1B) the scatter of points is curvilinear, indicating that the rates both of shape change and of size increase fall off with increasing maturity. In contrast the plot of allometry (Figure 5.1C) is linear, indicating that the rate (and the anatomical nature) of ontogenetic shape change with increasing size remains constant throughout the range of ontogeny we sample. The plots for the other populations (Figures 5.3 and 5.4) show similar features as do plots for several other primates (e.g., Cobb & O'Higgins, 2002; O'Higgins & Collard, 2002; O'Higgins *et al.*, 2001), indicating that decrease in the rate of shape and size change is common.

This constant increase of size with shape throughout ontogeny in turn points to incremental increases in the size and shape of the functional matrices controlling this allometry, which in themselves are constant in effect across the entire ontogenetic period. Thus, major morphological changes in the facial skeleton during ontogeny are very closely linked to facial size, irrespective of growth

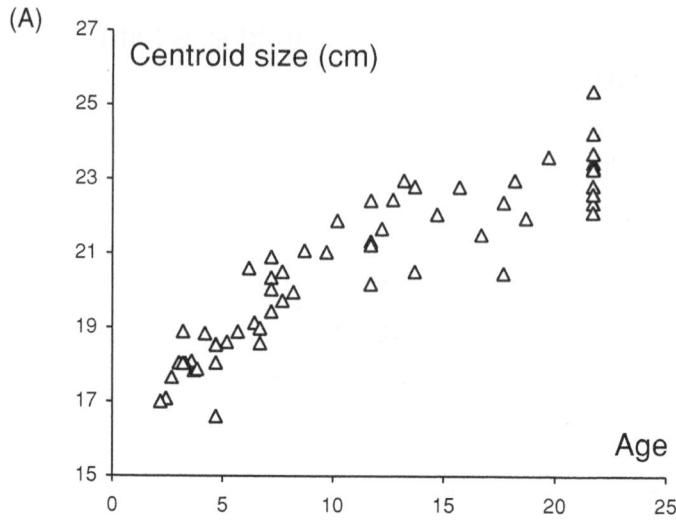

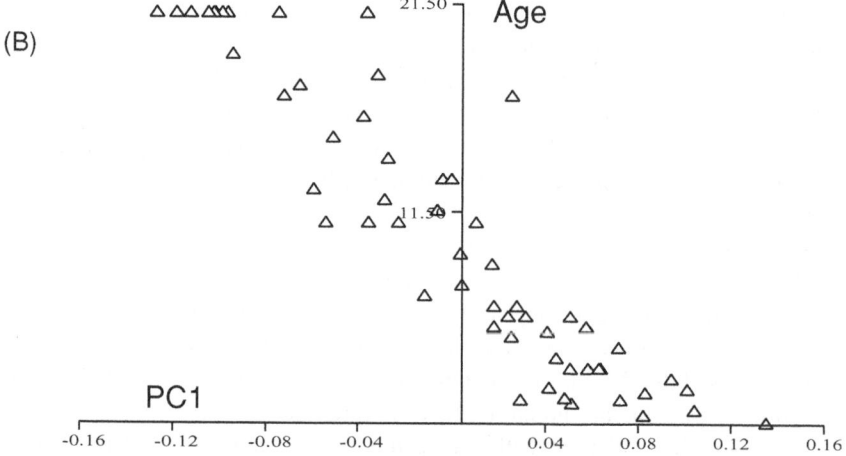

Figure 5.4   Caucasians: (A) Biological age vs. centroid size; (B) biological age vs. PC1; (C) centroid size vs. PC1.

rate. In turn, this finding predicts that the effect on form of facial modeling and remodeling remains more or less constant in anatomical location and rate throughout much, if not all, of the postnatal period. This finding concurs with that reported for the postnatal period up to eruption of the last permanent maxillary molar in *Cercocebus torquatus* (O'Higgins & Jones, 1998). However, in many remodeling studies (e.g., Kurihara *et al.*, 1980) there are subtle variations

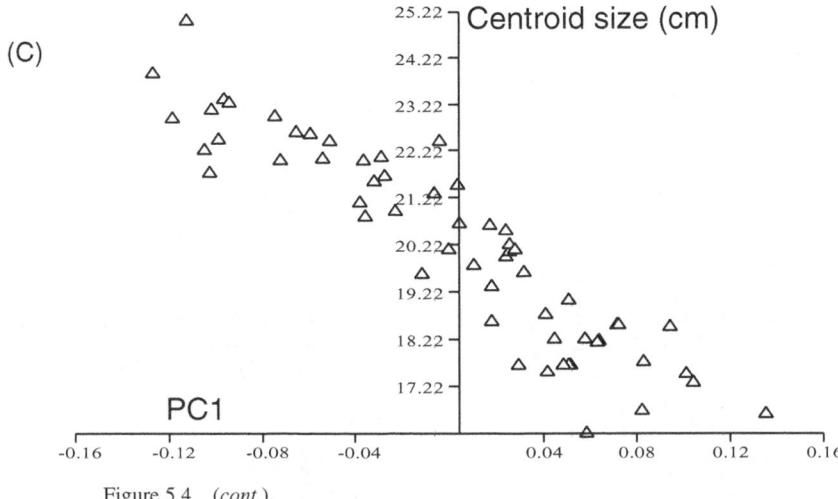

Figure 5.4   (*cont.*).

observed in the nature and location of remodeling fields during ontogeny. The findings of this study therefore indicate that these perturbations of remodeling must make only a small contribution to the ontogeny of facial form, at least of insufficient magnitude to present a detectable signal in cross-sectional studies of populations such as the present one. It will therefore be of interest to examine ontogeny within populations using longitudinal data in future studies.

Within the Arikara, the differences in facial shape between the small (Figure 5.2A and 5.2B) and large (Figure 5.2C and 5.2D) individuals include a relative reduction in the length of the frontal (Figure 5.2, label 1), a relative increase in maxillary height (Figure 5.2, label 2), a decrease in relative orbit size (Figure 5.2, label 3), a relative expansion of the zygomatic (Figure 5.2, label 4), a relative reduction in frontal breadth (Figure 5.2, label 5), relatively more superoposterior positioning of stephanion (Figure 5.2, label 6) representing the temporal muscle attachment, and an increase in relative alveolar prognathism (Figure 5.2, label 7). Are these features of ontogenetic shape change common to all the groups?

The answer from the analyses of ontogenetic shape trajectories is that some modern human populations grow along trajectories that are significantly different from one another. The magnitude of these divergences in terms of the size of the angle is comparable to that documented between species of non-human primates (Cobb & O'Higgins, 2002; O'Higgins *et al.*, 2001). We should be careful, however in how we interpret the magnitude of the angle since the degree of divergence of shape vectors says nothing about the actual features of anatomy that come to differ. This information is readily gleaned by careful comparison of visualizations of ontogenetic shape changes (examples of which are given in

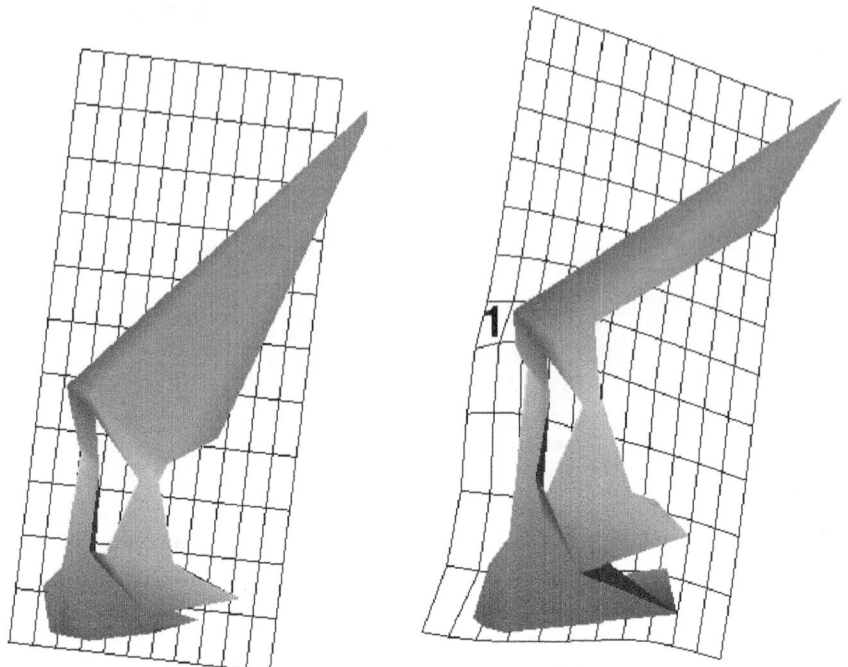

Figure 5.5   Caucasians: Morphological representations and transformation grids
showing the variation in facial shape represented by PC1, from the positive (left,
reference: PC1 0.14) to the negative extremes (right, target: PC1 −0.14), lateral view.
See text for explanation of numbered label.

Figures 5.2 and 5.5). They show, for example, that unlike the other populations,
Caucasians display a marked relative contraction of the supraorbital area with
increasing size, which may be reflected in the statistically significant angle
between the ontogenetic shape trajectory of the Caucasians and the other two
populations. Although this study does not set out specifically to explore the
underlying reasons for these observed differences in morphology, variations in
supraorbital morphology have in the past been put down to differences in degree
of neuroorbital disjunction (Hylander *et al.*, 1992; Ravosa, 1991; Weidenreich,
1941), or variation in supraorbital stress related to masticatory forces (see
review of this issue by Leiberman, 2000). Although there is no consensus
on this issue, the former is presently thought to be more likely, as bone in
this region is often more robust than is needed to counter masticatory loads
(Hylander *et al.*, 1992). Thus in the case of the Caucasians, it could be proposed
that the divergence of their ontogenetic allometry from the other groups may,
at least in part, be due to the development of a population-specific pattern of
neuroorbital conjunction.

The fact that all the populations can be distinguished on the basis of facial shape, irrespective of the age of the individuals analyzed, is important in the context of the study of ontogeny in fossil hominins as well as for the sciences of archaeology and forensic anthropology. A larger study on a greater number of populations (Strand Viðarsdóttir *et al.*, 2002) has revealed that the relative relationships (calculated using Mahalanobis' distances) between the combined adult and subadult populations closely reflect those calculated from adults alone. This relationship illustrates that similar morphological features are likely to distinguish the adults and the subadults, although these may be accentuated and modified throughout ontogeny as a result of the divergence of the ontogenetic shape trajectories.

Overall, our findings illustrate the general plasticity of the modern human facial skeleton, and how our facial shape can be relatively easily adapted through minor shifts in the ontogenetic process. The population-specific morphologies documented here also imply that a great deal of caution should be called for in comparative morphological studies with subadults of other hominin species. Given the significant interpopulation differences in adult and subadult facial shape, studies of comparative development between modern humans and fossils that use "conglomerate" age series based on a number of populations (e.g., the recent example of Ponce de León & Zollikofer, 2001) may obscure the subtleties of possible growth similarities and differences in diverse human groups.

Our study indicates new possibilities for workers in paleoanthropology and paleontology as a whole. Once we have a greater understanding of ontogenetic variation within our own species, we may be in a better position to assess and evaluate possible levels of species distinction between other hominin subadults. The findings of this study also point to a tantalizing possibility with regard to the forensic identification of subadult skeletal remains. At present there is no way of reliably assigning subadult remains to ancestral groups on the basis of skeletal morphology. The cross-validation study of the full data set reveals that in the "classic" test case of differentiating Caucasians, African Americans, and Native Americans, the success of ancestry categorization is on par with that of form functions commonly used for adults (e.g., Giles & Elliot, 1962). It is our intention that these types of studies will eventually form the basis of an identification system for subadults similar to the adult based system of CRANID (Wright, 1992).

Conclusions

In this chapter we have discussed growth of the facial skeleton and demonstrated the large amount of variation that exists in the facial ontogeny of the one species of the genus *Homo* that we can study comparatively. These results highlight the caution with which one should approach the study of ontogeny, but we hope that they have stimulated the reader's interest in the fascinating topic that is comparative ontogeny.

Acknowledgments

We thank Jennifer Thompson, Gail Krovitz, and Andrew Nelson for inviting us to participate in the symposium on "Growth in the Genus *Homo*" which formed the foundation for this book. We also thank the curators of the collections analyzed for access to the specimens in their care. Nick Jones deserves special thanks for his computing support, as do our colleagues in Durham and University College London for stimulating discussions. Chris Stringer deserves special thanks for his continued support of this work.

References

Bass, W. (1986). *Human Osteology: A Laboratory and Field Manual*. Columbia: Missouri Archaeological Society.

Batzer, M. A., Stoneking, M., Alegriahartman, M., Bazan, H., Kass, D. H., Shaikh, T. H., Novick, G. E., Ioannou, P. A., Scheer, W. D., Herrera, R. J., & Deininger, P. L. (1994). African origin of human-specific polymorphic Alu insertions. *Proceedings of the National Academy of Sciences of the USA*, **91**, 12288–12292.

Bookstein, F. L. (1989). Principal warps: Thin-plate splines and the decomposition of deformations. *Institute of Electrical and Electronics Engineers Transactions on Pattern Analysis and Machine Intelligence*, **11**, 567–585.

Bromage, T. (1986). A comparative scanning electron study of facial growth remodeling in early hominids. PhD dissertation, University of Toronto.

Brooks, S., & Suchey, J. (1990). Skeletal age determination based on the *os pubis*: A comparison of the Acsadi–Nemeskeri and Suchey–Brooks methods. *Human Evolution*, **5**, 227–238.

Brues, A. (1990). *People and Races*. Prospect Heights: Waveland Press.

Burrows, A. M., Richtsmeier, J. T., Mooney, M. P., Smith, T. D., Losken, H. W., & Siegel, M. I. (1999). Three-dimensional analysis of craniofacial form in a familial rabbit model of nonsyndromic coronal suture synostosis using Euclidean distance matrix analysis. *Cleft Palate Craniofacial Journal*, **36**, 196–206.

Canalis, E. (1993). Regulation of bone remodelling. In *Primer on Metabolic Bone Diseases and Disorders of Mineral Metabolisms*, ed. M. Favus, pp. 33–37. New York: Raven Press.

Cann, R. L., Stoneking, M., & Wilson, A. C. (1987). Mitochondrial DNA and human evolution. *Nature*, **325**, 31–36.

Cavalli-Sforza, L., Menozzi, P., & Piazza, A. (1994). *The History and Geography of Human Genes*. Princeton: Princeton University Press.

Cheverud, J. M., & Richtsmeier, J. T. (1986). Finite-element scaling applied to sexual dimorphism in rhesus macaque (*Macaca mulatta*) facial growth. *Systematic Zoology,* **35**, 381–399.

Cobb, S., & O'Higgins, P. (2002). The ontogeny of form variation in the African ape facial skeleton: A hierarchical approach to the interspecific comparison of ontogenetic trajectories. *American Journal of Physical Anthropology,* **34** (Suppl.), 55.

Collard, M., & O'Higgins, P. (2001). Ontogeny and homoplasy in the papionin monkey face. *Evolution and Development,* **3**, 322–331.

Corner, B. D., & Richtsmeier, J. T. (1991). Morphometric analysis of craniofacial growth in *Cebus apella. American Journal of Physical Anthropology,* **84**, 323–342.

Corner, B. D., & Richtsmeier, J. T. (1992). Cranial growth in the squirrel monkey (*Saimiri sciureus*): A quantitative analysis using three-dimensional coordinate data. *American Journal of Physical Anthropology,* **87**, 67–82.

Corner, B. D., & Richtsmeier, J. T. (1993). Cranial growth and growth dimorphism in *Ateles geoffroyi. American Journal of Physical Anthropology,* **92**, 371–394.

Djorovic, A., & Kalezic, M. L. (2000). Paedogenesis in European newts (*Triturus:* Salamandridae): Cranial morphology during ontogeny. *Journal of Morphology,* **243**, 127–139.

Dryden, I. L., & Mardia, K. V. (1992). Size and shape analysis of landmark data. *Biometrika,* **79**, 57–68.

Dryden, I. L., & Mardia, K. (1998). *Statistical Shape Analysis.* Chichester: John Wiley & Sons.

Eriksen, E., Vesterby, A., Kassen, M., Melsen, F., & Mosekilde, L. (1993). Bone remodeling and bone structure. In *Physiology and Pharmacology of Bone,* eds. G. Mundy & T. Martin, pp. 67–109. New York: Springer-Verlag.

Fehling, P., Alekel, L., Clasey, J., Rector, A., & Stillman, R. (1995). A comparison of bone mineral densities among female athletes in impact loading and active loading sports. *Bone,* **17**, 205–210.

Fink, W. L., & Zelditch, M. L. (1995). Phylogenetic analysis of ontogenetic shape transformations: A reassessment of the piranha genus *Pygocentrus* (Teleostei). *Systematic Biology,* **44**, 343–360.

Froment, A. (1992). Correspondence between anatomical differentiation and geographic distribution of modern humans studied by craniometry. *Comptes Rendus de l'Académie de Sciences de Paris,* **315**, 323–329.

Giles, E., & Elliot, O. (1962). Race identification from cranial measurements. *Journal of Forensic Science,* **7**, 147–157.

Gill, G., & Rhine, S. (eds.) (1990). *Skeletal Attribution of Race: Methods for Forensic Anthropology,* Anthropological Papers no. 4. Albuquerque: Maxwell Museum of Anthropology.

Good, P. (1993). *Permutation Tests: A Practical Guide to Resampling Methods for Testing Hypotheses.* New York: Springer-Verlag.

Guglielmino-Matessi, C., Gluckman, P., & Cavalli-Sforza, L. (1977). Climate and the evolution of skull metrics in man. *American Journal of Physical Anthropology,* **50**, 549–564.

Hanihara, T. (1992). Population prehistory of East Asia and the Pacific as viewed from craniofacial morphology. *American Journal of Physical Anthropology,* **91**, 173–187.

Hanihara, T. (1996). Comparison of craniofacial features of major human groups. *American Journal of Physical Anthropology*, **99**, 389–412.

Heinonen, A., Oja, P., Kannus, P., Sievanen, H., Haapasalo, H., Manttari, A., & Vuori, I. (1995). Bone mineral density in female athletes representing sports with different loading characteristics of the skeleton. *Bone*, **17**, 197–203.

Hennessy, R., & Stringer, C. B. (2002). Geometric morphometric study of the regional variation of modern human craniofacial form. *American Journal of Physical Anthropology*, **117**, 37–48.

Hernandez, M., Fox, C. L., & Garcia Moro, C. (1997). Fueguian cranial morphology: The adaptation to a cold, harsh environment. *American Journal of Physical Anthropology*, **103**, 103–117.

Howells, W. W. (1973). *Cranial Variation in Man: A Study by Multivariate Analysis of Patterns of Difference among Recent Human Populations*. Cambridge: Harvard University Press.

Howells, W. W. (1986). Physical anthropology of the prehistoric Japanese. In *Windows on the Japanese Past: Studies in Archaeology and Prehistory*, eds. R. Pearson, G. Barnes, & K. Hutterer, pp. 85–99. Ann Arbor: Center for Japanese Studies, University of Michigan.

Howells, W. W. (1989). *Skull Shapes and the Map: Craniometric Analyses in the Dispersion of Modern* Homo. Cambridge: Harvard University Press.

Howells, W. W. (1995). *Who's Who in Skulls: Ethnic Identification of Crania from Measurements*. Cambridge: Harvard University Press.

Hunt, N. (1998). Muscle function and the control of facial form. In *Clinical Oral Science*, eds. M. Harris, M. Edgar, & S. Meghji, pp. 120–133. Oxford: Wright.

Hylander, W., Picq, P., & Johnson, K. (1992). Bone strain and the supraorbital region in primates. In *Bone Biodynamics in Orthodontic and Orthopaedic Treatment*, eds. D. Carlson & S. Goldstein, pp. 315–349. Ann Arbor: Center for Human Growth and Development.

Kendall, D. G. (1984). Shape manifolds, Procrustean metrics and complex projective spaces. *Bulletin of the London Mathematical Society*, **16**, 81–121.

Kurihara, S., Enlow, D., & Rangel, R. (1980). Remodeling reversals in anterior parts of the human mandible and maxilla. *Angle Orthodontics*, **50**, 98–106.

Lahr, M. M. (1995). Patterns of modern human diversification: Implications for Amerindian origins. *Yearbook of Physical Anthropology*, **38**, 163–198.

Lanyon, L. (1993). Skeletal responses to physical loading. In *Physiology and Pharmacology of Bone*, eds. G. Mundy & T. Martin, pp. 485–505. New York: Springer-Verlag.

Lanyon, L., & Rubin, C. (1984). Static versus dynamic loads as an influence on bone remodeling. *Journal of Biomechanics*, **17**, 892–905.

Leigh, S. R., & Cheverud, J. M. (1991). Sexual dimorphism in the baboon facial skeleton. *American Journal of Physical Anthropology*, **84**, 193–208.

Lieberman, D. E. (1996). How and why humans grow thin skulls: Experimental evidence for systemic cortical robusticity. *American Journal of Physical Anthropology*, **101**, 217–236.

Lieberman, D. E. (2000). Ontogeny, homology and phylogeny in the hominid craniofacial skeleton: The problem of the browridge. In *Development, Growth and*

Evolution: Implications for the Study of Hominid Skeletal Evolution, eds. P. O'Higgins & M. Cohn, pp. 85–112. London: Academic Press.

Lovejoy, C., Meindl, R., Mensforth, R., & Barton, T. (1985). Multifactoral determination of skeletal age at death: A method and blind test of its accuracy. *American Journal of Physical Anthropology*, **68**, 1–14.

Martin, R., Burr, D., & Sharkey, N. (1998). *Skeletal Tissue Mechanics*. New York: Springer-Verlag.

McLachlan, J. (1994). *Medical Embryology*. Wokingham: Addison-Wesley.

Meindl, R. & Lovejoy, C. (1985). Ectocranial suture closure: A revised method for the determination of skeletal age at death and blind tests of its accuracy. *American Journal of Physical Anthropology*, **68**, 57–66.

Monteiro, L. R., Cavalcanti, M. J., & Sommer, H. J. iii (1997). Comparative ontogenetic shape changes in the skull of Caiman species (Crocodylia, Alligatoridae). *Journal of Morphology*, **231**, 53–62.

Moss, M. (1964). Vertical growth of the human face. *American Journal of Orthodontics*, **50**, 359–376.

Moss, M. (1968). The primacy of functional matrices on orofacial growth. *Dental Practice*, **19**, 65–73.

Moss, M. (1997a). The functional matrix hypothesis revisited. I: The role of mechanotransduction. *American Journal of Orthodontic and Dentofacial Orthopedics*, **112**, 8–11.

Moss, M. (1997b). The functional matrix hypothesis revisited. II: The role of an osseous connected cellular network. *American Journal of Orthodontic and Dentofacial Orthopedics*, **112**, 221–226.

Moss, M. (1997c). The functional matrix hypothesis revisited. III: The genomic thesis. *American Journal of Orthodontic and Dentofacial Orthopedics*, **112**, 338–342.

Moss, M. (1997d). The functional matrix hypothesis revisited. IV: The epigenetic antithesis and the resolving synthesis. *American Journal of Orthodontic and Dentofacial Orthopedics*, **112**, 410–417.

Moss, M., & Salentijn, L. (1969a). The primary role of functional matrices in facial growth. *American Journal of Orthodontics*, **55**, 566–577.

Moss, M., & Saletijn, L. (1969b). The capsular matrix. *American Journal of Orthodontics*, **56**, 474–490.

Nei, M., & Roychoudhury, A. K. (1993). Evolutionary relationships of human populations on a global scale. *Molecular Biology and Evolution*, **10**, 927–943.

Nur, N., & Hasson, O. (1984). Phenotypic plasticity and the handicap principle. *Journal of Theoretical Biology*, **110**, 275–297.

O'Higgins, P. (2000). The study of morphological variation in the hominid fossil record: Biology, landmarks, and geometry. *Journal of Anatomy*, **197**, 103–120.

O'Higgins, P., & Collard, M. (1999). Facial growth and sexual dimorphism in some papionin species. *American Journal of Physical Anthropology*, **28** (Suppl.), 214.

O'Higgins, P., & Collard, M. (2002). Sexual dimorphism and facial growth in papionin monkeys. *Journal of Zoology*, **257**, 255–272.

O'Higgins, P., & Jones, N. (1998). Facial growth in *Cercocebus torquatus*: An application of three-dimensional geometric morphometric techniques to the study of morphological variation. *Journal of Anatomy*, **193** (P2), 251–272.

O'Higgins, P., & Strand Viðarsdóttir, U. (1999). New approaches to the quantitative analysis of craniofacial growth and variation. In *Human Growth in the Past: Studies from Bones and Teeth*, eds. R. Hoppa & C. Fitzgerald, pp. 128–160. Cambridge: Cambridge University Press.

O'Higgins, P., Chadfield, P., & Jones, N. (2001). Facial growth and the ontogeny of morphological variation within and between *Cebus apella* and *Cercocebus torquatus*. *Journal of Zoology*, **254**, 337–357.

Parfitt, A. M., Mathews, C. H. E., Villanueva, A. R., Kleerekoper, M., Frame, B., & Rao, D. S. (1983). Relationships between surface, volume, and thickness of iliac trabecular bone in aging and in osteoporosis: Implications for the microanatomic and cellular mechanisms of bone loss. *Journal of Clinical Investigation*, **72**, 1396–1409.

Ponce de León, M., & Zollikofer, C. (2001). Neanderthal cranial ontogeny and its implications for late hominid diversity. *Nature*, **412**, 534–538.

Ranly, D. (1980). *A Synopsis of Facial Growth*. Norwalk: Appleton & Lange.

Ravosa, M. J. (1991). The ontogeny of cranial sexual dimorphism in two Old-World monkeys – *Macaca fascicularis* (Cercopithecinae) and *Nasalis larvatus* (Colobinae). *International Journal of Primatology*, **12**, 403–426.

Reilly, S. M. (1990). Comparative ontogeny of cranial shape in salamanders using resistant-fit theta rho analysis. In *Proceedings of the Michigan Morphometrics Workshop*, eds. F. J. Rohlf & F. L. Bookstein, pp. 311–321. Ann Arbor: University of Michigan Museum of Zoology.

Relethford, J. H. (1994). Craniometric variation among modern human populations. *American Journal of Physical Anthropology*, **95**, 53–62.

Relethford, J. H., & Harpending, H. (1994). Craniometric variation, genetic theory, and modern human origins. *American Journal of Physical Anthropology*, **95**, 249–270.

Rhine, S. (1990). Non metric skull racing. In *Skeletal Attributions of Race: Methods for Forensic Anthropology*, Anthropological Papers no. 4, eds. G. Gill & S. Rhine, pp. 9–20. Albuquerque: Maxwell Museum of Anthropology.

Richtsmeier, J. T., Cheverud, J. M., Danahey, S. E., Corner, B. D., & Lele, S. (1993a). Sexual dimorphism of ontogeny in the crab-eating macaque (*Macaca fascicularis*). *Journal of Human Evolution*, **25**, 1–30.

Richtsmeier, J. T., Corner, B. D., Grausz, H. M., Cheverud, J. M., & Danahey, S. E. (1993b). The role of postnatal growth pattern in the production of facial morphology. *Systematic Biology*, **42**, 307–330.

Rohlf, F. J. (1998). On applications of geometric morphometrics to studies of ontogeny and phylogeny. *Systematic Biology*, **47**, 147–158.

Rohlf, F. J. (1999). Shape statistics: Procrustes superimpositions and tangent spaces. *Journal of Classification*, **16**, 197–223.

Scheuer, L., & Black, S. (2000). *Developmental Juvenile Osteology*. London: Academic Press.

Schour, I., & Massler, M. (1941). The development of the human dentition. *Journal of the American Dental Association*, **28**, 1153–1160.

Shea, B. T. (1983). Allometry and heterochrony in the African apes. *American Journal of Physical Anthropology*, **62**, 275–289.

Shea, B. T. (1986). Ontogenetic approaches to sexual dimorphism in the anthropoids. *Human Evolution*, **1**, 97–110.

Slice, D. E., Bookstein, F. L., Marcus, L. F., & Rohlf, F. J. (1996). A glossary for geometric morphometrics. In *Advances in Morphometrics*, eds. L. F. Marcus, M. Corti, A. Loy, G. J. P. Naylor, & D. Slice, pp. 531–551. New York: Plenum Press.

Snow, C., Hartman, S., Giles, E., & Young, F. (1979). Sex and race determination of crania by calipers and computers: A test of the Giles and Elliot discriminant function in 52 forensic science cases. *Journal of Forensic Science*, **24**, 448–460.

Sokal, R., & Rohlf, F. J. (1995). *Biometry*. New York: W. H. Freeman.

Sperber, G. (1989). *Craniofacial Embryology*. Oxford: Wright.

Strand Viðarsdóttir, U. (1999). Changes in the form of the facial skeleton during growth: A comparative morphometric study of modern humans and Neanderthals. PhD dissertation, University College London.

Strand Viðarsdóttir, U., & O'Higgins, P. (2001). Geometric morphometrics and the analysis of variations in facial form: Robusticity of biological findings in relation to bilateral versus unilateral and missing landmarks. *Statistica*, **LXI**, 315–333.

Strand Viðarsdóttir, U.,O'Higgins, P., & Stringer, C. B. (2002). A geometric morphometric study of regional differences in the growth of the modern human facial skeleton. *Journal of Anatomy*, **201**, 211–229.

Stringer, C. B., Dean, M. C., & Humphrey, L. T. (1999). Regional variation in human mandibular morphology. *American Journal of Physical Anthropology*, **28** (Suppl.), 260.

Turner, C. I. (1992). The dental bridge between Australia and Asia: Following Macintosh into the East Asian heart of humanity. *Archaeology of Oceania*, **27**, 143–152.

Ubelaker, D. (1989). *Human Skeletal Remains: Excavation, Analysis, Interpretation*. Washington, DC: Smithsonian Institution Press.

Van Limborgh, J. (1970). A new view on the control of the morphogenesis of the skull. *Acta Morphologica Scandinavica*, **8**, 143.

Weidenreich, F. (1941). The brain and its role in the phylogenetic transformation of the human skull. *Transactions of the American Philosophical Society*, **31**, 321–442.

Wong, L., Dufresne, C. R., Richtsmeier, J. T., & Manson, P. N. (1991). The effect of rigid fixation on growth of the neurocranium. *Plastic and Reconstructive Surgery*, **88**, 395–403.

Wright, R. (1992). Correlation between cranial form and geography in *Homo sapiens*: CRANID – a computer program for forensic and other applications. *Archaeology of Oceania*, **27**, 105–112.

6 Linear growth variation in the archaeological record

L. T. HUMPHREY

Natural History Museum, London

Introduction

The evaluation of physical development in terms of percentage of mature state attained at a given age measures the tempo of development (Eveleth & Tanner, 1990). For some parameters such as tooth crown and root formation, or fusion of the diaphysis and epiphyses of a long bone, the final adult state is comparable for all individuals and the main difficulty is to define consistent intermediate states of maturation in terms of percentage of final adult state. For measures of skeletal growth the problem lies with the fact that the endpoint, or final adult size, is not always known. In cross-sectional studies, such as those that are conducted on any past population, whether fossil or more recent, the final growth attainment of any immature individual in the sample is both hypothetical and unknown. The mean value for mature individuals in the population under study can be used as an estimate of final growth attainment if this information is available. If other statistical parameters are known for the distribution of size in the adult population concerned, an estimate of error can also be made.

The evaluation of growth in terms of percentage of adult size has several advantages over straightforward measurement of absolute size attainment, particularly in a comparative context. The method adjusts for differences in adult size, and places the emphasis on the rate of progress towards adult size rather than actual size attained at any given age. Additionally, the effects of differences in measurement technique between studies are negated, provided each study is internally consistent. This is particularly relevant when making comparisons between data sets derived from measurements from radiographs and direct measurements on bones, because the problem of magnification introduced by radiographic techniques is reduced. It is not entirely eliminated because with

Patterns of Growth and Development in the Genus Homo, ed. J. L. Thompson, G. E. Krovitz and A. J. Nelson. Published by Cambridge University Press. © Cambridge University Press 2003.

144

some radiographic techniques the degree of magnification is slightly greater in large individuals than in small individuals (Feldesman, 1992).

Several studies have employed analysis of growth tempo to evaluate growth in a skeletal sample from a past population relative to a modern reference sample (Humphrey, 2000; Littleton, 1998; Lovejoy *et al.*, 1990; Saunders *et al.*, 1993; Wall, 1991). Typically, results are presented for individual skeletal measurements, although Lovejoy *et al.* (1990) calculated average percentage of adult size attained in successive age groups across all six long bones to provide an overall measure of linear growth attainment. This approach could be particularly important when dealing with partial or fragmentary skeletal remains that yield incomplete data sets, but should only be used if the variables combined in this way exhibit similar growth patterns (see Humphrey, 1998). Recently the technique was used to evaluate growth in a European Neandertal sample relative to modern reference samples derived from osteological and radiographic studies (Thompson & Nelson, 2000).

In this chapter, I review evidence for the range of modern human variation in progress towards adult size attainment in a single developmental system – growth of the femoral diaphysis – using studies of past populations represented by archaeological samples. In this study, data for each archaeological sample are reanalyzed in terms of residuals from the percentage of growth attained in a modern reference sample in successive age classes. The reference sample is derived from a mixed longitudinal analysis of children living in North America in the last century and can be used to represent the growth process of a healthy and well-nourished population (Maresh, 1955, 1970). The method used here is a powerful tool for visualizing fluctuations in growth tempo relative to this reference sample, which may represent periods of growth faltering and catch-up growth. Catch-up growth describes a rapid increase in growth velocity that may follow recovery from a disease or an improvement in nutritional status (Bogin, 1999), whereas growth faltering refers to a decrease in growth velocity resulting from disruption of the normal growth process. Although these terms typically describe changes in the growth velocity of a single child that are revealed by a longitudinal growth study, they are used here to describe general trends within a cross-sectional sample.

Material and methods

Reference sample

Data on the growth of the long bones in several recent populations have been compiled from measurements from radiographs of living children of known

146 L. T. Humphrey

chronological age, but the data presented by Maresh (1955, 1970) are those most often used as a modern reference sample or standard in studies of past populations. These data are from a mixed longitudinal radiographic study of children aged between two months and 18 years from Denver, Colorado. In total 121 females and 123 males were studied between 1935 and 1967, although not all individuals are represented by complete sets of radiographs. The earliest-born children were aged between 1 and 11.5 years at the onset of the study, and the latest-born children were still infants when the study was terminated. As a result, measurements for the youngest children reflect growth attainment in a more recent period than those of the older children. This could result in a slight distortion of the pattern of growth implied by the Denver sample if growth attainment improved over the period represented, but any such bias is likely to be relatively minor. In these publications, measurements of femur length are presented without epiphyses from two months to 12 years and with epiphyses from 10 to18 years. Separate data are presented for males and females. The first publication (Maresh, 1955) gives the 10th, 25th, 50th, 75th, and 90th percentiles and the range of values for successive age categories. A more recent compilation (Maresh, 1970) presents means and standard deviations for successive age categories as well as updated 10th, 50th, and 90th percentiles, but does not include the range.

Data from the Denver growth study are used here as a reference sample, against which to evaluate the growth of past populations. Mean femur length in each age group was calculated as the midpoint of the mean male and female values (data from Maresh, 1970). Males and females were analysed as a single sample since the sex of juveniles in past populations is typically unknown. The results presented here are based on calculation of residuals from the mean value for the Denver sample at a given age. In order to calculate accurate residuals at intermediate ages, curves were fitted to the data to describe the relationship between femur length and age in the Denver sample. Separate polynomial curves were fitted to data for individuals aged 2 years and under and for individuals aged 2 to 12 years because this allowed a closer fit to the original data. The standard deviation was calculated as the mean of the male and female standard deviation values (data from Maresh, 1970) at each age, and these values were used to calculate values for plus and minus one and two standard deviations from the mean.

The 10th and 90th percentile value for both males and females in the Denver sample fall within the range of values defined by ± 2 standard deviations from the combined mean, indicating that at least 80% of girls and boys fall within 2 standard deviations from the combined mean. The minimum and maximum values (data from Maresh, 1955) are close to but typically just outside the values for ± 2 standard deviations from the mean until about 6 years. After 6 years, the difference between the maximum values and the values for 2 standard deviations above the mean tend to be slightly higher for both males and females. This is particularly noticeable in females over the age of 10 and may reflect the onset of the adolescent growth spurt in some of the girls from the Denver sample.

Archaeological samples

Data on the growth of the femoral diaphysis in 11 past populations were compiled from published sources (Table 6.1). Published studies on long-bone growth in past populations have a patchy geographical coverage, with a preponderance of studies in North America (mostly of Native American groups) and Western Europe. The number of samples available for use in this study is further restricted because many studies of femoral growth in past populations do not give data on final adult size attainment. For example, although there are at least three published studies of human long-bone growth from archaeological sites in Africa (Armelagos *et al.*, 1972; Hummert & Van Gerven, 1983; Steyn & Henneberg, 1996), the data on adult size required to ascertain the percentage of adult size attained at specific ages were only available for the Wadi Halfa sample (Hummert & Van Gerven, 1983).

In this study Africa is represented by the Wadi Halfa sample (Armelagos *et al.*, 1972), which is from a number of cemeteries in Lower Nubia, Sudan, and incorporates individuals living in a period that spans nearly two millennia (350 BC – AD 1400). Western Asia is represented by a sample from two cemeteries in Bahrain that were used by small agricultural communities between 300 BC and AD 200 (Littleton, 1998). Eastern Asia is represented by a sample from a prehistoric site in central Thailand, Khok Phanom Di, that was occupied from 2000 to 1500 BC. The economy was centered on fishing and collecting, with rice cultivation in the middle period of occupation (Tayles, 1999).

Four European samples are used in this study, as well as a Canadian sample from a population of European origin. The Altenerding sample is from a site dating to the sixth and seventh centuries near Munich in Germany (Sundick, 1978). Most of the Ancient Slavic sample is from the ninth-century site of Mikulčice, but part of the sample is from two smaller seventh- and eighth-century cemetery samples (Stloukal & Hanáková, 1978). The Wharram Percy sample is from a deserted medieval village in Yorkshire, Northern England (Mays, 1999). The London sample represents an eighteenth- and nineteenth-century urban English population and comprises individuals buried in the crypts of Christ Church, Spitalfields, and the churches of St Brides and St Barnabas (Humphrey, 2000; Molleson & Cox, 1993; Scheuer & Bowman, 1995). The sample from Belleville, Ontario comprises individuals of European origin who

Table 6.1 *Details of archaeological samples*

Data sources and other relevant references	Sample	Period	Method of age determination	Sample size	Type of site/economy
North Africa/Middle East					
Armelagos et al., 1972	Wadi Halfa, Lower Nubia, Sudan	350 BC – AD 1400	Dental development (including emergence and probably formation)	55	Economy of region developed from irrigation farming to urbanization
Littleton, 1998	DS3 and Saar, Bahrain	300 BC – AD 200	Dental formation and eruption	73	Two cemeteries used by small agricultural communities
Asia					
Tayles, 1999	Khok Phanom Di, Thailand	2000–1500 BC	Tooth and crown formation and tooth emergence	25	Economy based on fishing, collecting, and rice cultivation
Europe/European origin					
Stloukal & Hanáková, 1978	Mikulčice, Nové Zámky, and Virt	7th–9th centuries	Tooth crown and root formation	190	No information provided for any of these three "Ancient Slavic" sites
Sundick, 1978	Altenerding, Germany	6th–7th centuries	Individuals placed into dental age stages based on emergence and formation	46	No information provided

Reference	Site	Period	Method	n	Description
Mays, 1999	Wharram Percy, North Yorkshire, England	10th–16th centuries	Dental calcification	104	Village cemetery used by rural medieval population
Humphrey, 2000 (see also Molleson & Cox, 1993; Scheuer & Bowman, 1995)	Christ Church, Spitalfields, St Brides, and St Barnabas, London, England	18th–19th centuries	Age at death indicated by coffin plates and further documentary evidence	66	Church crypts used by an urban population of relatively high socioeconomic status
Saunders et al., 1993	St Thomas's Church, Belleville, Ontario	19th century	Tooth crown and root formation	156	Settlement developed from village to market town to urban industrial "city" during the period of use of this cemetery
North America					
Ryan, 1976	San Cristóbal, New Mexico	AD 700–1500	Tooth and root formation, emergence and attrition	68	Population dependent on subsistence agriculture
Sundick, 1978 (see also Johnston, 1962)	Indian Knoll, Kentucky	Eastern Archaic, 3000 BC	Individuals placed into dental age stages based on emergence and formation	86	Shell midden site occupied by a pre-agricultural, pre-pottery population
Lovejoy et al., 1990 (see also Menforth, 1985)	Libben, Ohio	Late Woodland, AD 800–1100	Tooth crown and root development (adjusted)	131	Sedentary hunter–gatherers

were buried in the cemetery of St Thomas's Anglican Church from 1821 to 1874 (Saunders *et al.*, 1993).

Three Native American samples are included. The Indian Knoll sample is from a shell midden in Kentucky and dates from the pre-agricultural Eastern Archaic period (Johnston, 1962; data from Sundick, 1978). The Libben sample is from a Late Woodland (800–1100 AD) cemetery site on the banks of Portage River, Ohio (Lovejoy *et al.*, 1990). The site was occupied on a year-round basis by hunter–gatherers who relied heavily on local resources (Menforth, 1985). The San Cristóbal sample from New Mexico represents a Pueblo Indian population (700–1500 AD) dependent on subsistence agriculture (Ryan, 1976).

Data were recalculated as percentage of adult size attained for all of the analyses presented here. Percentage of adult size attained was calculated by dividing femoral length in each age group by estimated adult femoral length. In each case, measurements of the subadults are of the diaphysis only whereas measurements of adult femur length include the fused distal and proximal epiphyses. Comparisons between sites are restricted to children aged under 12 years. For the Denver sample, adult size was estimated as the mean of male femur length at age 18 years and female femur length at age 16 years (Maresh, 1970). For the Altenerding and Indian Knoll samples (Sundick, 1978) adult femur length was calculated as the mean femur length of males and females aged 17 and 18 years measured with epiphyses. Adult femur length for the Libben sample was calculated from data presented by Lovejoy *et al.* (1990) on diaphysis length and the percentage of adult size attained for successive age classes. For all other samples (Armelagos *et al.*, 1972; Ryan, 1976; Stloukal & Hanáková, 1978; Tayles, 1999), adult femur length was calculated as the mean of male and female femur length for adults aged over 20 years. Simon Mays, Judith Littleton, and Shelley Saunders kindly provided data on femur length in adult males and females from the Wharram Percy, Bahrain, and Belleville sites respectively.

Estimation of age at death

The age at death of individuals in the London crypt sample was derived from information on associated coffin plates, supplemented in many cases by additional documentary evidence. The sample was separated into half-yearly age classes until age 1 and yearly age classes from 1 to 5 years. Children aged between 5 and 12 years were divided into two evenly sized groups due to small sample size in this age range. The mean age at death and mean femoral length was calculated for individuals in each group. Nancy Tayles kindly provided individual dental age estimates for juveniles from Khok Phanom Di. The sample includes a large number of neonates but only 25 individuals aged between

1 month and 12 years with dental age estimates. These 25 infants and children were placed in five age classes. The first age group includes nine infants (under 1 year) and successive age groups for 1–2 years, 2–4.5 years, 8–11 years, and 12 years each include four children. Data points represent the mean age at death and the mean femoral length of individuals belonging to a particular age group.

The age grouping used for the other populations are those presented by the authors of each study. Sundick (1978) avoided assigning age at death to individual skeletons by placing individuals into dental age stages, with intervals chosen to reflect distinctive stages of tooth emergence and formation. In order to permit comparisons with other studies, these dental stages were translated into approximate age categories using the conversion table provided by Sundick. For all other samples, the age at death estimates used are those presented by the authors of each study. Age at death is normally determined from dental development, typically using developmental standards for individuals of European ancestry (see Table 6.1), but some studies take osteological development or dental wear into account (Johnston, 1962; Ryan, 1976). In most studies, ages are presented as the midpoint of a given dental age class and femoral length is presented as the mean value of individuals within that age class. For the Belleville sample, the mean age of individuals within each age class and the mean femoral length of individuals within that age class were used (Saunders *et al.*, 1993).

Accurate age estimation is fundamental to any direct study of skeletal growth in a past population. The use of inappropriate techniques for the evaluation of age at death can have a significant effect on the evaluation of skeletal growth. If age at death is systematically overestimated due to the use of an inappropriate developmental standard, both actual size and the percentage of adult size attained at a given age will be underestimated. Conversely, if age at death is underestimated, both size and the percentage of adult size attained at a given age will be overestimated. Apparent variations in growth trajectories could therefore arise due to variation in the methods used to assign age at death or differences among populations in the relationship between physiological age and chronological age in the developmental system used to assign age at death.

Tooth formation is recognized as a more useful and accurate indicator of subadult age than tooth emergence (Demirjian, 1986; Saunders, 1992; Smith, 1991), but since most studies of tooth formation document European-derived populations, there is very little information available on potential variation between populations. Information on tooth and root formation of the deciduous teeth that is critical for age evaluation in the first few years of life is particularly poor. Smith (1991) suggested there may be patterned, but not necessarily large, differences in the timing of tooth formation between populations. Population differences in the timing of tooth emergence and tooth crown and root formation are reviewed by Liversidge (this volume).

For their study of the Libben population, Lovejoy *et al.* (1990) adjusted dental development standards to allow for population differences in dental development. Their correction was based on studies indicating that the average age of emergence of the permanent teeth is earlier in Native Americans than in Americans of European ancestry (see references in Lovejoy *et al.*, 1990). Lovejoy *et al.* (1990) suggested a correction factor of 0.0575 years per year of growth, calculated using the age of emergence of the canine, premolars, and second molar. This corresponds to a downward age adjustment of approximately 3.5 months at age 5 years and nearly 7 months at age 10 years. This adjustment results in a cumulatively upward shift of the growth trajectory and a cumulatively earlier age for changes in the growth trajectory that might reflect significant biological events, such as the age at onset and completion of a period of growth faltering.

The underlying assumption that dental development is advanced in a uniformly cumulative manner in the Libben population is an oversimplification. Eveleth & Tanner (1990) consider emergence of the permanent dentition in terms of two active phases and a quiescent phase, and compare ages for the start and end of each phase in different population groups. In their sample of populations, the second phase (emergence of the canine, premolars, and second molar) was earlier in the Asiatic groups than in Europeans or Africans. However, completion of the first phase (emergence of incisors and first molar) occurred later on average in Asiatic groups than in European or African groups, indicating that relative advancement at a given stage (or stages) of tooth emergence should not be taken to imply that development is advanced at all stages of emergence. An additional concern is that the method developed by Lovejoy *et al.* (1990) is derived from studies of tooth emergence and applied to age estimates based on tooth formation (Saunders, 1992). As with tooth emergence, Native American groups do not appear to exhibit a consistent pattern of advanced tooth formation over Europeans. Tompkins (1996) found that Native Americans tended to be advanced in relative development of the second and third permanent molars compared to French Canadians, but delayed in their relative canine and incisor calcification.

The extent to which variations in the growth trajectories of the populations examined here are influenced by differences in methodology is difficult to quantify. Studies that have compared growth trajectories produced using different aging techniques on a single skeletal sample have demonstrated that the method used to assign age at death can have a marked impact on the resultant growth trajectory (Merchant & Ubelaker, 1977; Saunders *et al.*, 1993). Additionally, the shape of a growth trajectory constructed from age-grouped data could be an artefact of the way in which data are presented, since changes in a growth trajectory are constrained to occur at the specific ages (or midpoints of a specified age range) for which mean diaphysis lengths are available. If the age interval between these points is large, changes in growth trajectory may be over-looked. However, the data should not be over-divided because artificial changes in growth trajectory could be introduced due to random error caused by small sample size in some age groups. Typically, infants and young children are well represented in archaeological samples (although they may be underrepresented relative to the number of deaths in this age category) and problems caused by small sample size arise during mid to late childhood and adolescence. This is the case for most of the samples included in this study (with the exception of the Altenerding sample, which includes only three individuals aged under 2.5 years). Although it is more appropriate to calculate the mean age of individuals within an age class rather than use the midpoint of the age range, particularly when sample size is small and there is a risk that data points are distributed unevenly across the age range, this information is typically unavailable.

The effect of methodological factors on the growth trajectory for Indian Knoll is illustrated in Figure 6.1, which presents alternative growth trajectories

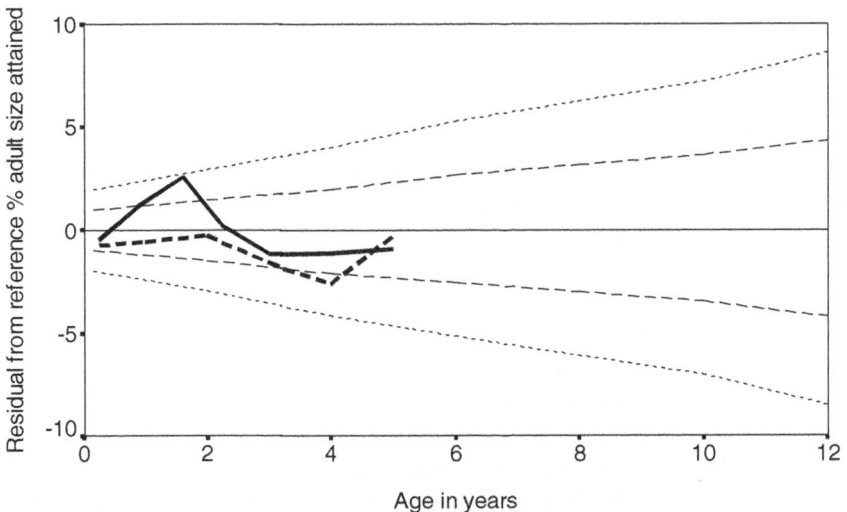

Figure 6.1   Comparison of relative femoral growth trajectories generated by Johnston (1962, thick dashed line) and Sundick (1978, thick solid line) for the Indian Knoll sample using different methods of sampling and estimation of age at death. The line $y = 0$ represents the mean percentage of adult size attained at a given age in the Denver sample. The growth trajectories for the Indian Knoll population illustrate deviations from this pattern. Each trajectory interpolates between individual points that represent the difference between the percentage of size attained in the Denver sample at a given age and the actual percentage of adult size attained in the archaeological populations. Thin dashed and dotted lines represent -2, -1, $+1$, and $+2$ standard deviations from the Denver mean.

derived from data presented by Johnston (1962) and Sundick (1978). These differences could reflect several different methodological factors including technique to assign age at death, selection of age groups and the number of individuals considered suitable for measurement within each age group. The trajectory produced using Sundick's data indicates a higher percentage of adult size attained between the age of 3 months and 4 years, but differences in the shape of the two trajectories are even more marked than differences in the percentage of adult size attained. Sundick's data imply a dramatic improvement in relative growth attainment between 3 and 20 months, followed by a steep decline until age 3 when the trajectory levels out. Johnston's data imply a slight improvement between birth and 2 years followed by a gradual decline from 2 to 4 years, and then a marked improvement. Despite these differences, there is a basic agreement between the two results of the two studies, with both trajectories indicating an initial improvement in growth attainment followed by a downward turn in early childhood, and lower levels of growth attainment than the Denver reference population after the age of 2 years.

Results

The growth trajectories displayed for each past population (Figure 6.2) illustrate deviations from the 'typical' (mean) Denver pattern. Each trajectory interpolates between individual points that represent the difference between the percentage of adult size attained in the Denver sample at a given age and the actual percentage of adult size attained in the archaeological population. The line $y = 0$ represents the mean percentage of adult size attained at a given age in the Denver sample. A growth trajectory that followed this line would indicate a population that pursued the same tempo of growth as the Denver sample throughout the growth period. The dashed and dotted lines illustrate -2, -1, $+1$, and $+2$ standard deviations from the mean for the Denver sample. A trajectory that tracked one of these lines, or maintained a fairly consistent position between lines would also indicate a population that was pursuing a pattern of growth that is consistent with that of the Denver sample, but with an average tempo of growth that was higher or lower than the mean for the Denver sample. A growth trajectory that does not maintain a relatively constant position within this framework indicates a population that exhibits a different pattern of growth. A steep downward deflection reflects a period in which the tempo of growth is markedly lower than the Denver sample whereas an upward deflection reflects a period in which the tempo of growth is markedly higher. Observed changes in growth trajectory may reflect periods of growth faltering or catch-up growth. It should be remembered, however, that methodological factors, such as sample size and statistical treatment of the data, could contribute to these apparent changes in growth tempo (see above).

North Africa and Western Asia

The youngest age category for which data are available for the Wadi Halfa sample (Figure 6.2A) is 6 months to 1 year. Femur length of infants in this age range is comparable to that of the Denver sample in terms of percentage of adult size attained. There is a decline in the percentage of adult femur length attained for age relative to the Denver sample between 1.5 and 3 years, indicating growth faltering in early childhood. For the rest of the childhood period, the values for the Wadi Halfa sample fluctuate at around 1–2 standard deviations below the

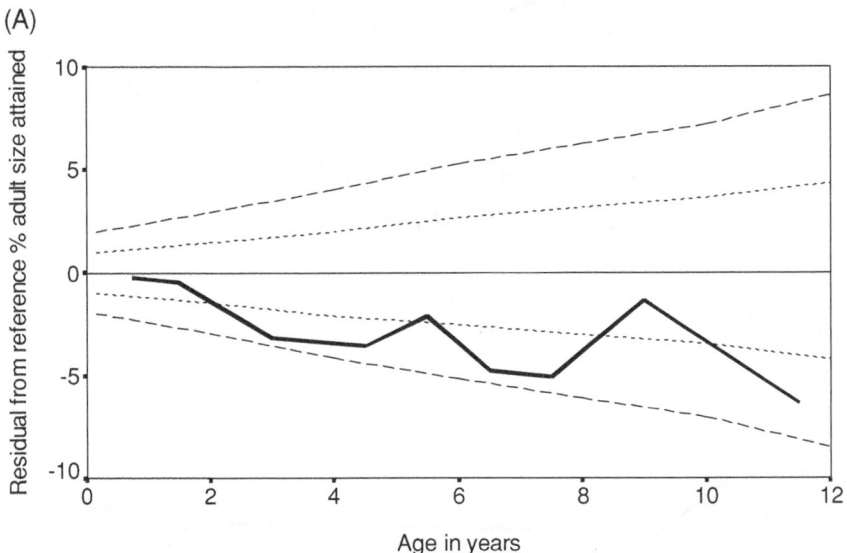

Figure 6.2   Relative femoral growth trajectories of samples from (A) Wadi Halfa, (B) Bahrain, (C) Khok Phanom Di, (D) London, (E) Ancient Slavic sites, (F) Wharram Percy, (G) Altenerding, (H) Belleville, (I) Indian Knoll, (J) San Cristóbal, (K) Libben. The line $y = 0$ represents the mean percentage of adult size attained at a given age in the Denver sample. The growth trajectories illustrate deviations from this pattern. Each trajectory interpolates between individual points that represent the difference between the percentage of size attained in the Denver sample at a given age and the actual percentage of adult size attained in the archaeological population. Thin dashed and dotted lines represent -2, -1, $+1$, and $+2$ standard deviations from the Denver mean.

mean value for the modern reference sample. The trajectory for the Bahrain sample falls steeply from an initial value above the Denver mean to more than 1 standard deviation below between the ages of 3 and 9 months (Figure 6.2B). A steady decline continues until 2.5 years, with values markedly lower than 2 standard deviations below the Denver mean throughout childhood.

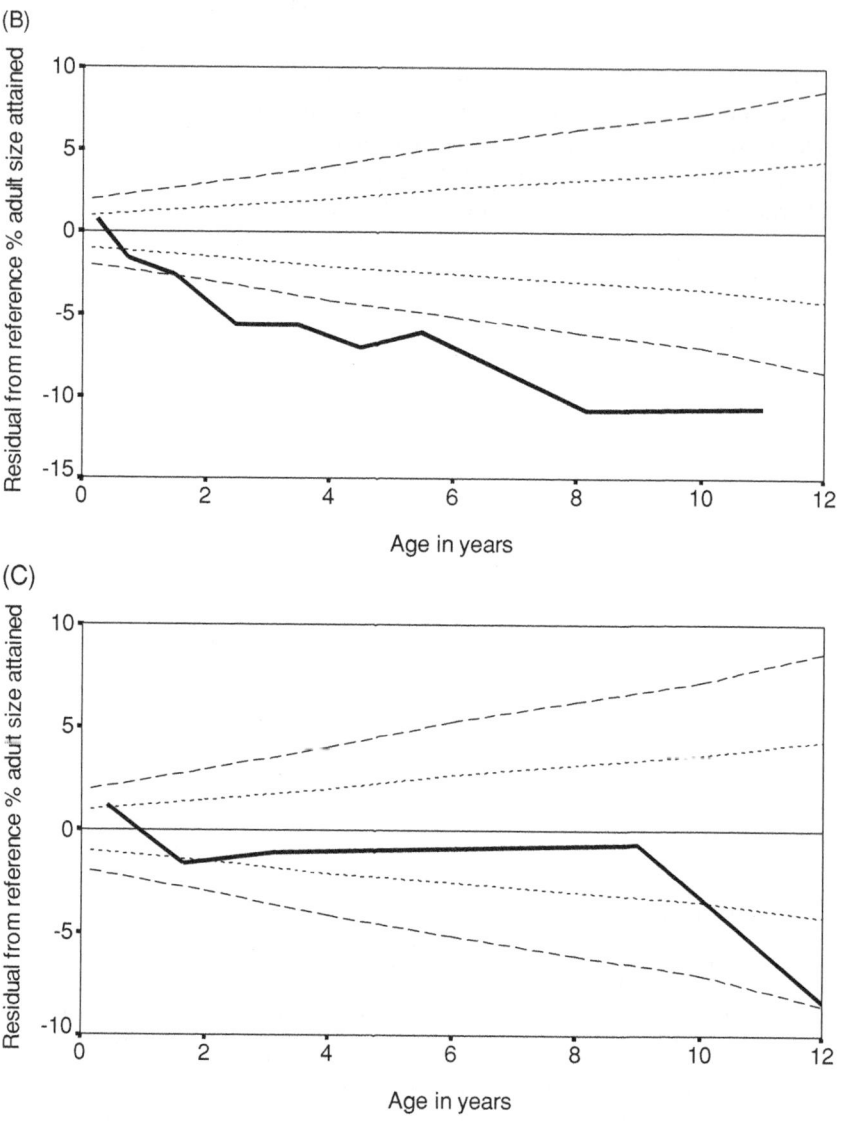

Figure 6.2   (*cont.*)

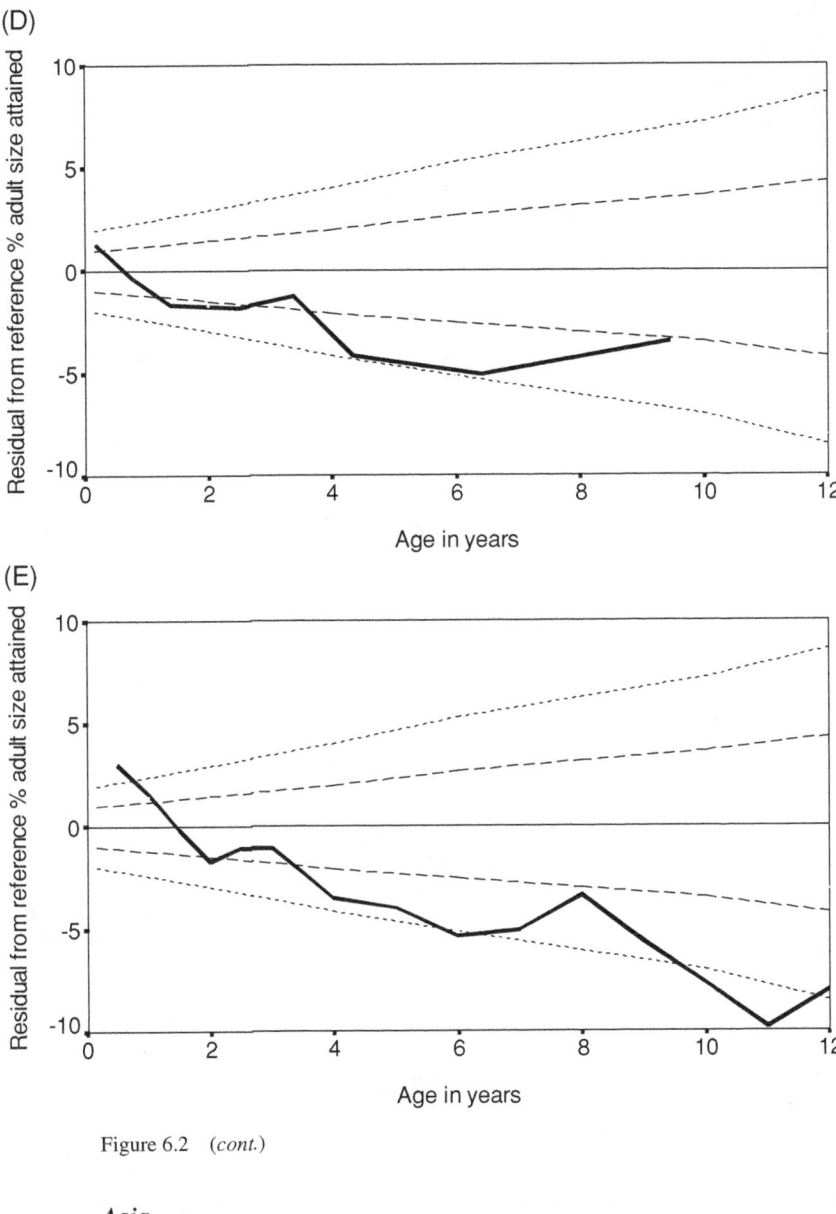

Figure 6.2   (*cont.*)

Asia

The trajectory for the Khok Phanom Di sample falls steeply from an initial value 1 standard deviation above the Denver mean to more than 1 standard deviation below the Denver mean between the first and second age group,

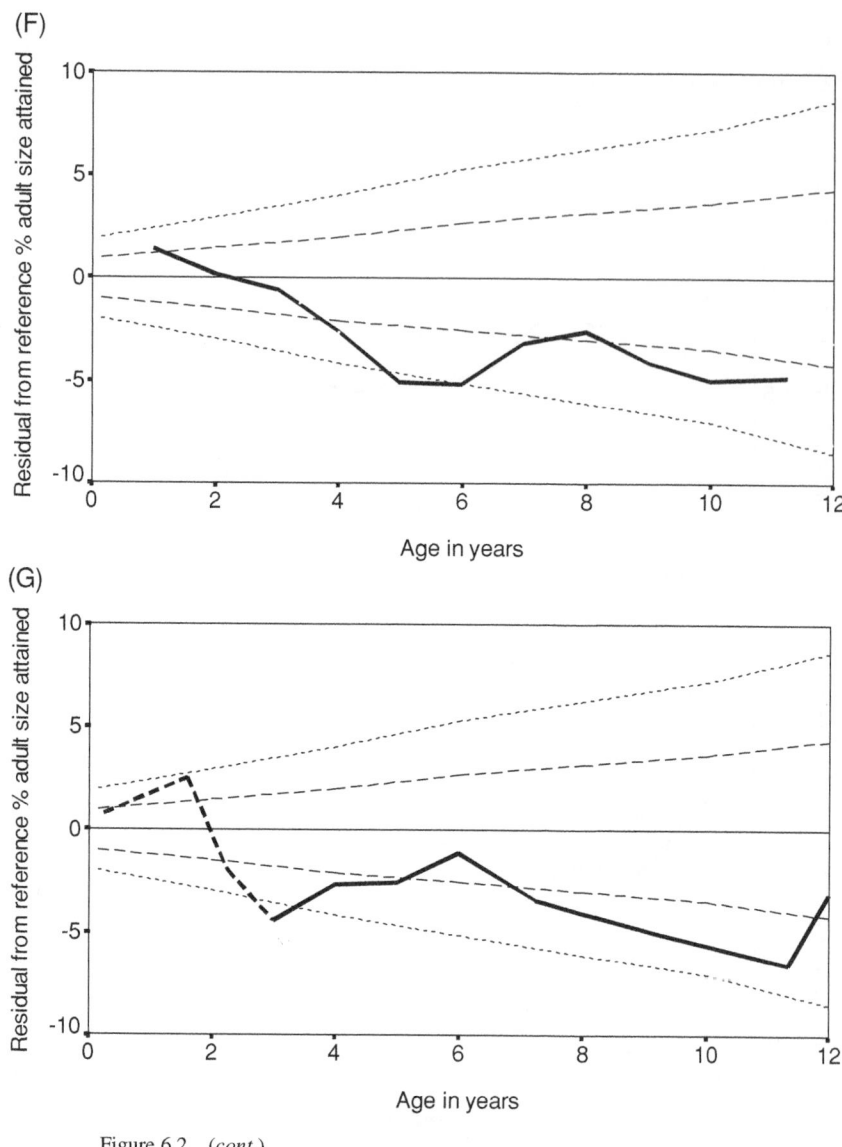

Figure 6.2   *(cont.)*

indicating growth faltering in late infancy or early childhood. Children in the third and fourth age groups (averaging 3 and 9 years) are within −1 standard deviation of the Denver mean in terms of percentage of adult size attained. After age 9, the growth trajectory turns steeply downwards, reaching almost 2 standard deviations below the mean for the Denver sample at age 12.

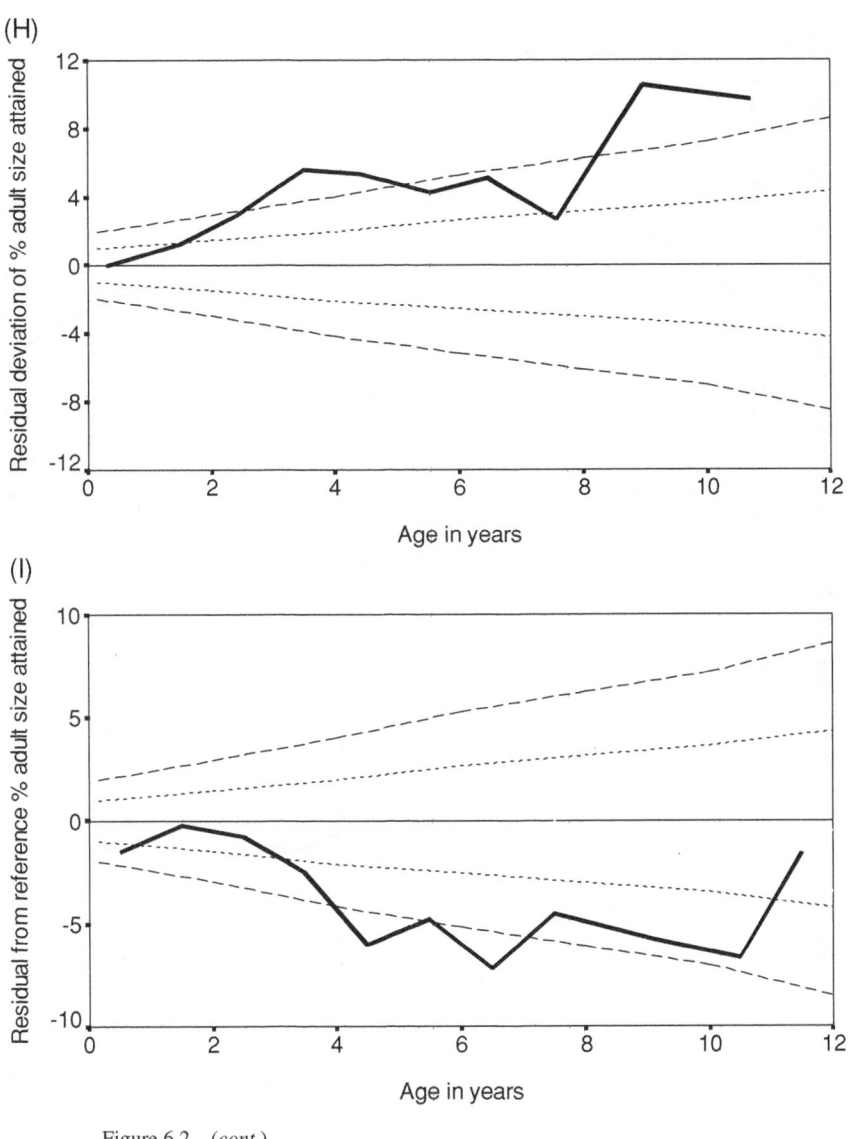

Figure 6.2 (*cont.*)

Europe/European origin

The growth trajectories for the London (Figure 6.2D) and Ancient Slavic (Figure 6.2E) samples are similar from infancy to mid childhood, although the starting-point differs. The value for the first age class of the London sample

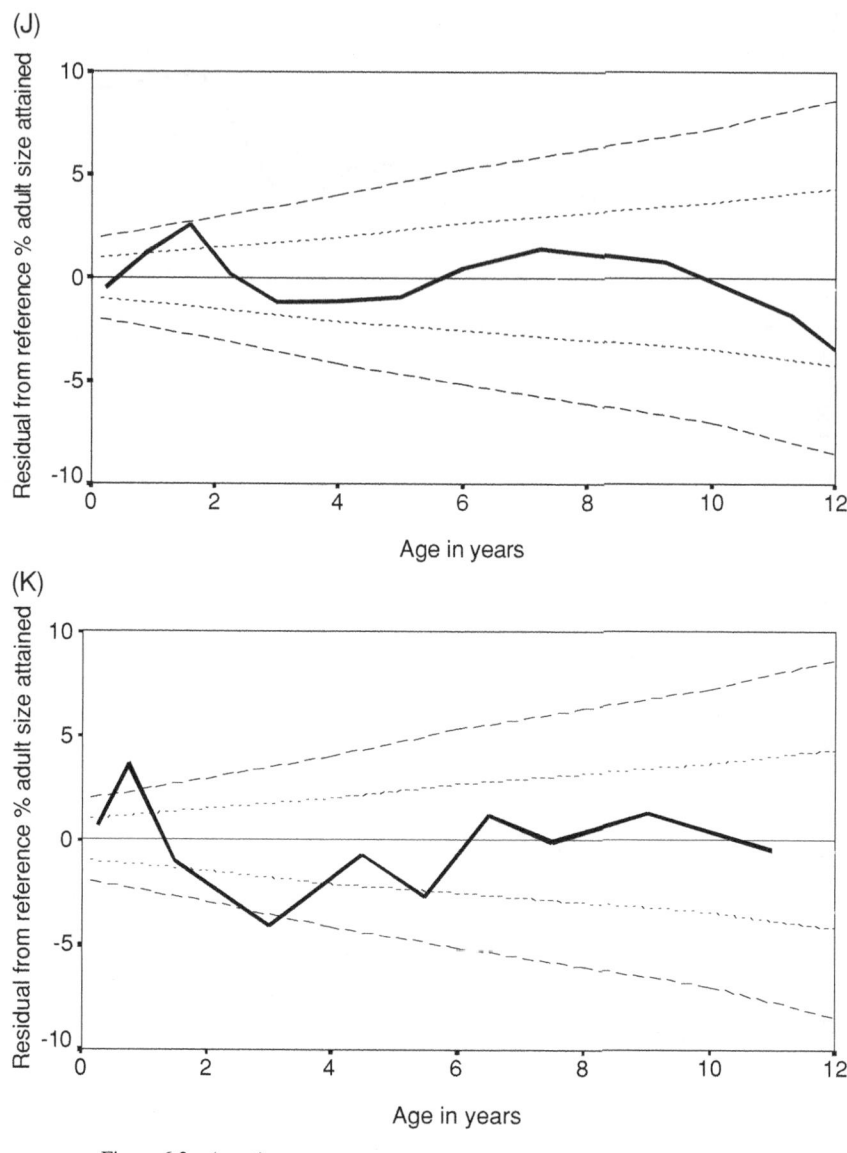

Figure 6.2   (*cont.*)

is about 1 standard deviation above the Denver mean whereas the first value for
the Ancient Slavic sample is more than 2 standard deviations above the mean
of the Denver sample. Both trajectories exhibit a marked decline in the first
two years of life and then level off temporarily at about 1 standard deviation

below the mean value for the modern reference sample in early childhood. The decline then continues, with both samples reaching about 2 standard deviations below the mean for the Denver sample at 4 to 6 years. The Wharram Percy sample (Figure 6.2F) exhibits a steady decline during early and mid childhood with values falling from 1 standard deviation above the Denver mean at age 1 to more than 2 standard deviations below the Denver mean at 5 years. After 6 years, there appears to be an improvement in terms of growth attainment in the London, Wharram Percy, and Ancient Slavic samples, with values reaching around 1 standard deviation below the mean for the Denver sample. This recovery is short-lived in the Ancient Slavic sample, with values dropping more than 2 standard deviations below the Denver mean at 10 and 11 years. The trajectory joining the three youngest age groups for the Altenerding sample (Figure 6.2G) is shown here as a dotted line because each point represents a single individual. At age 3, the value for the Altenerding sample is more than 2 standard deviations below the mean value for the Denver sample. In contrast to the other European samples, there is some evidence of an improvement in terms of percentage growth attainment in early childhood, but this is followed by a steady decline from age 6 to 11 years.

The trajectory for the Belleville sample shows an increasing trend from an initial value that is very close to the Denver mean, to a value more than 2 standard deviations above the Denver mean by the age of 3.5 years. Over the next four years, the trajectory converges towards the Denver mean reaching around 1 standard deviation above the Denver mean at age 7.5 years, but the values of subsequent age groups return to more than 2 standard deviations above the Denver mean (Figure 6.2H).

North America

The values for the youngest age group in each of the three North American samples are within 2 standard deviations of the mean value for the Denver sample, although uniquely among the samples compared for this study, the youngest age group for Libben (Figure 6.2I) has a value more than 1 standard deviation below the modern reference sample. All three North American samples initially exhibit an increase in terms of the percentage of adult size attained relative to the Denver sample, but the trajectories reverse between 9 months and 2 years of age, and indicate growth faltering in the second and third years of life. The growth trajectory for the Indian Knoll sample (Figure 6.2J) levels off just below the Denver mean at around 3 years of age and thereafter follows a level that is close to the Denver mean throughout childhood. Growth faltering in the San Cristóbal sample (Figure 6.2K) is more dramatic but after the age of 3 years there is a steep recovery and by the age of 6 years, this sample also follows a trajectory that is close to or above the mean for the Denver sample. Growth faltering in the Libben sample (Figure 6.2I) starts at around 18 months and continues into mid childhood, with values falling more than 2 standard deviations below the Denver mean at 4.5 and 6.5 years. In later childhood, the values lie between 1 and 2 standard deviations below the Denver mean, and there is further catch-up after age 10. The growth trajectory of the Libben sample would appear to lie even further below the Denver mean if Lovejoy and colleagues (1990) had not adjusted dental age estimates of the sample in an attempt to offset population differences in the timing of dental development.

Discussion

The trajectories for the past populations shown here typically fall below the mean of the Denver sample for most of the childhood period, indicating that they have achieved a lower percentage of adult size than this modern reference sample during this period of life. A continuation of this trend for the whole of the developmental period would imply a delayed adolescent growth spurt and a delay in the age of attainment of adult size. Prolongation of growth and the delay of final stature attainment under adverse environmental circumstances are outcomes that have been well documented in recent and historical populations (Bogin, 1999; Steegmann 1985), but there may also be a genetic component to population differences in the duration of linear growth (Eveleth & Tanner, 1990).

Marked fluctuations in the growth trajectory relative to the Denver mean suggest a different pattern of growth. The populations shown here vary in the extent to which they exhibit relative growth retardation, but the most consistent fluctuations occur in the first three years of life. A notable similarity between the growth trajectories of the three Native American samples is an initial increase in percentage of adult size attained relative to the Denver sample. There may be a genetic component to this pattern since it does not occur in the other population groups that have reasonable sample sizes in this age range. The upward trajectory suggests a more rapid rate of femoral growth attainment during infancy. However, systematic underestimation of the age of infants at the peak of the growth trajectory would produce the same effect. This could result from an overall delay in dental development at this stage or a more specific delay affecting the emergence or formation of one or more teeth that are critical for evaluation of age at death in this age group. If the upward trajectory genuinely reflects a rapid rate of femoral growth attainment, cultural factors affecting the treatment of infants could play a significant role. In all three populations, the period of rapid growth is followed by a dramatic reversal in the growth trajectory, but the age at which this occurs differs markedly, indicating that a multifactorial explanation for this pattern is required.

All of the archaeological populations shown here apart from Belleville exhibit a decline in the percentage of adult size attained relative to the modern reference sample during late infancy and early childhood. They differ in the age of onset and duration of this event. The period of rapid decline typically starts between 6 and 20 months of age and continues until around 3 years. This decline may reflect growth faltering during the weaning period. Exclusively breast-fed infants are shielded from infectious disease, in part due to the transfer of immunological components from breast milk (Garofalo & Goldman, 1998; King & Ulijaszek, 1999). The weaning process is initiated with the introduction of supplementary foods, and the age at supplementation is strongly influenced by prevailing cultural norms and household circumstances (Stuart-Macadam, 1995). The maximum age at which supplementary foods are likely to have been introduced is around 6 months, since the nutritional and caloric value of a mother's milk is generally insufficient to sustain an infant after this age. The contribution of supplementary foods to an infant or child's total dietary intake increases with age and these foods eventually entirely replace breast-milk. The rate of increasing supplementation, the suitability of supplementary foods, and the age of an infant or child at complete cessation of breast-feeding are also culturally mediated (Fildes, 1995; Katzenberg *et al.*, 1996; Stuart-Macadam, 1995), and are not known for these past populations.

The weaning period can be hazardous for several reasons. Supplementary foods can supply the additional nutrients required for a growing infant, but in cases where the weaning diet is deficient in protein or other essential nutrients, an infant or child may experience malnutrition. Supplementary foods can also represent a novel source of infection at this stage of life, because weaning foods are frequently contaminated due to poor water quality and hygiene (Katzenberg *et al.*, 1996; King & Ulijaszek, 1999). During the weaning process, the immunological protection provided by the mother's milk is reduced and eventually withdrawn. Older infants and children may be exposed to a greater and novel range of pathogens resulting from increased social interaction and independent physical mobility. Growth faltering during weaning typically results from complex interactions between undernutrition and infection (Calder & Jackson, 2000; King & Ulijaszek, 1999).

In some of the past populations shown here (Ancient Slav, San Cristóbal) the onset of growth faltering takes place between the ages of 6 months and 1 year. In these groups, it is possible the onset of the weaning process initiated a cycle of infection and malnutrition that resulted in growth faltering. The onset of growth faltering in the Bahrain sample takes place between the ages of 3 and

9 months. An early onset of growth faltering (before 6 months) could be the result of unsuitable early supplementation or poor nutritional status in the mother. Growth faltering is also initiated at an early age in the Spitalfields sample, which may correlate with historical accounts of artificial feeding practices in eighteenth- and nineteenth-century London (Molleson & Cox, 1993). In other populations, (Indian Knoll, Libben, and Wadi Halfa) marked growth faltering occurs only after the age of 18 months. In these groups growth faltering cannot be related to the introduction of supplementary foods since the maximum age at which dietary supplementation would be expected is around 6 months.

The duration and severity of growth faltering and the pattern of recovery also differs between populations. The trajectories for London, Wharram Percy, and the Ancient Slav sample are extremely similar between the age of 3 and 8 years, with values falling to around 2 standard deviations below the mean for the Denver sample at 4 to 6 years and then recovering. The Libben trajectory exhibits a similar trend and demonstrates that this pattern is not specific to European samples. In contrast, samples from San Cristóbal, Altenerding, and to a lesser extent Indian Knoll exhibit an upward growth trajectory between the ages of 3 and 6 years. Despite slight signs of recovery, the majority of populations analyzed continue to follow a pattern of growth attainment that is below that of the Denver mean throughout childhood. Only two of the populations achieve an average growth tempo that is not markedly different from that of the Denver sample in later childhood. From the age of approximately 6 years, the San Cristóbal growth trajectory lies very close to the mean trajectory for the Denver sample. The trajectory for the Indian Knoll sample fluctuates within 1 standard deviation of the mean trajectory for the Denver sample at all ages shown, apart from a brief period in early childhood when it is nearly 2 standard deviations above the Denver mean.

The femoral growth trajectory for the nineteenth-century cemetery sample from St Thomas's Church, Belleville, Ontario is advanced over the Denver sample in terms of percentage of adult size attained in all but the first age group. Interestingly, children from this sample are similar in size to those of the Denver sample, and this apparent anomaly is explained by the fact that final growth attainment is lower in the Belleville sample than in the Denver sample (Saunders *et al.*, 1993). The difference in adult size could be the result of reduced growth attainment in adolescence in the Belleville sample relative to the Denver population (Saunders *et al.*, 1993), but another explanation is possible that reflects the particular circumstances of the sample. The cemetery of St Thomas' Church from which the Belleville sample was drawn was in use for only 53 years during a period of rapid socioeconomic change. On average, measurements of infants and children in this sample reflect growth attainment in a more recent period and under a different set of environmental circumstances to measurements of the adult sample. It is therefore possible that the very high levels for percentage growth attainment revealed by this analysis are an artefact caused by a population undergoing rapid secular change in growth rate and final growth attainment.

In living populations, it is sometimes possible to make carefully controlled comparisons between the growth of genetically different population groups living under similar environmental circumstances or genetically similar populations living under very different circumstances in order to distinguish between genetic and environmental causes of observed growth differences (Bogin, 1999). In past populations, it is usually not possible to distinguish between environmental and genetic causes of differential growth when comparing genetically dissimilar groups. Numerous cultural practices could potentially have an impact on the exposure of infants and young children to pathogens and on their energy intake and expenditure, and hence contribute to these variable patterns of growth faltering and catch up. These include differences in breast-feeding (spacing of feeding episodes, age at cessation), supplementary feeding (nutritional value, digestibility, and quantity given at different ages), culturally determined patterns of food allocation (gender-related preferences, response to shortage), potential for food contamination (methods of food storage and preparation), amount of contact with other members of the community, freedom or constraint of infant mobility, as well as population density and waste management. The climate of a region could influence both nutritional intake, due to its effect on the predictability and seasonality of food availability, and pathogen load.

When working with archaeological samples, an additional set of factors can confound accurate interpretation of growth in a past population (Humphrey, 2000; Saunders, 1992). Methodological concerns include small samples, incomplete or biased preservation and recovery, poor temporal resolution, and the reliability of age-at-death determinations. Analysis of cross-sectional growth results in an average trend that does not reveal individual variation in the timing or severity of changes in growth tempo, and will result in a smoothing out of growth events that are imperfectly synchronized between individuals (see for example Tanner *et al.* (1966) in relation to the adolescent growth spurt). Furthermore, data collected from a mortality sample may not accurately reflect the parameters of the living population from which the sample was drawn (Wood *et al.*, 1992). Particular emphasis has been placed on the effect of mortality bias on linear growth and other non-specific indicators of childhood health, and based on extensive review and analysis, Saunders & Hoppa (1993) concluded that although the linear growth of survivors is often greater than that of non-survivors, the effect of this on resultant (average) growth trajectories is relatively minor.

166   L. T. Humphrey

Conclusion

The 11 samples presented here demonstrate a range of different patterns of growth tempo during infancy and childhood. These differences are shaped by the interaction of genetic, cultural, and environmental influences and may be influenced by random sampling effects and other methodological factors. A decline in growth attainment, relative to the Denver sample, during infancy and childhood is an almost universal feature of these past populations, although the populations differ in the age, severity, and duration of this process. The archaeological samples included in this study are all from sedentary populations, and therefore reflect a comparatively recent mode of human existence. Population size in most of these communities was probably large enough to sustain density-dependent diseases, and sedentism carries other risks for disease transmission including infestation by disease-carrying pests and the effects of poor sanitation. The data presented here do not provide evidence of clear-cut geographical or chronological patterns in population differences in the tempo of femoral growth, or differences related to socioeconomic factors. A thorough evaluation of regional variation in growth patterns and temporal variation within and between regions would require a more comprehensive representation of past populations from different temporal and cultural horizons than is currently available.

Acknowledgments

I thank the editors for inviting me to participate in this volume, and for their helpful comments on the paper. I would especially like to thank Nancy Tayles, Simon Mays, Judith Littleton, and Shelley Saunders for providing me with additional data on the skeletal samples from Khok Phanom Di, Wharram Percy, Bahrain, and Belleville, which enabled me to include these sites in this overview.

References

Armelagos, G. J., Mielke, J. H., Owen, K. H., Van Gerven, D. P., Dewey, J. R., and Mahler, P. E. (1972). Bone growth and development in prehistoric populations from Sudanese Nubia. *Journal of Human Evolution*, 1, 89–119.
Bogin, B. (1999). *Patterns of Human Growth*, 2nd edn. Cambridge: Cambridge University Press.
Calder, P. C., & Jackson, A. A. (2000). Undernutrition, infection and immune function. *Nutrition Research Reviews*, 13, 3–29.

Demirjian, A. (1986). Dentition. In *Human Growth: A Comprehensive Treatise*, vol. 2, *Postnatal Growth and Neurobiology*, 2nd edn., eds. F. Falkner & J. M. Tanner, pp. 269–298. New York: Plenum Press.

Eveleth, P. B., & Tanner, J. M. (1990). *Worldwide Variation in Human Growth*, 2nd edn. Cambridge: Cambridge University Press.

Feldesman, M. R. (1992). Femur/stature ratio and estimates of stature in children. *American Journal of Physical Anthropology*, **87**, 447–459.

Fildes, V. A. (1995). The culture and biology of breastfeeding: An historical review of Western Europe. In *Breastfeeding: Biocultural Perspectives*, eds. P. Stuart-Macadam & K. A. Dettwyler, pp. 1–37. New York: Aldine de Gruyter.

Garofalo, R. P., & Goldman, A. S. (1998). Cytokines, chemokines, and colony-stimulating factors in human milk: The 1997 update. *Biology of the Neonate*, **74**, 134–142.

Hummert, J. R., & Van Gerven, D. P. (1983). Skeletal growth in a medieval population from Sudanese Nubia. *American Journal of Physical Anthropology*, **60**, 471–478.

Humphrey, L. T. (1998). Patterns of growth in the modern human skeleton. *American Journal of Physical Anthropology*, **105**, 57–72.

Humphrey, L. T. (2000). Growth studies of past populations: An overview and an example. In *Human Osteology in Archaeology and Forensic Science*, eds. S. Mays & M. Cox, pp. 25–38. Greenwich: Medical Media Ltd.

Johnston, F. E. (1962). Growth of the long bones of infants and young children at Indian Knoll. *American Journal of Physical Anthropology*, **20**, 249–254.

Katzenberg, M. A., Herring, D. A., & Saunders, S. R. (1996). Weaning and infant mortality: Evaluating the skeletal evidence. *Yearbook of Physical Anthropology*, **39**, 177–199.

King, S. E., & Ulijaszek, J. (1999). Invisible insults during growth and development: Contemporary theories and past populations. In *Human Growth in the Past: Studies from Bones and Teeth*, eds. R. Hoppa & C. Fitzgerald, pp. 161–182. Cambridge: Cambridge University Press.

Littleton, J. (1998). *Skeletons and Social Composition: Bahrain 300* BC – AD *250*, British Archaeological Reports International Series no. 703. Oxford: British Archaeological Reports.

Lovejoy, C. O., Russell, K. F., & Harrison, M. L. (1990). Long bone growth velocity in the Libben population. *American Journal of Human Biology*, **2**, 533–542.

Maresh, M. M. (1955). Linear growth of long bones of extremities from infancy through adolescence. *American Journal of Diseases of Childhood*, **89**, 725–742.

Maresh, M. M. (1970). Measurements from roentgenograms. In *Human Growth and Development*, ed. R. W. McCammon, pp. 157–200. Springfield: Charles C. Thomas.

Mays, S. A. (1999). Linear and appositional long bone growth in earlier human populations: A case study from Mediaeval England. In *Human Growth in the Past: Studies from Bones and Teeth*, eds. R. Hoppa & C. Fitzgerald, pp. 290–312. Cambridge: Cambridge University Press.

Menforth, R. P. (1985). Relative tibia long bone growth in the Libben and Bt-5 prehistoric skeletal populations. *American Journal of Physical Anthropology*, **68**, 247–262.

Merchant, V. L., & Ubelaker, D. H. (1977). Skeletal growth of the protohistoric Arikara. *American Journal of Physical Anthropology*, **46**, 61–72.

168   *L. T. Humphrey*

Molleson, T., & Cox, M. (1993). *The Spitalfields Project*, vol. 2, *The Middling Sort*, Council for British Archaeology Research Report no. 86. York: Council for British Archaeology.

Ryan, A. S. (1976). Long bone growth in a prehistoric population from San Cristóbal, New Mexico. *Michigan Discussions in Anthropology*, **2**, 55–75.

Saunders, S. R. (1992). Subadult skeletons and growth related studies. In *Skeletal Biology of Past Peoples: Research Methods*, eds. S. R. Saunders & M. A. Katzenberg, pp. 1–19. New York: Wiley-Liss.

Saunders, S. R., & Hoppa, R. D. (1993). Growth deficit in survivors and non-survivors: Biological mortality bias in subadult skeletal samples. *Yearbook of Physical Anthropology*, **36**, 127–151.

Saunders, S., Hoppa, R., & Southern, R. (1993). Diaphyseal growth in a nineteenth-century skeletal sample of subadults from St Thomas's Church, Belleville, Ontario. *International Journal of Osteoarcheology*, **3**, 265–281.

Scheuer, J. L., & Bowman, J. E. (1995). Correlation of documentary and skeletal evidence in the St. Brides's crypt population. In: *Grave Reflections: Portraying the Past through Cemetery Studies*, eds. S. R. Saunders & A. Herring, pp. 49–70. Toronto: Canadian Scholars' Press.

Smith, B. H. (1991). Standards of human tooth formation and dental age assessment. In *Advances in Dental Anthropology*, eds. M. A. Kelley & C. S. Larson, pp. 143–168. New York: Wiley-Liss.

Steegmann, A. T. (1985). 18th-century British military stature: Growth cessation, selective recruiting, selective trends, nutrition at birth, cold and occupation. *Human Biology*, **57**, 77–95.

Steyn, M., & Henneberg, M. (1996). Skeletal growth of children from the Iron Age site at K2 (South Africa). *American Journal of Physical Anthropology*, **100**, 389–396.

Stloukal, M., & Hanáková, H. (1978). Die Länge der Längsknochen altslawischer Bevölkerungen – unter besonderer Berücksichtigung von wachstumsfragen. *Homo*, **29**, 53–69.

Stuart-Macadam, P. (1995). Biocultural perspectives on breastfeeding. In *Breastfeeding: Biocultural Perspectives*, eds. P. Stuart-Macadam & K. A. Dettwyler, pp. 101–126. New York: Aldine de Gruyter.

Sundick, R. I. (1978). Human skeletal growth and age determination. *Homo*, **29**, 228–248.

Tanner, J. M., Whitehouse, R. H., & Takaishi, M. (1966). Standards from birth to maturity for height, weight, height velocity and weight velocity: British children 1965. *Archives of Disease in Childhood*, **41**, 454–471, 613–635.

Tayles, N. (1999). *The People*, vol. 5, *The Excavation of Khok Phanom Di, a Prehistoric Site in Central Thailand*, Reports of the Research Committee no. LXI. London: Society of Antiquaries.

Thompson, J. L., & Nelson, A. J. (2000). The place of Neandertals in the evolution of hominid patterns of growth and development. *Journal of Human Evolution*, **38**, 475–495.

Tompkins, R. L. (1996). Human population variability in relative dental development. *American Journal of Physical Anthropology*, **99**, 79–102.

Wall, C. E. (1991). Evidence of weaning stress and catch-up growth in the long bones of a central Californian Amerindian sample. *Annals of Human Biology*, **18**, 9–22.

Wood, J. W., Milner, G. R., Harpending, H. C., & Weiss, K. M. (1992). The osteological paradox: Problems of inferring prehistoric health from skeletal samples. *Current Anthropology*, **3**, 343–370.

7 Hominid growth and development: The modern context

J. L. THOMPSON
University of Nevada, Las Vegas

A. J. NELSON
University of Western Ontario

G. E. KROVITZ
Pennsylvania State University

> Arm length varies among humans, and some people must have longer arms
> than others. The average chimp has a longer arm than the average human,
> but this doesn't mean that a relatively long-armed human is genetically
> similar to apes. Normal variation *within* a population is a different biological
> phenomenon from differences in average values *between* populations.
>
> (Gould, 1981: 127)

Introduction

The above quotation, from Stephen Jay Gould, is a good starting-point for
any discussion of the evolution of human ontogeny. This is because we must
integrate an understanding of within- and between-population variation in addressing the questions of when and how the modern human pattern of growth
and development first appeared. However, before we can even attempt to address this question, we must first consider: *What is the modern human pattern
of growth and development? What aspects of growth and development make
modern humans unique?* Inevitably, what we attempt to do is to characterize
what constitutes the average pattern. Such an approach, to typify a species,
is a common practice in paleoanthropology. We use similarities, differences,
and unique features to distinguish between different species, and even genera.
Hence the practice of using a "type specimen," an individual fossil against
which all others are compared to determine their species attribution, in the

Patterns of Growth and Development in the Genus Homo, ed. J. L. Thompson, G. E. Krovitz and
A. J. Nelson. Published by Cambridge University Press. © Cambridge University Press 2003.

170

naming of a new fossil species. A similar approach is used in auxological pale-ontology (Bogin, this volume; Tillier, 2000), a term used to describe research into patterns of growth and development of fossil species. However, if we are to determine when the modern human pattern of growth and development first evolved, we must first establish what we actually mean by that phrase and, in fact, provide a characterization of the modern human pattern against which we compare those of other extant primate as well as extinct hominid species. In this way researchers attempt to define what may be equated to an "average" pattern for populations and species. This approach is used to determine if the differences in average values *between* populations reveal consistent patterns. Normal variation *within* populations is then used to compare *within* population variation to *between* population variation in order to assess the significance of observed interpopulation differences. How the resulting comparison is inter-preted can differ between researchers depending upon their theoretical approach (see below and Nelson *et al.,* Chapter 17, this volume).

To contextualize the modern pattern, most researchers commonly make com-parisons with other living primates, like *Pan troglodytes*, in order to determine what is unique about our genus and species in adult individuals that are at the end point of the ontogenetic trajectory. From the point of view of morphol-ogy, comparisons allow us to characterize modern humans as a large-bodied primate, trunchally erect, with skull situated atop a vertebral column that takes on a sigmoid curve. The modern human skull is large, rounded, with high forehead and a relatively small, broad but short, orthognathic face retracted underneath the skull, with a relatively small nose, and a mandible exhibiting a bony protrusion we commonly call a chin. The dentition is relatively small, with a small, incisiform canine and molars with low, rounded cusps. Our arms are short relative to our legs and the pelvis is short and broad with efficient transport of weight from the trunk to the lower appendicular skeleton, where a relatively long femoral neck and a valgus knee are well positioned for effi-cient bipedal locomotion; a locomotor strategy enhanced by a stable foot able to distribute weight in a pattern that is distinct from that seen in occasion-ally bipedal apes. Adult humans are therefore easily distinguishable from adult apes in their overall skeletal morphology, but to achieve this modern skele-ton, human growth has had to change in significant ways from that of other apes. For example, we continue to grow well in the second decade of life as opposed to apes whose growth terminates about the age of 12 years. The propor-tions and length of growth stages as well as the overall duration of the growth period are all unique to our species. Therefore establishing the ontogeny of particular defining morphological features of modern humans as well as the evolution of the modern life-history pattern, are important aims of auxological paleontology.

This summary will first define what is meant by the modern human pattern of ontogeny, distinguishing between the two processes of growth and development. The life-history stages, which make the overall modern human pattern of ontogeny distinguishable from that of non-human primates, are then outlined and discussed. This aspect of development is important, since these life stages, as defined here and by Bogin (this volume, and references therein), are tied to specific dental and skeletal maturational events and so can be used to evaluate when these stages appear in the fossil record.

In analyses of modern humans, as well as fossil samples, available specimens are treated as a single population and the immature individuals graded by their estimated age at death to provide a cross-sectional sample for analysis. This type of analysis must ignore, at least in the first instance, the variables of individual and geographical variation and sexual dimorphism. However, many morphological characteristics are ontogenetically variable, some features appearing early or late in the growth of an individual. Thus it is pertinent to the discussion of the modern human pattern of growth and development to outline potential causes of within-species variability as well as how this variation may be interpreted by individual researchers. A review of the papers in this section follows, with particular attention to how these papers further our knowledge of modern patterns of growth and development.

Modern human pattern of growth and development

Definition of pattern of growth and development

What is meant by the modern human pattern of growth and development? Modern human *growth* refers to changes in "total body size, body proportions, length and size of segments" (Roche, 1992: 18) and also includes dimensions and proportions of the bones of the skull and face. The length, breadth, proportions, shape, and/or robusticity of bones relative to their adult form can be used to assess the rate of growth and the amount of growth remaining (e.g., Humphrey, this volume; Nelson & Thompson, 1999, 2002, in press). Therefore, we have several ways to assess the modern human pattern of growth in living and extinct hominids, even if we restrict ourselves to the growth of the bony elements of the body.

Development involves changing from an immature to a mature stage, and can be marked in skeletal samples by skeletal and dental maturation, including ages at peak height velocity (Bogin, 1999; Roche, 1992). The fact that growth events occur in a certain order allows assessment of how "developed" or mature an individual was at the time of death. Dentally, the number of teeth formed and erupted in the jaws, and their stage of calcification, can be examined in order to estimate age at death, using appropriate standards (see Liversidge, this volume). Skeletally, the degree of maturation can be assessed by examining how ossified a bone is, or if the ends of bones, the epiphyses, have or have not yet fused to the shaft (e.g., Saunders, 1992). This information, considered as a whole, tells us how close an individual is to the adult form, and can be used to tell us something about the nature of development of a particular species (Thompson, 1998). A new way of examining development builds on the work of D'Arcy Thompson, by using three-dimensional geometric morphometric techniques (e.g., Euclidean distance matrix analysis, Procrustes analysis, and thin plate splines) to characterize shape change in larger regions of the craniofacial skeleton (see Krovitz, 2000, this volume; Strand Viðarsdóttir & O'Higgins, this volume).

The *pattern of growth* relates to the relative sequence, or order, of certain events over the course of one's life. Generally speaking, the development of the modern human pattern of growth and development has involved the addition and extension of developmental stages and the delay of somatic growth, relative to the ape pattern (Schultz, 1960; Watts, 1986). The characterization of these phases helps to clarify the unique aspects of the human growth program, highlighting how that program differs from that of non-human primates. Furthermore, the definitions of these stages combine aspects of morphology that are visible in the skeleton, with aspects of behavior that are, for the most part, invisible in the fossil record.

Life stages

Old World monkeys and apes have only three basic stages in their growth and development: *Infancy, juvenile*, and *adulthood* (Bogin & Smith, 1996; Smith & Tompkins, 1995). According to Bogin (1999, this volume) the modern human pattern demonstrates five identifiable stages: *Infancy, childhood, juvenile, adolescent*, and *adult* stages (Bogin & Smith, 1996; Smith & Tompkins, 1995).

Infancy has been defined as the period from immediately after birth (0–28 days is referred to as the neonatal stage in Bogin, 1999, this volume) to the time of weaning (Bogin & Smith, 1996). The stage of infancy ends with weaning. According to Lee *et al.* (1991), while there is an isometric relationship between birth weight and weaning weight, the timing of this threshold event varies according to a number of factors, including food availability and quality. This relationship between birth weight and weaning was confirmed by Dettwyler (1995), who found that, in humans, weaning (complete cessation of breast-feeding) takes place at around 3 years of age in well-nourished populations, but later, between 3 and 4 years of age, in less well-nourished ones. This is about the same time that weaning takes place in large-bodied apes. In general, first molar eruption is strongly correlated with the time of weaning in non-human primates (see Bogin, 1999, this volume). However, the fact that humans typically erupt their first molar about 6 years of age, while the first molar of non-human primates (e.g., chimpanzees) erupts at about 3 years is particularly significant. This means that, in humans, this relationship between M1 eruption and time of weaning has become decoupled (see discussion of childhood below).

Human infancy is also distinguished from that of apes by the very high rates of brain growth that continue from the fetal period into the first year of life (Dienske, 1986), leading to a condition referred to as secondary altriciality. Portmann (1941) demonstrated the difference between altricial and precocial mammals. Altricial mammals usually have short gestation periods and produce helpless, poorly developed young that require prolonged care in a protected area like a nest or den. In contrast, precocial mammals have a relatively long gestation period and their young are born in a more advanced state of development with no nest or den required. Humans are unusual in that like precocial mammals, they have a long gestation period (Martin & MacLarnon, 1990) but like other altricial mammals their young are relatively helpless, hence the term secondary altriciality. The state of secondary altriciality, which is one of the unique features of modern human infancy, can be estimated based on the ratio of adult/newborn brain weight and so its evolutionary history can be assessed (Rosenberg, 1992; Smith & Tompkins, 1995).

Childhood follows *infancy*, and is defined by Bogin (1999) as the time when the individual has been weaned, but remains dependent on adults for food and protection. Among modern primates this period is unique to modern humans, allowing for a delay in sexual development and an opportunity for further brain and somatic growth, but at a rate much reduced from that seen in *infancy*. Mann *et al.* (1996) have argued that the childhood developmental stage is marked by a period of stasis in dental eruption that occurs between the eruption of the lateral incisors and the second molars in modern humans (Hellman, 1943) but not in apes (Mann *et al.*, 1996).

The *juvenile* stage, occurring between infancy and puberty, is seen in modern human and non-human primates and represents a delay in sexual maturation and growth (Bogin, 1999). This stage begins with the eruption of the first molar and the cessation of brain growth and ends with puberty (Bogin & Smith, 1996). While apes and humans both share a juvenile phase, this stage is relatively shorter in humans due to the insertion of the extra stage of childhood into the growth program.

The *adolescent* phase is marked by the onset of puberty and ceases with the termination of growth. It is characterized by the attainment of full sexual maturity, the development of secondary sexual characteristics, a growth spurt, and the adoption of adult behaviors (Bogin, 1993, 1994a, 1994b, 1995, 1999, this volume; Bogin & Smith, 1996; Tanner, 1962, 1979, 1990). Puberty in African apes begins at the age of 7–10 years of age with the onset of menarche and the appearance of other sexual characteristics in females and the acceleration of testes growth in males (Leigh & Shea, 1996). Puberty occurs in human girls and boys about the age 10 and 12 years, respectively. During adolescence, humans undergo a period of growth called the adolescent growth spurt, which is apparently unique to humans in that it involves a rapid acceleration in the rate of growth soon after the eruption of the second molar, resulting in an increase in linear dimensions of all parts of the skeleton including the face. While both human and non-human primates demonstrate some sort of acceleration in growth velocity during this time (Laird, 1967; Leigh, 1996), this "long absolute delay in the initiation of the human growth spurt" (Leigh 1996: 455) as well as the magnitude and late timing of the human growth spurt (and thus the nature of adolescence) appear to be unique to modern humans (Bogin, this volume). However, this is an area of some controversy (see Leigh (1996) for references, and see Krovitz *et al.*, this volume; also Nelson *et al.*, Chapter 17, this volume for further discussion).

Adulthood in all primates begins with the cessation of growth and the attainment of full sexual maturity, and ends with the death of the individual (Bogin, 1997). Therefore, modern human life stages are unique compared to that of apes in exhibiting: (1) secondary altriciality; (2) the childhood stage of development; and (3) the adolescent stage with its accompanying growth spurt.

Life history

Additional life-history events, like menarche, age at first reproduction, and longevity, are also correlated with aspects of the skeleton and dentition (i.e., dental eruption, brain size, and/or body size) and give insights into the comparative life histories of modern apes (including humans) and early hominids. Menarche is strongly correlated with the eruption of the M2 (Smith, 1992), while age of first reproduction is strongly correlated with the value of life expectation at birth and is highly correlated with body size (Harvey & Zammuto, 1985). Longevity is strongly correlated with brain weight and body weight (Cutler, 1980; Hofman, 1984; Martin, 1981, 1983; Sacher, 1975). Age of menarche and age at first reproduction are broadly similar in apes (including humans) (Richard, 1985), but longevity estimates of modern humans far exceed those of apes (Thompson, 1998). Following an ape rate of growth and development, the australopithecines and early *Homo* likely lived no longer than about

30 years of age (Thompson, 1986). Thus another milestone of the modern human pattern – the extension of lifespan – evolved after the appearance of the genus *Homo* (Bogin, 1999).

Causes of within-species variation

It is relatively easy to characterize the average modern human pattern of growth and development, and this helps us establish major differences in growth patterns seen between genera. However, it is much more difficult to characterize the within-population, or within-species, variability. Yet, our documentation and treatment of within-species variation is essential to our interpretation of the fossil record.

Within-species variation includes individual variation, sexual dimorphism, geographic variation, and microevolutionary change over time. Variation is always present within species with no two individuals being exactly alike. Some of this variation results from environmental influences during growth and development and some from an individual's genetic program. It is the latter that provides the crucial "raw material" for long-term evolution. Major changes over long periods of time (i.e., speciation) depend on variation, and variation is responsible for both the evolutionary divergence of populations and the increased diversity within a population. Variation ultimately comes from mutations, and is reshuffled by sexual recombination, migration, and genetic drift, while natural selection edits and shapes the available mutations into the non-random patterns that we see in the world around us (Dawkins, 1986; Mayr, 2001).

Individual differences in morphology, for example, can arise through differential rates of growth of certain parts of the body relative to others, but since variation is affected by the interaction of functional complexes, a small change in one area can result in small-to-significant changes in another, contiguous area. The evolution of morphological characters that differentiate species from each other is achieved through alterations in the inherited pattern of growth and development (see McBratney-Owen & Lieberman, this volume). Since functional systems can evolve at different rates, fossils sampled from different time periods may appear largely similar, except with some differences in certain areas of the skeleton. Conversely, genetic changes influencing the timing of gene expression during growth and development may produce substantial changes in the phenotype. Thus, individual differences may occur because of subtle variations in the rate of development of certain elements of the skeleton, but this same process can also produce species- or genus-level differences.

An additional source of variation that needs to be considered is that resulting from allometric growth. Allometry describes the changes in shape that accompany changes in size. The changes in shape come about because of

differential rates of growth in body parts. Thus, some within-species variation occurs because of changes of size resulting in an alteration of relationships
between different parts of an organism. Differences between species can involve the extension or truncation of growth of any or all parts of the organism.
Therefore allometric growth, involving both changes in size and shape, plays an
important part in the development of individual differences but also in the evolution of morphological characters and must be accounted for in the assessment
of within-species variation (see Leigh, 2001 and references therein).

Assessments of the juvenile fossil record typically use modern human and/or
ape dental and skeletal development to provide a chronological baseline and as
models or a framework against which to compare various lines of ontogenetic
information. Variation among the modern forms is usually used as a guide for
the interpretation of the level of distinction (genus, species, subspecies) between
fossil samples, but the perspective of the researcher can be influenced by their
interpretation of that variation. For example, some researchers who seek central
tendencies in the data may interpret consistent patterns of differences between
samples as species level distinctions (e.g., Krovitz, this volume; Miller, 1991;
Tattersall, 1986, 1992; Thompson & Nelson, 2000) while other researchers
that emphasize outliers might interpret overlap in patterns between living humans and recent hominids as merely constituting within-species variation (e.g.,
Clegg & Aiello, 1999; see also Kondo & Ishida, this volume). This difference
in perspective has implications both for interpreting the number of species in
the genus *Homo* (see Krovitz *et al.*, this volume, for discussion) and for the
debate regarding origin of modern humans. Considering the latter debate, some
researchers in paleoanthropology see recent and ancient anatomically modern
humans as a separate species from other archaic hominids (Stringer, 1994;
Stringer & Andrews, 1988; Stringer *et al.*, 1984), and would retain sole use of
the taxon *Homo sapiens* to highlight those species differences. Others support
the view that anatomically modern humans are simply one of several subspecies
of *Homo sapiens*, and so further subdivide the taxon (e.g., Wolpoff, 1980, 1996;
Wolpoff & Caspari, 1997; Wolpoff *et al.*, 1984). Thus how one interprets within-
and between-species variation is obviously of great importance to the study of
human evolution and the evolution of the modern human pattern of growth and
development (see also Nelson *et al.*, Chapter 17, this volume).

Evidence for modern human growth as presented in this volume

This section will characterize the modern human pattern of growth and development, explicitly recognizing the nature of the variations in that pattern, while
also summarizing the contributions of the authors in this section.

The study of the pattern of growth and development of extant modern humans has had a long history (Eveleth & Tanner, 1991; Roche, 1992; Tanner, 1960, 1962) and continues today as a vibrant and growing field of study (e.g., Bogin, 1999, this volume, and references therein) as reviewed and discussed in papers in this volume. Bogin (this volume) outlines the major life-history stages of modern humans and how they are used in auxology. He follows this discussion with the major contrasts between the stages of growth seen in humans and apes in some detail, and ends with a detailed presentation of his interpretation of the adaptive value of the delayed growth period and when it may have evolved. The life-history stages outlined by Bogin are gradually becoming "the standard scheme" used within auxological paleontology. The benefit of such clearly defined stages, beyond distinguishing our species from that of other apes, is that the stages are tied to biological events that can be detected using the dentition and/or skeleton (see discussion above).

Bogin's chapter is unique in this section since he presents an explanation of *why* the extra growth stages of childhood and adolescence evolved, drawing on a detailed review of the literature (Bogin, 1999). It is likely that these stages of growth appeared within the evolution of the genus *Homo*. For example, Bogin proposes that childhood would have begun to appear at the time of *H. habilis* based on moderate brain size increase in this species. By maintaining an ape-like length of weaning of about 3 years, decreasing the interbirth interval relative to that seen in apes, but maintaining strong parental care, *H. habilis* females would increase their reproductive fitness and this pattern would be selected for over time. In contrast, the adolescent stage, with its characteristic growth spurt, would have evolved to increase the survival, and ultimately reproductive fitness, of the youthful individuals, as they could practice adult skills without adult responsibilities. These are plausible arguments given what we know about the increase in cultural complexity, the committed shift to a hunting–gathering subsistence mode at least by *H. erectus* times, as well as the change in environmental pressures acting on these populations as they migrated within and beyond Africa (see Krovitz *et al.*, this volume, for further discussion).

The chapter by McBratney-Owen & Lieberman (this volume) provides an excellent example of how to isolate morphological aspects of the modern human ontogenetic pattern. The authors concentrate on the modern human feature of facial retraction, which they define as an anteroposteriorly short and superoinferiorly vertical face that is tucked almost entirely underneath the anterior cranial fossa. Previous research demonstrates that while variation in facial position is influenced by the same structural variables (cranial base angle, length and width of the cranial base and fossa, and facial length) in extant primates, archaic *Homo*, and anatomically modern humans (AMHS), facial retraction is found only in AMHS. The authors assert that this a unique feature, supporting a distinct phylogenetic species designation for modern humans

(*H. sapiens*) (also see Lieberman *et al.*, 2002). This paper explores the ontogenetic origin of facial projection, and the degree to which it interacts with the structural variables mentioned above during growth and development. Three-dimensional landmark coordinate data are collected from ontogenetic samples of *H. sapiens* and *P. troglodytes,* and analyzed using Euclidean distance matrix analysis.

The variables that McBratney-Owen & Lieberman consider do contribute to different patterns of facial position, and exhibit different growth patterns in humans and chimpanzees. The morphological pattern of facial retraction in *H. sapiens* likely occurs via developmental shifts that take place prenatally or very early in postnatal ontogeny (also see Krovitz, 2000, this volume; Lieberman *et al.*, 2002). They also predict that the accelerated growth of the cranial base seen in humans will not be found in archaic *Homo*. Instead, the degree and structure of facial projection, and hence the underlying ontogenetic pattern, is more similar between archaic *Homo* and *P. troglodytes*. The authors note that the relationship between facial position and cranial base proportions may well be related to brain shape. Thus, facial retraction may be a consequence of the relative expansion of the temporal lobes, in combination with a reduction in facial size, in the recent evolution of *H. sapiens*. This hypothesis needs further investigation from an ontogenetic perspective. This comparison between living taxa contributes to our understanding of cranial growth differences between *H. sapiens* and archaic *Homo*. It does so by establishing key elements of ontogeny leading to structural differences between *H. sapiens,* with its characteristic facial retraction, and species with facial projection, such as Neandertals. In other words, contrasts between the two extant species in facial growth may shed light on growth processes in recent members of the genus *Homo*. These results also highlight important differences in the evolution of complex morphological structures within the genus *Homo*. Clearly, the growing face is a complicated, multifaceted anatomical complex in which individual components have evolved in a mosaic pattern over the course of human evolution.

The first section of this volume also contains three chapters that explore variability in modern human growth patterns in the dentition, face, and postcranial skeleton (chapters by Liversidge, Strand Viðarsdóttir & O'Higgins, and Humphrey, respectively). The chapter by Liversidge (this volume) is of importance to paleoauxological studies since it provides a detailed review of the literature pertaining to variation in modern human dental development. She begins with a discussion of the various methods used to quantify dental formation, followed by a discussion of methods used to document dental growth data. She highlights the variability of the frequently used methods, the way dental age and dental development stages are recorded, how age ranges are defined, and how dental eruption is defined and quantified. Liversidge documents differences in quantification, collection, documentation, and presentation of the data between studies, as well as differences in terminology used, which make the assessment of patterns within and between populations a very difficult undertaking. Her examination of dental eruption studies reveals little evidence for population differences in the eruption of deciduous teeth. While interpopulation variation in the eruption of permanent teeth is greater than in deciduous teeth, no clear pattern of population differences is apparent. Her review indicates that in a tooth-by-tooth comparison between human populations, there is some variation. However, there is no consistent pattern of early vs. late eruption of teeth across populations. In other words, a population that erupts their M1 earlier than other populations may have other teeth that erupt relatively late. This means that no one population studied had a consistent pattern of early-, average-, or late-erupting teeth. Liversidge notes, however, that several studies comparing eruption of the whole dentition have found population differences of between 3 to 6 months in the mandibular dentition, particularly in early-emerging teeth. This observation is of importance when interpreting the age of individuals from the later fossil record based on dental emergence data alone.

In a review of tooth formation research, Liversidge (this volume) discusses some of the most familiar studies of modern human dental formation, and finds that differences in methodology and sampling may introduce some of the observed variability between results, as was seen in her review of dental eruption. She reviews crown and root formation data from cross-sectional and longitudinal studies, compares results derived from a variety of methodologies (e.g., timing of development, maturity scales, scoring methods, etc.), and concludes that population differences in both dental eruption and tooth formation are small (less than 1 standard deviation). Her summary of modern human dental studies emphasizes several points: (1) clinical eruption varies more than dental formation; (2) dental formation varies less in deciduous teeth than in permanent teeth; (3) later-forming permanent teeth (e.g., M3) vary more than earlier-forming teeth (e.g., M1); and (4) population differences in formation and eruption are small in early childhood, slightly more variable during the eruption of the mixed dentition, and are most variable in later forming teeth. While Liversidge calls for further and more standardized research, she concludes that age estimates based on the developing dentition in humans should not differ substantially between populations. This finding is of particular importance for paleoauxological studies, which depend heavily upon modern dental development standards to estimate age at death of fossil specimens, emphasizing the importance of using crown and root formation data, whenever possible.

In contrast, the paper by Strand Viðarsdóttir & O'Higgins (this volume) demonstrates that modern human facial morphology is ontogenetically variable, with clear morphological differences arising between populations, even at early ages. This work is important in the context of this volume for it warns against using single populations, or even averaging populations, to represent "the modern human pattern" of facial ontogeny. While there are plenty of studies that document population differences in modern human craniofacial form, this study differs by concentrating on the ontogenetic basis of those differences (see also Krovitz, 2000). The use of three-dimensional geometric morphometrics allows for the isolation of variables contributing to overall variation (see above) including allometry and sexual dimorphism. The authors focus on the ontogeny of facial shape differences in Caucasians, African-Americans, and Native Americans, since the facial morphology of adults of these populations is readily differentiated using forensic techniques. Analyses concentrate on the relationships between age, facial size, and facial shape. The results indicate that differences in facial shape, both in population-specific morphologies as well as ontogenetic allometries and shape trajectories, arise early in development, possibly even prenatally (similar to Krovitz, 2000). The fact that these population-specific adult morphs may have distinguishable underlying ontogenetic patterns adds value to this research since it is pertinent in a modern as well as fossil context. They also discuss environmental influences on morphology that include external selective pressures (such as geography), as well as functional matrices of the surrounding skull that influence the growth and development of the facial architecture.

The last chapter in this section (Humphrey) examines linear growth of the femoral diaphysis to document between-population variation. Available data from archaeological samples representing populations from North Africa and Western Asia, Asia, Europe (or European origin), and North America (Native American) are compared against a reference sample using dental age as a proxy for chronological age. The reference sample (Maresh, 1955, 1970) is derived from a population of children of European descent and models the femoral growth process of a healthy and well-nourished population. The purpose of this study is to visualize fluctuations in growth tempo relative to this reference sample, which may represent periods of growth faltering and catch-up growth.

All of the archaeological samples display varying degrees of apparent growth retardation, or growth faltering, at different ages of onset and duration, but growth trajectories of all samples fall below that of the reference sample for late infancy and most of childhood. This indicates that a smaller percentage of adult size is attained during these stages for all archaeological populations and may be the result of growth faltering during weaning. The early onset of growth faltering may indicate that weaning is initiated earlier in some populations relative to others. Humphrey concludes that the introduction of foods into the diet to supplement breast-milk may explain growth faltering, particularly in samples that demonstrate early onset of growth retardation (relative to the comparative sample), due to possible contamination of food or water, as well as other factors.

In other samples (Indian Knoll, Libben, and Wadi Halfa) growth faltering occurs after 18 months, so is not likely due to the introduction of supplementary foods, but perhaps is due to weaning (total cessation of breast-feeding: see Bogin, 1999). A period of recovery follows this initial period of growth faltering. Both the period of faltering and recovery differ in degree and duration between samples, and relative to the reference sample. Humphrey rightly cautions that there are numerous factors that can affect the variation in the pattern of growth and development in living, as well as past populations, including environmental (including cultural practices relating to breast-feeding, supplementary foods, and food allocation and/or contamination among others) and genetic causes, as well as methodological variability in sample size, sample recovery, and time depth, as well as accuracy of age-at-death estimations. Humphrey's results indicate a common pattern of growth in most of her archaeological samples relative to her comparative sample, with a decline of growth attainment during infancy and childhood. The difference in age of onset, degree, and length of this decline varies between populations, but not in a systematic way that would indicate clear geographic or temporal patterns. Humphrey concludes that this result needs to be interpreted with caution because there are not enough data yet available to thoroughly document regional and/or temporal variation in this aspect of skeletal growth. Despite this caveat, this chapter summarizes data from the available archaeological record and presents a strong foundation for further research on this topic. With respect to this volume, her work indicates that there appears to be a common pattern of growth retardation in archaeological modern populations, making it likely that paleontological populations will also exhibit the same pattern and so has significance for studies attempting the interpretation of femoral growth patterns in fossil children.

Conclusions

This summary chapter synthesizes the papers in this section and some of the current literature on the topic of modern human patterns of growth and development. Bogin's work (this volume) shows that the modern human life-history schedule, as contrasted against that of living primates, includes the feature of secondary altriciality, the insertion of the childhood stage of growth, and an exaggerated skeletal growth spurt during adolescence, thus resulting in relatively delayed somatic and sexual development. According to Bogin, this pattern results in a shortening of the infancy period (relative to apes), providing the opportunity for a shorter interbirth interval, and more offspring to be produced over an individual's lifespan (which is also elongated compared with the lifespan of apes). These biological features, as outlined above, coordinate with particular dental and skeletal developmental events and can therefore be used to interpret the evolution of the modern pattern of ontogeny (see discussions in Krovitz *et al.*, this volume, and Nelson *et al.,* Chapter 17, this volume). Since auxological paleontology depends upon the fossil record, this link between skeletal maturation and life-history biology is paramount.

Examination of morphological features from an ontological perspective is central to the interpretation of phylogenetic relationships between species. The adult modern human morphological pattern, as characterized at the beginning of this chapter, is the endpoint of various growth trajectories affecting each bone as well as integrated functional complexes of bones (e.g. skull and face). McBratney-Owen & Lieberman's study is an important example of this approach, as they focus on the ontogeny of facial retraction, an aspect of modern morphology that appears to be unique to anatomically modern humans. Their research isolates linear measures that contribute to facial projection over the growth period, and test several structural hypotheses about the relationship between the face and cranial base. By comparing modern humans with another living species, *P. troglodytes*, they establish that facial retraction is achieved in a particular ontogenetic fashion in modern humans, while facial projection is achieved in *P. troglodytes* through another process. These results are suggested to be similar to the growth of facial projection in non-modern fossil hominids. While investigation of the modern pattern of growth and development is the purpose of this section, these results make an equally important contribution to the investigation of facial growth in non-human primates. Establishing these important contrasts in ontogenetic processes allows a deeper understanding of the similarities and differences in facial growth in extinct fossil hominids.

The remaining papers in this section explicitly address the normal variation between modern human populations seen in the modern human dentition (Liversidge, this volume), face (Strand Viðarsdóttir & O'Higgins, this volume), and postcranial skeleton (Humphrey, this volume). These papers measure variation within and between modern human populations, and thus provide a comparative context for variation seen in fossil species. The chapter on dental variation (Liversidge, this volume) provides an exhaustive review of the literature, and supports previous research showing that age estimation in future paleoauxological studies should preferentially rely on dental formation and calcification instead of dental eruption in modern humans. While formation and eruption are more variable in permanent teeth than in the deciduous dentition, permanent teeth still provide a reasonably accurate estimation of age since they vary comparatively little between modern human populations. The paper on facial variation (Strand Viðarsdóttir & O'Higgins, this volume) shows that population-specific facial features found fairly predictably in certain adult populations are present in young individuals, and thus are set early in life. Because of the population variation in facial morphology, paleoauxological comparisons should use either multiple samples of modern humans, or a sample from a single population with the caveat that any contrasts/similarities do not encompass the modern range of variability in this feature. Finally, variation in femoral growth in modern human archaeological samples (Humphrey, this volume) gives us fresh insight into the complications of research in this area. Many of the modern human ontogenetic samples frequently used for comparative purposes are derived from well-fed populations (primarily of European origin), which are not ideal as a basis for comparison with the environmentally challenged populations of fossil hominids. Collectively, the chapters in this section outline aspects of growth and development in modern humans, and set the stage for the following sections that discuss the evolutionary history of these modern human patterns of growth and development in the fossil record.

References

Bogin, B. (1993). Why must I be a teenager at all? *New Scientist,* **137**, 34–38.
Bogin, B. (1994a). Human learning: Evolution of anthropological perspectives. In *The International Encyclopedia of Education,* vol. 5, 2nd edn, eds. T. Husén & T. N. Postlethwaite, pp. 2681–2685. New York: Pergamon Press.
Bogin, B. (1994b). Adolescence in evolutionary perspective. *Acta Paediatrica* Suppl. **406**, 29–35.
Bogin, B. (1995). Growth and development: Recent evolutionary and biocultural research. In *Biological Anthropology: The State of the Science,* eds. N. T. Boaz & L. D. Wolfe, pp. 49–70. Bend, Oregon: International Institute for Human Evolutionary Research.
Bogin, B. (1997). Evolutionary hypotheses for human childhood. *Yearbook of Physical Anthropology,* **40**, 63–89.
Bogin, B. (1999). *Patterns of Human Growth,* 2nd edn. Cambridge: Cambridge University Press.
Bogin, B., & Smith, B. H. (1996). Evolution of the human life cycle. *American Journal of Human Biology,* **8**, 703–716.
Clegg, M., & Aiello, L. C. (1999). A comparison of the Nariokotome *Homo erectus* with juveniles from a modern human population. *American Journal of Physical Anthropology,* **110**, 81–93.
Cutler, R. G. (1980). Evolution of human longevity. In *Aging, Cancer and Cell Membranes,* eds. C. Borek, C. Fenoglio, & D. King, pp. 43–79. New York: Thieme–Stratton.
Dawkins, R. (1986). *The Blind Watchmaker.* New York: W. W. Norton.
Dettwyler, K. A. (1995). A time to wean: The hominid blueprint for the natural age of weaning in modern human populations. In: *Breastfeeding: Biocultural Perspectives,* eds. P. Stuart-Macadam & K. A. Detwyller, pp. 39–73. New York: Aldine de Gruyter.

Dienske, H. (1986). A comparative approach to the question of why human infants develop so slowly. In *Primate Ontogeny, Cognition and Social Behaviour,* eds. J. G. Else & P. C. Lee, pp. 145–154. Cambridge: Cambridge University Press.

Eveleth, P. B., & Tanner, J. M. (1991). *Worldwide Variation in Human Growth*, 2nd edn. Cambridge: Cambridge University Press.

Gould, S. J. (1981). *The Mismeasure of Man.* New York: W. W. Norton.

Harvey, P. H., & Zammuto, R. M. (1985). Patterns of mortality and age at first reproduction in natural populations of animals. *Nature*, **315**, 319–320.

Hellman, M. (1943). The phase of development concerned with erupting the permanent teeth. *American Journal of Orthodontics,* **29**, 507–526.

Hofman, M. A. (1984). On the presumed coevolution of brain size and longevity in hominids. *Journal of Human Evolution*, **13**, 371–376.

Krovitz, G. E. (2000). Three-dimensional comparisons of craniofacial morphology and growth patterns in Neandertals and modern humans. PhD thesis, Johns Hopkins University, Baltimore.

Laird, A. K. (1967). Evolution of the human growth curve. *Growth,* **31**, 345–355.

Lee, P. C., Majluf, P., & Gordon, I. J. (1991). Growth, weaning and maternal investment from a comparative perspective. *Journal of Zoology, London*, **225**, 99–114.

Leigh, S. R. (1996). Evolution of human growth spurts. *American Journal of Physical Anthropology,* **101**, 455–474.

Leigh, S. R. (2001). Evolution of human growth. *Evolutionary Anthropology*, **10**, 233–236.

Leigh, S. R., & Shea, B. T. (1996). Ontogeny of body size variation in African apes. *American Journal of Physical Anthropology,* **99**, 43–65.

Lieberman, D. E., McBratney, B. M., & Krovitz, G. E. (2002). The evolution and development of cranial form in *Homo sapiens. Proceedings of the National Academy of Sciences of the USA*, **99**, 1134–1139.

Mann, A., Lampl, M., & Monge, J. M. (1996). The evolution of childhood: Dental evidence for the appearance of human maturation patterns. *American Journal Physical Anthropology* (Suppl.), **21**, 156.

Maresh, M. M. (1955). Linear growth of long bones of extremities from infancy through adolescence. *American Journal of the Diseases of Childhood*, **89**, 725–742.

Maresh, M. M. (1970). Measurements from roentgenograms. In *Human Growth and Development*, ed., R. W. McCammon, pp. 157–200. Springfield: Charles C. Thomas.

Martin, R. D. (1981). Relative brain size and basal metabolic rate in terrestrial vertebrates. *Nature*, **293**, 57–60.

Martin, R. D. (1983). *Human Brain Evolution in an Ecological Context*, 52nd James Arthur Lecture on the Evolution of the Human Brain. New York: American Museum of Natural History.

Martin, R. D., & MacLarnon, A. M. (1990). Reproductive patterns in primates and other mammals: The dichotomy between altricial and precocial offspring. In *Primate Life History and Evolution,* ed. C. J. DeRousseau, pp. 47–79. New York: Wiley-Liss.

Mayr, E. (2001). *What Evolution Is.* New York: Basic Books.

Miller, J. A. (1991). Does brain size variability provide evidence of multiple species in *Homo habilis? American Journal of Physical Anthropology*, **84**, 385–398.

Nelson, A. J., & Thompson, J. L. (in press). Le Moustier 1 and the interpretation of stages in Neandertal growth and development. In *The Neandertal Adolescent Le Moustier 1: New Aspects, New Results*, ed. H. Ullrich. Berlin: Staatliche Museen.

Nelson, A. J., & Thompson, J. L. (1999). Growth and development in Neandertals and other fossil hominids: Implications for hominid phylogeny and the evolution of hominid ontogeny. In *Growth in the Past: Studies from Bones and Teeth*, eds. R. D. Hoppa & C. M. FitzGerald, pp. 88–110. Cambridge: Cambridge University Press.

Portmann, A. (1941) Die Tragzeiten der Primaten und die Dauer der Schwangerschaft beim Menschen: Ein Problem der vergleichenden Biologie. *Revue Suisse de Zoologie*, **48**, 511–518.

Rak, Y., Ginzburg, A., & Geffen, E. (2002). Does *Homo neanderthalensis* play a role in modern human ancestry? The mandibular evidence. *American Journal of Physical Anthropology*, **119**, 199–204.

Richard, A. (1985). *Primates in Nature*. New York: W. H. Freeman.

Roche, A. F. (1992). *Growth, Maturation and Body Composition: The Fels Longitudinal Study 1929–1991*. Cambridge: Cambridge University Press.

Rosenberg, K. R. (1992). The evolution of modern human childbirth. *Yearbook of Physical Anthropology*, **35**, 89–124.

Sacher, G. A. (1975). Maturation and longevity in relation to cranial capacity. In *Primate Functional Morphology and Evolution*, ed. R. H. Tuttle, pp. 413–441. Paris: Mouton.

Saunders, S. R. (1992). Subadult skeletons and growth related studies. In *Skeletal Biology of Past Peoples: Research Methods*, eds. S. R. Saunders & M. A. Katzenberg, pp. 1–20. New York: Wiley-Liss.

Schultz, A. H. (1960). Age changes in primates and their modification in man. In *Human Growth*, ed. J. M. Tanner, pp. 1–20. Oxford: Pergamon Press.

Smith, B. H. (1993). The physiological age of KNM-WT 15000. In *The Nariokotome* Homo erectus *Skeleton*, eds. A. Walker & R. Leakey, pp. 196–220. Cambridge: Harvard University Press.

Smith, B. H. (1992). Life history and the evolution of human maturation. *Evolutionary Anthropology*, **1**, 134–142.

Smith, B. H., & Tompkins, R. L. (1995). Toward a life history of the Hominidae. *Annual Review of Anthropology*, **24**, 257–279.

Stringer, C. B. (1994). Out of Africa: A personal history. In *Origins of Anatomically Modern Humans*, eds. M. H. Nitecki & D. V. Nitecki, pp. 149–172. New York: Plenum Press.

Stringer, C. B., & Andrews, P. (1988). Genetic and fossil evidence for the origin of modern humans. *Science*, **239**, 1263–1268.

Stringer, C. B., Hublin, J.-J., & Vandermeersch, B. (1984). The origin of anatomically modern humans in Western Europe. In *The Origins of Modern Humans*, eds. F. H. Smith & F. Spencer, pp. 51–135. New York: Alan R. Liss.

Tanner, J. M. (1960). *Human Growth*. New York: Pergamon Press.

Tanner, J. M. (1962). *Growth at Adolescence*, 2nd edn. Oxford: Blackwell Scientific Publications.

Tanner, J. M. (1979). A concise history of growth studies from Buffon to Boas. In *Human Growth*, vol. 3, *Neurobiology and Nutrition*, eds. F. Falkner & J. M. Tanner, pp. 515–593. New York: Plenum Press.

Tanner, J. M. (1990). *Foetus into Man*. Cambridge: Harvard University Press. Cambridge.

Tattersall, I. (1986). Species recognition in human paleontology. *Journal of Human Evolution*, **15**, 165–175.

Tattersall, I. (1992). Species concepts and species identification in human evolution. *Journal of Human Evolution*, **22**, 341–349.

Thompson, J. L. (1986). A paleodemographic analysis of the East African Plio-Pleistocene Hominidae. MA thesis, Trent University, Peterborough, Canada.

Thompson, J. L. (1998). Neanderthal growth and development. In *The Cambridge Encyclopedia of Human Growth and Development,* eds. S. J. Ulijaszek, F. E. Johnston, & M. A. Preece, pp. 106–107. Cambridge: Cambridge University Press.

Thompson, J. L., & Nelson, A. J. (2000). The place of Neandertals in the evolution of hominid patterns of growth and development. *Journal of Human Evolution*, **38**, 475–495.

Tillier, A-m. (2000). Palaeoauxology applied to Neanderthals: Similarities and contrasts between Neanderthal and modern children. In *Children in the Past: Paleoauxology, Demographic Anomalies, Taphonomy and Mortuary Practices*, ed. A-m. Tillier, *Anthropologie (Brno)*, **38**, 109–120.

Watts, E. S. (1986). The evolution of the human growth curve. In *Human Growth*, vol. 1, 2nd edn., eds. F. Faulkner & J. M. Tanner, pp. 153–156. New York: Plenum Press.

Wolpoff, M. H. (1980). *Paleoanthropology*. New York: Alfred A. Knopf.

Wolpoff, M. H. (1996). *Human Evolution*. New York: McGraw-Hill.

Wolpoff, M., & Caspari, R. (1997). *Race and Human Evolution*. New York: Simon & Schuster.

Wolpoff, M. H., Xin Zhi, W., & Thorne, A. G. (1984). Modern *Homo sapiens*: A general theory of hominid evolution involving the fossil evidence from East Asia. In *The Origins of Modern Humans: A World Survey of the Fossil Evidence*, eds. F. Smith & F. Spencer, pp. 411–483. New York: Alan R. Liss.

Part II
The first steps: From australopithecines to Middle Pleistocene Homo

Part II

8   *Reconstructing australopithecine growth and development: What do we think we know?*

University of the Witwatersrand

> ... distinctive patterns of dental development characterize *A. africanus* and *A. robustus* ... This is not to imply that we necessarily know the cause of these pattern differences – only that they are apparently real and need some biological explanation.
>
> (Conroy & Kuykendall, 1995: 128)

> It is perhaps time we characterize early hominids more for what they are than what they are like, insofar as this is possible. We should be cognizant of the fact that early hominids were not apes as we know them today. They were unique creatures varying in morphology, behaviour and ecology, their ontogenetic strategies reflecting life history variation through evolutionary time.
>
> (Bromage, 1987: 271)

Introduction

Paleoanthropologists have always taken note of the maturational status of fossil hominid specimens as an important comparative feature. In Dart's (1925) original description of the Taung child, he inferred an age at death of approximately 6 years based on comparisons with the emergence status of the M1 in modern human children. Over time, the notion that australopithecines followed a prolonged human-like schedule of growth and development took firm hold in the paleoanthropological literature (Kyauka, 1994; Mann, 1975). Prolonged "human-like" growth and development was considered to be essential to the evolution of notably "human" traits such as intelligence, language, tool production and use, social behavior, and culture itself (Dobzhansky, 1962; Isaac,

Patterns of Growth and Development in the Genus Homo, ed. J. L. Thompson, G. E. Krovitz and A. J. Nelson. Published by Cambridge University Press. © Cambridge University Press, 2003.

1972; Lancaster, 1975; Lovejoy, 1981; Mann, 1975). Retrospectively, the focus of many of these models was to document the evolutionary origin of modern *human* adaptive traits – i.e., on "human" rather than "hominid" evolution (Macho, 2001).

Leaving aside our inherent specific interest in *human* evolution, there are several interrelated reasons for pursuing comparative studies about growth and development in extinct (fossil) taxa (see also Wood, 1996): (1) fossils of subadults preserve information about the pattern and process of growth and maturation, and thus provide a basis for reconstruction and comparison of the evolutionary history of growth and development in extinct lineages; (2) morphological evolution is ultimately a result of modifications to growth and development over time (Gould, 1977; Hall, 2002), and so our knowledge of growth differences among species potentially improves our understanding of variation in adult morphology and structure (McNamara, 2002); and (3) an individual's survival to adulthood is in large part dependent on successful, "normal" growth and development through critical periods of life history (Smith, 1989a, 1991), and we can thus improve our understanding of an organism's overall adaptive strategy by studying species-specific aspects of growth and development. Synthesizing all such considerations, taxonomic and phylogenetic models rely on the assessment of recognizable morphological patterns that are the end product of the growth and development process under the influence of adaptive, selective, and other evolutionary factors.

Life history is a "big picture" concept referring to the scheduling of developmental and reproductive stages during an organism's lifespan (Kelley, 2002; Shea, 1990). It involves an organism's allocation of resources toward different survival factors, and particularly toward reproductive success (Harvey *et al.*, 1987; Smith, 1989a). For the reasons outlined above, accurate reconstruction of growth patterns and processes in extinct species is informative for refining and testing adaptive models to clarify what we think we know about evolution and paleobiology. In paleoanthropology, growth and development is one aspect of life history for which we have perhaps the best direct fossil evidence in the form of subadult skeletal and, particularly, dental remains of early hominids.

This volume focuses on the evolution of growth and development in the genus *Homo*, but this contribution aims to review the information presently known regarding the patterns and schedules of growth and development in Plio-Pleistocene early hominids commonly known as the "gracile" and "robust" australopithecines. This will assist in understanding the earlier evolutionary history of hominid growth and development by establishing what is known (and what is not) about growth, development, and life history of the australopithecines.

In addition, refinement of our reconstructions of australopithecine life history will provide the basis for a more informed discussion about evolutionary pattern and process in later hominid taxa (i.e., within the genus *Homo*), particularly with respect to issues such as heterochrony. In many cases, evolutionary analyses are reliant on modern apes to suggest "primitive" conditions for hominids, but modern apes clearly do not represent hominid ancestors, and have their own (largely unknown) evolutionary history. For certain questions, it now seems possible to establish the actual primitive (ancestral) states from which later growth and life history adaptations evolved.

The generic names *Australopithecus* and *Paranthropus* are used in this paper to distinguish between these extinct taxa, while retaining the conventional usage of the common term "australopithecines" for both. This decision was taken in recognition of the consistent developmental differences documented between "gracile" and "robust" australopithecines (Bromage, 1985, 1989; Bromage & Dean, 1985; Smith, 1986; and others), and the potential significance of such observations in the context of life history.

Background

From about the mid 1980s, the notion that early hominids followed a pro-longed, so-called "human-like" schedule of growth and maturation met with serious challenges (Beynon & Dean, 1988; Bromage, 1987; Bromage & Dean, 1985; Conroy & Vannier, 1987, 1988; Dean, 1987a; Smith, 1989a, 1989b). The arguments focused almost exclusively on comparisons of dental development patterns, and on interpretations of microanatomical structures in tooth enamel (see Mann, 1988; Mann *et al.*, 1987, 1990a, 1990b, 1991; Wolpoff *et al.*, 1988 for critiques of those studies) because of the relative abundance of fossilized teeth and jaws in contrast to postcranial skeletal remains. In general, the dentition provides a highly versatile and stable means of assessing maturity status for comparative purposes (Demirjian, 1986; Miles, 1963, 1978), and such studies primarily utilized radiographic and histological techniques that could be applied to both extant (skeletal) and extinct (fossil) samples.

Perhaps the most consequential question initially was whether a prolonged "human-like" schedule of growth and development was a universal (derived) hominid feature – did even the earliest hominids grow up like humans? In pursuit of an answer, initial studies of rates of dental development in modern hominids and hominoids also documented variation in dental developmental patterns among samples of extant taxa to construct comparative standards for assessment of fossil specimens representing extinct taxa. It became clear from early studies (Beynon & Dean, 1988; Beynon & Wood, 1987; Bromage & Dean, 1985; Conroy & Vannier, 1991a, 1991b; Smith, 1986) that early hominids demonstrated a mosaic of "ape-like" and "human-like" developmental features, and that early hominid dental developmental patterns were thus unique. Additional studies using a variety of approaches have since documented differences among fossil hominid and modern taxa in growth of both the facial (e.g., Ackerman & Krovitz, 2002; Braga, 1998; Bromage, 1985, 1987, 1989; Kuykendall *et al.*, 2002; Shea, 1983) and the postcranial skeleton (e.g., Berge, 1998, 2002; Tardieu, 1998, 1999).

Comparative life-history studies have drawn on the mass of data available from a variety of modern primates and other mammals to produce predictions about early hominid life history from factors that can be documented (or at least estimated) from the fossil record – such as brain and body size, and the timing and pattern of dental development. Life-history and dental development studies together suggest that the evolutionary origin of the modern human growth and development period (and life-history strategy) lies well within our own genus, probably only arising in populations of *Homo erectus* (Beynon & Dean, 1988; Bogin & Smith, 1996; Bromage & Dean, 1985; Smith, 1986) or even later (Dean *et al.*, 2001), and coinciding with a cranial capacity exceeding 1000 cm^3 (Smith, 1991; Smith & Tompkins, 1995; Smith *et al.*, 1995). Thus, a general explanation is that the prolongation of development is tied in some way to increased cranial capacity (Martin, 1981, 1983; Sacher & Staffeldt, 1974; Smith, 1986; Smith *et al.*, 1995), and to an adaptive strategy involving successful acquisition of rich energy resources, low juvenile mortality, and other factors leading to increased reproductive success (see Bogin, 1997; Bogin & Smith, 1996; Smith & Tompkins, 1995).

In general, variation in life history strategies can be organized along a *fast–slow life-history continuum* (Harvey *et al.*, 1989; Promislow & Harvey, 1990; Read & Harvey, 1989) according to the overall pace of their maturity schedule – and specifically, the timing of reproductive maturity. A variety of life-history variables are highly correlated, and both the timing of tooth emergence (especially that of M1) (Smith, 1986, 1989a) and of tooth crown formation times during dental development (Macho, 2001) have been shown to be strongly correlated with life-history parameters such as adult brain weight, maternal body weight, gestation length, birth weight, age at weaning, and overall lifespan. This means that development is similarly patterned in different species regardless of the overall timing of maturity – "life history as a whole is expressed as distinct suites or packages" (Kelley, 2002: 225). The trick is in unraveling the complex linkages between brain and body size, reproductive and developmental parameters, ecological strategies, and so on, in order to produce informative and testable evolutionary models regarding life history and adaptation in different early hominid species.

It is clear that particular details of development – the exact sequence of tooth development, the relative timing of crown versus root formation, the relationship of tooth development to facial growth and structure, and so on – vary in species with generally similar life-history schedules (i.e., that mature at similar ages). Interpretation of such developmental differences is by no means straightforward, but probably can be discussed in relation to variation in eco- logical and dietary factors (Godfrey *et al.*, 2001; Macho, 2001), and may also be interpreted in the context of ontogenetic scaling resulting from variation in body size among closely related species (Shea, 1983). At least in a general sense, such factors correspond with current ideas about ecological and dietary differences between gracile and robust australopithecines (see Macho, 2001, and discussion below).

So instead of "When did the modern human pattern and schedule of life history evolve?," the relevant question at this stage is "What do observed vari- ations in the pattern and schedule of growth and development events among (extinct) hominid taxa tell us about the paleobiology, life history and evolution of early hominids?"

A number of detailed reviews and discussions of evidence relating to differ- ent aspects of early hominid growth and life history are available (Anemone, 2002; Bromage, 1987; Dean, 1987b, 2000; Kelley, 2002; Kyauka, 1994; Macho & Wood, 1995; Mann *et al.*, 1990a; Smith, 1989b, 1992; and others); this con- tribution will first review and clarify the observed interspecific differences in chronology and timing of dental development, patterns of dental development, and growth and development of the australopithecine face. A few relevant stud- ies of postcrania (cited above) are available, but are not included here primarily because they require detailed considerations of heterochronic processes that are beyond the scope of the present discussion.

Chronology and timing of australopithecine dental development

Since fossil samples are by nature both fragmentary and cross-sectional, most conventional (longitudinal) approaches to determining the timing of growth and development in living populations are not directly applicable for extinct hominid species (e.g., Bogin, 1988; Eveleth & Tanner, 1990). However, mineralized dental tissues preserve incremental growth markings observable either on the crown surface (perikymata), or in sectioned enamel (striae of Retzius), that allow retrieval of longitudinal data even from isolated teeth. Histological analysis of such microstructural features produces estimates for the absolute chronology of variables such as crown and root formation periods, rates of enamel secretion, age at death, and the total period of tooth development. Dean (1987b, 1989, 2000) and Ramirez-Rozzi (1993, 1995) have provided essential reviews of the theoretical and practical aspects of this research and the application to life-history reconstruction in fossil hominids, and Mann *et al.* (1990a) offered a more critical review of these issues.

The first problem was to reconstruct dental development schedules in different early hominid species for comparison with modern ape and human standards. Initial studies utilized perikymata counts from the labial surface of incisor crowns (Beynon & Dean, 1988; Bromage & Dean, 1985), and striae of Retzius counts from fractured molar tooth surfaces (Beynon & Wood, 1987) in fossil hominid species to produce estimates for crown formation times and for age at death of the individuals represented. Thus, the chronological age for attainment of specific dental maturity events (e.g., incisor and molar crown completion) could be reconstructed for *Australopithecus* and *Paranthropus*, and compared to modern ape and human dental development schedules (Figure 8.1). These studies concluded that dental maturity estimates for *Australopithecus* and *Paranthropus* were more in line with modern ape – not human – dental developmental schedules. Rather than achieving dental maturity at 20 years of age or older, on a similar scale as in modern humans (following Demirjian *et al.*, 1973; Eveleth & Tanner, 1990; Moorrees *et al.*, 1963), early hominids most likely achieved dental maturity on the order of 10–12 years of age on a schedule similar to modern chimpanzees (following Anemone *et al.*, 1991; Kuykendall, 1996; Nissen & Riesen, 1964; Reid *et al.*, 1998).

Despite this general similarity between *Australopithecus* and *Paranthropus* in the overall timing of dental (and by inference, somatic) maturity, other details of dental development – such as crown formation times and the attendant details of enamel microstructure – have been shown to differ, and may even have taxonomic utility (Bromage & Dean, 1985; Ramirez-Rozzi, 1993). Table 8.1 presents published ranges for crown formation times (in years) calculated from surface perikymata or striae of Retzius counts from broken tooth crowns in extinct hominids, modern African apes, and humans (references given in Table 8.1).

It is immediately obvious from Table 8.1 that available samples of fossil hominid teeth are small (or even lacking completely), and this hinders rigorous statistical comparison between fossil taxa. However, the data that are available suggest important general differences in crown formation times between early (extinct) hominids and modern apes and humans. First, certain tooth types demonstrate considerably shorter crown formation times in any early hominid taxon compared to any modern taxon, and such differences are particularly observable for canine and incisor crown formation times, which may differ by more than a year (incisors) or several years (canines) in different species. Molar crowns are the exception, as they demonstrate broadly similar (and overlapping) crown formation times among taxa (Dean, 2000; Dean & Wood, 1981; Schwartz & Dean, 2000). Second, crown formation times in *Australopithecus*

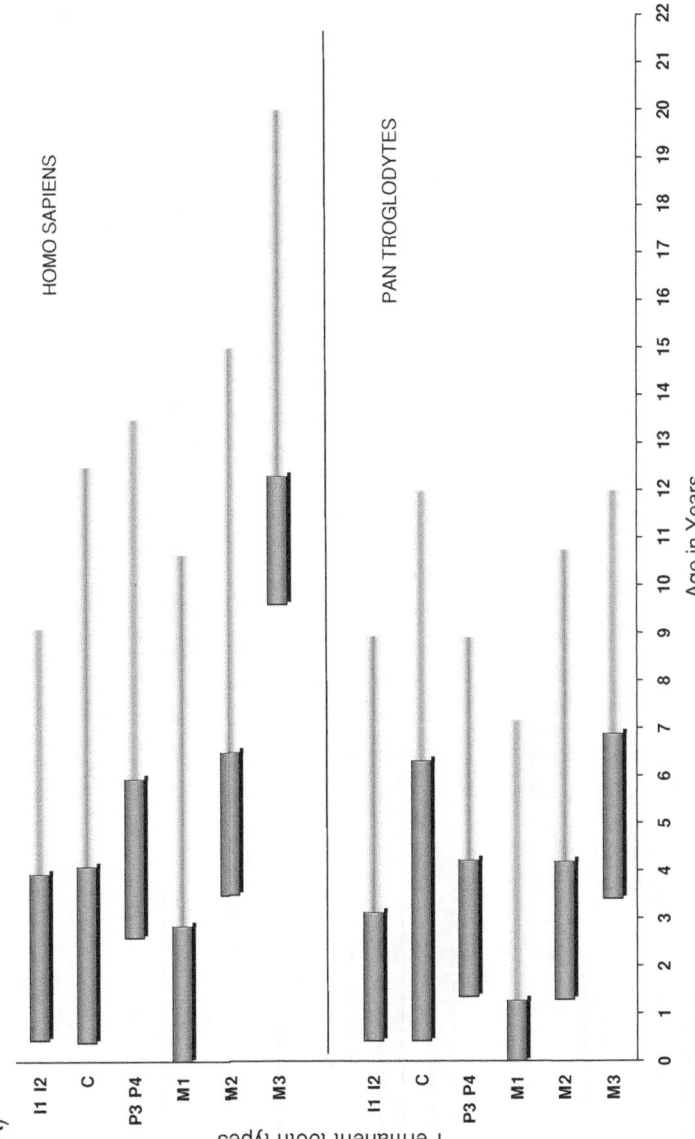

Figure 8.1   Dental development schedules for (A) modern humans and chimpanzees and (B) australopithecines. Thick light gray bars (to the left) represent periods of crown formation, and thin dark gray bars (to the right) represent periods of root formation. Estimates for modern taxa are based on radiographic studies in humans (Moorrees *et al.*, 1963), and both radiographic and histological studies in chimpanzees (Kuykendall, 1996; Reid *et al.*, 1998). Estimates for australopithecines are calculated from incremental features in tooth enamel after Beynon & Dean (1988).

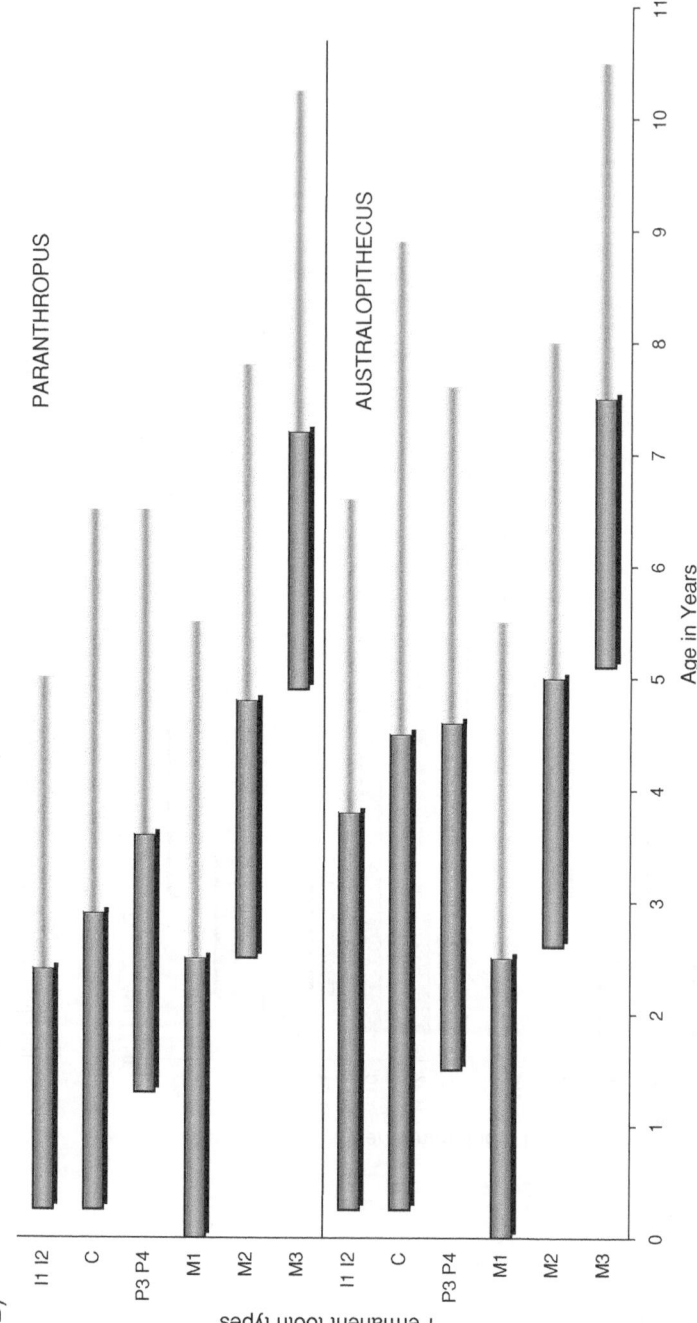

Figure 8.1 (cont.)

Table 8.1 *Crown formation times in years for different permanent tooth types in early hominid species, modern African apes, and humans. Estimates are from histological studies (sources given below). Data consist of estimated sample ranges (minimum–maximum for all available teeth), means from several specimens indicated by curved brackets (), or estimates from single specimens indicated by square brackets []. All estimates rounded to the nearest 0.1 year. Note that certain discrepancies are evident among studies due to sampling and perhaps methodological differences*

Tooth type	A. afarensis	A. africanus	P. aethiopicus	P. boisei	P. robustus	Early Homo	Pan	Gorilla	H. sapiens
Incisors	3.0–3.6	3.19	?	2.5–3.2 (2.2)	1.6–2.2 (1.9)	2.5–2.6 (2.6)	4.5–5.6	3.6–4.5	4.2
Canine	[3.6]	?	?	(2.4)	?	?	7.1	>5.0	?
Premolars	[3.0]	?	?	[2.4]	?	?	3.6–5.0	3.4–5.1	2.6–4.1
Molars	?	?	1.9–2.5 (2.4)	2.1–2.6[a] (2.4) 2.7–3.4[b] (2.9)	?	2.4–2.6 (2.5)	2.7–4.0	2.6–3.3	1.7–3.4

Sources for estimates: *A. afarensis*, LH 2, Bromage & Dean (1985), LH2, LH3, LH6, Beynon & Dean (1988); *A. africanus*, Sts 24a, Bromage & Dean (1985); *P. robustus*, SK 62, 63, 71, 73, Bromage & Dean (1985); Early *Homo*, KNM-ER 820, SK 74b, Bromage & Dean (1985); *P. aethiopicus*, Omo 84-100, 57.6-244, 33-3325, 33-6172, Ramirez-Rozzi (1993); *P. boisei*, KNM-ER 801a, 801c, 802e, 816b, 1171c, 1479a, 1819, 3737, OH 30, Beynon & Wood (1987), KNM-ER 1477, 812, 1820, OH 30, Dean (1987), KNM-ER 733D, Beynon & Dean (1987), Omo 33-65, 76-37, 141-1, 141-2, Ramirez-Rozzi (1993); *Pan*, Beynon et al. (1998); *Gorilla*, Beynon et al. (1991); *H. sapiens*, Moorees et al. (1963), Ramirez-Rozzi (1993), Bromage & Dean (1985).

[a] Estimates for *P. boisei* molars from Beynon & Dean (1988).

[b] Estimates for *P. boisei* molars from Ramirez-Rozzi (1993).

and early *Homo* are roughly similar, while *Paranthropus* tooth crowns tend to develop over distinctly shorter periods. In fact, *P. robustus* incisor crown formation times are shorter than in any other taxon, probably achieving completion in 60–70% of the period required for *Australopithecus* incisors (Table 8.1) (Beynon & Dean, 1988; Bromage & Dean, 1985). Additionally, crown formation times differ between South African and East African robust australopithecines, and Ramirez-Rozzi (1993, 1995) has also reported differences in crown formation times between *P. aethiopicus* and *P. boisei* in East Africa, estimated from isolated and fractured molar teeth attributed to these taxa.

At a developmental level, such differences in crown formation times are explained in terms of interspecific variation in at least three microstructural factors (Beynon & Dean, 1988; Dean, 2000; Dean & Reid, 2001; Ramirez-Rozzi, 1993, 1995; Schwartz & Dean, 2000). First, robust australopithecine teeth possess fewer surface perikymata (and striae in cross-section) than other species, and perikymata in these specimens are thus more widely spaced across the surface of the tooth crown. Having fewer, more widely spaced perikymata is indicative of a faster enamel extension rate (representing the number of newly differentiated ameloblasts secreting enamel over a standard period of time). Second, the angulation of the striae as they course from the enamel–dentine junction to the enamel surface is more acute in *Paranthropus,* indicating a more rapid ameloblast differentiation rate (i.e., a relatively greater number of active ameloblasts during crown formation). Third, the proportion of striae of Retzius in appositional (cuspal) versus imbricational (lateral or cervical) enamel also differs among taxa, suggesting that the progress of crown formation varies not only between different hominids (e.g., between *Paranthropus* and *Homo*), but also in different parts of the tooth crown. As a result of these factors, even the larger robust australopithecine postcanine teeth form in equivalent or less time compared to other taxa, and the smaller robust australopithecine incisor teeth take considerably less time for crown formation, as illustrated in Table 8.1.

Recent studies have shown that slow enamel formation rates and longer crown formation times associated with the prolonged modern human growth period are a very late evolutionary trait, probably arising only after *H. erectus/ergaster* (Dean *et al.*, 2001), and thus after a substantial increase in cranial capacity. Thus, early hominids (including australopithecines, *H. habilis*, and perhaps other species) might all be characterized by broadly similar overall periods of dental development, though variation in the underlying growth processes of tooth enamel results in species-specific crown formation times for individual teeth (Dean *et al.*, 2001; Moggi-Cecchi, 2001a), apart from variation in gross factors such as tooth size, morphology, and enamel thickness. Such differences in the absolute timing of crown formation result in variation in relative developmental relationships (patterns) among tooth types in different taxa.

Patterns of australopithecine dental development

Differences in the timing of developmental events between taxa (such as the time required for crown formation in different tooth types, as discussed above) are translated into variation in the sequence or pattern of development. Like comparisons of developmental timing, research on the pattern of dental development in fossil hominids has usually focused on comparisons between modern humans and our closest extant relative, *Pan troglodytes*, in order to determine whether particular early hominid specimens and taxa were "human-like" or "ape-like" in their dental development pattern. The primary comparative studies in this area either utilized tooth emergence data (e.g., Conroy & Mahoney, 1991; Dean, 1985; Garn *et al.*, 1957; Koski & Garn, 1957; Kuykendall *et al.*, 1992; Smith, 1994), or documented dental development patterns of permanent tooth crowns and roots (e.g., Anemone, 1995; Anemone *et al.*, 1991, 1996; Bromage, 1987; Dean, 1987a; Dean & Wood, 1981; Kuykendall, 1996; Kuykendall & Conroy, 1996; Simpson *et al.*, 1990, 1991, 1992; Smith, 1986, 1991). There are no detailed assessments of deciduous or mixed dental development patterns in fossil hominids, although descriptions of some specimens are available (e.g., Conroy & Vannier, 1991b; Grine, 1981; Moggi-Cecchi *et al.*, 1998).

Reviews of research in this area have been provided by Anemone (2002), Beynon *et al.* (1998), Kuykendall (2002), Macho & Wood (1995), Reid *et al.* (1998), and others. Three commonly cited differences in the pattern of permanent tooth emergence and development between humans and chimpanzees have been utilized in assessments of permanent tooth development in fossil hominids (as discussed below; see also Anemone *et al.*, 1996; Conroy & Vannier, 1991a; Dean & Wood, 1981; Garn *et al.*, 1957; Kuykendall & Conroy, 1996):

1. Compared to chimpanzees, human anterior tooth development is advanced relative to the molars. The permanent incisors and first molar in humans typically demonstrate very similar stages of development, while in chimpanzees the incisors demonstrate considerably delayed developmental status compared to the M1.

2. Although the incisors and canine in both chimpanzees and humans demonstrate early (and roughly concurrent) crown initiation, the greater proportion and longer duration of crown formation required for large chimpanzee canines results in delayed crown completion, emergence, and root formation relative to the M2 and premolars in comparison to humans; only the M3 is consistently delayed in development relative to the canine in chimpanzees, whereas in humans the canine tooth usually completes development before the premolars and M2.

3. Development of the molar teeth in humans involves relative delays compared to chimpanzee molar tooth development. Typically, in most human samples studied, crown mineralization between successive molars (e.g., M2 versus M1) involves delays so that in radiographic or computerized tomography (CT) evaluation, initial crown mineralization of more distal molars is not observed until the more mesial molar crown is complete and some root formation is also detectable. In apes however, molar crown development overlaps so that the crowns of more distal teeth demonstrate some degree of crown formation prior to completion of the more mesial molar crown.

Interspecific differences in features such as tooth size, space within the bony jaws for growth and development of tooth crypts (which may relate to overall size and prognathism of the facial skeleton, among other factors), the timing of social and physical maturity in males (with regard to canine tooth size and development), and the amount of developmental time available to growing individuals, are thought to contribute to variation in the pattern, or relative timing, of dental development (Beynon & Dean, 1988; Bromage, 1987; Conroy & Kuykendall, 1995; Conroy & Vannier, 1991a; Dean, 1987a, 1988a; Simpson *et al.*, 1990; Smith, 1991). Thus, differences in patterns of development reflect differences in life history of the species involved – even if the details of life history are not clearly understood.

Using standards for dental development in modern humans and apes as a comparative baseline, available samples of fossil hominid teeth and jaws have been assessed using radiographic, CT and histological methods to describe the developmental patterns typical of different early hominid species. Table 8.2 summarizes the published observations documenting dental stage assessments for early hominid specimens, in accordance with dental development scoring systems such as Demirjian *et al.* (1973). Assessments presented in Table 8.2 were derived from written descriptions, figures, and radiographic or CT-based assessments in the literature (references given in Table 8.2); data are clearly lacking for some teeth and some taxa.

Documented patterns of dental development in "gracile" early hominid species (*A. afarensis* and *A. africanus*) are generally similar to those observed in modern chimpanzees (Beynon & Dean, 1988; Bromage, 1987; Conroy & Kuykendall, 1995; Conroy & Vannier, 1991a, 1991b; Dean, 1989; Smith, 1986, 1991). In particular, incisor crown mineralization is consistently delayed by one or two stages compared to the M1. Developmental stages for canines are similar to the incisors, but delayed compared to M1 and some other postcanine teeth (as in chimpanzees) even though canine crowns in hominid species may be reduced in size compared to modern apes. However, there are differences from chimpanzees – gracile australopithecine canines tend to be advanced in

Table 8.2 *Dental development stages and inferred perikymata ages at death for available australopithecine fossil specimens*

Specimen	I1	I2	C	P3	P4	M1	M2	M3	Perikymata age (years)
Australopithecus africanus									
Sts 2	–	–	3	3	2	5	–	–	2.5–3.0
Taung	4	4	4	3–4	3	6	2	–	3.5
Sts 24	4–5	4–5	4	3–4	3	6	2	–	3.5–4.0
Sts 18	–	–	–	4	3–4	6–7	4	–	4.0–4.5
MLD 2	–	–	5–6	6	5	7	6	–	6.6
Sts 8	–	–	–	–	–	7–8	7–8	4	7.75
Stw 327	–	–	–	–	7	7–8	7	4	7.0–8.0
Stw 151	6	5–6	3–4	4–5	4	7	4	0?	5.2–5.3
Paranthropus boisei									
KNM ER 1477	4	4	3	3?	3?	4	–	–	2.5–3.0
KNM ER 812	5	4?	3?	3?	3?	5	–	–	2.5–3.0
KNM ER 1820	5	4?	3?	3	3	5–6	–	–	2.5–3.1
OH 30	4	3	3	3	3	5	–	–	2.7–3.2
Paranthropus robustus									
SK 438	–	–	–	–	0	3–4	–	–	1.0–1.5
SK 64	–	–	–	2	0	3–4	0	–	1.0–2.0
SK 3978	3	3	3	2	0	3–4	–	–	1.0–2.0
SK 62	5	5	4	4	2–3	6	2–3	–	2.5–3.0
SK 61	5	5	4	4	3	6	–	–	3.0
SK 63	5–6?	5–6?	4	4	3	6–7	3	–	3.0

Sources: Most data taken from Conroy & Vannier (1991b, Table 1: 141). STW 151 tooth mineralization stages taken from descriptions provided in Moggi-Cecchi *et al.* (1998). East African robust australopithecines from Dean (1987a). Dental development stages as in Demirjian *et al.* (1973).

development relative to the M2 (Table 8.2), while development of the large chimpanzee canine crown is delayed relative to all other teeth (Kuykendall & Conroy, 1996). Moggi-Cecchi (2001b), using the coefficient of variation (CV)-based comparison of dental development scores developed by Smith (1986), found that *Australopithecus* CVs for estimated dental age were typically below the ape mean CV, but within the range of both apes and modern humans. In other words, dental development patterns in *Australopithecus* are not clearly "ape-like," even if they matured at a similar age.

Establishment of the dental development pattern in *Paranthropus* has been controversial because of ambiguities with reconstructed specimens (SK 63), misidentification of deciduous incisors as permanent teeth (SK 61), and observed intraspecific variation in dental development patterns (Conroy, 1988; Dean, 1985; Grine, 1987). Nonetheless, it is commonly reported that robust australopithecines shared a "superficially human-like" pattern of development involving the incisors and M1 (Beynon & Dean, 1988; Bromage, 1987; Conroy & Kuykendall, 1995; Conroy & Vannier, 1991a, 1991b; Dean, 1985, 1988a, 1989; Smith, 1986, 1991). In fact, permanent incisor crown formation in *Paranthropus* is relatively advanced compared to M1 crown mineralization in humans (Smith, 1986, 1989b), and *Paranthropus* premolars are developmentally advanced compared to modern humans (Beynon & Dean, 1987; Dean, 1988a). These factors constitute a unique dental development pattern in *Paranthropus* (see also Conroy & Kuykendall, 1995; Smith, 1991), despite a general similarity to humans in the M1–I1 developmental pattern.

The difficulty in characterizing dental development in robust australopithecine specimens may be due to the combination of very brief estimated crown formation times for the canine, premolars, and M2 (Table 8.1), the developmental requirements of large and thick-enamelled postcanine teeth (Schwartz & Dean, 2000), the relationships between crown and root formation with respect to tooth emergence timing (Dean, 1985), and the tendency to make dichotomous comparisons between "ape-like" and "human-like" dental development patterns (Bromage, 1987; Smith, 1991). The best way to characterize robust australopithecine dental development (as well as that of gracile australopithecines) is that it is unique in comparison to modern apes or humans.

Other factors are problematic in assessment of developmental patterns in fossil taxa. The omnipresence of small and fragmentary fossil assemblages prevents adequate recognition and assessment of intra- and interspecific variation – in this case, of developmental variability. This problem may be exacerbated by the conventional use of assessment and scoring methods for dental development that are incapable of detecting subtle developmental variation (Beynon *et al.*, 1998; Kuykendall, 2002). For example, the M1 in SK 61 and SK 62 has been assessed at similar stages of development, with significant root formation complete (stage 6 according to Conroy & Vannier, 1991b; see Table 8.2). However, Conroy & Vannier (1991b) described advanced incisor root development (longer roots) for the incisors of SK 62 compared to the incisors of SK 61, though all incisors in these specimens were scored at stage 5 (see their Table 1:141; Table 8.2 here). Other assessment methods are also problematic; the CV-based comparison designed by Smith (1986) and used by others (e.g., Moggi-Cecchi, 2001b) clearly discriminates between *Australopithecus* and *Paranthropus*, but also places the latter well within the modern human range. While this importantly underscores the parallel M1–I1 developmental pattern, the measure in itself is not informative about unique and subtler aspects of *Paranthropus* development. It is thus necessary to consider results obtained using different techniques, and to examine multiple aspects of dental development, in order to effectively describe and compare dental developmental patterns in closely related species.

Another factor pertinent to this discussion is the age-at-death distribution of fossils attributed to different taxa. From Table 8.2, the estimated ranges for age at death (from perikymata estimates) for the *A. africanus* sample extend from 2.0 to 8.0 years. The estimated ranges for age at death among the robust australopithecine sample are comparatively younger than estimates for *A. africanus*, extending from 1.0 to 3.2 years (see Table 8.2 and Smith, 1991). Such differences in age-at-death distributions of the available fossil samples certainly affect analysis and interpretation of results in comparisons between taxa, and may potentially reflect differences in taphonomy, causes of mortality, or other biological/life-history factors (they may also simply reflect some unrecognized sampling factor regarding size or completeness of specimens, etc.).

In any event, the differences in age-at-death distribution are noteworthy in light of the finding by Smith (1989b; 1991) and Conroy & Kuykendall (1995) that dental development patterns between apes and humans, and between early hominids and either extant taxon, do not reliably distinguish among taxa at younger ages – e.g., identification of an individual's dental developmental pattern as "ape" or "human" is reliable only after M1 emergence. It is therefore essential that future studies incorporate both the deciduous teeth and the mixed dentition (rather than only the permanent teeth) into comparisons of dental developmental patterns.

Growth and development of the australopithecine face

Many researchers have utilized comparative aspects of facial morphology, such as varying degrees of prognathism in different hominid species, to explain features such as the similarity in the M1–I1 dental development pattern in (orthognathic) *Paranthropus* and modern *Homo*, or the similarities noted above between (prognathic) *Australopithecus* and *Pan* (Anemone, 1995; Bromage, 1985; Conroy & Kuykendall, 1995; Conroy & Vannier, 1987; Dean, 1985, 1988a; Simpson *et al.*, 1990; Smith, 1991). However certain we can be that dental development and facial growth are intimately coordinated (Lavelle *et al.*, 1977), our understanding of these complex growth relationships remains rather "immature." Nonetheless, certain general patterns of facial growth appear to correspond with the distinctions among hominid taxa already noted for dental development.

Bromage (1985, 1987, 1989) examined patterns of cranial bone remodeling (cortical bone resorption and deposition) during ontogeny in 29 fossil hominid specimens representing the genera *Australopithecus*, *Paranthropus*, and early *Homo*. This research utilized scanning electron microscopy (SEM) imaging analysis, and compared the hominid data to patterns observed in modern humans (from Enlow, 1966, 1975) and chimpanzees (Bromage's own data). The detailed description of facial bone growth remodeling patterns is complex (see Bromage, 1989), but two basic ontogenetic patterns were identified and associated with prognathic versus orthognathic facial structure in early hominids.

Australopithecus and early *Homo* demonstrate the first pattern, in which most bony surfaces of the face, orbit, and mandible are depository, reflecting the general anteriorly directed vector of growth resulting in a more prognathic facial profile (Bromage, 1985, 1989). This pattern is generally similar to that observed in modern chimpanzees.

The second pattern of facial bone remodeling is associated with *Paranthropus*, and is similar to that documented for modern *Homo*. This pattern involves extensive areas of resorption in the anterior temporal fossa, nasoalveolar clivus, canine–molar region, and in some anteriorly facing surfaces of the mandible (Bromage, 1987, 1989). These resorptive regions result in a generally inferiorly directed vector of growth associated with the more orthognathic facial profile typical of these taxa.

However, Sts 52 preserves resorptive patterns for the mandible that do not have modern analogs – they were not observed in either modern humans or chimpanzees (Bromage, 1989). It is relevant that the most extensive comparative data available document bone remodeling patterns only for modern humans and macaques (Dutterloo & Enlow, 1970; Enlow, 1966, 1975), while equivalent data for chimpanzees are limited to a small sample provided by Bromage (1989). A more informative interpretation of such "anomalous" specimens, and of intra- and interspecific variation in both modern and fossil taxa, would be enhanced by further comparative work on modern ape facial growth, and obviously, by an expanded fossil sample.

In any event, the distribution pattern of such resorptive and depositional surfaces suggests synchrony with other aspects of facial growth. For example, in the Taung juvenile, the identification of depositional surfaces on the medial wall of the orbit indicates that these surfaces are expanding laterally to accommodate expansive growth of the ethmoidal sinuses (Bromage, 1985). Conroy & Vannier (1987) also noted that the developing maxillary sinus in Taung is relatively large, and demonstrates pneumatized extensions into the hard palate and zygoma. Both features are typical of chimpanzees, and distinct from humans, at similar stages of dental and facial development (see Koppe & Ohkawa, 1999). Following up on these observations, Kuykendall *et al.* (2002) reported that though Taung exhibits similarities in maxillary sinus configuration with chimpanzees at a comparable developmental stage, adult specimens attributed to either *Australopithecus* (Sts 5, Stw 73, AL 200) or *Paranthropus* (SK 12) do not demonstrate maxillary sinus morphology (e.g., intrapalatal extensions) typical of adult chimpanzees. Though the details are clearly not yet understood, these observations allow the inference to be made that the developmental pattern of the maxillary sinus in australopithecines – and the related aspects of facial growth – must be distinct from either modern apes or humans.

Bromage (1985, 1989) also proposed that depositional surfaces on the premaxillary and lower lateral maxillary surfaces in *Australopithecus*, combined with mid-palatal and incisive sutural growth, are necessary to fulfill the space requirements of the large incisor crowns that develop in the premaxillary region in early phases of facial growth (see also Conroy & Vannier, 1987). In support of this observation, Braga identified two developmental patterns relating to incisive suture fusion in chimpanzees, modern humans and fossil hominids (Braga, 1998; Maureille & Braga, 2002). The first pattern can be generalised as "late fusion," and was observed in chimpanzees and *Australopithecus*. The incisive suture, which has both an anterior or facial and a palatal component, is thought to remain patent later in ontogeny in order to accommodate development of the prognathic faces, large anterior dentition, and diastemata characterizing these species. The second pattern of "early fusion" was observed in *Paranthropus* and modern *Homo*, which share orthognathic facial structure and small incisors, and lack diastemata.

However, there remain many questions about the comparative ontogeny of this suture and related aspects of facial growth. For example, the palatal component of the incisive suture is completely unfused in Taung, whereas a chimpanzee at a similar developmental stage would already demonstrate partial fusion (Braga, 1998). In addition, the available *Paranthropus* sample includes only older subadults, in which the palatal component was already partially fused, and the anterior component was completely fused. In fact, the available hominid fossils do not constitute a complete enough assemblage to document the complete sequence of incisive suture fusion for any early hominid species. Additionally, there are differences in the timing of incisive suture fusion in *Pan* and *Gorilla* skulls (Schultz, 1948), though both hominoids are clearly prognathic. Clarification of these questions will require more complete documentation of incisive suture fusion in different taxa, and supplementation of the available comparative sample.

Similarities in facial morphology (Dean & Wood, 1982), facial architecture (Bromage, 1992), and dental development (Dean, 1986, 1988a; see above) between *Paranthropus* and *Homo* have been discussed for some time, and are usually considered in terms of homoplasy (but see Bromage, 1992; Dean & Wood, 1982). These structural and developmental associations are the basis for a consistent association in growth patterns between species that are more orthognathic (*Paranthropus* and modern *Homo*) and those that are more prognathic (*Australopithecus*, *Pan*, and early *Homo*). In an alternate model, Dean discounts the influence of factors such as facial prognathism and cranial base flexion since these structures are highly variable in both fossil hominids and Old World monkeys (Dean, 1988b; Dean & Wood, 1984). Instead, he suggested that similarities in certain aspects of growth and development of the cranial base (e.g., coronal orientation of the petrous bones) in both early and later *Homo* and *Paranthropus* are established at birth, and relate to similarities in growth and morphology of the cerebellum and the posterior cranial fossa in which it sits.

An alternate approach is to employ morphometric techniques in the reconstruction and comparison of growth trajectories among modern and fossil species. Ackermann & Krovitz (2002) compared reconstructed facial growth trajectories in *A. africanus*, modern apes, and humans. Their results also support the conclusion that patterns of ontogenetic divergence are established very early in development, and that "juvenile growth plays a relatively minor role in determining final adult facial morphology" (p. 146). Although interspecific differences in growth pattern were noted in the maxilla, zygomatic region, and midface, their analysis indicates that after M1 emergence, growth patterns were remarkably similar in all taxa, and do not account for observed differences in adult facial morphology. In fact, in simulations in which they "grew" Taung according to chimpanzee, gorilla, and human growth models, the resulting adult morphology remained similar to the adult *A. africanus* specimen Sts 5. One acknowledged difficulty with such research is that the fossil record only provides data for the beginning (Taung) and end (Sts 5) points for reconstruction of growth trajectories in *A. africanus*.

One of the most interesting aspects of such research is that it appears to contradict assessments of patterns of permanent tooth development, in which discrimination based on patterns of tooth mineralization in chimpanzees, humans, and fossil hominids becomes much clearer *only after* M1 emergence (see section above; Conroy & Kuykendall, 1995; Smith, 1989b, 1991). It is also puzzling in light of the numerous differences between *Australopithecus* and modern *Homo* noted above in features such as patterns of tooth mineralization and facial bone remodeling – these lines of evidence suggest that developmental patterns become *more divergent* with age. In any case, there is a clear need for further research regarding the complex developmental interactions between facial growth, bone remodeling, dental development, and related factors during hominid evolution.

Summary and discussion

Based on chronologies derived from dental microstructure and predictions relating brain size to developmental timing, both *Australopithecus* and *Paranthropus* would have achieved developmental maturity at about 10–12 years of age. In addition, *Australopithecus* can be characterized by late development of the permanent incisors and canines relative to the M1, similar crown formation times for the incisors, canines, and premolars on the order of 3 to 3.5 years, complete fusion of the incisive suture only after M1 emergence, sagittally oriented petrous bones, relatively large maxillary sinuses that extend across the hard palate at about the time that the M1 achieves functional occlusion, and facial bone remodeling and growth resulting in a marked degree of facial prognathism. In comparison, *Paranthropus* is characterized by relatively early development of the anterior permanent dentition – more or less concurrent with M1 development, more rapid crown formation times for all permanent teeth (incisors to molars) roughly between 2 and 3.5 years, early fusion of the incisive suture, coronally oriented petrous bones, and facial bone remodeling and growth resulting in a relatively flat, orthognathic face.

As presented throughout this paper, *Australopithecus* is frequently compared with *Pan* regarding such developmental characteristics (although *Australopithecus* does demonstrate unique features), while *Paranthropus* demonstrates a rather puzzling combination of modern human and other unique developmental patterns. The situation is further complicated by observations of *differences* between East and South African *Paranthropus*, apparent *similarities* between *A. africanus* and *A. afarensis* (Kyauka, 1994), and anomalous specimens such as Sts 52 (Bromage, 1989) and Stw 183 (Moggi-Cecchi *et al.*, 1998) that are not neatly characterized by these descriptions.

The australopithecines were comparatively small-brained, quasi-bipedal hominids with no perfectly suited modern analogs. It is unlikely that the low levels of interspecific variation indicated by either brain or body size estimates for gracile and robust australopithecines (McHenry, 1988, 1992) can be informative on their own about differences in life-history schedules (Smith *et al.*, 1995) – even though both are highly correlated with many life-history traits (Harvey & Clutton-Brock, 1985; Harvey *et al.*, 1987; Smith, 1986).

Thus, the relevant questions stemming from comparisons between gracile and robust australopithecines are different from those concerning the evolution of "human life history," or from comparisons between taxa that clearly differ in the duration of growth and development or factors such as brain and body size. Answers to these questions would involve issues identified earlier, such as the adaptive significance of interspecific variation in developmental patterns, the evolutionary history of hominid growth and development, the ontogenetic basis of structural variation, and taxonomic and phylogenetic aspects of ontogeny and life history.

To add to the confusion, at least some aspects of australopithecine ontogenetic variation probably resulted indirectly from natural selection relating to other, less obvious, environmental or adaptive factors. For example, developmental variation between gracile and robust australopithecines results in differences in tooth size, prognathism, and other structural features related to masticatory and dietary adaptations. It is generally maintained that dietary differences between *Australopithecus* and *Paranthropus* played a central role in the evolution of their different masticatory adaptations/structures (e.g., Conroy, 1990; Grine, 1988) – but could diet also play a role in the observed ontogenetic differences between robust and gracile australopithecines?

Recent analyses of the relationship between different aspects of dental development and life-history evolution in primates (Macho, 2001) and bovids (Macho & Williamson, 2002) provide one plausible model that is consistent with estimates of more rapid developmental timing in robust australopithecines. Essentially, mammals that occupy open habitats (as attributed to *Paranthropus*) and subsist on low-quality diets (also attributed to *Paranthropus*) tend to mature relatively early compared to closely related species of similar body size (*also* attributed to *Paranthropus*), and finally, tend to attain larger adult body size. This last point may or may not be true of *Paranthropus* generally, but is not inconsistent with body size estimates for East African species of this genus (McHenry, 1992). Accordingly, Godfrey *et al.* (2001: 204) concluded that "diet affects the absolute pace of dental development, independent of body size, age at first breeding, and cranial capacity," finding that folivorous primates (low-quality diet) have accelerated dental development schedules compared to frugivores (high-quality diet) of similar body size, and that dental development itself is more advanced at weaning in folivores. It is therefore interesting to note that Aiello *et al.* (1991) observed tooth wear on deciduous molars (suggestive of weaning) prior to M1 emergence in specimens of robust australopithecines. Separate studies of life-history variables in extant non-human primates obtained high correlations between age at weaning and both age at M1 eruption (Smith, 1986; $r = 0.93$) and average molar crown formation time (Macho, 2001; $r = 0.88$).

Thus, the evidence for shorter crown formation times and generally more rapid dental development in *Paranthropus* may be regarded as consistent with several interrelated aspects of robust australopithecine paleobiological models – occupation of open habitats, subsistence on a lower-quality diet, and perhaps slightly larger body sizes than *Australopithecus*. In comparison, reconstructions of *Australopithecus* paleobiology involve closed or partially wooded habitats, more varied and higher-quality diets, perhaps slightly smaller body sizes, and comparatively longer periods of dental development and crown formation times. However, it is realized that these models for robust and gracile australopithecines do not necessarily provide "solutions" to the puzzling observations regarding australopithecine dental development and life history; rather, it is more likely that they provide fertile fodder for future research efforts.

On a more general note, reconstructions of growth and development in *Australopithecus* and *Paranthropus* are both unique and mosaic – this review has outlined some general similarities with either modern chimpanzees or humans (respectively), but emphasizes that early hominids followed life-history schedules that are not observed in any modern primate. In particular, modern chimpanzees do not reflect primitive evolutionary characters for early hominid growth and development, even if overall growth and development periods were similar. The extensive and consistent similarities in various aspects of growth and development between *Australopithecus* and early *Homo* certainly lend additional support to hypotheses of close phylogenetic relatedness, but so far do not seem capable of identifying the "true" ancestral species for *Homo*. On the other hand, the growing list of similarities between *Paranthropus* and modern *Homo* remain enigmatic – do they all really reflect homoplasy?

In the past couple of decades, an abundance of data and many essential insights have resulted from research concerning the evolution of hominid life history, yet it remains for all the pieces of the puzzle to be identified and put into place. Much of this information has resulted from the innovative application of new technology allowing us to non-invasively extract previously unimaginable data from the scarce fossil assemblages (Schwartz & Kuykendall, 1996; Wood, 2000). The effective use of new technology in combination with ongoing advances in analytical techniques in life-history research will yield not only new information, but also new and unimagined questions concerning the evolution of hominid life history.

Acknowledgments

I would like to express my gratitude to the editors of this volume for inviting me to contribute, and for their continued support and patience throughout. In addition, Jacopo Moggi-Cecchi and Gary Schwartz provided valuable comments on earlier drafts that definitely improved the paper. Finally, my thanks to the Head of the School of Anatomical Sciences at the University of the Witwatersrand, Beverley Kramer, for her friendship and continued support for paleoanthropological research in the School.

References

Ackermann, R. R., & Krovitz, G. E. (2002). Common patterns of facial ontogeny in the hominid lineage. *Anatomical Record*, **269**, 142–147.
Aeillo, L. C., Montgomery, C., & Dean, M. C. (1991). The natural history of tooth attrition in hominoids. *Journal of Human Evolution*, **21**, 397–412.

Anemone, R. L. (1995). Dental development in chimpanzees of known chronological age: Implications for understanding the age at death of Plio-Pleistocene hominids. In *Aspects of Dental Biology: Palaeontology, Anthropology and Evolution*, ed. J. Moggi-Cecchi, pp. 201–215. Florence: International Institute for the Study of Man.

Anemone, R. L. (2002). Dental development and life history in hominid evolution. In *Human Evolution through Developmental Change*, eds. N. Minugh-Purvis & K. J. McNamara, pp. 249–280. Baltimore: Johns Hopkins University Press.

Anemone, R. L., Watts, E. S., & Swindler, D. S. (1991). Dental development of known-age chimpanzees, *Pan troglodytes* (Primates, Pongidae). *American Journal of Physical Anthropology*, **86**, 229–241.

Anemone, R. L., Mooney, M. P., & Siegel, M. I. (1996). Longitudinal study of dental development in chimpanzees of known chronological age: Implications for understanding the age at death of Plio-Pleistocene hominids. *American Journal of Physical Anthropology*, **99**, 119–133.

Berge, C. (1998). Heterochronic processes in human evolution: An ontogenetic analysis of the hominid pelvis. *American Journal of Physical Anthropology*, **105**, 441–459.

Berge, C. (2002). Peramorphic processes in the evolution of the hominid pelvis and femur. In *Human Evolution through Developmental Change*, eds. N. Minugh-Purvis & K. J. McNamara, pp. 381–404. Baltimore: Johns Hopkins University Press.

Beynon, A. D., & Dean, M. C. (1987). Crown-formation time of a fossil hominid premolar tooth. *Archives of Oral Biology*, **32**, 773–780.

Beynon, A. D., & Dean, M. C. (1988). Distinct dental development patterns in early fossil hominids. *Nature*, **335**, 509–514.

Beynon, A. D., & Wood, B. A. (1987). Patterns and rates of enamel growth in the molar teeth of early hominids. *Nature*, **326**, 493–496.

Beynon, A. D., Dean, M. C., & Reid, D. J. (1991). Histological study on the chronology of the developing dentition in gorilla and orang-utan. *American Journal of Physical Anthropology*, **86**, 189–203.

Beynon, A. D., Clayton, C. B., Ramirez Rozzi, F. V., & Reid, D. J. (1998). Radiographic and histological methodologies in estimating the chronology of crown development in modern humans and great apes: A review, with some applications for studies on juvenile hominids. *Journal of Human Evolution*, **35**, 351–370.

Bogin, B. (1988). *Patterns of Human Growth*. Cambridge: Cambridge University Press.

Bogin, B. (1997). Evolutionary hypotheses for human childhood. *Yearbook of Physical Anthropology*, **40**, 63–89.

Bogin, B., & Smith, B. H. (1996). Evolution of the human life cycle. *American Journal of Human Biology*, **8**, 703–716.

Braga, J. (1998). Chimpanzee variation facilitates the interpretation of the incisive suture closure in South African Plio-Pleistocene hominids. *American Journal of Physical Anthropology*, **105**, 121–135.

Bromage, T. G. (1985). Taung facial remodeling: A growth and development study. In *Hominid Evolution: Past, Present, and Future*, ed. P. V. Tobias, pp. 239–245. New York: Academic Press.

Bromage, T. G. (1987). The biological and chronological maturation of early hominids. *Journal of Human Evolution*, **16**, 257–272.

Bromage, T. G. (1989). Ontogeny of the early hominid face. *Journal of Human Evolution*, **18**, 751–773.

Bromage, T. G. (1992). The ontogeny of *Pan troglodytes* craniofacial architectural relationships and implications for early hominids. *Journal of Human Evolution*, **23**, 235–251.

Bromage, T. G., & Dean, M. C. (1985). Re-evaluation of the age at death of immature fossil hominids. *Nature*, **317**, 525–527.

Conroy, G. C. (1988). Alleged synapomorphy of the M1/I1 eruption pattern in robust australopithecines and *Homo*: Evidence from high-resolution computed tomography. *American Journal of Physical Anthropology*, **75**, 487–492.

Conroy, G. C. (1990). *Primate Evolution*. New York: W.W. Norton.

Conroy, G. C., & Kuykendall, K. L. (1995). Paleopediatrics: Or when did human infants really become human? *American Journal of Physical Anthropology*, **98**, 121–131.

Conroy, G. C., & Mahoney, C. J. (1991). Mixed longitudinal study of dental emergence in the chimpanzee, *Pan troglodytes* (Primates, Pongidae). *American Journal of Physical Anthropology*, **86**, 243–254.

Conroy, G. C., & Vannier, M. W. (1987). Dental development of the Taung skull from computerized tomography. *Nature*, **329**, 625–627.

Conroy, G. C., & Vannier, M. W. (1988). The nature of Taung dental maturation continued. *Nature*, **333**, 808.

Conroy, G. C., & Vannier, M. W. (1991a). Dental development in South African Australopithecines. I: Problems of pattern and chronology. *American Journal of Physical Anthropology*, **86**, 121–136.

Conroy, G. C., & Vannier, M. W. (1991b). Dental development in South African Australopithecines. II: Dental Stage Assessment. *American Journal of Physical Anthropology*, **86**, 121–156.

Dart, R. A. (1925). *Australopithecus africanus*: The man–ape of South Africa. *Nature*, **115**, 195–199.

Dean, M. C. (1985). The eruption pattern of the permanent incisors and first permanent molars in *Australopithecus (Paranthropus) robustus*. *American Journal of Physical Anthropology*, **67**, 251–257.

Dean, M. C. (1986). *Homo* and *Paranthropus*: Similarities in the cranial base and developing dentition. In *Major Topics in Primate and Human Evolution*, eds. B. Wood, L. Martin, & P. Andrews, pp. 249–265. Cambridge: Cambridge University Press.

Dean, M. C. (1987a). The dental developmental status of six East African juvenile fossil hominids. *Journal of Human Evolution*, **16**, 197–213.

Dean, M. C. (1987b). Growth layers and incremental markings in hard tissues: A review of the literature and some preliminary observations about enamel structure in *Paranthropus boisei*. *Journal of Human Evolution*, **16**, 157–172.

Dean, M. C. (1988a). Growth of teeth and development of the dentition in *Paranthropus*. In *Evolutionary History of the "Robust" Australopithecines*, ed. F. E. Grine, pp. 43–54. New York: Aldine de Gruyter.

Dean, M. C. (1988b). Growth processes in the cranial base of hominoids and their bearing on morphological similarities that exist in the cranial base of *Homo* and

Paranthropus. In *Evolutionary History of the "Robust" Australopithecines*, ed. F. E. Grine, pp. 107–112. New York: Aldine de Gruyter.

Dean, M. C. (1989). The developing dentition and tooth structure in hominoids. *Folia Primatologica*, **53**, 160–176.

Dean, M. C. (2000). Progress in understanding hominoid dental development. *Journal of Anatomy*, **197**, 77–101.

Dean, M. C., & Reid, D. J. (2001). Perikymata spacing and distribution on hominid anterior teeth. *American Journal of Physical Anthropology*, **116**, 209–215.

Dean, M. C., & Wood, B. A. (1981). Developing pongid dentition and its use for ageing individual crania in comparative cross-sectional growth studies. *Folia Primatologica*, **36**, 111–127.

Dean, M. C., & Wood, B. A. (1982). Basicranial anatomy of Plio-Pleistocene hominids from East and South Africa. *American Journal of Physical Anthropology*, **59**, 157–174.

Dean, M. C., & Wood, B. A. (1984). Phylogeny, neoteny and growth of the cranial base in hominoids. *Folia Primatologica*, **43**, 157–180.

Dean, M. C., Leakey, M. G., Reid, D., Schrenk, F., Schwartz, G. T., Stringer, C., & Walker, A. (2001). Growth processes in teeth distinguish modern humans from *Homo erectus* and earlier hominins. *Nature*, **414**, 628–631.

Demirjian, A. (1986). Dentition. In *Human Growth: A Comprehensive Treatise*, eds. F. Falkner & J. M. Tanner, pp. 269–295. New York: Plenum Press.

Demirjian, A., Goldstein, H., & Tanner, J. M. (1973). A new system of dental age assessment. *Human Biology*, **45**, 211–227.

Dobzhansky, T. (1962). *Mankind Evolving.* New Haven: Yale University Press.

Dutterloo, H. S., & Enlow, D. H. (1970). A comparative study of cranial growth in *Homo* and *Macaca*. *American Journal of Anatomy*, **127**, 357–367.

Enlow, D. (1966). A comparative study of facial growth in *Homo* and *Macaca*. *American Journal of Physical Anthropology*, **24**, 293–307.

Enlow, D. (1975). *Handbook of Facial Growth.* Toronto: W.B. Saunders.

Eveleth, P. B., & Tanner, J. M. (1990). *Worldwide Variation in Human Growth*, 2nd edn. Cambridge: Cambridge University Press.

Garn, S. M., Koski, K., & Lewis, A. B. (1957). Problems in determining the tooth eruption sequence in fossil and modern man. *American Journal of Physical Anthropology*, **15**, 313–331.

Godfrey, L. R., Samonds, K. E., Jungers, W. L., & Sutherland, M. R. (2001). Teeth, brains, and primate life histories. *American Journal of Physical Anthropology*, **114**, 192–214.

Gould, S. J. (1977). *Ontogeny and Phylogeny.* Cambridge: Harvard University Press.

Grine, F. E. (1981). A new composite juvenile specimen of *Australopithecus africanus* (Mammalia, Primates) from Member 4 Sterkfontein Formation, Transvaal. *Annals of the South African Museums*, **84**, 169–201.

Grine, F. E. (1987). On the eruption pattern of the permanent incisors and first permanent molars in *Paranthropus*. *American Journal of Physical Anthropology*, **72**, 353–359.

Grine, F. E. (1988). Evolutionary history of the "robust" australopithecines: A summary and historical perspective. In *Evolutionary History of the "Robust" Australopithecines*, ed. F. E. Grine, pp. 509–520. New York: Aldine de Gruyter.

Hall, B. K. (2002). Evolutionary developmental biology: Where embryos and fossils meet. In *Human Evolution through Developmental Change*, eds. N. Minugh-Purvis & K. J. McNamara, pp. 7–27. Baltimore: Johns Hopkins University Press.

Harvey, P. H., & Clutton-Brock, T. H. (1985). Life-history variation in primates. *Evolution*, **39**, 559–581.

Harvey, P. H., Martin, R. D., & Clutton-Brock, T. H. (1987). Life histories in comparative perspective. In *Primate Societies*, eds. B. B. Smuts, D. L. Cheney, R. M. Seyfarth, R. W. Wrangham, & T. T. Struhsaker, pp. 181–196. Chicago: University of Chicago Press.

Harvey, P. H., Read, A. F., & Promislow, D. E. L. (1989). Life history variation in placental mammals: Unifying the data with the theory. *Oxford Surveys of Evolutionary Biology*, **6**, 13–31.

Isaac, G. L. (1972). Chronology and the tempo of cultural change during the Pleistocene. In *Calibration of Hominid Evolution*, eds. W. Bishop & J. Miller, pp. 381–430. Edinburgh: Scottish Academic Press.

Kelley, J. (2002). Life-history evolution in Miocene and extant apes. In *Human Evolution through Developmental Change*, eds. N. Minugh-Purvis & K. J. McNamara, pp. 223–248. Baltimore: Johns Hopkins University Press.

Koppe, T., & Ohkawa, Y. (1999). Pneumatization of the facial skeleton in Catarrhine primates. In *The Paranasal Sinuses of Higher Primates: Development, Function, and Evolution*, eds. T. Koppe, H. Nagai, & K. W. Alt, pp. 77–119. Chicago: Quintessence Publishing Co.

Koski, K., & Garn, S. M. (1957). Tooth eruption sequences in fossil and modern man. *American Journal of Physical Anthropology*, **15**, 469–488.

Kuykendall, K. L. (1996). Dental development in chimpanzees (*Pan troglodytes*): The timing of tooth calcification stages. *American Journal of Physical Anthropology*, **99**, 135–157.

Kuykendall, K. L. (2002). An assessment of radiographic and histological standards of dental development in chimpanzees. In *Human Evolution through Developmental Change*, eds. N. Minugh-Purvis & K. J. McNamara, pp. 281–304. Baltimore: Johns Hopkins University Press.

Kuykendall, K. L., & Conroy, G. C. (1996). Permanent tooth calcification in chimpanzees (*Pan troglodytes*): Patterns and polymorphisms. *American Journal of Physical Anthropology*, **99**, 159–174.

Kuykendall, K. L., Mahoney, C. J., & Conroy, G. C. (1992). Probit and survival analysis of tooth emergence ages in a mixed-longitudinal sample of chimpanzees (*Pan troglodytes*). *American Journal of Physical Anthropology*, **89**, 379–399.

Kuykendall, K. L., Bozic, J., & Conroy, G. C. (2002). A comparative analysis of tooth mineralization and paranasal sinus development of the Taung child. *American Journal of Physical Anthropology*, **34** (Suppl.), 98–99.

Kyauka, P. S. (1994). Developmental patterns of the earliest hominids: A morphological perspective. In *Integrative Paths to the Past: Palaeoanthropological Advances in Honor of F. Clark Howell*, eds. R. S. Corrucini & R. L. Ciochon, pp. 229–250. New York: Prentice Hall.

Lancaster, J. B. (1975). *Primate Behavior and the Emergence of Human Culture*. New York: Holt, Rinehart, & Winston.

216   *K. L. Kuykendall*

Lavelle, C. L. B., Shellis, R. P., & Poole, D. F. G. (1977). *Evolutionary Changes to the Primate Skull and Dentition*. Springfield: Charles C Thomas.

Lovejoy, C. O. (1981). The origin of man. *Science*, **211**, 341–350.

Macho, G. A. (2001). Primate molar crown formation times and life history evolution revisited. *American Journal of Primatology*, **55**, 189–201.

Macho, G. A., & Williamson, D. K. (2002). The effects of ecology on life history strategies and metabolic disturbances during development: An example from the African bovids. *Biological Journal of the Linnean Society*, **75**, 271–279.

Macho, G. A., & Wood, B. A. (1995). The role of time and timing in hominid dental evolution. *Evolutionary Anthropology*, **4**, 17–31.

Mann, A. E. (1975). *Some Paleodemographic Aspects of the South African Australopithecines*. Philadelphia: University of Pennsylvania.

Mann, A. E. (1988). The nature of Taung dental maturation. *Nature*, **333**, 123.

Mann, A. E., Lampl, M., & Monge, J. M. (1987). Maturational patterns in early hominids. *Nature*, **328**, 673–674.

Mann, A. E., Lampl, M., & Monge, J. M. (1990a). Patterns of ontogeny in human evolution: Evidence from dental development. *Yearbook of Physical Anthropology*, **33**, 111–150.

Mann, A. E., Monge, J. M., & Lampl, M. (1990b). Dental caution. *Nature*, **348**, 202.

Mann, A. E., Monge, J. M., & Lampl, M. (1991). Investigation into the relationship between perikymata counts and crown formation times. *American Journal of Physical Anthropology*, **86**, 175–188.

Martin, R. D. (1981). Relative brain size and basal metabolic rate in terrestrial vertebrates. *Nature*, **293**, 57–60.

Martin, R. D. (1983). *Human Brain Evolution in an Ecological Context*, 52nd James Arthur Lecture on the Evolution of the Human Brain. New York: American Museum of Natural History.

Maureille, B., & Braga, J. (2002). Between the incisive bone and premaxilla: From African apes to *Homo sapiens*. In *Human Evolution through Developmental Change*, eds. N. Minugh-Purvis & K. J. McNamara, pp. 464–478. Baltimore: Johns Hopkins University Press.

McHenry, H. (1988). New estimates of body weight in early hominids and their significance to encephalization and megadontia in "robust" australopithecines. In *Evolutionary History of the "Robust" Australopithecines*, ed. F. E. Grine, pp. 133–148. New Yok: Aldine de Gruyter.

McHenry, H. (1992). How big were early hominids? *Evolutionary Anthropology*, **1**, 15–20.

McNamara, K. J. (2002). What is heterochrony? In *Human Evolution through Developmental Change*, eds. N. Minugh-Purvis & K. J. McNamara, pp. 1–4. Baltimore: Johns Hopkins University Press.

Miles, A. E. W. (1963). The dentition in the assessment of individual age in skeletal material. In *Dental Anthropology*, ed. D. R. Brothwell, pp. 191–209. New York: Pergamon Press.

Miles, A. E. W. (1978). Teeth as an indicator of age in man. In *Development, Function, and Evolution of Teeth*, eds. P. M. Butler & K. A. Joysey, pp. 455–464. New York: Academic Press.

Moggi-Cecchi, J. (2001a). Questions of growth. *Nature*, **414**, 595–597.

Moggi-Cecchi, J. (2001b). Patterns of dental development of *Australopithecus africanus*, with some inferences on their evolution with the origin of the genus *Homo*. In *Humanity from African Naissance to Coming Millennia*, eds. P. V. Tobias, M. A. Raath, J. Moggi-Cecchi, & G. A. Doyle, pp. 125–133. Firenze: Firenze University Press, and Johannesburg: Witwatersrand University Press.

Moggi-Cecchi, J., Tobias, P. V., & Beynon, A. D. (1998). The mixed dentition and associated skull fragments of a juvenile fossil hominid from Sterkfontein, South Africa. *American Journal of Physical Anthropology*, **106**, 425–465.

Moorrees, C. F. A., Fanning, E. A., & Hunt, E. E. J (1963). Age variation of formation stages for ten permanent teeth. *Journal of Dental Research*, **42**, 1490–1502.

Nissen, H. W., & Riesen, A. H. (1964). The eruption of the permanent dentition in the chimpanzee. *American Journal of Physical Anthropology*, **22**, 285–294.

Promislow, D. E. L., & Harvey, P. H. (1990). Living fast and dying young: A comparative analysis of life-history variation among mammals. *Journal of Zoology*, **220**, 417–437.

Ramirez-Rozzi, F. V. (1993). Tooth development in East African *Paranthropus*. *Journal of Human Evolution*, **24**, 429–454.

Ramirez-Rozzi, F. V. (1995). Time of crown formation in Plio-Pleistocene hominid teeth. In *Aspects of Dental Biology: Palaeontology, Anthropology and Evolution*, ed. J. Moggi-Cecchi, pp. 217–238. Florence: International Institute for the Study of Man.

Read, A. F., & Harvey, P. H. (1989). Life history differences among eutherian radiations. *Journal of Zoology*, **219**, 329–353.

Reid, D. J., Schwartz, G. T., Dean, C., & Chandrasekera, M. S. (1998). A histological reconstruction of dental development in the common chimpanzee, *Pan troglodytes*. *Journal of Human Evolution*, **35**, 427–448.

Sacher, G. A., & Staffeldt, E. F. (1974). Relation of gestation time to brain weight for placental mammals: Implications for the theory of vertebrate growth. *American Naturalist*, **108**, 593–615.

Schultz, A. H. (1948). The relation in size between premaxilla, diastema and canine. *American Journal of Physical Anthropology*, **6**, 163–180.

Schwartz, G. T., & Dean, M. C. (2000). Interpreting the hominid dentition: Ontogenetic and phylogenetic aspects. In *Development, Growth and Evolution: Implications for the Study of the Hominid Skeleton*, eds. P. O'Higgins, & M. Cohen, pp. 207–233. London: Academic Press.

Schwartz, G. T., & Kuykendall, K. L. (1996). Enamel structure and development. *Evolutionary Anthropology*, **5**, 150–151.

Shea, B. T. (1983). Size and diet in the evolution of African ape craniodental form. *Folia Primatologica*, **40**, 32–68.

Shea, B. T. (1990). Dynamic morphology: Growth, life history, and ecology in primate evolution In *Primate Life History and Evolution*, Monographs in Primatology, vol. 14, ed. C. J. De Rousseau, pp. 325–352. New York: Wiley-Liss.

Simpson, S. W., Lovejoy, C. O., & Meindl, R. S. (1990). Hominoid dental maturation. *Journal of Human Evolution*, **19**, 285–297.

Simpson, S. W., Lovejoy, C. O., & Meindl, R. S. (1991). Relative dental development in hominoids and its failure to predict somatic growth velocity. *American Journal of Physical Anthropology*, **86**, 113–120.

Simpson, S. W., Lovejoy, C. O., & Meindl, R. S. (1992). Further evidence on relative dental maturation and somatic developmental rate in hominoids. *American Journal of Physical Anthropology*, **87**, 29–38.

Smith, B. H. (1986). Dental development in *Australopithecus* and early *Homo*. *Nature*, **323**, 327–330.

Smith, B. H. (1989a). Dental development as a measure of life history in primates. *Evolution*, **43**, 683–688.

Smith, B. H. (1989b). Growth and development and its significance for early hominid behaviour. *OSSA*, **14**, 63–96.

Smith, B. H. (1991). Dental development and the evolution of life history in Hominidae. *American Journal of Physical Anthropology*, **86**, 157–174.

Smith, B. H. (1992). Life history and the evolution of human maturation. *Evolutionary Anthropology*, **1**, 134–142.

Smith, B. H. (1994). Sequence of emergence of the permanent teeth in *Macaca*, *Pan*, *Homo*, and *Australopithecus*: Its evolutionary significance. *American Journal of Human Biology*, **6**, 61–76.

Smith, B. H., & Tompkins, R. L. (1995). Toward a life history of the hominidae. *Annual Review of Anthropology*, **24**, 257–279.

Smith, R. J., Gannon, P. J., & Smith, B. H. (1995). Ontogeny of australopithecines and early *Homo*: Evidence from cranial capacity and dental eruption. *Journal of Human Evolution*, **29**, 155–168.

Tardieu, C. (1998). Short adolescence in early hominids: Infantile and adolescent growth of the human femur. *American Journal of Physical Anthropology*, **107**, 163–178.

Tardieu, C. (1999). Ontogeny and phylogeny of femoro-tibial characters in humans and hominid fossils: Functional influence and genetic determinism. *American Journal of Physical Anthropology*, **110**, 365–377.

Wolpoff, M., Monge, J., & Lampl, M. (1988). Was Taung human or an ape? *Nature*, **335**, 501.

Wood, B. A. (1996). Hominid palaeobiology: Have studies of comparative development come of age? *American Journal of Physical Anthropology*, **99**, 9–15.

Wood, B. A. (2000). Investigating human evolutionary history. *Journal of Anatomy*, **197**, 3–17.

9   *Growth and life history in* Homo erectus

S. C. ANTÓN
Rutgers University

S. R. LEIGH
University of Illinois

Introduction

Evolution modifies the developmental pattern and thus understanding the evolution of development is critical to identifying how and when the presumed descendant morphological patterns, such as our own, originated. Likewise, understanding how and when these patterns changed ultimately assists in answering why they changed; that is, what evolutionary problems and solutions they reflect. Here we address two aspects of ontogeny in *Homo erectus* in order to define developmental shifts that have characterized later human evolution. First, we undertake a preliminary heterochronic comparative analysis of cranial ontogeny in *H. erectus* and *H. sapiens*. This investigation focuses on alterations in the relations between size, shape, and age at maturation between ancestral and descendant species. Heterochronic transformations can thus be taken to refer, rather narrowly, to shifts in allometric or relative growth trajectories between an ancestor and a descendant species. Using this approach requires both juvenile and adult fossils, although it does not require knowledge of the specific developmental age of each juvenile fossil (Shea, 2000). Second we explore differences in how these taxa grew to reach adult size, with special emphasis on whether or not a growth spurt characterized *H. erectus*. This component of the research investigates the possibility that growth in *H. erectus*, like that in *H. sapiens*, was subdivided into relatively discrete time periods, including a period prior to an adolescent growth spurt and a period after the initiation of the growth spurt. This information complements heterochronic investigations, and enables the examination of possible life history correlates of growth in both *H. erectus* and *H. sapiens*.

Patterns of Growth and Development in the Genus Homo, ed. J. L. Thompson, G. E. Krovitz and A. J. Nelson. Published by Cambridge University Press. © Cambridge University Press 2003.

Theoretical context

A number of recent contributions address questions regarding the evolution of human growth and development from a variety of perspectives, including growth in time (Antón, 2002; Bogin, 1994a, 1999; Leigh, 1996, 2001; Minugh-Purvis & McNamara, 2002) and in terms of heterochrony (Godfrey *et al.*, 1998; Gould, 1977; Shea, 1989, 2002). Unfortunately, there currently exist several renditions of heterochronic theory (Leigh, 2002; Nehm, 2001; Shea, 2000, 2002), that often find application to problems in human evolution, and these studies often are not linked closely with analyses of growth in time. Specifically, a number of authors apply heterochrony, which includes allometric techniques, to understand shifts in size and shape ontogeny between ancestor and descendant species (Godfrey *et al.*, 1998; Shea, 1983, 2000, 2002). Alternatively, variation in the duration of growth phases has been discussed using heterochronic terminology (McKinney & McNamara, 1991). This version of heterochrony explores shifts in the allocation of growth to different time periods. These alternative views of heterochrony (as well as others: Zelditch, 2001) are important in understanding the evolution of development. Unfortunately, most authors do not distinguish between these forms of heterochrony, and terminologies from each branch are, problematically, often used interchangeably. This has resulted in a fair degree of basic terminological confusion, particularly in human evolutionary studies, since the evolution of the human brain is often a major research focus of heterochronic analyses (cf. McKinney, 2002; Rice, 2002; Shea, 1989).

We contend that these versions of heterochrony require analytical separation (*sensu* Shea, 2000, 2002). First, we advocate analyzing morphological aspects of changes in ontogeny through the context developed by Gould (1977), which emphasizes ontogenetic allometric dimensions of growth and development (i.e., changes in size and shape). Second, we suggest that a separate conceptual framework be applied to how these species allocate time to different periods of ontogeny. In the first case, our objective is to test the hypothesis that paedomorphosis, or the evolution of a juvenilized cranial morphology, characterized later changes in human evolution. In the second, we evaluate the hypothesis that *H. erectus* exhibited a skeletal growth spurt comparable to that observed in virtually all contemporary *H. sapiens* populations. We must emphasize that our analyses are very preliminary, serving mainly to define the value of different approaches, whether or not these approaches should be considered in the analysis of future questions, and to offer directions for future research.

Heterochrony

Heterochrony investigates evolutionary modifications of development in terms of size, shape, and age of developmental events between ancestor and descendant. This definition derives directly from Gould's (1977) refinements of de Beer's (1971) heterochronic terminology. In this framework, heterochrony is explicitly tied to shifts in size and shape relations, or allometries between ancestors and descendants. Perhaps paradoxically, given the name heterochrony, this approach emphasizes size and shape relations to a greater degree than timing shifts.

The analysis of allometric heterochronic transformations requires a two-step diagnostic process of first evaluating how the shape of the descendant adult compares with that of ancestral juveniles and adults, and testing for the presence of size/shape dissociation between ancestor and descendant ontogenies (Leigh *et al.*, 2003; Shea, 2002). The second step involves defining which of several developmental processes might have produced a given morphological result. In terms of heterochrony, several authors have argued that paedomorphosis, or the evolution of a descendant adult that resembles a juvenile of the ancestor, has characterized human evolution (Gould, 1977). This is one of several possible outcomes of heterochronic transformations (see definitions by Leigh *et al.*, 2003, and Shea, 2002). Here we formally test the hypothesis that cranial form in extant humans is paedomorphic (or juvenilized) relative to *H. erectus*. In addition, we attempt to define which of several heterochronic processes may have produced this result. We focus particularly on neoteny, which is defined as a paedomorphic shape produced through the process of allometric dissociation. In other words, neoteny requires that the shape of the descendant adult resemble the shape of an ancestral juvenile. Such a resemblance indicates only that the descendant is paedomorphic, a result which can be produced either by neoteny or hypomorphosis (Leigh *et al.*, 2003; Shea, 1989, 2000, 2002). Neoteny occurs when allometric dissociation (non-overlapping regression lines) produces a shape like that of the juvenile ancestor.

Temporal patterning of growth and the human growth spurt

A complementary approach to considering the evolution of hominin growth patterns involves the comparison of life-history patterns in living apes and humans in order to consider when differences in life history may have originated in one lineage or the other. Most of this history is based on informal comparisons. For example, Weidenreich (1941, 1943) based his view of accelerated growth on ideas about the relationship between sutural fusion and dental eruption in humans and great apes, and a general idea that Asian *H. erectus* was more ape-like than human-like in its cranial development. More recently, a separate analytical tradition that derives from heterochrony has been applied to this problem. Specifically, several authors have argued for an exclusively "time-based" version of heterochronic theory that predicts morphological change as a result of changes in the duration of various growth periods (McKinney & McNamara, 1991; see also Minugh-Purvis & McNamara, 2002). For example, McKinney and McNamara have explicitly argued that a process termed "sequential hypermorphosis" is responsible for the evolution of the human brain. This hypothesis predicts that each of several brain growth periods (however these might be defined) has undergone a prolongation in duration. They argue that this has resulted in a relatively large brain in human evolution, and thus reject the notion that paedomorphosis can account for evolutionary changes in the human brain (see further critique by Godfrey *et al.*, 1998; Shea, 2000).

This view of heterochrony is divorced from size and shape change, so this approach is of limited utility unless certain conditions are met. Although prolongation of growth periods can produce relatively large structures, to do so requires either increases in growth rates or maintenance of growth rates at ancestral levels. If anything, current data based on fossil teeth argue for a decrease in dental developmental growth rates in *H. sapiens* over those seen in *H. erectus* (Dean *et al.*, 2001). Furthermore, research on the length and composition of human growth periods indicates extension of an early period of growth that is coupled with relative reductions in body mass growth rates (Leigh, 2001). These kinds of changes pose problems for the notion of sequential hypermorphosis, because the increased length of early growth periods is coupled with reduced growth rates. The nature of evolutionary change in later growth periods may be more straightforward because humans present growth spurts of durations and rate comparable to expectations based on body size.

Further complexity in debates concerning later growth periods revolves around the adolescent growth spurt. For example, Bogin (1994a, 1994b, 1999) argues that humans uniquely possess a skeletal growth spurt, particularly when measured by stature. This attribute reflects an "inserted" period of ontogeny unique to humans and entirely absent in early hominins, including *H. erectus*. On the other hand, the attributes of human mass growth spurts are consistent with those expected for primates of our body size (Leigh, 2001). Adolescent skeletal growth spurts are known in non-human primates, particularly in the face (S. R. Leigh, unpublished data), but their presence remains to be convincingly demonstrated in dimensions that might be homologized with human stature. Thus, the presence of an adolescent growth spurt may not be unique to humans (contra Bogin, 1994a, 1994b, 1999).

Here we focus specifically on whether or not an adolescent skeletal growth spurt was present in the skull of *H. erectus*. Trying to determine the timing of origin of these differences requires both juvenile and adult fossils and additionally requires some independent estimate of developmental age of the juvenile fossils. This latter requirement poses something of a quandary, as we shall see below. The only direct evidence for when a growth spurt might have originated must be the fossil record. Several authors have tried to use the rather remarkable KNM-WT 15000 skeleton (the "Nariokotome Boy") to address whether early *H. erectus* underwent an adolescent growth spurt. Smith (1993) suggested there was not a growth spurt on the basis of the disagreement between dental, postcranial, and statural developmental ages for KNM-WT 15000 and it is in part these data on which Bogin relies when making his assessments of growth in *H. erectus* (Bogin, 1994a; Bogin & Smith, 1996). Others have suggested that the data are not inconsistent with the presence of a growth spurt either because the lack of agreement in developmental ages is not outside the bounds of human variation (Clegg & Aiello, 1999), or because adult *H. erectus* has specific femoral morphology that is similar to that which is acquired in modern humans only during the adolescent growth spurt (Tardieu, 1998), or because comparisons of cranial size and shape in KNM-WT 15000 and adult *H. erectus* seem to imply a substantial amount of growth left to be accomplished before adult morphology is attained (Antón, 2002). However, none of these studies has attempted to reconstruct growth velocity or rate of growth (change is size per unit time). Since the adolescent growth spurt is a peak in growth velocity relative to both earlier and later periods, it is necessary to examine growth rates when trying to determine if such growth spurts existed in *H. erectus*, thus supporting the contention of subdivision of growth periods in *H. erectus*. Here we attempt such preliminary reconstructions in order to test whether an adolescent growth spurt can be detected in *H. erectus*.

Materials and methods

Materials

Allometric heterochronic analyses
Diagnoses of heterochronic results and processes are based on comparative morphometric analyses of ontogeny in *H. erectus* and anatomically modern *H. sapiens*. Data for analyses of *H. erectus* are derived from primary measurements (SCA) and from literature sources (see Antón, 1997 for details). The most important specimen in this sample is the subadult from Mojokerto. This specimen defines the small end of the size range for the *H. erectus* sample, and

thus exerts a heavy influence on the position of regression lines calculated for scaling analyses. The subadult KNM-WT 15000 is also included in parietal and occipital analyses, but due to damage of the frontal cannot be included in the three-dimensional analyses. The remainder of the sample follows Antón (1997) including both African and Asian specimens typically classified as *H. erectus* as well as late *H. erectus* from Ngandong, Indonesia.

The anatomically modern *H. sapiens* sample represents a variety of world-wide populations including crania from individuals of all ages from South Africa, Australia, Europe, and North America (*n* = 307). Seventy-four of the South African and Native American individuals were measured by SRL with a Microscribe 3D digitizer, the remainder were measured by SCA using standard linear measurements and spreading and sliding calipers. The 74 crania measured by digitizer and a limited sample of *H. erectus* crania are used in comparisons of vault shape based on angular data from the vault. We did not have the opportunity to assess differences between caliper and digitizer measures. However, previous analyses by one of us (SRL) have shown no statistically significant differences between comparable collected data sets on monkey skulls. The entire modern human and *H. erectus* samples are used in analyses of linear measurements of individual vault bones (i.e., frontal, parietal, and occipital chords), which focus on scaling analyses of *H. erectus* and anatomically modern *H. sapiens*.

Growth spurt analyses

As noted above, the adolescent growth spurt represents an increase in the *velocity* of growth that results in an increase in linear dimensions. In living humans, growth velocity increases dramatically for stature during adolescence, resulting in increases of as much as 31% in linear height (Marshall & Tanner, 1986; Smith, 1993). Peak velocities of facial growth, although less extreme, also occur at this time resulting in increases of mandibular height of about 25% from 12 to 20 years (Buschang *et al.*, 1983; Goldstein, 1936; Marshall & Tanner, 1986). Upper facial height increases somewhat less dramatically. Our first task must be to explore whether we can see these same changes in growth velocity that we know to occur in modern humans based on longitudinal somatic data, using skeletal measures of the modern human face. If we cannot visualize these changes in samples in which we know they occur, there is little reason to look for them in other samples. In light of the preservation common in the fossil record, we attempt to visualize these changes using measures of upper facial height (nasion–prosthion) and mandibular corpus height (at M1).

Because of significant geographic variation in facial size and shape among modern human groups (e.g., Howells, 1973: Figure 3), the anatomically modern human samples are analyzed separately by geographic region. For this analysis we include two groups of widely different geographic location and cranial form, a dolichocephalic Australian ($n = 26$) sample and a brachycephalic Alaskan ($n = 32$) sample. Each sample consists of subadults and adult females. Using adult females should be a conservative test for growth spurts since females attain smaller absolute facial measurements and mixed-sex juvenile samples should, if anything, tend to overestimate percentage of adult height achieved. Sample sizes are severely limited by the very small number of adolescent crania available in collections. Subadult human samples are grouped into the following age classes: infant (~18 months), early childhood (~2–3 years), late childhood (~4–6 years), juvenile (~8–10 years), beginning adolescence (11/12 years), mid adolescence (14–17 years), and adult. Sample sizes vary by geographic group and age class; individual data points are plotted in the figures. Age classes from infant to beginning adolescence are attributed on the basis of the dental developmental age of each specimen following dental eruption sequences outlined by Ubelaker (1984). Beginning adolescents are further recognized by open spheno-occipital synchondroses and billowing of the occipital condyles indicative of juvenile status. Mid adolescence is recognized by the presence of an adult dentition without M3 in occlusion, an open spheno-occipital synchondrosis, and no billowing of the occipital condyles. Mature individuals all have M3 in occlusion and the spheno-occipital synchondrosis fused. Age determinations and measurements were made by SCA. The *H. erectus* specimens used in these analyses are limited by preservation to the African samples and include the subadult KNM-WT 15000 for both facial and mandibular analyses, subadults KNM-ER 820 and 1507 and adults KNM-ER 730 and 992 for mandibular analysis and adult KNM-ER 3733 for the facial analysis.

Methods

Heterochronic analyses

The two-stage procedure discussed previously is used to calibrate heterochronic transformations given the ancestor–descendant relationship represented by these data. In general, the first stage of this diagnostic process involves determining whether or not the descendant adult shape is juvenilized or paedomorphic in relation to an ancestral condition. For this analysis, frontal, parietal, and occipital angles, derived from three-dimensional landmark coordinates, are employed to evaluate whether or not modern human adult calvaria approximate the shape of juvenile *H. erectus*. These angles were calculated directly from three-dimensional coordinates using the Law of Cosines. Our results are somewhat preliminary because three-dimensional landmark coordinates were available only for the 74 Native American and South African crania and for two

Table 9.1 *Reduced major axis regression results for* H. erectus *analyses. The independent variable is overall cranial size (geometric mean of cranial dimensions). All variables are expressed as base-10 logarithms*

Dependent variable	Intercept	95% Lower confidence interval	95% Upper confidence interval	Slope	95% Lower confidence interval	95% Upper confidence interval	Pearson correlation coefficient
Frontal chord	−0.888	−1.428	−0.349	1.473	1.198	1.748	0.952
Parietal chord	−0.179	−1.064	0.706	1.096	0.646	1.547	0.765
Occipital chord	0.048	−0.659	0.754	0.95	0.590	1.309	0.800

adult and one juvenile *H. erectus* (Sangiran 2, Ngandong 6, and Mojokerto). The calculated angles provide a measure of skull roundness, enabling comparisons of shapes between the *H. erectus* and anatomically modern *H. sapiens* samples. This method fulfills the first part of the diagnostic procedure. If Mojokerto falls within or very near the range of shapes for anatomically modern *H. sapiens*, then we can suggest that modern human cranial shapes resemble those of juvenile *H. erectus*.

Next we evaluate the relationship between size and shape during ontogeny in both samples based on investigations of relative growth trajectories for each taxon. Thus, we present a second stage of the heterochronic analysis, which evaluates whether or not ontogenetic regression lines differ in their position and slopes. Specifically, we plot natural logarithms for the frontal chord, parietal chord, and occipital cord for each individual against a measure of overall size, which is defined as the geometric mean of these three chords. Relative growth plots provide information about size, shape and size/shape dissociation. Non-overlapping and non-parallel regression lines reflect size and shape dissociation between taxa, leading to identification of neoteny as the process that produces the heterochronic result of paedomorphosis. A linear regression line describes ontogenetic allometry in *H. erectus*, whereas a lowess regression line is employed for anatomically modern *H. sapiens* (Table 9.1). We opt for lowess regression in the latter case given obvious curvilinearity in the data.

Diagnosing neoteny (or other heterochronic process) requires that we use these plots to evaluate shape similarities between taxa. Specifically, an isometric line in log space (slope = 1) describes a range of geometrically identical data points (Gould, 1977; Jungers *et al.*, 1995; Leigh *et al.*, 2003). In order to assess whether or not adult anatomically modern *H. sapiens* exhibits shapes like those of juvenile *H. erectus*, we pass an isometric line through data points for Mojokerto (the smallest *H. erectus* skull in the sample) and Sangiran 2 (the smallest adult skull in the sample). If a regression line for anatomically modern *H. sapiens* overlaps or falls between these isometric lines, then the average for the anatomically modern *H. sapiens* sample can be considered paedomorphic. If separate regression lines describe ontogenetic allometries for each taxon, then neoteny is implied as the process that produces paedomorphosis. We must emphasize that other processes can produce paedomorphosis (Gould, 1977; Leigh *et al.*, 2003; Shea, 2000, 2002). However, these processes involve similar or overlapping regression lines. In the interest of clarity, we refer the reader to several references that spell these protocols out in detail (Gould, 1977; Leigh *et al.*, 2003; Shea, 2000, 2002). Moreover, given the small sample size for *H. erectus*, we treat these results only in a preliminary manner, and do not apply formal statistical tests of the regression lines.

Growth spurt analyses

In cross-sectional samples, such as human skeletons and the fossil record, we can only measure growth velocity indirectly by considering the "amount of growth" (differences in measurements) between individuals of different ages. Such cross-sectional samples are known to underestimate growth spurt magnitudes and thus should be conservative tests of whether growth spurts existed in a sample (e.g., Boas, 1892; Leigh, 1996). Clearly, the extremely small sample size for fossil upper facial data may not conform to this trend. We estimate growth rates in each of our samples by dividing the difference between successive mean facial values by the difference in successive mean age values. We then plot these velocities (*y*-axis) against the average age of the two age classes used to compare the values (*x*-axis), creating an arithmetic-velocity curve (corresponding to the "pseudo-velocity curve" of Hamill *et al.*, 1973 and Coelho, 1985). While juvenile and adolescent ages can be identified with some certainty, adult age estimates have no such limitations. The ages assigned for mature individuals are critical to determining the adolescent growth velocity. In the human model of the *H. erectus* analyses we estimate mature ages based on the average age of first birth in human hunter–gatherers of between 18 and 20 years (Hill & Hurtado, 1996; Howell, 1979; Pennington, 2001) and the age of third molar eruption and spheno-occipital synchondrosis closure (Buikstra & Ubelaker, 1994). In the ape model we estimate mature ages based on the age of first birth in the African apes of between 10 and 15 (Leigh & Shea, 1996; Watts & Pusey, 1993) and the age of third molar eruption and spheno-occipital synchondrosis fusion, signaling cessation of craniofacial growth, between 10 and 12 years of age (e.g., Kuykendall & Conroy, 1996). To visualize an adolescent growth spurt we must see an increase in the magnitude of these velocities in the later part of growth.

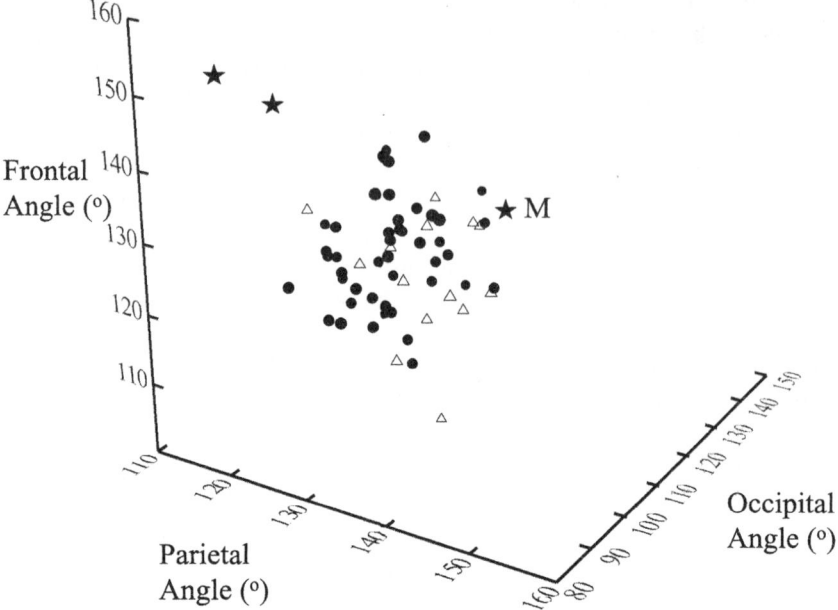

Figure 9.1   Three-dimensional plot of cranial angles. *H. erectus* specimens are designated by filled stars (Mojokerto is identified by the letter M). Adult *H. sapiens* are identified by filled circles, while juveniles are presented as unfilled triangles.

Results

Heterochrony

In comparison of frontal, parietal, and occipital angles, the presumed descendant, adult human, shape resembles the shape of an ancestral juvenile, Mojokerto (Figure 9.1). The skull shape of Mojokerto is closely associated with, if slightly outside of, the data points representing anatomically modern *H. sapiens*. In contrast, data points for adult *H. erectus* Sangiran 2 and Ngandong 6 are outside the scatter for anatomically modern *H. sapiens*. A few adult Native American crania do fall within a shape-space for *H. erectus* as delineated by the plane connecting Mojokerto and the adult *H. erectus* data points. If we had data from subadult *H. erectus* specimens older than Mojokerto we would be better able to define the range of the juvenile *H. erectus* shape-space and we might find that the modern humans plot only in the juvenile portion of the *H. erectus* space. This remains to be tested when larger datasets of *H. erectus* are available. The cranium of anatomically modern *H. sapiens* is shaped much like a juvenile *H. erectus*, and thus can be considered paeodomorphic. Additionally, the degree of shape change during ontogeny in *H. erectus* is substantially greater than the degree of shape change in anatomically modern *H. sapiens*. As recognized by Gould (1977), human cranial shape changes are limited during ontogeny. We can add that ontogenetic shape changes are quite substantial in *H. erectus* (Antón, 1997, 1999; Antón & Franzen, 1997). In addition, the juvenile skull shape of *H. sapiens* is evident despite the fact that adult anatomically modern *H. sapiens* possess very large brains compared to *H. erectus* (Antón & Leigh, 1998). In other words, the juvenile shape of the modern human brain, as reflected in vault shape, is preserved through a massive ontogenetic and phylogenetic size increase. This implies that size and shape are dissociated between this presumed ancestor–descendant pair.

Size and shape dissociation

Plots of relative growth for the frontal, parietal, and occipital chords suggest that size and shape dissociation are responsible for the paedomorphosis observed in shape comparisons (Figure 9.2). It is important to note that, for single variables, there is considerable overlap in the shape between taxa. This arises as a consequence of estimating shape based on single linear dimensions. However, shape information from bivariate plots is consistent with analyses based on angular data above.

The majority of adult anatomically modern *H. sapiens* data points lie between the isometric lines for the smallest *H. erectus* skulls; as noted above, we are focusing on adults – the adult data points need to fall in this space. Most importantly, predicted lowess regression lines for adults fall within the boundaries of the isometric lines defined by *H. erectus*. Complex or curvilinear allometry seems to characterize the anatomically modern *H. sapiens* regression line, but in the adult size regions, this regression line is either well within or overlaps the boundaries set by the two smallest *H. erectus* crania. The differences between the lines between taxa reflect a dissociation between size and shape. Thus, for the frontal chord, neoteny (defined as paedomorphic or juvenilized shape in the descendant produced through allometric dissociation) is the implied heterochronic process that produces a juvenilized shape in anatomically modern *H. sapiens*. The parietal chord is consistent with this pattern, with the majority of adult data points lying within the shape lines, but the result is less clear. Once again, size and shape are dissociated, but there is substantial overlap in shapes towards the upper end of the size ranges. *Homo erectus* presents relatively low values for the parietal chord during early parts of ontogeny, while the largest anatomically modern *H. sapiens* far exceed *H. erectus* in the parietal

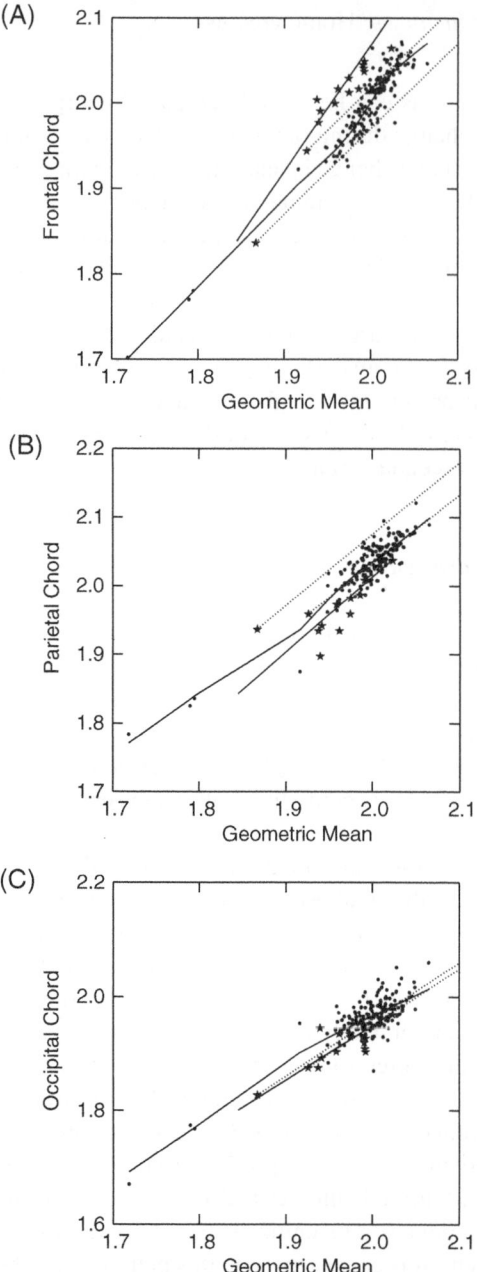

Figure 9.2   Log–log plots of cranial chords and geometric means of frontal, parietal, and occipital chords. *Homo erectus* fossils are identified by stars, *H. sapiens* juveniles and adults by circles. Parallel isometric lines through the Mojokerto and Sangiran 2 data points identify the *H. erectus* subadult morphospace. A lowess regression line describes the relationship between variables in *H. sapiens* and a linear regression line describes the relationship in *H. erectus*.

chord measure. Even though several adult modern humans are very large, some of these individuals are nearly identical in shape to Mojokerto. The occipital chord resembles the parietal chord in that size and shape are dissociated. The average human shape is between the isometric lines for *H. erectus*, while the largest anatomically modern humans present shapes that may not be seen in *H. erectus* and are thus outside the shape envelope defined by juvenile fossils.

Growth through time and growth spurts

Analysis of various facial measurements for *H. sapiens* shows that it is often possible to visualize the adolescent growth spurt from skeletal measures of small cross-sectional samples. Figure 9.3 shows raw data for facial height measurements relative to dental age for Alaskan and Australian modern humans, and *H. erectus*. These graphs are fairly typical of the diverse geographic samples we considered, and show a sizable gap between the 11/12-year-olds and the later adolescents and adults. There is a 15–20% difference between the 11/12-year-old and adult measurements, with a lesser degree of difference between the 15-year-olds and adults. The raw data for mandibular measures are largely similar, although differences between 11/12-year-olds and adult measurements are somewhat smaller than those seen for upper facial height. It should be noted, however, that some samples showed no "break" between these ages in either facial or mandibular heights.

When we convert these raw data into arithmetic-velocity curves we see a pattern of changing growth velocity that is similar to, although less pronounced than, that derived from modern human longitudinal data (Table 9.2; Figure 9.4). For mandibular data, both Alaskan and Australian samples show marked increases in growth velocities beginning between early and mid adolescence. These velocities are double those of the preceding juvenile to adolescent time period in the Alaskans (0.74 vs. 0.37 mm/year) and the Australians are more than three times those of the earlier period (1.53 vs. 0.42 mm/year). The Alaskan sample continues this strong velocity increase by again doubling its velocity in the adolescent to adult period to 1.37 mm/year, whereas in the Australian sample, arithmetic-velocity between the mid adolescent and adult periods falls off to just slightly higher (0.63 mm/year) than pre-adolescent levels. Only the Alaskan sample was sufficiently preserved to examine growth velocities in upper facial height. Again there is a strong increase in the beginning to mid-adolescent period (1.91 mm/year vs 0.68 mm/year) and in the mid-adolescent to adult period (1.19 mm/year). The time-scales are broad and the data coarse-grained, yet a clear increase in velocity is visible during adolescence in the human samples.

232   *S. C. Antón & S. R. Leigh*

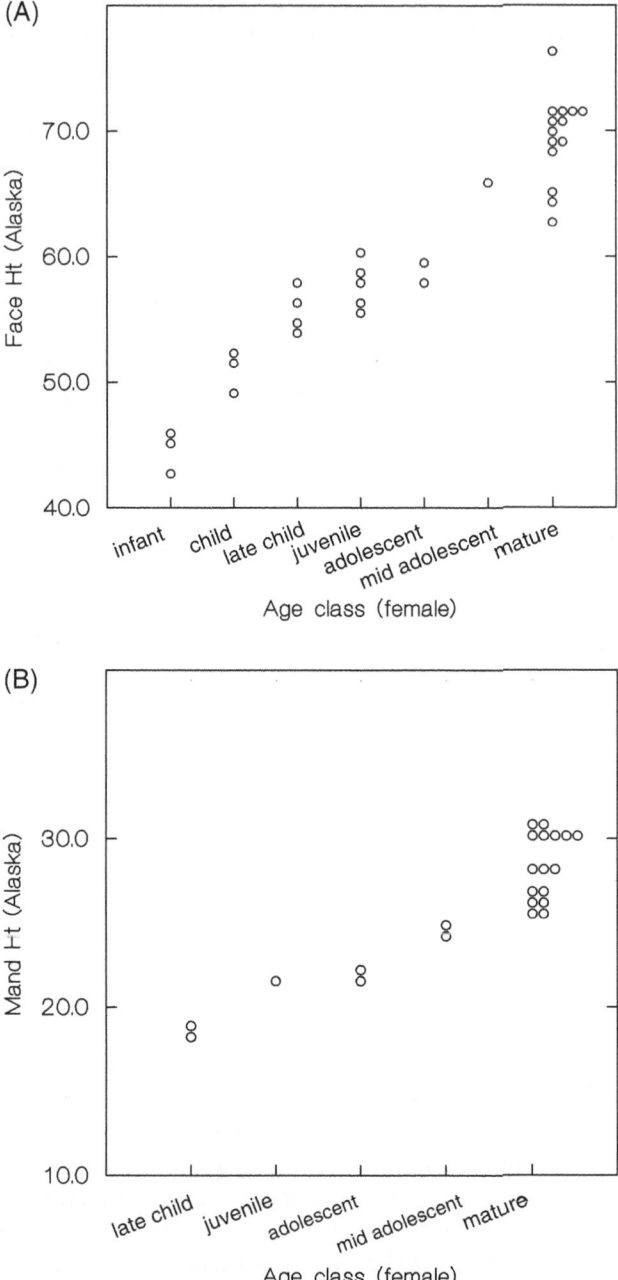

Figure 9.3   Raw data by age class for *H. sapiens* from Alaska for upper facial height (A), and mandibular height (B), *H. sapiens* from Australia mandibular height (C), and *H. erectus* mandibular height (D).

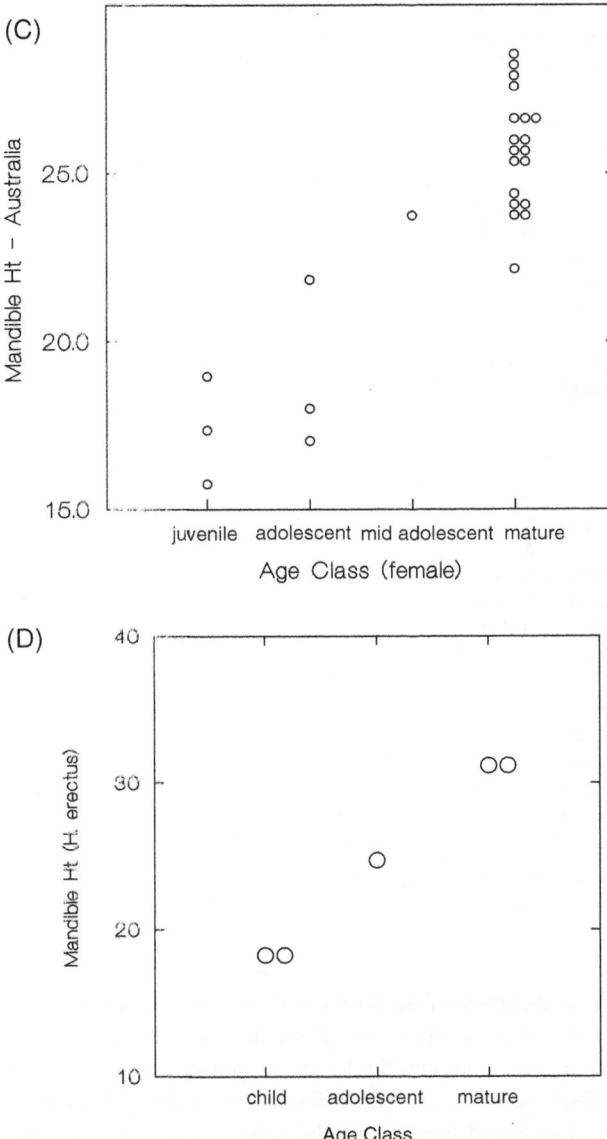

Figure 9.3   (*cont.*)

Table 9.2 *Arithmetic velocities for* H. sapiens *and* H. erectus *skeletal samples based on human age standards*[a]

	Alaska	Australia	H. erectus
Mandible growth velocity			
Child – juvenile (7.5 yrs)	0.74	–	–
Juvenile – adolescent I (10.5 yrs)	0.37	0.42	–
Adolescent I – adolescent II (13 yrs)	0.74	1.53	–
Adolescent II – adult[b]			
(mean = 16.5 yrs)	1.37	0.63	–
(mean = 17.5 yrs)	0.81	0.38	–
Child – adolescent I (8.5 yrs)	0.57	–	1.06
Adolescent I – adult[b]			
(mean = 14.5 yrs)	0.91	0.97	0.98
(mean = 15.5 yrs)	0.71	0.76	0.76
Upper face growth velocity			
Infant – early child (3 yrs)	3.48	–	–
Early child – late child (5 yrs)	2.9	–	–
Late child – juvenile (7.5 yrs)	0.63	–	–
Juvenile – adolescent I (10.5 yrs)	0.68	–	–
Adolescent I – adolescent II (13 yrs)	1.91	–	–
Adolescent II – adult[b]			
(mean = 16.5 yrs)	1.19	–	–
(mean = 17.5 yrs)	0.72	–	–
Adolescent I – adult[b]			
(mean = 14.5 yrs)	1.57	–	0.72
(mean = 15.5 yrs)	1.19	–	0.55
Child – adolescent I (8.5 yrs)	1.27	–	–

[a] Age classes as defined in text. Mean age between classes in parentheses.
[b] Adult ages varied between 18 and 20 years, as per text. Mean age between classes in parentheses.

Because the age classes of the fossil samples are even more coarse-grained than the modern human samples, we also calculate arithmetic-velocities over multiple age classes, comparable to those preserved in the fossil record (Table 9.2). Thus we calculate velocities for the modern human sample between late-childhood and beginning-adolescent values, skipping the juvenile data points, and we calculate velocities between beginning-adolescent and mature values, skipping the mid-adolescent values. The effect of combining age classes depresses the differences between growth velocities. As a result, although it is still possible in some cases to see increases between the earlier velocities of childhood to adolescence and adolescence to adult periods, the differences are not very great. For mandibular values depending on the age at maturity

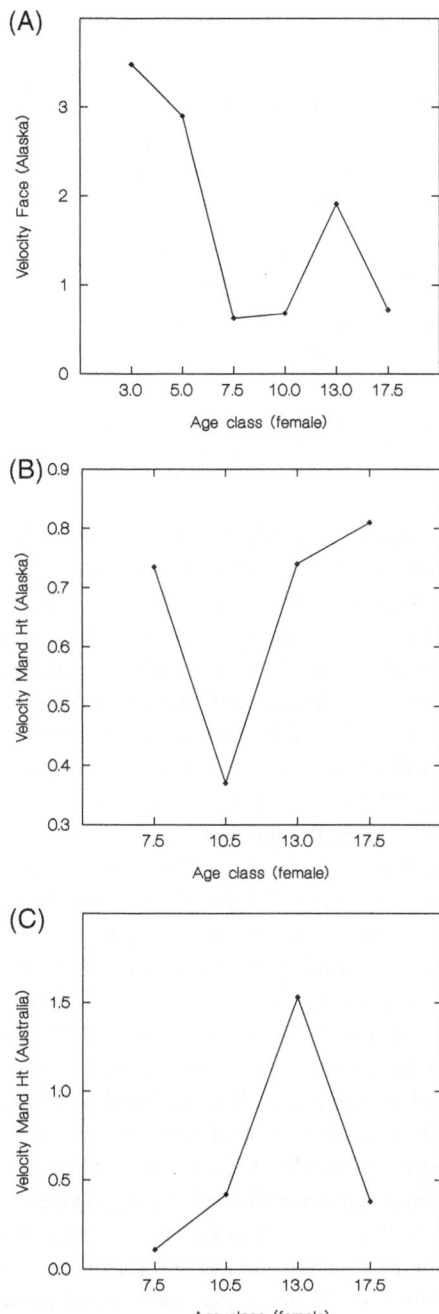

Figure 9.4   Arithmetic-velocity curves for *H. sapiens* from Alaska for upper facial height (A) and mandibular height (B), and *H. sapiens* from Australia mandibular height (C).

Table 9.3 *Arithmetic velocities for* H. erectus *using human and ape standards in mm/year*

	Human (maturity 18 years)	Human (maturity 20 years)	Ape (maturity 10 years)	Ape (maturity 12 years)	*Pan* (maturity 12 years)
Mandible growth velocity					
Adolescent I – adult	0.98	0.76	3.42	1.71	1.36
Child – adolescent I	1.06	1.06	1.73	1.73	1.00
Upper facial growth velocity					
Adolescent I – adult	0.72	0.55	2.50	1.25	2.78

used, the adolescent to adult values are as little as 0.14 mm/year different, and for upper facial height the child to adolescent values are even slightly greater (0.08 mm/year) than the adolescent to adult value, implying no growth spurt.

Even in a single geographic region (Africa) the picture is a bit muddier in the fossil samples. As with the cranial contour data the mandibular data suggest that the KNM-WT 15000 mandible would have had to undergo substantial growth in height to reach adult size. The KNM-WT 15000 mandible is about 29% of the adult mandibular height of KNM-ER 992 or 730. However, we do not have any juvenile individuals older than KNM-WT 15000 with which to show whether there was a gap similar to that seen in the modern human samples. When we consider upper facial height we have only the single KNM-WT 15000 specimen to compare with the East Turkana adult KNM-ER 3733. Facial height differs about 10–12% between the two. This is not as great a value as the 20% typically seen in modern humans (Buschang *et al.*, 1983; Goldstein, 1936; Marshall & Tanner, 1986), although if we include the possibility that KNM-WT 15000 is male and KNM-ER 3733 may be female the discrepancy may be explained.

When we convert the data into arithmetic-velocities the results vary depending upon the age standards used. In the first analysis we use developmental ages based on modern human standards (Tables 9.2 and 9.3; Figure 9.5). We ran two trials, both using an age of 11 years for KNM-WT 15000 (Smith *et al.*, 1995) and one using an adult age of 18 years, the other an adult age of 20 years for mature specimens from Koobi Fora. Both analyses yield relatively small mandibular growth velocities nearly identical to the absolute values for the modern human samples. Using an age at maturity of 18 years, pre-adolescent and adolescent growth velocities are essentially identical (1.06 and 0.98 mm/year). Using an age at maturity of 20 years, adolescent growth velocity falls to 0.76 mm/year. In contrast to the results for the human samples, the adolescent growth velocities for upper facial height are much smaller than those seen for mandibular height,

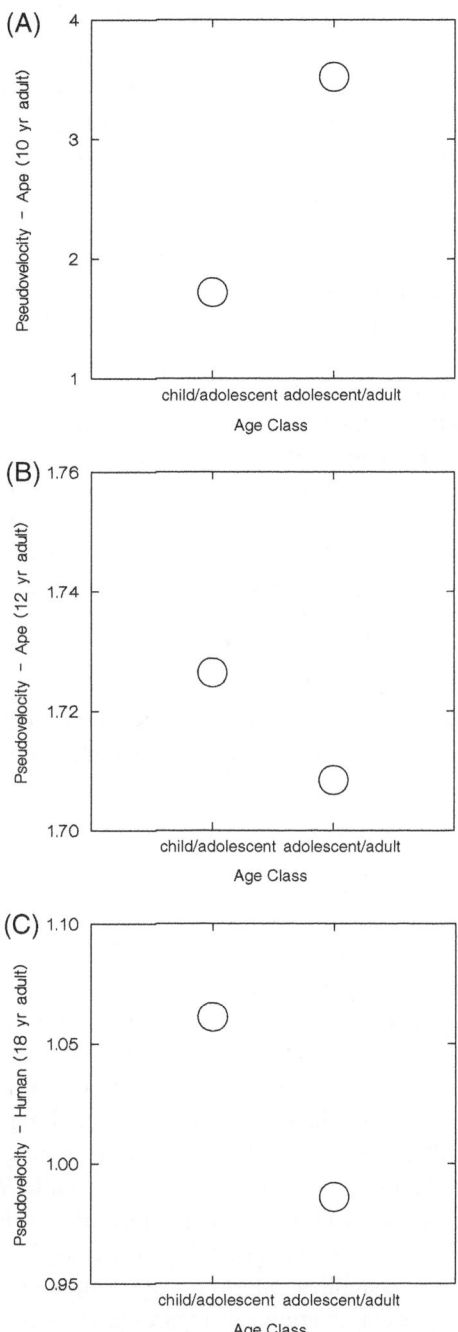

Figure 9.5   Arithmetic-velocity curves for mandibular data for *H. erectus* using ape (A, B) and human (C) models for estimating adult and subadult ages.

regardless of the age used for maturity (0.55 to 0.72 mm/year) (Table 9.2). Because there are no known *H. erectus* children's faces, we cannot compare between childhood and adolescent upper facial growth velocities.

In the second analysis using developmental ages based on ape standards growth velocity is substantially greater than when using the human standards because time scales are relatively shorter (Table 9.3; Figure 9.5). The most recent data on dental microstructure suggest that ape standards are the most appropriate standards for aging the African *H. erectus* specimens (Dean *et al.*, 2001). Again we ran two trials, both using an age of 8 years for KNM-WT 15000 and one using an adult age of 10 years, the other an adult age of 12 years for the mature specimens. Using a 10-year age of maturity, mandibular height shows a substantial increase in growth velocity during adolescence, from 1.73 mm/year from childhood to adolescence to 3.42 mm/year from adolescence to maturity. This adolescent growth velocity represents about 10% of the total mandibular height per year, which is comparable to the relationship seen between the magnitude of weight gain per year and the average adult weight in the African apes (Leigh, 1996). Alternatively, using a 12-year age of maturity, mandibular growth velocity is essentially the same during both preadolescent and adolescent periods at 1.73 and 1.71 mm/year, respectively. Again facial growth velocities are slower than mandibular velocities varying between 2.5 and 1.25 mm/year when using 10-year-old and 12-year-old maturity values, respectively.

Discussion

Heterochrony, size, and shape

Our two-step heterochronic analysis indicates that anatomically modern *H. sapiens* are paedomorphs relative to *H. erectus*. Furthermore, we cannot reject neoteny as a causal process of paedomorphosis, particularly with regard to the frontal and parietal chord measures. The pattern in the occipital region is less clear, but a tendency towards neoteny is evident. Modern humans present an adult skull shape that is very similar to skull shapes seen in juvenile *H. erectus*. This similarity in shape is apparent despite a major increase in the size of the brain in modern humans. We must emphasize that the heterochronic transformation that we have observed applies only to a very limited region of the skull. We obviously cannot present an argument regarding other morphological or behavioral systems. Nevertheless, we must also emphasize that human neoteny occurs in an extremely important derived trait, the shape of the neurocranium. Consequently, neoteny is either directly or indirectly associated with selective factors that have influenced the size and shape of the modern human brain.

Ontogenetic shape change is quite marked in *H. erectus*, but is very limited in *H. sapiens* (Gould, 1977). Thus, cranial shape change dominates the ontogeny of *H. erectus*, but cranial and presumably brain size changes, coupled with conservative shape change, prevails in the ontogeny of *H. sapiens*. In anatomically modern *H. sapiens* the conservation of a juvenilized brain shape through a comparatively large size increase may have important implications for improving our understanding of three kinds of evolutionary changes. These include maternal metabolic costs, brain and neural function, and cranial functional morphology.

Disruptions of size/shape trajectories occur because, at early postnatal ages, the modern human brain is much larger than the brain of our youngest *H. erectus* individual. Consequently, we can relate size/shape dissociations to greater levels of prenatal and early postnatal brain growth in *H. sapiens*. This implies that human neoteny must involve adaptations responsible for allocating greater amounts of metabolic resources to offspring in support of early brain growth, thus involving increased costs of offspring to modern human mothers relative to *H. erectus*. This cost is exaggerated during the postnatal period by the increased rate of brain growth (and increased infant dependency) resulting from the secondary altriciality of human neonates (*sensu* Martin, 1990: 426). Regardless of whether these costs may be met by reducing other "expensive tissues" (*sensu* Aiello & Wheeler, 1995), by increasing diet quality (*sensu* Leonard & Robertson, 1994, 1997), or by some combination of the two, the implication of increased maternal costs in *H. sapiens* remains. It should be noted that paedomorphosis in and of itself need not require increased costs of brain growth if it is produced through size and shape association (hypomorphosis). However, this implies maintenance of ancestral connections between size and shape that our data do not support.

We are reluctant to move from gross cranial shape to inferences regarding brain function. However, given the obvious cognitive differences between humans and other living primates, it would be premature to eschew discussion of possible neural correlates of our findings. Thus, we have previously hypothesized that the paedomorphic shape of the modern human brain may be related to human neural plasticity (Antón, 2002; Antón & Leigh, 1998). Specifically, adult modern humans possess a brain that may be structured much like a juvenile *H. erectus* and this may imply that modern human brains retain the attributes of a juvenile brain well into the developmental period. If gross proportions provide insight into brain structure and function, then paedomorphosis might help explain why the human brain exhibits a remarkable ability to reconstruct synaptic connections despite both focal trauma and radical surgeries (Vining *et al.*,

1997). For example, human brains are sufficiently plastic to regain language abilities that are within normal limits even after left-side hemispherectomy that eliminates Broca's area (Vining *et al.*, 1997). This example clearly illustrates the level of plasticity in the human brain, but it may also reflect selection for paedomorphosis through neoteny as a mechanism that sustains high levels of plasticity. While adult modern human brains are clearly less plastic than those of modern juveniles, adult brains display at least some plasticity (Rauschecker, 1995), which may be comparable to levels seen in young *H. erectus*. We infer that paedomorphosis may grant our own species a higher degree of neural plasticity than was available to adult *H. erectus*.

In this view, the heterochronic shifts that we have measured reflect a comparatively "simple" way to respond to selection for neural plasticity. Retention of juvenile shapes implies that selection for novel brain structures or proportions need not be invoked to explain the initial evolution of human neural plasticity. If we accept the idea that juvenile *H. erectus* individuals exhibited enhanced neural plasticity relative to adult *H. erectus*, then selection for the retention of neural plasticity might involve the retention of a juvenile brain shape. In other words, the relations among brain components that are observed in juvenile *H. erectus* are present at new sizes in *H. sapiens*. Obviously, retention of a paedomorphic morphology is coupled with a substantial evolutionary increase in size. It is very difficult to determine how selection may have affected the relations between size and shape, but dissociation may suggest that paedomorphosis was selectively favored independent of increases in brain size.

At an ecological level, paedomorphosis is often conceptualized as a heterochronic adaptation that limits the degree of evolutionary specialization. This idea accords well with suggestions that behavioral flexibility is a "hallmark" of human evolution (Potts, 1996). Early investments in brain size growth coupled with retention of ancestral juvenile cranial shapes may reflect the major adaptation to selection for behavioral flexibility. Thus, paedomorphosis through the process of neoteny may have been a response to selection favoring high levels of neural plasticity.

Alternatively, we might also reasonably argue that the heterochronic shifts that we measured simply reflect a conservative way to increase brain size (with the attendant "advantages") while also maintaining functional relations within the skull. That is, unlike the neuroplasticity view, paedomorphosis may in fact have been directly linked to size increase as a conservative way to maintain biomechanical functions. In this view, paedomorphosis would not be equated with flexibility but with conservatism. Because our skull shape worked for a juvenile of an ancestral form, retaining similar shape relations into a larger size range may actually be just a conservative way of accommodating the size increase of our brain and thus the shape change may be driven by size increases.

Because of differences in the strength of a sphere versus more angular forms, retention of the juvenile, more globular, vault shape has the added advantage of allowing maximal size increase without requiring increases in cranial thickness. In fact, such a shift permits the reduction of cranial thickness and the energy and resources necessary to grow thick cranial walls. Of course this advantage accrues regardless of the underlying reason for the change in cranial shape. Brain size increase, at least beyond a critical threshold, may require a less angular, more juvenile, vault form for mechanical reasons. The globular, juvenile form of the modern human brain, then, may bear no relation to juvenilized function or increased neural plasticity.

Growth through time and adolescent growth spurts

Opinions differ as to the significance (adaptive advantage) of an adolescent growth spurt, from the more behavioral/cultural advantages of allowing a period for practice of adult economic, social, and sexual behaviors without the full consequences accruing when mistakes are made (Bogin, 1994b), to the more physical advantages of delaying the increased metabolic risk of larger body size to a later time period (Leigh, 1996). However, the strong adaptive advantages suggested for such a spurt provide good reason to want to assess when that spurt began in our lineage.

Because of the rather severe limitations of the fossil record, the results of the growth velocity analyses are equivocal as to whether an adolescent growth spurt occurred in *H. erectus*. However, we have sufficient data to urge caution in dismissing the idea of an adolescent growth spurt in early *H. erectus*. If we accept that ape standards are more appropriate for determining chronological age in early *H. erectus* as Dean and colleagues (2001) have suggested, then the case for the growth spurt is made much stronger. However, even if we do not, at best, we cannot distinguish unequivocally whether or not early *H. erectus* lacked an adolescent growth spurt. Given the relatively ubiquitous nature of somatic adolescent growth spurts in the great apes (Leigh, 1996) and the attendant difficulties of visualizing growth spurts in cranial skeletal measures, even in animals in which we know quite extreme adolescent growth spurts exist somatically, it seems conservative to accept that some type of growth spurt existed in early *H. erectus*. If we then add the even greater difficulty of visualizing these spurts in animals such as the great apes who exhibit less exaggerated changes in growth velocity than in modern humans, it seems almost prudent to accept that a growth spurt likely occurred in early *H. erectus*. However, only better fossil data and clearer means of assessing developmental age will tell.

Conclusions

Comparative analyses of ontogeny in *H. erectus* and in anatomically modern *H. sapiens* suggest that the shape of the modern human neurocranium is paedomorphic relative to an ancestral condition represented by *H. erectus*. Paedomorphosis is a result of the heterochronic process of neoteny, which involves major differences in allometric trajectories. Most specifically, paedomorphosis is produced through size and shape dissociation. Dissociation of size and shape indicates substantial increases in brain size during early growth that would have substantially elevated maternal metabolic costs of pregnancy and lactation. We suggest that human neurocranial paedomorphosis reflects either selection for high levels of neural plasticity or a conservative solution to accommodating extreme increases in brain size. Neotenic shifts in development obviate major restructuring of the human brain, but simply involve retention of juvenile shapes.

Analyses of life-history patterns are equivocal as to the presence of an adolescent growth spurt in *H. erectus*. However, even the sparse data that are available cannot rule out the possibility that such a spurt existed. Additional research to increase human and non-human primate comparative samples and to discover additional juvenile *H. erectus* is needed.

Acknowledgments

We thank the volume editors for the invitation to participate. This paper is the amalgamation of portions of two separate papers presented at the original symposium of the American Association of Physical Anthropologists in 2001. We are grateful to the following individuals for access to specimens in their care: Dr. T. Jacob, Gadjah Mada University, Yogyakarta, Indonesia; Dr F. Aziz, Quaternary Research Laboratory, Geological Research and Development Center, Bandung, Indonesia; Dr J. L. Franzen, Senckenberg Museum, Frankfurt, Germany; Dr D. Lordkipanidze, Georgian State Museum, Tbilisi, Georgia; Drs I Tattersall and K. Mowbray, American Museum of Natural History, New York, USA; Dr M. Wiant, Illinois State Museum, UIUC, USA; Dr Francis Thackery, Transvaal Museum, South Africa. A. Taylor and W. R. Leonard provided helpful discussion. This study was partially supported by NSF BCS-9804861 and the Ford Foundation (SCA).

References

Aiello, L., & Wheeler, P. (1995). The expensive tissue hypothesis: The brain and digestive system in human and primate evolution. *Current Anthropology*, **36**, 199–221.

Antón, S. C. (1997). Developmental age and taxonomic affinity of the Mojokerto child, Java, Indonesia. *American Journal of Physical Anthropology*, **102**, 497–514.

Antón, S. C. (1999). Cranial growth in *Homo erectus*: How credible are the Ngandong juveniles? *American Journal of Physical Anthropology*, **108**, 223–236.

Antón, S. C. (2002). Cranial growth in *Homo erectus*. In *Human Evolution through Developmental Change*, eds. N. Minugh-Purvis & K. McNamara, pp. 349–380. Baltimore: Johns Hopkins University Press.

Antón, S. C., & Franzen, J. L. (1997). The occipital torus and developmental age of Sangiran-3. *Journal of Human Evolution*, **33**, 599–610.

Antón, S. C., & Leigh, S. R. (1998). Paedomorphosis and neoteny in human evolution. *Journal of Human Evolution*, **34**, A2.

Boas, F. (1892). The growth of children. *Science*, **19**, 281–282.

Bogin, B. (1994a). *Patterns of Human Growth*. Cambridge: Cambridge University Press.

Bogin, B. (1994b). Adolescence in evolutionary perspective. *Acta Paediatrica* Suppl., **406**, 29–35.

Bogin, B. (1999). *Patterns of Human Growth*, 2nd edn. Cambridge: Cambridge University Press.

Bogin, B., & Smith, B. H. (1996). Evolution of the human life cycle. *American Journal of Human Biology*, **8**, 703–716.

Buikstra, J. E., & Ubelaker, D. H. (1994). *Standards for Data Collection from Human Skeletal Remains*, Arkansas Archeological Survey Research Series no. 44. Fayetteville: Arkansas Archeological Survey.

Buschang, P. H., Baume, R. M., & Nass, G. G. (1983). Craniofacial growth maturity gradient for males and females between 4 and 16 years of age. *American Journal of Physical Anthropology*, **61**, 373–381.

Clegg, M., & Aiello, L. C. (1999). A comparison of the Nariokotome *Homo erectus* with juveniles from a modern human population. *American Journal of Physical Anthropology*, **110**, 81–94.

Coelho, A. M. (1985). Baboon dimorphism: Growth in weight, length, and adiposity from birth to 8 years of age. In *Nonhuman Primate Models for Human Growth and Development*, ed. E. S. Watts, pp. 125–159. New York: Alan R. Liss.

Dean, C., Leakey, M. G., Reid, D., Schrenk, F., Schwartz, G. T., Stringer, C., & Walker, A. (2001). Growth processes in teeth distinguish modern humans from *Homo erectus* and earlier hominins. *Nature*, **414**, 628–631.

de Beer, G. (1971). *Embryos and Ancestors*. Oxford: Clarendon Press.

Godfrey, L. R., King, S. J., & Sutherland, M. R. (1998). Heterochronic approaches to the study of locomotion. In *Primate Locomotion: Recent Advances*, eds. E. Strasser, J. Fleagle, A. Rosenberger, & H. McHenry, pp. 177–207. New York: Plenum Press.

Goldstein, M. S. (1936). Changes in dimensions and form of the face and head with age. *American Journal of Physical Anthropology*, **22**, 37–89.

Gould, S. J. (1977). *Ontogeny and Phylogeny*. Cambridge: Harvard University Press.

Hamill, P. V. V., Johnston, F. E., & Lemeshow, S. (1973). *Body Weight, Stature and Sitting Height: White and Negro Youths 12–17 Years*, Vital and Health Statistics, Series 11, Data from the National Health Survey no. 126. Rockville: National Center for Health Statistics.

Hill, K. R., & Hurtado, A. M. (1996). *Ache Life History*. New York: Aldine de Gruyter.

Howell, N. (1979). *Demography of the Dobe !Kung*. New York: Academic Press.

Howells, W. W. (1973). *Cranial Variation in Man*. Cambridge: Peabody Museum of Archeology and Ethnology.

Jungers, W. L., Falsetti, A. B., & Wall, C. E. (1995). Shape, relative size, and size-adjustments in morphometrics. *Yearbook of Physical Anthropology*, **38**, 137–162.

Kuykendall, K. L., & Conroy, G. C. (1996). Permanent tooth calcification in chimpanzees *(Pan troglodytes)*: Patterns and polymorphisms. *American Journal of Physical Anthropology*, **99**, 159–174.

Leigh, S. R. (1996). Evolution of human growth spurts. *American Journal of Physical Anthropology*, **101**, 455–474.

Leigh, S. R. (2001). Evolution of human growth. *Evolutionary Anthropology*, **10**, 223–236.

Leigh, S. R. (2002). Book review: *Human Evolution through Developmental Change*. *Journal of Human Evolution*, **43**, 768–770.

Leigh, S. R., & Shea, B. T. (1996). Ontogeny of body size variation in African apes. *American Journal of Physical Anthropology*, **99**, 43–66.

Leigh, S. R., Shah, N., & Buchanan, L. S. (2003). Ontogeny and phylogeny in papionin primates. *Journal of Human Evolution* (in press).

Leonard, W. R., & Robertson, M. L. (1994). Evolutionary perspectives on human nutrition: The influence of brain and body size on diet and metabolism. *American Journal of Human Biology*, **6**, 77–88.

Leonard, W. R., & Robertson, M. L. (1997). Comparative primate energetics and hominid evolution. *American Journal of Physical Anthropology*, **102**, 265–281.

Marshall, W. A., & Tanner, J. M. (1986). Puberty. In *Human Growth: A Comprehensive Treatise,* vol. 2, eds. F. Falkner & J. M. Tanner, pp. 171–209. New York: Plenum Press.

Martin, R. D. (1990). *Primate Origins and Evolution*. Princeton: Princeton University Press.

McKinney, M. L. (2002). Brain evolution by stretching the global mitotic clock of development. In *Human Evolution through Developmental Change*, eds. N. Minugh-Purvis & K. J. McNamara, pp. 173–188. Baltimore: Johns Hopkins University Press.

McKinney, M. L., & McNamara, K. J. (1991). *Heterochrony: The Evolution of Ontogeny*. New York: Plenum Press.

Minugh-Purvis, N., & McNamara, K. J. (eds.) (2002). *Human Evolution through Developmental Change*. Baltimore: Johns Hopkins University Press.

Nehm, R. H. (2001). The developmental basis of morphological disarmament in *Prunum* (Neogastropoda: Marginellidae). In *Beyond Heterochrony: The Evolution of Development*, ed. M. Zelditch, pp. 1–26. New York: John Wiley & Sons.

Pennington, R. (2001). Hunter–gatherer demography. In *Hunter–Gatherers: An Interdisciplinary Perspective*, eds. C. Panter-Brick, R. H. Layton, & P. Rowley-Conwy, pp. 170–204. Cambridge: Cambridge University Press.

Potts, R. (1996). *Humanity's Descent: The Consequences of Ecological Instability*. New York: William Morrow.

Rauschecker, J. P. (1995). Developmental plasticity and memory. *Behavior and Brain Research*, **66**, 7–12.

Rice, S. H. (2002). The role of heterochrony in primate brain evolution. In *Human Evolution through Developmental Change*, eds. N. Minugh-Purvis & K. J. McNamara, pp. 154–172. Baltimore: Johns Hopkins University Press.

Shea, B. T. (1983). Allometry and heterochrony in the African apes. *American Journal of Physical Anthropology*, **62**, 275–290.

Shea, B. T. (1989). Heterochrony in human evolution: The case for neoteny reconsidered. *Yearbook of Physical Anthropology*, **32**, 690–702.

Shea, B. T. (2000). Current issues in the investigation of evolution by heterochrony, with emphasis on the debate over human neoteny. In *Biology, Brains and Behavior: The Evolution of Human Development*, eds. S. Taylor Parker, J. Langer, & M. L. McKinney, pp. 181–214. Santa Fe: School of American Research Press.

Shea, B. T. (2002). Are some heterochronic transformations likelier than others? In *Human Evolution through Developmental Change*, eds. N. Minugh-Purvis & K. McNamara, pp. 79–101. Baltimore: Johns Hopkins University Press.

Smith, B. H. (1993). The physiological age of KNM-WT 15000. In *The Narioko-tome* Homo erectus *Skeleton*, eds. A. Walker & R. Leakey, pp.195–220. Cambridge: Harvard University Press.

Smith, R. J., Gannon, P. J., & Smith, B. H. (1995). Ontogeny of australopithecines and early *Homo*: evidence from cranial capacity and dental eruption. *Journal of Human Evolution*, **29**, 155–168.

Tardieu, C. (1998). Short adolescence in early hominids: Infantile and adolescent growth of the human femur. *American Journal of Physical Anthropology*, **107**, 163–178.

Ubelaker, D. H. (1984). *Human Skeletal Remains: Excavation, Analysis, Interpretation*, revised edn. Washington, DC: Taraxacum Press.

Vining, E. P., Freeman, J. M., Pillas, D. J., Uematsu, S., Carson, B. S., Brandt, J., Boatman, D., Pulsifer, M. B., & Zuckerberg, A. (1997). Why would you remove half a brain? The outcome of 58 children after hemispherectomy: The Johns Hopkins experience 1968 to 1996. *Pediatrics*, **100**, 163–171.

Watts, D. P., & Pusey, A. E. (1993). Behavior of juvenile and adolescent great apes. In *Juvenile Primates*, eds. M. E. Pereira & L. A. Fairbanks, pp. 148–167. New York: Oxford University Press.

Weidenreich, F. (1941). The brain and its role in the phylogenetic transformation of the human skull. *Transactions of the American Philosophical Society*, **31**, 321–442.

Weidenreich, F. (1943). The skull of *Sinanthropus pekinensis*: A comparative study on a primitive hominid skull. *Paleontologia Sinica*, New Series D, **10**, 1–298.

Zelditch, M. L. (ed.) (2001). *Beyond Heterochrony: The Evolution of Development*. New York: John Wiley & Sons.

10 Patterns of dental development in Lower and Middle Pleistocene hominins from Atapuerca (Spain)

J. M. BERMÚDEZ DE CASTRO
Museo Nacional de Ciencias Naturales, Madrid

F. RAMÍREZ ROZZI
Centre National des Recherches Scientifiques

M. MARTINÓN-TORRES
Museo Nacional de Ciencias Naturales, Madrid

S. SARMIENTO PÉREZ
Museo Nacional de Ciencias Naturales, Madrid

A. ROSAS
Museo Nacional de Ciencias Naturales, Madrid

Introduction

Knowledge about the evolution of life history in hominins is one of the most important challenges in palaeoanthropology. A list of traits, such as a relatively short birth interval, helpless newborns, a high rate of postnatal brain growth, extended period of offspring dependency, intense maternal and paternal care, prolonged period of maturation, marked adolescent growth spurt, and delayed reproduction cycles characterize our unique life strategy, differentiating us from living great apes. The questions of how, when, and in which chronological order these traits were acquired have attracted the attention of paleoanthropologists throughout the twentieth century (Bolk, 1926; Dart, 1948; Etkin, 1954; Isaac, 1978; Keith, 1949; Lancaster, 1978; Le Gros Clark, 1947; Lovejoy, 1981; Mann, 1968; Montagu, 1961; Washburn, 1960).

In the last 20 years there has been an intense debate concerning whether early hominins had an ape- or human-like life-history pattern. Information obtained from the study of tooth development and eruption patterns, as well as the dental tissues (especially enamel) has been used as the basic argument to defend the

Patterns of Growth and Development in the Genus Homo, ed. J. L. Thompson, G. E. Krovitz and A. J. Nelson. Published by Cambridge University Press. © Cambridge University Press 2003.

246

opposing views in this debate (Bromage & Dean, 1985; Conroy & Vannier, 1987; Mann, 1975, 1988; Mann *et al.*, 1987, 1990a, 1990b, 1991; Simpson *et al.*, 1990; Smith, 1986, 1994). Recently, the general consensus is that a short life history characterized *Australopithecus, Paranthropus*, and early *Homo*. The subsequent evolution of the genus *Homo* included an overall lengthening of development, as well as the appearance of new life stages (Bogin & Smith, 1996). Late Pleistocene hominins may have been characterized by a life-history pattern similar to that of modern humans (Dean *et al.*, 2001; Smith & Tompkins, 1995; Tompkins, 1996). However, there is inadequate information available between 1.5 and 0.1 million years (Ma) that can be used to investigate the evolution of life history. Exactly when and how did a fully modern human maturational pattern appear in the evolutionary history of the genus *Homo*?

Dental development and life history

Dental development can be considered a fitting measure of life history in primates (Smith, 1989) as its study enables the inference of other life-history traits. The time and timing of those formation events, as well as tooth eruption, constitute the dental development pattern. This process is closely integrated into the overall plan of somatic growth and development and is characteristic of each species (Smith, 1991a). Dental development is highly heritable and relatively resistant to extreme malnutrition and disease processes (Lewis & Garn, 1960). According to these authors, there is less variation in the pattern of dental development (as measured by the coefficient of variation) than in skeletal maturation parameters.

There is a high correlation among different markers of life history across primate species, including birth weight, gestation length, age at weaning, inter-birth interval, and age at sexual maturity or lifespan (Harvey & Clutton-Brock, 1985). Certainly, the development of a species is the execution of a program under strict genetic control with some influence from environmental factors. The different variables of dental development also have a high correlation with other life-history variables, as has been shown by Smith (1989) in a studied sample of 21 primate species. This author illustrated the strong correlation between two dental development markers, the age at first permanent molar (M1) eruption and the age at complete dentition, with brain weight for newborns (0.99 in both cases) and adults (0.98 and 0.97 respectively). Thus, brain development is an essential reference point for the definition of species' life-history patterns. Brain metabolism and energy processing comprise the pacemaker of vertebrate growth and aging (Sacher & Staffeldt, 1974). Unlike brain growth that is highly influenced by environmental factors, dental development is very resistant to environmental perturbations and has a relatively low variance (Lewis & Garn, 1960). Therefore, brain weight on one hand, and the pattern of dental development and eruption on the other, comprise two strongly related aspects defining the life history of primate species.

Time and timing: Two different approaches

Some life-history features, such as the total duration of the growth process in extinct species, can be ascertained through the determination of time and timing in dental development. The wide fossil record of available hominin teeth allows a valuable approach to this purpose. The crown formation times can be calculated through the study of incremental growth lines of the hard dental tissues (e.g., Beynon & Dean, 1988; Bromage, 1987; Bromage & Dean, 1985; Dean, 1987a, 1987b; Dean *et al.*, 1986). However, determination of the timing of dental development processes does not allow direct calculation of the times of somatic development. Some dental growth events have very similar timing among living great apes and modern humans, especially with teeth belonging to the same dental field, i.e., P3 and P4. However, other events exhibit differences in timing between species as will be shown later. These differences can be used to test whether the dental developmental pattern of a fossil species is more similar to that of modern humans or living great apes. The application of the timing of dental development as a method for inferring features of hominin life history has its supporters (Conroy & Vannier, 1987; Mann, 1968, 1975; Smith & Tompkins, 1995) and its detractors (Simpson *et al.*, 1990, 1992). There are also those who recommend caution when drawing conclusions about somatic maturation through the application of this method (Dean *et al.*, 2001). It is known for example, that members from the genus *Paranthropus* share some similarities with *H. sapiens* in the timing of dental development (Bromage, 1987; Conroy & Vannier, 1991; Smith, 1986, 1994). Those similarities led to conclusions about maturation periods in early hominins (Mann, 1975) that we now believe to be incorrect (i.e., Beynon & Dean, 1988; Bromage, 1987; Bromage & Dean, 1985; Dean, 1987a, 1987b; Conroy & Vannier, 1991; Smith, 1991a). However, *Paranthropus* also presents very significant differences from *H. sapiens* in other aspects of dental development, such as the rate and pattern of enamel growth (Beynon & Wood, 1987; Dean *et al.*, 2001). When comparing relatively distant or different groups, as in the case of different genera (*Paranthropus* and *Homo*), interpretation of pattern similarities requires caution. It is important to bear in mind that a similarity might simply be a homoplasy. However, when considering the dental development pattern of specimens included in the same genus, e.g., *Homo*, we are working in the same adaptive range or within a

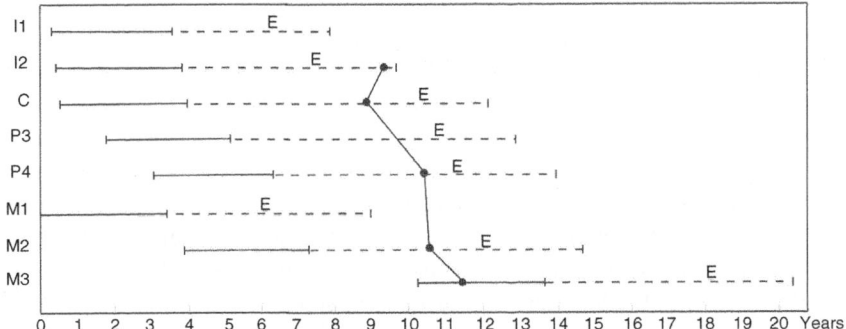

Figure 10.1   Development of mandibular teeth of Hominid 18 from Sima de los Huesos charted on standards for modern humans. Solid lines, duration of crown formation; dotted lines, root formation; E, emergence of teeth. (Chart design from Smith (1993).)

more homogeneous context. Therefore, differences in the timing of dental development events, if detectable, may have a similar meaning in the time and rate of somatic development of those species.

Smith (1986) determined the state of dental development in several specimens of *Australopithecus*, *Paranthropus*, and *Homo*. Their respective states were then plotted on summary charts of dental development of the living great apes and modern humans. The degree of similarity of the developmental state of each studied specimen (measured by the proximity of the plotted points to a straight line) with the ape and human reference charts was measured by the fitness of their profile with one or another chart (see Figure 10.1 for an example of this method application). Smith observed that living great apes, *Australopithecus*, and some early *Homo* specimens share an S-shaped pattern when plotted on modern human standards. This means that the anterior teeth, from I1 to P3, appear delayed in formation relative to M1 (Smith, 1986, 1991b).

On the other hand, Smith (1986) estimated the dental age of every tooth from the hominin specimens, according to the standard of modern human populations. The coefficient of variation (CV) of the data for each specimen was obtained. The analysis was performed with all dental fields (incisors, canines, premolars, and molars) and at least six of the seven maxillary or mandibular teeth (I1–M2) represented. Smith (1986, 1991b) observed that the CV of *H. sapiens* was considerably lower than the CV calculated for *H. habilis*, *Australopithecus*, and living great apes. *Paranthropus* specimens yield CV values intermediate between *H. sapiens* and *Australopithecus*, although closer to this last group. In this context, Smith (1986) suggested that *Paranthropus* was characterized by a unique condition in terms of dental development, where anterior teeth present very advanced development in comparison to M1.

The greatest potential of the dental development pattern as a discriminating tool among species is approached when comparing the development state of teeth belonging to different tooth classes. This comparison reveals the relative independence of the evolutionary trajectories of the different dental fields (Smith, 1994). In closely related hominins, the particular timing of developmental events of the anterior teeth relative to the posterior teeth may give information about the timing of somatic growth and development (Smith, 1994). Therefore, through the determination of the mineralization stages of fossil teeth it is possible to make inferences about the general growth processes of our ancestors. Studies based on the histology of hard tissues have shown that dental formation can be accurately characterized from the analysis of enamel microstructure. Previous studies on dental formation based on enamel microstructure have been carried out on Plio-Pleistocene hominins. The purpose of this analysis was to determine if a prolonged growth period was already present in the oldest hominins. These studies demonstrated that Plio-Pleistocene hominins were characterized by a short time of dental maturation, suggesting that patterns of growth and development in these fossil hominins were closer to those of living great apes than modern humans (Beynon & Dean, 1988; Dean, 1987a, 1987b). A number of studies on enamel microstructure of the most recent fossil hominin species, *H. neanderthalensis*, have not yielded conclusive results yet. Some findings point to rapid anterior tooth formation in concordance with studies on other anatomical parts which propose precocious development for this species (Dean *et al.*, 1986; Ramirez Rozzi, 1993a, 1996). However, molar crown formation does not confirm these previous observations (Dean *et al.*, 2001). Further work needs to be done on other Pleistocene *Homo* species to establish time and patterns of crown formation and, in turn, to obtain information about growth processes in these species.

It is well established that the time of crown formation is influenced by the extension rate of enamel (Dean & Reid, 2001; Ramirez Rozzi, 1993b). The extension rate of enamel is the number of ameloblasts becoming active per day and as a result, this parameter reflects the advance of the enamel-matrix front. In the inner layers of the enamel, the extension rate can be estimated by the slope of the incremental lines (striae of Retzius) at their intersection with the enamel–dentine junction. The more acute the slope, the higher the extension rate. At the enamel surface, the extension rate of enamel can be deduced from the spacing between the perikymata (smooth depressions on the enamel surface produced by contact with the striae of Retzius). In modern humans, the extension rate becomes lower from the cusp tip towards the cervix. At the enamel surface, the slowing of the enamel extension rate can be observed by the reduction of the space between perikymata. This distinctive number and distribution defines a characteristic perikymata packing pattern in modern humans: perikymata are spaced farther apart near the cusp tip and become closer toward the cervix (Beynon & Dean, 1988; Dean, 1987a; Dean & Reid, 2001; Ramirez Rozzi, 1993a, 1996). Thus, changes in the extension rate during crown formation can be estimated from the number and the spacing variation of perikymata. Since the extension rate of enamel influences the time of crown formation, a similar extension rate between two hominid species, e.g., a similar perikymata packing pattern, may suggest analogous crown formation times.

Materials: The Atapuerca hominins

We present here evidence derived from Lower and Middle Pleistocene human fossil remains recovered from two localities of the Sierra de Atapuerca, northern Spain. The Lower Pleistocene fossils come from the TD6 level (Aurora stratum) of the Gran Dolina site, in the Railway Trench which cuts the south-east slope of the Sierra de Atapuerca (Carbonell *et al.*, 1995). These fossils were found in sediments located about 1 meter below the Matuyama–Brunhes boundary (Parés & Pérez-González, 1995) and assigned to a new *Homo* species, *H. antecessor* (Bermúdez de Castro *et al.*, 1997). The history of the archaeological investigations of Gran Dolina, the studies of biochronology, faunal remains, lithic industry, magnetochronology, paleoecology, stratigraphy, taphonomy, and zooarchaeology, as well as the results of the electron spin resonance and uranium (U) series dating methods of the TD6 level, can be found in a special issue of the *Journal of Human Evolution* (Bermúdez de Castro *et al.*, 1999a).

The Middle Pleistocene fossils come from the Sima de los Huesos (SH) site in the Cueva Mayor-Cueva del Silo karst system of the Sierra de Atapuerca (Arsuaga *et al.*, 1997a), about 1 kilometer away from the Railway Trench. By 2001, the human bone breccias of this site have yielded a large hominin sample of nearly 4000 fossils belonging to a minimum of 28 individuals (J. M. Bermúdez de Castro, personal observations). All human bones were deposited during the same sedimentation period (Bischoff *et al.*, 1997), and because of their relative morphological homogeneity they probably belong to the same biological population (Arsuaga *et al.*, 1997b; Bermúdez de Castro, 1988; Carretero *et al.*, 1997; Rosas, 1997). The SH hominins have been included in the species *H. heidelbergensis* (Arsuaga *et al.*, 1997b). The combination of the existing U-series (Bischoff *et al.*, 1997), the macro- and microfauna evidence (Cuenca-Bescós *et al.*, 1997; García *et al.*, 1997), and the paleomagnetic study (Parés *et al.*, 2000) suggest an age between 325 and 200 ka for the Atapuerca SH hominins. However, recent radiometric studies (U-series) of a 14-cm-thick *in situ* speleothem overlying the mud-breccia containing the human bones has provided a minimum age of 350 ka for the SH hominins (Bischoff *et al.*, 2002).

Estimations of the speleothem growth rate, correlation of the fauna from SH (micro- and macromammals) to other Atapuerca sites (e.g., TD6, TD8, TD10, and TD11 levels of Gran Dolina), as well as the normal magnetization of the SH fossiliferous mud give an interval of 400 to 500 ka (oxygen isotope stages 12 to 14) for these hominins (Bischoff *et al.*, 2002). Additional information concerning the SH site can be found in a special issue of the *Journal of Human Evolution* (Arsuaga *et al.*, 1997a).

The total of 85 human remains recovered from the Aurora stratum belong to a minimum of six individuals, and three of them offer information concerning their pattern of dental development. Hominid 1 (the holotype of *H. antecessor*: Figure 3 of Carbonell *et al.*, 1995 and Figures 1, 4, and 5 of Bermúdez de Castro *et al.*, 1999b) suffered a stress episode during early childhood, which produced a disturbance in the formation of the dental tissues. A line of enamel hypoplasia is observed surrounding the crown of the maxillary and mandibular C, P3, P4, and M2, whereas the mandibular I2 and the mandibular and maxillary M1 exhibit a line of dentine fault at the root level, roughly in the same relative position. The right mandibular M3 of this individual did not erupt so he/she died before reaching maturity. Hominid 2 died during early childhood. This individual is represented by a left maxillary fragment with dc and dm1 in place (Figure 5 of Carbonell *et al.*, 1995 and Figure 2 of Bermúdez de Castro *et al.*, 1999b). These teeth are completely formed. Hominid 3, who is represented by a partial face (Figure 1 of Bermúdez de Castro *et al.*, 1997 and Figure 3 of Bermúdez de Castro *et al.*, 1999b), also died before reaching maturity since the M3 crown is not fully formed.

One individual from the SH site offers the opportunity to study patterns of dental development in this sample. Hominid 18 of the SH site died at between 8 and 11 years of age (according to modern human standards), and is represented by 33 isolated teeth, including the upper and lower dm2 and 29 permanent teeth (Bermúdez de Castro & Rosas, 2001). To determine that all these teeth belong to the same individual, the fit of interproximal wear facets of the fully erupted teeth (upper and lower dm2, permanent M1, and central and lateral incisors), as well as similarity in size, were taken into account.

Methods

The mineralization stages of each tooth class (seen by computerized tomography-scan, conventional radiographic observation or directly with the naked eye in the case of loose teeth) of the TD6 and SH hominins were scored using the method of Moorrees *et al.* (1963), with additional stages following Smith (1991b). The mean age of attainment of the different mineralization stages was interpolated from a sample of mandibular and maxillary teeth of Caucasian children recorded by Anderson *et al.* (1976). The ages in modern humans by which stages of root mineralization are comparable to that seen in TD6 and SH hominins were then recorded. Since it is not our purpose to make an age prediction for these hominids, the conversion of Anderson *et al.* (1976) data into age predictions as in Smith (1991b) was not employed.

In order to estimate the fraction (percentage) of root formed in the teeth of Hominid 18 from SH, measurements of completely formed root length were first taken in other complete SH isolated teeth. Root length was defined as the maximum distance in projection from the enamel–cementum junction on the buccal face to the apex of the root (in the case of single-rooted teeth) or to the apex of the longest root (in teeth with multiple roots). The percentage of root formed in Hominid 18 was obtained with reference to the mean length of the complete roots recorded in other SH teeth. The measurements are displayed in Table 2 of Bermúdez de Castro & Rosas (2001).

To ascertain whether the TD6 and SH hominins exhibit a pattern of dental development closer to that of living great apes or modern humans, the criteria of Smith (1994) were followed. We found a high degree of discrimination when both molar and incisor/canine fields are considered, particularly any combination involving M2 and an anterior tooth. The tooth class observed is more critical than tooth number (Smith, 1994). In this study, the relative development of the teeth was investigated by means of cluster analysis, performed with Euclidean distances and complete linkage (furthest neighbor) amalgamation rule. Age scores of every tooth (interpolated from human standards) were divided by the mean age of the individual (Smith, 1994). Each standardized score was subsequently considered as an independent variable. In order to include the largest possible number of specimens, matrix distances were computed with two variables (teeth), under the condition that the incisor/canine and molar fields were both represented. The comparative data were obtained from Smith (1994), who estimated the ages for the corresponding mineralization stages of I2 to M2 in living great apes, *H. sapiens* and different fossil hominin specimens.

In order to know if either the extension rate of enamel or crown formation time in *H. heidelbergensis* was close to that of modern humans, the perikymata packing pattern of some of the SH teeth was studied. The dental fossils were replicated using the addition-curing resin Coltene President in its variants putty and light (Beynon, 1987). Epoxy resin casts were made from each silicone replica and examined in reflected light with a Wild M8 stereomicroscope. Teeth presenting wear equal to or above 25% of total crown height were excluded from the study. Teeth that failed to show perikymata over an area higher than 20% of the buccal enamel surface were not included either. The resultant subsample of SH teeth comprises 11 lower I1, 28 lower I2, 22 lower C, 13 upper I1, 12 upper I2, and 17 upper C. Each tooth replica was then positioned with the buccal face orthogonal to the optical axis of the microscope. A measure of buccal enamel height was taken using a vernier micrometer eyepiece connected to a digital ocular measure linked, in turn, to a calculator–meter–printer RZD-DO (Leica). Buccal enamel height was divided into 10 equal divisions (10th percentiles) from the first formed enamel at the cusp to the last formed at the cervix (Dean & Reid, 2001; Reid & Dean, 2000). Perikymata counts were made in each of the divisions of crown height. The concavity of the division corresponding to the last-formed enamel (cervix) makes the count of perikymata difficult and thus this section was excluded from the study. Divisions that failed to show perikymata over the total height were not included in the study. The variation of perikymata number along the 10 percentiles of the crown height allows the establishment of the perikymata packing pattern. Correlation between the height of the buccal face and the perikymata number in the last divisions was tested to observe if perikymata number, and so crown formation, is related to crown height. Results from SH teeth were compared with data from modern humans obtained from ground sections of 16 lower I1, 14 lower I2, 14 lower C, 19 upper I1, 16 upper I2, and 39 upper C previously prepared for histological examination (Dean & Reid, 2001; Reid *et al.*, 1998). In ground sections, the same methodology applies, the buccal face is divided in percentiles and the incremental lines (striae of Retzius) are counted in each 10th percentile (see Dean & Reid, 2001). *T*-tests were used to assess the significance of the differences in the number of perikymata in each division and for each tooth type, between the Atapuerca and *H. sapiens* samples.

Results

Table 10.1 shows the stages of mineralization of TD6 and SH hominin teeth, along with the chronological ages associated with these stages in modern humans. Table 10.2 shows the resultant values obtained from dividing the age score of I2, C, M1, and M2 of each specimen by its mean dental age. Some specimens have similar values (generally close to 1.0) for I2 and M1 formation. This is the case of Hominin 1 from TD6, Hominin 18 from SH, specimen Zhou B-I (late *H. erectus*), and both specimens of *H. sapiens*. These results indicate similarity between both I2 and M1 ages of formation and tooth age and the average dental age of the individual. The Gibraltar 2 Neandertal is a special case, where the I2 and M1 formation ages coincide, but the average dental age of the individual is higher than the ages for those teeth. For the gorilla sample, one chimpanzee specimen and the Sts 24, Taung, and LH 3 australopithecines, I2 formation appears considerably delayed in comparison with M1 formation.

Table 10.1 *Mineralization stages (MS) of teeth of Hominins 1 and 3 from level TD6 (Aurora stratum) of the Gran Dolina site and Hominin 18 from the Sima de los Huesos (SH) site and chronological ages associated with those stages in living humans*

	TD6 Hominin 1		TD6 Hominin 3		SH Hominin 18	
	MS[a]	Age[b]	MS	Age[b]	MS	Age[b]
Maxilla						
I1					Ac	>10.6
I2			A 1/2-Ac	10.5	A1/2	10.0
C	Cr3/4−	3.8	R3/4-RC	10.3	R1/2–R3/4	8.8
P3	Cr3/4+	5.0	R3/4-Rc	10.5		
P4	Cr2/3+	5.2	R2/3	10.1	R1/2	9.7
M1	R1/4	4.9	Ac	>10.1	Ac	>10.1
M2	Cr1/2	5.3	R1/3	10.0	R1/2−	10.4
	Mean age 4.84		Mean age 10.25			
Mandible						
I1					Ac	>9.2
I2	Ri-R1/4	5.4			Ac	>9.9
C	Cr3/4	3.9			R1/2–R3/4	8.7
P3	Cr3/4	4.7				
P4	Cr1/2	4.8				
M1	R1/4+	5.0			Ac	>10.0
M2	Cr1/2−	5.2			R1/2	10.5

[a] Mineralization stages at the age (years) of formation of a conspicuous enamel (hypoplasia) and dentine development disturbance in Hominin 1 from TD6.
[b] Using standards in Anderson *et al.* (1976).

A third group formed by one *Pan* specimen, the Stw 151 and LH 6 australopithecines, and KNM-WT 15000 and KNM-ER 820 *H. ergaster* specimens shows an intermediate situation where the relative delay of I2 formation compared to M1 formation is less pronounced. Nevertheless, in the "Turkana Boy" hominin, the relative delay of I2 formation in relation to the average dental age is more pronounced than in the other members of this third group.

These data are illustrated in the cluster shown in Figure 10.2A. Hominin 1 from TD6 and Hominin 18 from SH cluster with *H. sapiens*, late *H. erectus*, and Neandertal specimens. A second main branch clearly groups great apes and australopithecines. Note the clustering of KNM-ER 820, LH 6, one specimen of chimpanzee and Stw 151 with Taung and LH 3 on one side, and the clustering of KNM-WT 15000 with specimens of *Pan* and *Gorilla* on the other side. Some authors (Bromage, 1987) assigned KNM-ER 820 to early *Homo*, whereas other authors suggest with caution that this specimen should be attributed either to

Table 10.2 *Values which result from dividing the age scores of I2, C, M1, and M2 of each specimen by the mean age of the specimens. TD6 and SH specimens from this study, other data from Smith (1986, 1994)*

	I2	C	M1	M2
H. sapiens	1.05	0.98	1.01	0.93
H. sapiens	0.97	0.87	0.97	1.15
Gibraltar 2	0.94	0.96	0.85	1.21
Teshik Tash		0.98		1.06
Ehringsdorf		0.97		1.09
SH-H18	1.01	0.89	1.05	1.05
Zhou B-I	1.17	0.85	1.15	0.89
TD6 H1 mx		0.78		1.09
TD6 H1 md	1.12	0.80	1.03	1.07
TD6 H3		1.00		0.97
KNM-WT 15000	0.84	0.87	1.06	1.13
KNM-ER 820	0.98	0.82	1.18	0.98
KNM-ER 1507		0.68		1.08
KNM-ER 1590		0.60		0.96
Stw 151	1.02	0.66	1.28	1.12
Taung	0.83	0.69	1.35	1.12
Sts 24	0.77		1.57	
LH 3	0.88		1.37	
LH 6	0.98		1.18	
Pan troglodytes	1.00	0.46	1.21	1.42
P. troglodytes	0.79	0.51	1.21	1.32
Gorilla gorilla	0.88	0.47	1.23	1.21
G. gorilla	0.78	0.46	1.45	1.27

Homo aff. *H. erectus* (or *H. ergaster*) (Wood, 1991), or early African *H. erectus* (Walker, 1993). The specimen KNM-WT 15000 is also considered either as an early African *H. erectus* (Walker, 1993) or as a representative of *H. ergaster* (Wood, 1992). Thus, from the data of Table 10.2 and Figure 10.2A it seems that either these *Homo* specimens still preserved the primitive condition observed in *Australopithecus* for I2/M1 relative development or they had not reached the derived condition observed in later *Homo* for this trait of dental development. Further findings of early *Homo* specimens probably will shed light on this question.

As shown in Table 10.2, *Pan* and *Gorilla* specimens present very different values for M2 and C formation, with a marked delay for C formation. These results are expected to some extent if we consider that values in Table 10.1 are obtained from modern human population standards. For this same reason, *H. sapiens* specimens display similar values for M2 and C formation. This is also

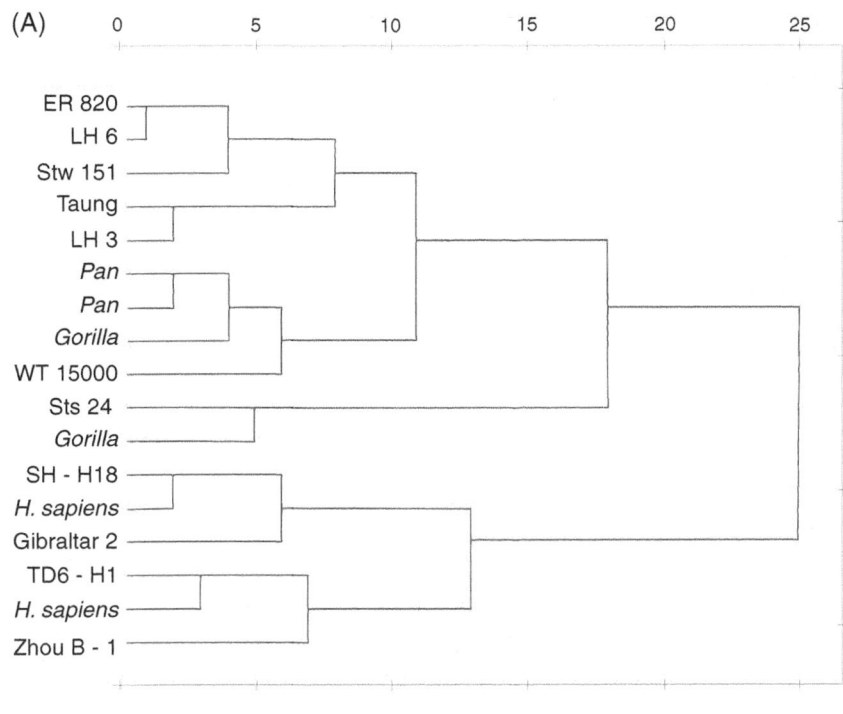

(A)

I2 & M1

Figure 10.2   Cluster analysis computed from dental development scores of (A) lateral
incisor and first molar, (B) canine and second molar of some fossil hominins. Data
from Smith (1986, 1994). Fossil specimens are classified as follows: LH 3 and LH 6
(Laetoli) are *A. afarensis*; Taung, Sts 24, and Stw 151 (Sterkfontein) are *A. africanus*;
ER 1590 is *H. habilis*; WT 15000 (West Turkana), ER 1507, and ER 820 (East
Rudolf) are *H. ergaster*; Zhou B-I (Zhoukoudian) is *H. erectus*; TD6-H1 (Hominid 1
from Gran Dolina–TD6, Atapuerca) and TD6-H3 (Hominid 3 from Gran Dolina–TD6,
Atapuerca) are *H. antecessor*; SH-H18 (Hominid 18 from Sima de los Huesos,
Atapuerca) is *H. heidelbergensis*; Gibraltar 2, Teshik Tash, and Ehringsdorf are *H.
neanderthalensis*. Although the M1 of KNM-WT 15000 is fully formed, we have
considered an age of 10.0 years for this specimen. This assumption is more
conservative than the dental age of 10.4–11.7 assigned to the Nariokotome specimen
using modern human standards (Smith, 1993).

apparent with other *Homo* specimens like TD6 and SH, as well as with Gibraltar
2, Teshik Tash, KNM-WT 15000, and KNM-ER 820 fossils, suggesting that they
fit human standards for the C/M2 relative development. Finally, Taung, Stw 151,
KNM-ER 1590 (*H. habilis*) and KNM-ER 1507 (classified as *Homo* aff.
H. erectus by Wood, 1991) show an intermediate situation where C formation
appears delayed compared to M2 formation, but to a lesser degree than in living
great apes. These findings are also reflected in the cluster shown in Figure 10.2B

(B)

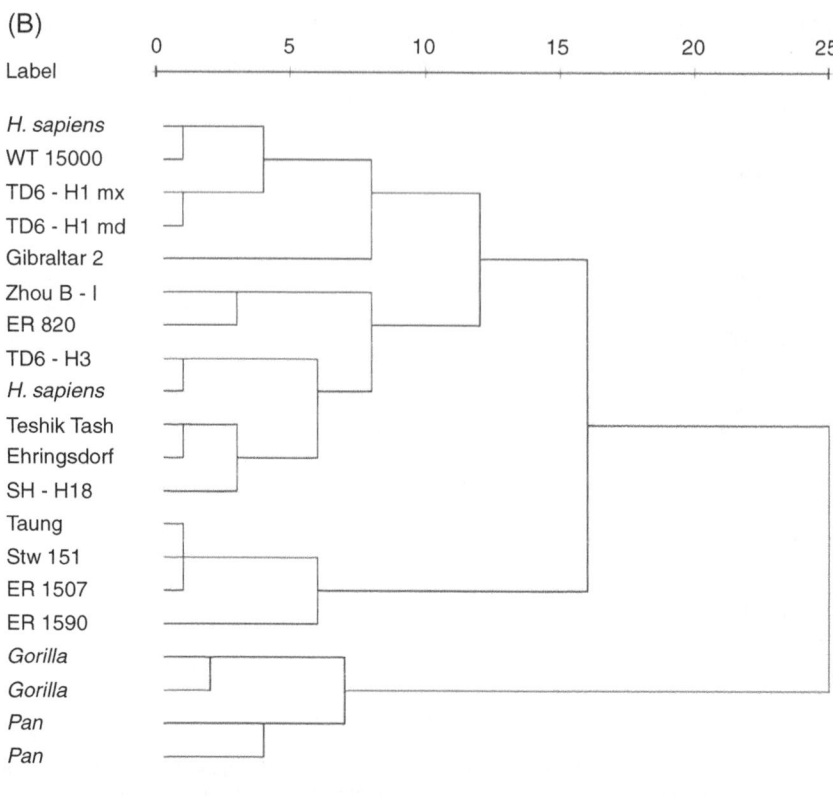

C & M2

Figure 10.2   *(cont.)*

where three main groups can be observed. The first branch assembles *Pan* and *Gorilla* samples. The second branch includes australopithecines, and specimens KNM-ER 1507 and KNM-1590. Finally, the remaining *Homo* representatives are located in a larger group where no further distinctions can be made in absence of chronological or geographical criteria for further grouping. In this case it seems that the genus *Homo* shows a derived condition for C/M2 relative development with regard to early hominins.

Perikymata counts in the divisions of the SH anterior teeth are shown in Table 10.3. The numbers of perikymata recorded in each division for each single individual are plotted in Figure 10.3. Perikymata are more widely spaced near the cusp tip (3rd, 4th, or 5th percentiles) and become closer toward the cervix (8th or 9th percentiles). No correlation was found between the number of perikymata registered in the last divisions and crown height, except for the

Table 10.3 *Perikymata count for each division of the crown height in H. heidelbergensis anterior teeth from the Sima de los Huesos site*

10th percentile	Lower I1			Lower I2			Lower C			Upper I1			Upper I2			Upper C		
	n	$\bar{x}$	SD	n	$\bar{x}$	SD	n	$\bar{x}$	SD	n	$\bar{x}$	SD	n	$\bar{x}$	SD	n	$\bar{x}$	SD
5th	2	9	3	10	10	2	12	13	4	10	10	2	7	9	1	9	14	2
6th	3	11	2	18	12	2	18	15	3	10	13	2	10	12	2	14	16	4
7th	9	14	3	25	15	3	21	17	3	12	14	3	11	15	3	17	18	4
8th	10	17	4	28	18	3	22	21	4	13	17	3	11	17	2	16	21	4
9th	11	19	3	28	22	4	22	24	4	13	19	3	12	20	3	15	22	3

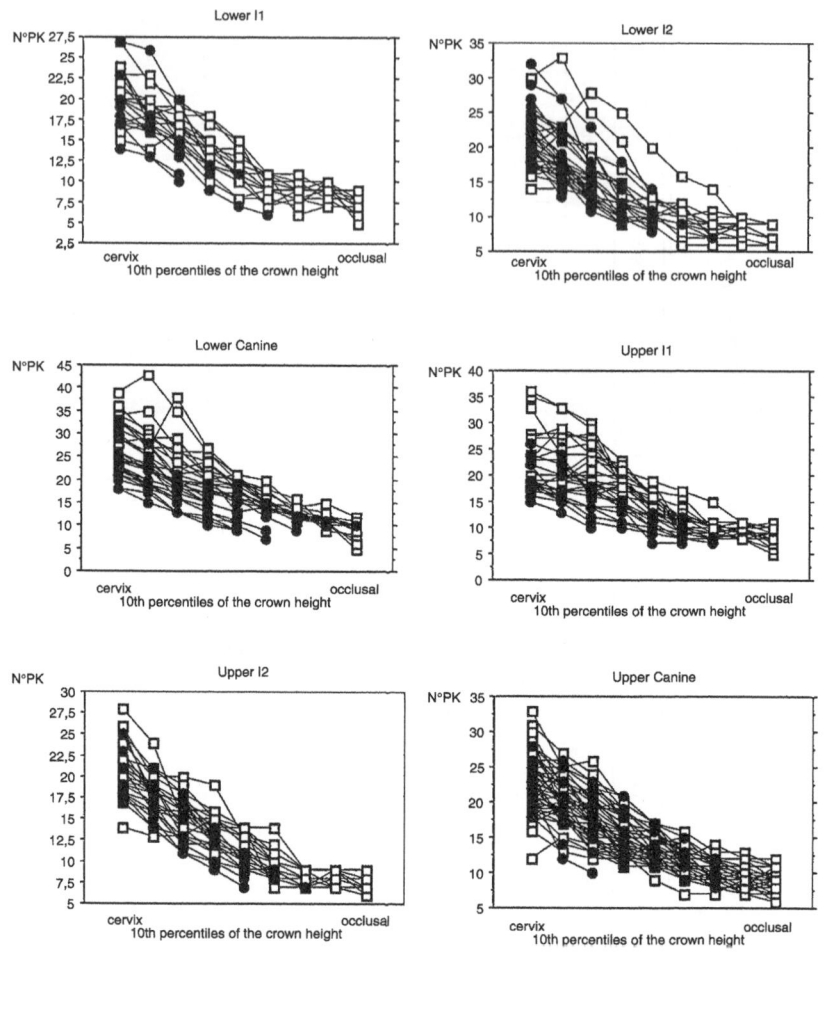

Figure 10.3   Perikymata number in each height crown division for each individual in modern humans and *H. heidelbergensis* from Sima de los Huesos. Both human species show a similar overall pattern. Perikymata become closer, and thus perikymata number increases, toward the cervix. It indicates that each successive height crown division takes longer to form during crown formation, as a result of a decreasing extension rate.

lower C where a correlation exists ($P < 0.01$) between crown height and the total perikymata number from the 5th to the 9th percentiles. In the modern human sample, the number of perikymata recorded is around 10 in the 3rd and 4th percentiles, reaching a little more than 20 perikymata in the 9th percentile.

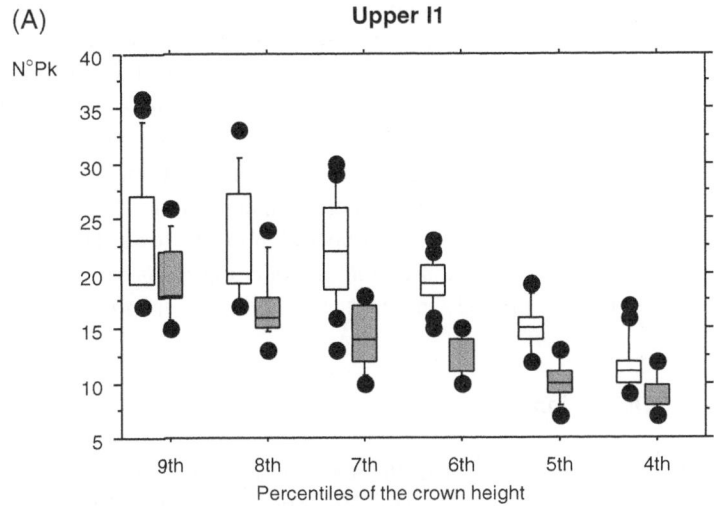

(A)

Upper I1

N°Pk

Percentiles of the crown height

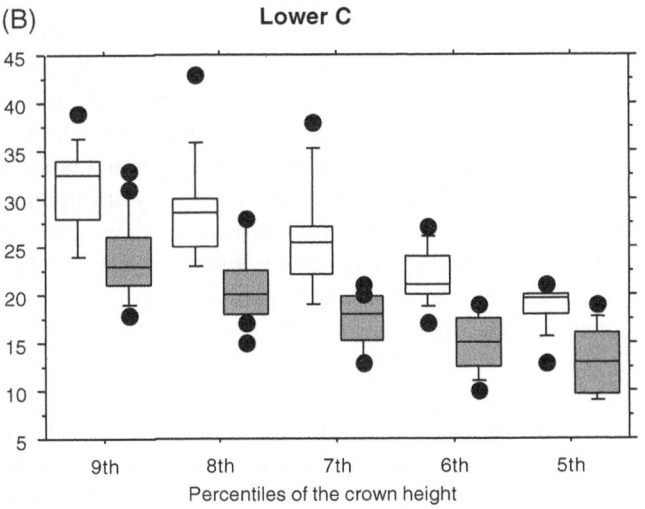

(B)

Lower C

Percentiles of the crown height

Homo sapiens

Homo heidelbergensis

Figure 10.4   Perikymata number in the height crown divisions in (A) the upper I1 and (B) the lower canine. *Homo heidelbergensis* presents a lower number of perikymata for each crown height division than modern humans. A significant difference ($P < 0.01$) is found from the 4th to the 8th percentiles in the upper I1 and from the 3rd to the 9th percentiles in the lower canine. It indicates that each 10th percentile of the UI1 and LC crown height in *H. heidelbergensis* were formed in a shorter time than their counterparts in modern humans.

Similar values are found in anterior teeth from SH. In both human species, the space between perikymata decreases toward the cervix in a similar manner. The perikymata packing pattern in the upper I2, upper C, lower I1, and lower I2 in the SH hominins is similar to that in modern humans. However, significant differences ($P < 0.01$) between species are found in the perikymata number of the last divisions of the lower C and the upper I1, with the number of perikymata being lower in SH than in modern humans (Figure 10.4). These results suggest that crown formation in the upper I2, upper C, lower I1, and lower I2 in *H. heidelbergensis* was similar to that in modern humans but crown formation in the lower C and upper I1 was shorter in the former than in the latter.

Discussion

The relative development of teeth from the anterior dental field (incisors and canines) in relation to teeth from the molar field is a useful factor in distinguishing living great apes from modern humans. This relative development can also be used to distinguish modern humans from extinct hominin species. In this context, australopithecines and early *Homo* (*H. habilis* and *H. ergaster*) show relative I2/M1 development similar to living great apes and, hence, different from modern humans. Moreover, australopithecines present relative C/M2 development that is intermediate between living great apes and modern humans, whereas the C/M2 relationship in early *Homo* does not yield significant differences among later *Homo* specimens of the Pleistocene and modern humans. It is possible to affirm therefore, that *H. ergaster* exhibits a dental development pattern intermediate between australopithecines and modern humans. Although with the available data it is not possible to be more precise, it seems reasonable to state that at least 800 ka a *Homo* species, *H. antecessor*, presented a dental developmental pattern similar to that of modern populations (Bermúdez de Castro *et al.*, 1999c). If our assumption that the timing of dental development patterns evolved in line with the lengthening of the hominin growth period is correct, the *H. antecessor* species would be characterized by a prolonged maturation period presumably similar to that of modern populations. Therefore, the evolution of a modern pattern of development would have taken place in the period between the late Pliocene and the late Lower Pleistocene. In this interval, the childhood and adolescence periods (with its characteristic growth spurt) would have emerged (Bogin & Smith, 1996).

In a recent paper, Dean *et al.* (2001) conclude that truly modern dental development and the related prolonged growth period of modern humans emerged relatively late in human evolution. These authors base their conclusion on the fact that modern humans and one Neandertal (Tabun C1) share a slow trajectory of enamel growth in permanent teeth which differs from australopithecines, *H. habilis*, *H. ergaster*, and *H. erectus*, who exhibit faster rates of enamel formation resembling modern and fossil African apes. Furthermore, they find that *Australopithecus*, one *H. ergaster* specimen (KNM-WT 15000) and one *H. erectus* specimen (Sangiran S7–37) show tooth formation times shorter than those observed in modern humans.

We believe that it may be premature to affirm that a slow rate of enamel growth in permanent teeth can be regarded as a life-history trait associated with the extended growth period of modern humans. The fossils studied by Dean *et al.* (2001), except Tabun C1, may be older than 1.5 Ma. However, we believe that for understanding the evolution of the genus *Homo* it will be essential to study fossils belonging to the period between this date and the beginning of the Upper Pleistocene. According to Dean *et al.* (2001), the first evidence of this slow trajectory of enamel growth in permanent teeth appeared with *H. neanderthalensis* and *H. sapiens*. If this is the case, two alternative hypotheses emerge: either both species have acquired the trait independently, or both species have inherited it from their common ancestor. The first hypothesis would imply a case of homoplasy in Neandertals and modern humans, and therefore, both *Homo* lineages would have prolonged their growth periods independently. Though this scenario is possible, we support the second more parsimonious hypothesis that this trajectory of dental growth is inherited from a common ancestor, *H. antecessor*. Using mitochondrial DNA obtained from the Neandertal type specimen Krings *et al.* (1997, 1999) have proposed that the lineages leading to modern humans and Neandertals diverged 465 ka ago (with confidence limits of 317 and 741 ka). The recent findings in the TD6 level from Gran Dolina (Bermúdez de Castro *et al.*, 1997) also support this scenario. Therefore, the present fossil evidence would point to the slow rate of enamel growth as a trait with an earlier appearance than that proposed by Dean *et al.* (2001).

Furthermore, and as stated above, there is a high correlation between brain weight and the time of dental eruption (Smith, 1989; Smith *et al.*, 1995). Based on this correlation, it is possible to predict that those hominids who reached a cranial capacity of about 1000 cm^3 approached the modern grade of life history (Bogin & Smith, 1996; Smith, 1991a; Smith *et al.*, 1995). The dimensions of the skull fragment ATD6-15 (probably belonging to the Hominid 3 from the Gran Dolina site) point to a cranial capacity above 1000 cm^3 for this individual (Bermúdez de Castro *et al.*, 1997; Carbonell *et al.*, 1995). This trait is in accordance with the clearly human-like dental development observed in the TD6 hominids. Moreover, the calvaria of Ceprano, recently assigned to *H. antecessor*, (Manzi *et al.*, 2001) yields a cranial capacity close to 1200 cm^3 (Ascenzi *et al.*, 1996). Consequently, this species and *H. heidelbergensis* (e.g., Arsuaga *et al.*, 1997) have a brain size within the known ranges for modern humans, further suggesting that they have a more human-like life-history pattern.

The perikymata packing pattern in hominins differs from that in great apes (Dean & Reid, 2001). In *H. heidelbergensis*, it appears to follow the same overall pattern observed in australopithecines and fossil *Homo* species: spacing between perikymata decreases towards the cervix, and the percentiles corresponding to the last-formed enamel present higher values in the perikymata count. Closer inspection reveals that in SH teeth, not only the perikymata packing pattern but also the number of perikymata present in each division is similar to that in modern humans, except for the upper I1 and lower C (see below). This suggests a high extension rate near the cusp decreasing towards the cervix, with very close values in both modern humans and *H. heidelbergensis*.

Since extension rate influences the time of crown formation, different crown formation times between species can be due to distinct extension rates, as seen in differences between *Homo* and *Paranthropus* (Beynon & Dean, 1988; Beynon & Wood, 1987). According to this line of reasoning, the times of crown formation can be estimated from a given extension rate of enamel. The same perikymata packing pattern with a similar number of perikymata in modern humans and *H. heidelbergensis* indicates that their extension rates were similar and, therefore, that their crown formation times were substantially the same.

Although the number of perikymata in each height division may be similar between species, it has been suggested that a different day-interval between adjacent striae of Retzius might result in different crown formation times (Beynon & Dean, 1988; Bromage & Dean, 1985). However, recent work on enamel microstructure of extant hominoids suggests that the day-interval between adjacent striae is similar in modern humans, chimpanzees, and gorillas (Reid & Dean, 2000; Schwartz & Dean, 2001; Schwartz *et al.*, 2001). On the basis of these results, Dean & Reid (2001) assume a similar interval of 9 days between adjacent striae in Plio-Pleistocene hominid species, Pleistocene *Homo* species, and modern humans. Therefore, the same 9 day-interval is also assumed for *H. heidelbergensis*. Since the number of perikymata is similar in each height division and the day-interval between striae is accepted as the same, it can be inferred that the time of lateral enamel formation in *H. heidelbergensis* was the same as in modern humans.

While it is true that the upper I1 and the lower C from Sima de los Huesos present a lower perikymata number than modern humans, the upper I1 and the lower C in modern humans appear to have proportionately longer crown formation times in comparison with other modern human anterior teeth. When crown formation times for all teeth are compared among the great apes, *Australopithecus*, and *Paranthropus*, the same pattern is observed in these three groups, with canines taking longer to form than the other anterior teeth (see Figure 3 in Dean & Reid, 2001). Crown formation time in modern humans follows the same pattern with the difference that the upper I1 and lower C take considerably longer to form than the other anterior teeth (Dean & Reid, 2001). The upper I1 and the lower C of *H. heidelbergensis*, with a shorter time of crown formation, keep proportionality with crown formation times of other teeth as expected according to the pattern observed in non-human hominoids. It is therefore possible to assume that relatively longer crown formation times in the upper I1 and lower C are a recent acquisition in human evolution, possibly characterizing *H. sapiens*.

Conclusions

On the basis of the data presented here, we suggest that the modern pattern of dental development timing may have emerged in the genus *Homo* at the end of the Lower Pleistocene. At present, *H. antecessor* is the earliest species to show a dental development pattern similar (although less derived in some aspects, e.g., delayed M3 calcification and long I2 and C crown formation times) to that of modern human groups such as European or European origin populations. If our hypothesis about *H. antecessor* being the common ancestor of modern *H. sapiens* populations and the European lineage of the Middle–Upper Pleistocene (formed by the chronospecies *H. heidelbergensis* and *H. neanderthalensis*) is correct, these species would have inherited the modern pattern of dental development timing from their common ancestor, *H. antecessor*.

However, information about crown formation times of *Homo* fossil species is still very limited. In this work, we present the first data on the perikymata pattern in *H. heidelbergensis*, as obtained from the large dental sample of the Sima de los Huesos. This species displays a perikymata pattern similar to that of *H. sapiens*, as well as a number of perikymata in each crown division, similar in both species except in the upper I1 and the lower C. Furthermore, it has been possible to describe the dental development pattern in one of the individuals from the human fossil hypodigm of the Sima de los Huesos, SH 18. As expected (see paragraph above) the dental development pattern of this hominin is similar to that of modern populations. Consequently, except for crown formation times of the upper I1 and lower C, it seems that the main aspects of modern human dental development were already present in *H. heidelbergensis* living 400 ka ago.

It is now known that crown formation times in *H. ergaster* and *H. erectus* were shorter than those of modern populations (Dean *et al.*, 2001). Further investigation on which *Homo* species approached modern human-like crown formation times is required. If the time and timing of dental development evolved relatively in parallel, we predict that crown formation times in *H. antecessor* will be approximately the same as in *H. sapiens*. Further studies will provide the necessary information to verify if a fully modern pattern of dental development was already present in *H. antecessor*.

As stated before, a modern-like pattern of dental development in a fossil species would be related to a prolonged pattern of maturation. Assuming that an essentially human life cycle appears once the threshold of $1000\,cm^3$ of cranial capacity is accomplished (Smith, 1991a), our data on the endocranial volume in *H. antecessor* and *H. heidelbergensis* would represent more strong evidence for a modern life-history pattern for these species. Consequently, our predictions are that, at least 800 ka ago, a *Homo* species would have acquired traits such as a relatively short birth interval, a high rate of postnatal brain growth, a childhood phase, an extended period of offspring dependency, a marked adolescent growth spurt, and delayed reproduction cycles.

Acknowledgments

We sincerely thank Jennifer Thompson, Gail Krovitz, and Andrew Nelson for their kind invitation to participate in the Symposium on "Patterns of Growth and Development in the Genus *Homo*" celebrated in Kansas City, Missouri (March 2001) as well to elaborate this chapter for this publication. We also thank Chris Dean and Don Reid for allowing us to use their data on modern humans. The authors are especially indebted to all people belonging to the Atapuerca research team, whose fieldwork and research effort have made possible the elaboration of this work. This research was supported by the Ministerio de Ciencia y Tecnología of the Junta de Castilla y León (Project no. BXX2000-1258-C03–01), the Atapuerca Foundation and Duques de Soria Foundation.

References

Anderson, D. L., Thompson, G. W., & Popovich, F. (1976). Age of attainment of mineralization stages of the permanent dentition. *Journal of Forensic Science*, 21, 191–200.
Arsuaga, J. L., Bermúdez de Castro, J. M., & Carbonell, E. (1997a). The Sima de los Huesos site. *Journal of Human Evolution*, 33.
Arsuaga, J. L., Martínez, I., Gracia, A., & Lorenzo, C. (1997b). The Sima de los Huesos crania (Sierra de Atapuerca, Spain). *Journal of Human Evolution*, 33, 219–281.
Ascenzi, A., Biddittu, I., Cassoli, P. F., Segre, A. G., & Segre-Naldini, E. (1996). A calvarium of late *Homo erectus* from Ceprano, Italy. *Journal of Human Evolution*, 31, 409–423.
Bermúdez de Castro, J. M. (1988). Dental remains from Atapuerca/Ibeas (Spain). II: Morphology. *Journal of Human Evolution*, 17, 279–304.

Bermúdez de Castro, J. M., & Rosas, A. (2001). Pattern of dental development in Hominid XVIII from the Middle Pleistocene Atapuerca-Sima de los Huesos site (Spain). *American Journal of Physical Anthropology*, **114**, 325–330.

Bermúdez de Castro, J. M., Arsuaga, J. L., Carbonell, E., Rosas, A., Martínez, I., & Mosquera, M. (1997). A hominid from the Lower Pleistocene of Atapuerca, Spain: Possible ancestor to Neandertals and modern humans. *Science*, **276**, 1392–1395.

Bermúdez de Castro, J. M., Carbonell, E., & Arsuaga, J. L. (1999a). Gran Dolina site: TD6 Aurora Stratum (Burgos, Spain). *Journal of Human Evolution*, **37**.

Bermúdez de Castro, J. M., Rosas, A., & Nicolás, M. E. (1999b). Dental remains from Atapuerca-TD6 (Gran Dolina site, Burgos, Spain). *Journal of Human Evolution*, **37**, 523–566.

Bermúdez de Castro, J. M., Rosas, A., Carbonell, E., Nicolás, E., Rodríguez, J., & Arsuaga, J. L. (1999c). A modern human pattern of dental development in Lower Pleistocene hominids from Atapuerca-TD6 (Spain). *Proceedings of the National Academy of Sciences of the USA*, **96**, 4210–4213.

Beynon, A. D. (1987). Replication technique for studying microstructure in fossil enamel. *Scanning Microscopy*, **1**, 663–669.

Beynon, A. D., & Dean, M. C. (1988). Distinct dental development patterns in early fossil hominids. *Nature*, **335**, 509–514.

Beynon, A. D., & Wood, B. A. (1987). Patterns and rates of enamel growth in the molar teeth of early hominids. *Nature*, **326**, 493–496.

Bischoff, J. L., Fitzpatrick, J. A., León, L., Arsuaga, J. L., Falgueres, C., Bahain, J. J., & Bullen, T. (1997). Geology and preliminary dating of the hominid-bearing sedimentary fill of the Sima de los Huesos Chamber, Cueva Mayor of the Sierra de Atapuerca, Burgos, Spain. *Journal of Human Evolution*, **33**, 129–154.

Bischoff, J. L., Shampa, D. D., Aramburu, A., Arsuaga, J. L., Carbonell, E., & Bermúdez de Castro, J. M. (2002). The Sima de los Huesos hominids date to beyond U/Th equilibrium (>350 kyrs) and perhaps to 400–600 kyrs: New radiometric dates. *Journal of Archaeological Science*, **30**, 275–280.

Bogin, B., & Smith, B. H. (1996). Evolution of the human life cycle. *American Journal of Human Biology*, **8**, 703–716.

Bolk, L. (1926). On the problem of anthropogenesis. *Proceedings of the Section of Sciences, Koninklijke Akademie van Wetenschappen te Amsterdam*, **29**, 465–475.

Bromage, T. G. (1987). The biological and chronological maturation of early hominids. *Journal of Human Evolution*, **16**, 257–272.

Bromage, T. G., & Dean, M. C. (1985). Re-evaluation of the age at death of immature fossil hominids. *Nature*, **317**, 525–527.

Carbonell, E., Bermúdez de Castro, J. M., Arsuaga, J. L., Díez, J. C., Rosas, A., Cuenca-Bescós, G., Sala, R., Mosquera, M., & Rodríguez, X. P. (1995). Lower Pleistocene hominids and artifacts from Atapuerca-TD6 (Spain). *Science*, **269**, 826–830.

Carretero, J. M., Arsuaga, J. L., & Lorenzo, C. (1997). Clavicles, scapulae and humeri from the Sima de los Huesos site (Sierra de Atapuerca, Spain). *Journal of Human Evolution*, **33**, 357–408.

Conroy, G. C., & Vannier, M. W. (1987). Dental development of the Taung skull from computerized tomography. *Nature*, **329**, 625–627.

Conroy, G. C., & Vannier, M. W. (1991). Dental development in South African australopithecines. I: Problems of pattern and chronology. *American Journal of Physical Anthropology*, **86**, 121–136.

Cuenca-Bescós, G., Laplana-Conesa, C., Canudo, J. I., & Arsuaga, J. L. (1997). Small mammals from Sima de los Huesos. *Journal of Human Evolution*, **33**, 175–190.

Dart, R. A. (1948). The infancy of *Australopithecus*. In *Robert Broom Commemorative Volume*, ed. A. L. du Toit, pp. 143–152. Cape Town: Royal Society of South Africa.

Dean, M. N. (1987a). The dental developmental status of six East African juvenile fossil hominids. *Journal of Human Evolution*, **16**, 197–213.

Dean, M. C. (1987b). Growth layers and incremental markings in hard tissues: A review of the literature and some preliminary observations about enamel structure in *Paranthropus boisei*. *Journal of Human Evolution*, **16**, 157–172.

Dean, M. C., & Reid, D. J. (2001). Perikymata spacing and distribution on hominid anterior teeth. *American Journal of Physical Anthropology*, **116**, 209–215.

Dean, M. C., Stringer, C. B., & Bromage, T. G. (1986). Age at death of the Neanderthal child from Devil's Tower, Gibraltar and the implications for studies of general growth and development in Neanderthals. *American Journal of Physical Anthropology*, **70**, 301–309.

Dean, M. C., Leakey, M. G., Reid, D., Schrenk, F., Schwartz, G. T., Stringer, C. B., & Walker, A. (2001). Growth processes in teeth distinguish modern humans from *Homo erectus* and earlier hominins. *Nature*, **414**, 628–631.

Etkin, W. (1954). Social behavior and the evolution of man's mental faculties. *American Naturalist*, **88**, 129–142.

García, N., Arsuaga, J. L., & Torres, T. (1997). The carnivore remains from the Sima de los Huesos Middle Pleistocene site (Sierra de Atapuerca, Spain). *Journal of Human Evolution*, **33**, 155–174.

Harvey, P. H., & Clutton-Brock, T. H. (1985). Life history variation in primates. *Evolution*, **39**, 559–581.

Isaac, G. L. (1978). The food-sharing behavior of protohuman hominids. *Scientific American*, **238**, 90–108.

Keith, A. (1949). *A New Theory of Human Evolution*. London: Watts.

Krings, M., Stone, A., Schmitz, R.-M., Krainitzki, H., Stoneking, M., & Pääbo, S. (1997). Neandertal DNA sequences and the origin of modern humans. *Cell*, **90**, 19–30.

Krings, M., Geisert, H., Schmitz, R. W., Krainitzki, H., & Pääbo, S. (1999). DNA sequence of the mitochondrial hypervariable region II from the Neandertal type specimen. *Proceedings of the National Academy of Sciences of the USA*, **96**, 5581–5585.

Lancaster, J. B. (1978). Carrying and sharing in human evolution. *Human Nature*, **1**, 82–89.

Le Gros Clark, W. E. (1947). Observations on the anatomy of the fossil Australopithecinae. *Journal of Anatomy*, **83**, 300–333.

Lewis, A. B., & Garn, S. M. (1960). The relationship between tooth formation and other maturational factors. *Angle Orthodontics*, **30**, 70–77.

Lovejoy, C. O. (1981). The origin of man. *Science*, **211**, 341–350.

Mann, A. (1968). The paleodemography of *Australopithecus*. PhD dissertation, University of California, Berkeley.

Mann, A. E. (1975). *Some Paleodemographic Aspects of the South African Australopithecines*, University of Pennsylvania Publications in Anthropology, no. 1. Philadelphia: University of Philadelphia.

Mann, A. E. (1988). The nature of Taung dental maturation. *Nature*, **333**, 123.

Mann, A., Lampl, M., & Monge, J. (1987). Maturational patterns in early hominids. *Nature*, **328**, 673–674.

Mann, A., Lampl, M., & Monge, J. (1990a). Patterns of ontogeny in human evolution: Evidence from dental development. *Yearbook of Physical Anthropology*, **33**, 111–150.

Mann, A., Monge, J. M., & Lampl, M. (1990b). Dental caution. *Nature*, **348**, 202.

Mann, A., Monge, J. M., & Lampl, M. (1991). Investigation into the relationship between perikymata counts and crown formation times. *American Journal of Physical Anthropology*, **86**, 175–188.

Manzi, G., Mallegni, F., & Ascenzi, A. (2001). A cranium for the earliest Europeans: Phylogenetic position of the hominid from Ceprano, Italy. *Proceedings of the National Academy of Sciences of the USA*, **98**, 10011–10016.

Montagu, M. F. A. (1961). Neonatal and infant immaturity in man. *Journal of the American Medical Association*, **178**, 56–57.

Moorrees, C. F. A., Fanning, E. A., & Hunt, E. E. (1963). Age variation of formation stages for ten permanent teeth. *Journal of Dental Research*, **42**, 1490–1502.

Parés, J. M., & Pérez-González, A. (1995). Paleomagnetic age for hominid fossils at Atapuerca archaeological site, Spain. *Science*, **269**, 830–832.

Parés, J. M., Pérez-González, A., Weil, A. B., & Arsuaga, J. L. (2000). On the age of the hominid fossils at the Sima de los Huesos, Sierra de Atapuerca, Spain: Paleomagnetic evidence. *American Journal of Physical Anthropology*, **111**, 451–461.

Ramirez Rozzi, F. V. (1993a). Microstructure et développement de l'émail dentaire du néandertalien de Zafarraya, Espagne: Temps de formation et hypocalcification de l'émail dentaire. *Comptes Rendus de l'Académie des Sciences de Paris*, **316**, 1635–1642.

Ramirez Rozzi, F. V. (1993b). Teeth development in East African *Paranthropus*. *Journal of Human Evolution*, **24**, 429–454.

Ramirez Rozzi, F. V. (1996). Comment on the causes of thin enamel in Neandertals. *American Journal of Physical Anthropology*, **99**, 625–626.

Reid, D. J., & Dean, M. C. (2000). Brief communication: The timing of linear hypoplasias on human anterior teeth. *American Journal of Physical Anthropology*, **113**, 135–139.

Reid, D. J., Beynon, A. D., & Ramirez Rozzi, F. V. (1998). Histological reconstruction of dental development in four individuals from a medieval site in Picardie, France. *Journal of Human Evolution*, **35**, 463–477.

Rosas, A. (1997). A gradient of size and shape for the Atapuerca sample and Middle Pleistocene hominid variability. *Journal of Human Evolution*, **33**, 319–331.

Sacher, G. A., & Staffeldt, E. F. (1974). Relation of gestation time to brain for placental mammals. *American Naturalist*, **108**, 593–616.

Schwartz, G. T., & Dean, M. C. (2001). The ontogeny of canine dimorphism in extant hominoids. *American Journal of Physical Anthropology*, **115**, 269–283.

Schwartz, G. T., Reid, D. J., & Dean, M. C. (2001). Development aspects of sexual dimorphism in hominoid canine. *International Journal of Primatology*, **22**, 837–860.

Simpson, S. W., Lovejoy, C. O., & Meindl, R. S. (1990). Hominoid dental maturation. *Journal of Human Evolution*, **19**, 285–297.

Simpson, S. W., Lovejoy, C. O., & Meindl, R. S. (1992). Further evidence on relative dental maturation and somatic developmental rate in hominoids. *American Journal of Physical Anthropology*, **87**, 29–38.

Smith, B. H. (1986). Dental development in *Australopithecus* and early *Homo*. *Nature*, **323**, 27–33.

Smith, B. H. (1989). Dental development as a measure of life history in primates. *Evolution*, **43**, 683–688.

Smith, B. H. (1991a). Dental development and the evolution of life history in Hominidae. *American Journal of Physical Anthropology*, **86**, 157–174.

Smith, B. H. (1991b). Standards of human tooth formation and dental age assessment. In *Advances in Dental Anthropology*, eds. M. A. Kelley & C. S. Larsen, pp. 143–168. New York: Wiley-Liss.

Smith, B. H. (1993). The physiological age of KNM-ER 15000. In *The Nariokotome* Homo erectus *Skeleton*, eds. A. Walker & R. Leakey, pp. 195–220. Berlin: Springer-Verlag.

Smith, B. H. (1994). Patterns of dental development in *Homo, Australopithecus, Pan*, and *Gorilla. American Journal of Physical Anthropology*, **94**, 307–325.

Smith, B. H., & Tompkins, R. L. (1995). Toward a life history of the Hominidae. *Annual Review of Anthropology*, **24**, 257–279.

Smith, R. J., Gannon, P. J., & Smith, B. H. (1995). Ontogeny of australopithecines and early *Homo*: Evidence from cranial capacity and dental eruption. *Journal of Human Evolution*, **29**, 155–168.

Tompkins, R. L. (1996). Relative dental development of Upper Pleistocene hominids compared to human population variation. *American Journal of Physical Anthropology*, **99**, 103–118.

Walker, A. (1993). Perspectives of the Nariokotome discovery. In *The Nariokotome* Homo erectus *Skeleton*, eds. A. Walker & R. Leakey, pp. 411–430. Berlin: Springer Verlag.

Washburn, S. L. (1960). Tools and human evolution. *Scientific American*, **203**, 63–75.

Wood, B. A. (1991). *Koobi Fora Research Project*, vol. 4, *Hominid Cranial Remains*. New York: Oxford University Press.

Wood, B. A. (1992). Origin and evolution of the genus *Homo*. *Nature*, **355**, 783–790.

11 *Hominid growth and development from australopithecines to Middle Pleistocene* Homo

G. E. KROVITZ
Pennsylvania State University

J. L. THOMPSON
University of Nevada, Las Vegas

A. J. NELSON
University of Western Ontario

Introduction

This chapter reviews studies of juvenile dental, cranial, and postcranial remains, along with aspects of life history or demography that relate to juvenile individuals, from Lower and Middle Pleistocene members of the genus *Homo* (i.e., *Homo habilis, Homo erectus, Homo antecessor*, and *Homo heidelbergensis*). It is understood that these species names are not uniformly accepted by all paleoanthropologists, and that there is disagreement about what fossil specimens belong in each taxon. However, use of these species names allows for a much clearer discussion than using vague terms like "early *Homo*," which has been variably used to refer to specimens of *H. habilis* and/or *H. erectus*. Additionally, as will be seen by the end of this review, there are large developmental differences between these groups that appear to justify discussion in these categories.

To set the stage for what we know about growth and development in the genus *Homo*, Kuykendall (this volume) reviews what is currently known about growth and development in robust and gracile australopithecines. Much of the research focuses on dental development, particularly aspects of timing and rate, and the relationship between dental development and life-history stages. Although it is clear that the australopithecines had unique developmental patterns that were unlike living apes or humans, on the whole they still matured following a faster,

Patterns of Growth and Development in the Genus Homo, ed. J. L. Thompson, G. E. Krovitz and A. J. Nelson. Published by Cambridge University Press. © Cambridge University Press 2003.

more "ape-like" life-history schedule and did not have the extended period of growth and development commonly associated with modern humans. The australopithecines probably also did not have the childhood and adolescent life-history stages that are seen in modern human ontogeny (Bogin, 1997, 1999b, this volume; Bogin & Smith, 1996; Smith & Tompkins, 1995). It is therefore within the genus *Homo* that modern human life-history patterns evolved. In the following discussion on Lower and Middle Pleistocene members of the genus *Homo*, the term "ape-like" will be used to describe elements of growth and development that are similar in timing, rate, or pattern to either australopithecines or fossil or modern apes. It is likely that many such features are retained from a great ape–hominid common ancestor, or are possibly primitive for primate growth in general. The term "human-like" refers to aspects of growth and development that are similar to elements of growth and development in modern humans.

Homo habilis

Some workers prefer to divide the material commonly referred to as *H. habilis* into two species, *H. habilis* and *H. rudolfensis* (e.g., Dunsworth & Walker, 2002; Wood, 1992; Wood & Richmond, 2000). However, this is a contentious subject, and it is unclear how many species are represented in this fossil material, and what particular fossil specimens should be allocated to these different species. Additionally, even though Olduvai Hominid (OH) 7, the type specimen of *H. habilis*, is a juvenile (although it is likely that more than one individual is represented by remains attributed to OH 7, as discussed in Day, 1986; Dunsworth & Walker, 2002), non-adult fossil remains from both putative species are scarce. Because of this limited sample size, it would be extremely difficult to explore any potential developmental differences between *H. habilis* and *H. rudolfensis* (assuming, of course, that there was agreement about what specimens belonged in these species). Thus all relevant specimens are discussed in this chapter as belonging to *H. habilis*.

Dental remains

Dental development can be separated into studies of the rate of developmental events (for example, the rate at which the enamel forms in a particular tooth crown), and the relative timing of developmental events (for example, the timing of particular stages of tooth formation). Separating these two types of studies is the approach taken by Kuykendall and Bermúdez de Castro *et al.* in this volume, and will be used in this review.

It is generally thought that crown formation times for canines and incisors are considerably shorter in "early *Homo*" (and in the australopithecines:
see Kuykendall, this volume) than in modern humans. However, many papers referring to dental development in "early *Homo*" include the specimens
KNM-ER 820 and KNM-ER 1507 (e.g., Bromage & Dean, 1985; Dean, 1985,
1987). These specimens are better described as *H. erectus* (Walker, 1993; Wood,
1992), dating to a time period (1.6–1.8 Ma) when *H. erectus* is known from the
same strata (Feibel *et al.*, 1989), and will therefore be discussed in the *H. erectus*
section below. Few studies of incremental enamel growth have actually considered *H. habilis* specimens. Two *H. habilis* juveniles, OH 7 and KNM-ER 1590,
were included in a recent study by Dean & Reid (2001) that used perikymata
counts on the tooth surface to determine crown formation times for the anterior
teeth. KNM-ER 1590 was found to be most similar to the australopithecines,
thus having faster, more ape-like developmental timing, while OH 7 was in the
modern human range for crown height formation (Dean & Reid, 2001). Another
recent study on enamel formation rates for anterior teeth (Dean *et al.*, 2001),
found that specimens attributed to *H. habilis* (and *H. rudolfensis*) resembled
fossil and modern apes in their developmental trajectories for enamel growth
(see further discussion of this study below). Thus, development of the anterior
dentition in *H. habilis* does appear to be faster and more ape-like, although
the measurements for OH 7 mentioned above are contrary to this conclusion.
Finally, crown formation times for molars are roughly similar between *H. habilis* and modern humans (Beynon & Wood, 1987).

Studies that examine the patterning of dental development also show more
ape-like dental development patterns for *H. habilis*. For example, Smith (1986,
1989a, 1991, 1994) found that dental development patterns for the specimen
KNM-ER 1590 had a more consistent fit with ape standards of dental development, suggesting that this specimen showed a more "primitive" pattern of
dental development along with gracile australopithecines. KNM-ER 1590 also
showed delayed development of the incisors relative to M1, as was found in
more ape-like dental development patterns (Smith, 1986, 1989a, 1991, 1994).
KNM-ER 1590 and OH 6 also show dental development patterns like those of
Australopithecus based on analysis of the coefficient of variation of dental ages
(Moggi-Cecchi, 2001).

Work has been done to show the high correlation between aspects of dental
development, such as M1 eruption, and many important aspects of life history,
including brain size (Smith, 1989a, 1989b). Using brain weight to predict life-
history measures in fossil hominids, B.H. Smith predicted that the *H. habilis* M1
would have erupted at 4.0 years, compared to a modern human eruption average
of 5.9 years (1989b, 1991, 1994). Additional work by R. J. Smith and colleagues
(1995) suggested that M1 eruption in *H. habilis* was intermediate between

australopithecines (with very ape-like M1 eruption times) and *H. erectus* (with a more human-like eruption time). These studies suggest that *H. habilis* had a slightly more prolonged maturational period than has been suggested by the other dental studies.

Cranial and mandibular remains

The type specimen of *H. habilis*, OH 7, is a juvenile, although the cranial remains from this specimen include only parts of both parietals, and a fragmentary mandible (much of the alveolar process and dentition, and little else) (Tobias, 1991). OH 13 and OH 16 are older subadult specimens, with the third molars erupting (Tobias, 1991). Bromage (1989) looked at microscopic bone resorption and deposition on the face and mandible, using OH 7, OH 13, SK 27, and the *H. erectus* mandible KNM-ER 820 (once again lumped in with "early *Homo*"). He found nothing in the remodeling patterns of the lower midfacial region or the mandible that differentiates *H. habilis* from *Australopithecus* (although they both differ from *Paranthropus* and modern humans). Thus, both *H. habilis* and *Australopithecus* share remodeling patterns that contribute to the development of a prognathic face. Additionally, the specimens OH 7, OH 13, and OH 16 have been used to represent juvenile *H. habilis* in a taxonomic comparison with an Asian *H. erectus* juvenile specimen (Mojokerto) (Antón, 1997).

Postcranial remains

Few definitive juvenile *H. habilis* postcranial remains have been recovered. The type specimen, OH 7, preserves 13 hand bones (although it is unclear how many individuals are represented by these bones: Day, 1986). Several studies have been carried out on adult "early *Homo*" postcranial specimens KNM-ER 1481A and B, KNM-ER 1472, KNM-ER 1476, and KNM-ER 3951 (Tardieu, 1998, 1999; Tardieu & Trinkaus, 1994), but there is no clear way to determine what species these elements belong to (Tardieu, 1998). Since it has been suggested that these specimens have *H. erectus*-like morphology (i.e., as distinct from OH 62: Ruff, 1995; and also see discussion in Kennedy, 1983, and Trinkaus, 1984), they will be discussed in the section below. Adult *H. habilis* postcrania are primarily known from OH 62, a fragmentary specimen that has been reconstructed as having an australopithecine-like body design (Richmond *et al.*, 2002; Wood & Richmond, 2000), although Asfaw *et al.* (1999) have noted that femoral length cannot be reliably estimated for this specimen. Regardless of whether or not that turns out to be true, juvenile postcrania (particularly long bones) from this group would advance our knowledge of ontogeny at the starting-point of our lineage.

Life-history issues

Bogin (1997, 1999b, this volume; Bogin & Smith, 1996) has tried to reconstruct when the childhood phase (i.e., the period following infancy when the youngster is weaned from nursing but is still dependent for food and protection) was added to the hominid life-history schedule. He has concluded that the slightly larger brain sizes in *H. habilis*, when compared to the australopithecines, could be due to either extended fetal and infancy periods, or to a reduced duration of infancy and insertion of a short childhood phase. An advantage of a reduced infancy period is that it reduces the duration (and thus, metabolic expense) of nursing and allows the mother to reproduce again sooner.

Homo erectus

Some workers prefer to divide the material commonly referred to as *H. erectus* into two species, *H. erectus* and *H. ergaster* (Dunsworth & Walker, 2002; Wood & Collard, 1999). It is understood that "if these two subsamples represent separate species, then pooling them to discuss developmental patterns, without first establishing that their specific-level growth patterns are similar, is unwarranted" (Antón, 2002: 370). However, since the small amount of juvenile *H. erectus/ H. ergaster* material makes it difficult to compare developmental patterns between these two groups, this discussion will consider all this material as *H. erectus*. There are many clear differences between *H. erectus* and *H. habilis*, including increased mean brain size, different cranial shape, larger body mass, and more human-like body proportions (when compared to the OH 62 *H. habilis* postcranial reconstruction), and a marked increase in geographic range.

Dental remains

A recent study on enamel formation rates for anterior teeth (Dean *et al.*, 2001), found that specimens attributed to *H. erectus* (along with australopithecines and *H. habilis*) resembled fossil and modern apes in their developmental trajectories for enamel growth, not modern humans. This study showed that the developmental formation of thick enamel in modern human teeth was different from the formation of thick enamel in earlier hominids (including *H. erectus*). Additionally, tooth crown and root formation times were shorter in two *H. erectus* specimens, KNM-WT 15000 and Sangiran S7-37, than in modern humans, revealing a faster rate of dental development. While it was not unexpected for *H. habilis* to show an ape-like enamel growth trajectory (see discussion above), it was somewhat surprising that *H. erectus* also showed such an ape-like pattern. Yet this study was not alone in its findings, as the *H. erectus* specimen KNM-ER 820 also showed an ape-like pattern and rate of dental formation (Bromage & Dean, 1985).

However, other studies reveal a different picture. When dental development patterns in *H. erectus* specimens were compared with modern human and ape dental development standards, specimens KNM-ER 820 and KNM-ER 1507 had an equivocal fit to both ape and modern human standards, while Zhoukoudian (ZH) B-I had a better fit to ape standards (Smith, 1986, 1989a, 1991). In a later paper (Smith, 1994), dental development in *H. erectus* specimens KNM-ER 820, KNM-WT 15000, and ZH B-I was shown to be most consistent with human dental development standards, although KNM-ER 1507 and KNM-ER 1590 were shown to best fit with ape dental development standards. Smith concluded that by *H. erectus*, the approach to modern human developmental patterns had begun (see also Moggi-Cecchi, 2001). The relationship between cranial capacity and M1 eruption also suggested a more extended period of development in *H. erectus* than in earlier hominids (Smith, 1989b, 1991, 1994; Smith *et al.*, 1995). Further disagreement over the appropriateness of ape-like or human-like developmental models for *H. erectus* (particularly relating to dental and skeletal age assessments for KNM-WT 15000, as initiated by Smith, 1993) is discussed below.

Cranial and mandibular remains

The available juvenile *H. erectus* cranial specimens include: Mojokerto (calvaria: age ~4–6 years), Ngandong 2 (frontal: age ~8–11), Zhoukoudian Skull VIII (occipital fragment: age >6 years), KNM-WT 15000 (skull and postcrania: age 8–13 years; see discussion below) (Antón, 1997, 1999, 2002; Walker & Leakey, 1993). Juvenile *H. erectus* mandible specimens include: KNM-WT 15000, KNM-ER 820, KNM-ER 1507, Zhoukoudian B-I, B-III, B-IV, B-V, C-I, F-I (see Antón, 2002; Dean, 1987; Walker & Leakey, 1993 for descriptions).

Juvenile *H. erectus* facial morphology is only preserved in KNM-WT 15000. Three-dimensional geometric morphometric methods were used to compare adolescent facial growth patterns in *H. erectus* (i.e., growth from KNM-WT 15000 to adult specimen KNM-ER 3733, and a simulated male adult *H. erectus*), modern humans, and chimpanzees (Richtsmeier & Walker, 1993). This study found great overall similarity in the adolescent facial growth patterns in these species, since species-specific features were probably present by adolescence, and the adolescent facial growth patterns might serve to maintain or amplify already present morphology. They suggested that there might be a generalized pattern of adolescent facial growth in all hominoids, with species-specific features developing early during ontogeny (presaging the results of Ackermann & Krovitz, 2002; Krovitz, 2000, this volume; Ponce de León & Zollikofer, 2001).

Neurocranial growth has been considered in *H. erectus* as well. The work of Antón has shown that *H. erectus* juveniles (1) have more rounded frontal and parietal bones and less angular occipital bones than do adult *H. erectus* specimens, and (2) have similar cranial vault contours to both juvenile and adult modern humans (Antón, 1997, 2002; Antón & Leigh, 1998, this volume). Differential growth of the cranial superstructures (i.e., localized thickening of ectocranial bone and diploë that develop as the cranial tables differentiate throughout the subadult years) creates the differences in cranial vault contours that are observable between *H. erectus* and modern human adults (Antón, 1997, 2002). Similarity between juvenile *H. erectus* and modern human cranial vault contours suggests that modern human neurocranial shape is paedomorphic (specifically, neotenic) relative to *H. erectus'* neurocranial shape (Antón, 1997; Antón & Leigh, 1998, this volume; Montagu, 1955). Differences in the rate or duration of growth are probably responsible for the observable adult differences, as modern human neurocranial form is not merely an extension of *H. erectus* growth patterns into an adult hominid with larger brain size (Antón, 1997; Antón & Leigh, this volume). Paedomorphosis of the neurocranium, and therefore by extension, the brain, may enable greater neurocranial flexibility in modern humans by allowing the human brain to retain the plasticity of a juvenile brain throughout the developmental period (Antón & Leigh, 1998, this volume). Interestingly, the work of Antón suggests *differences* in neurocranial growth (particularly in growth of neurocranial superstructures) between *H. erectus* and modern humans, while the work of Richtsmeier and Walker shows overall *similarity* in facial growth patterns between the two. This is evidence that different regions of the skull need to be considered independently, and that different anatomical regions evolve in different ways, so we cannot consider one process of "global heterochrony" to explain all changes in the skull (see also Krovitz, 1999, 2000, and Williams *et al.*, this volume).

Postcranial remains

The juvenile *H. erectus* specimen KNM-WT 15000 preserves a relatively complete postcranial skeleton, and therefore offers an extraordinary opportunity to study postcranial growth in *H. erectus*. Detailed descriptions and analyses of this specimen can be found in Walker & Leakey (1993). The initial physiological age assessment of KNM-WT 15000 is found in Smith (1993), and controversy regarding this subject is discussed below.

The relatively complete pelvis and associated femora in KNM-WT 15000 allowed analysis of the hip region and birth canal in *H. erectus* (Ruff, 1995; Walker & Ruff, 1993). *H. erectus* had a relatively mediolaterally broad and anteroposteriorly narrow true (lower) pelvis (similar to that of australopithecines). Since the mediolateral dimensions of the pelvis could only increase to a certain point due to the mechanical constraints of bipedalism, this morphology precluded the rotational birth process characteristic of modern humans, and constrained neonatal cranial capacity (Ruff, 1995). Ruff argued that this pelvic shape and birth mechanism, with resulting constraint on neonatal (and therefore adult) cranial capacity was in place until at least the Middle Pleistocene (see discussion below). However, Rosenberg & Trevathan (1996) suggest that *H. erectus* could have had a rotational form of birth even if the process differed from that observed in modern humans. The initial analysis of the KNM-WT 15000 pelvis suggested that the secondary altricial condition (i.e., neonates born with relatively smaller brain size, and the maintenance of high fetal brain growth rates for the first postnatal year) had already evolved in *H. erectus* (Walker, 1993; Walker & Ruff, 1993). However, a reconsideration of the error involved in this process (including "transforming a juvenile to an adult pelvis, a male to a female pelvis, using different human populations for the relationship between size of the birth canal and neonate head") should lead one to "accept the Dean *et al.* (2001) result as indicating faster development in *H. erectus* rather than the result of the complicated exercise carried out by Walker & Ruff (1993)" (A. Walker, pers. comm.).

In bicondylar angle, KNM-WT 15000 was fully modern, which was not surprising, since this feature (as well as many other aspects of this postcranial skeleton) reflects a habitual bipedal posture and locomotion (Tardieu & Trinkaus, 1994). An examination of ontogenetic reshaping of the distal femoral epiphysis in specimens KNM-ER 1481A, KNM-ER 1472, KNM-ER 3951, and KNM-WT 15000 revealed that, compared to *Australopithecus afarensis* (which had a short adolescent growth period and early epiphyseal closure), *H. erectus* had a prolonged adolescent growth period and delayed epiphyseal closure (Tardieu, 1998). It was suggested that the prolongation of peripubertal growth period was caused by the heterochronic process of time hypermorphosis (Tardieu, 1998).

Adolescent growth spurt?

The adolescent growth spurt is usually thought of as "a dramatic increase in skeletal growth due to an increase in the velocity of growth around the time of sexual maturity" (Antón, 2002: 371). An adolescent growth spurt (particularly in stature) has been suggested to be a unique character found only in modern humans (Bogin, 1999a, 1999b, this volume); although see Leigh (1996) for discussion of body mass growth spurts in non-human primates. Adolescence is a time when humans learn culturally important values and adult roles, suggesting that the presence or absence of this feature throughout the hominid record could have important implications for the evolution of social and cultural complexity (Antón, 2002; Bogin, this volume).

The initial physiological age estimate for KNM-WT 15000 (Smith, 1993) sparked a contentious discussion on the presence or absence of an adolescent growth spurt in *H. erectus*. Because of the remarkable preservation of dental and postcranial remains in this specimen, Smith (1993) was able to use a variety of maturational systems in her age assessment, and compared her findings with modern human and chimpanzee developmental standards. If KNM-WT 15000 was aged solely according to human standards, the dental age was 11.0 years, skeletal age was 13–13.5 years, and statural age was 15 years. However, if he was aged solely according to chimpanzee standards, the dental age was 7–8 years and skeletal age was 7.5 years. Although skeletal and dental age estimates are slightly more consistent when using a chimpanzee model, it was suggested that development of *H. erectus* matched neither species precisely, and that it was more intermediate between humans and chimpanzees. Indeed, a conceptual model based on the relationship between adult brain weight and various life history measures suggested an intermediate developmental pattern for *H. erectus*, and gave KNM-WT 15000 a developmental age of 9–10 years. However, the data implied that an adolescent growth spurt was not present in *H. erectus*, and that KNM-WT 15000 had reached more of his adult size by early adolescence, instead of suppressing growth in late childhood and catching up with an adolescent growth spurt (Smith, 1993).

Smith's (1993) assessment of KNM-WT 15000 was criticized by Clegg & Aiello (1999), who used a modern human skeletal sample of known age and the same age estimation methodology as Smith (1993), to examine variation in age as inferred from the maturity indicators. They showed that most of the modern human individuals examined also had a disjunction between dental and skeletal age estimates, as seen in KNM-WT 15000, and suggested that this pattern was within the normal range of human variation and did not indicate a different developmental pattern (i.e., implying that *H. erectus* did experience an adolescent growth spurt). Additionally, KNM-WT 15000 does not match adult *H. erectus* facial and mandibular dimensions and cranial contours, indicating that growth was still required to achieve adult dimensions and neurocranial shape, and suggesting that KNM-WT 15000 had yet to enter an adolescent growth spurt (Antón, 2002; Antón & Leigh, this volume). An adolescent growth spurt in *H. erectus* was also suggested by Tardieu's analysis of femoral morphology (1998), although it would have been less pronounced and of shorter duration than in modern humans.

At this point the data for an adolescent growth spurt in *H. erectus* are equivocal, although there is enough evidence for a growth spurt to "urge caution" in dismissing an adolescent growth spurt in this species (Antón, 2002; Antón & Leigh, this volume). However, it is clear that KNM-WT 15000 shows a delayed level of craniodental maturation relative to postcranial maturation, and that this difference is particularly evident when using modern human standards of age estimation. Thus, KNM-WT 15000 had yet to achieve adult craniofacial dimensions (Antón, 2002; Antón & Leigh, this volume), while postcranial measures show that he had already achieved most of his adult growth (Smith, 1993; Thompson & Nelson, 2000). This might suggest that the craniofacial and postcranial growth spurts had a different relationship in *H. erectus* than in modern humans. The recent study by Dean *et al.* (2001) showing that *H. erectus* resembled modern and fossil apes in dental development rate gave support to Smith's (1993) young age assessment for KNM-WT 15000, and suggested that an age closer to 8 years was better for this individual. If future work confirms the ape-like timing of *H. erectus*'s dental development, then the evidence for a more ape-like timing of adolescent postcranial growth becomes a lot stronger as well.

Life-history issues

There are clear biological and behavioral changes between *H. habilis* and *H. erectus*, many of which are potentially linked to a larger proportion of meat in the *H. erectus* diet (presumably from scavenging, not hunting) (Shipman, 1986; Shipman & Walker, 1989). The larger body mass estimated for *H. erectus* (in comparison with *H. habilis*) would lead to increased basal metabolic requirements and daily energy expenditures (Aiello & Wells, 2002). This increased expense was probably solved by an increased quality of food (i.e., more mammalian meat and fat), rather than increased quantity of lower-quality vegetable food (Aiello & Wells, 2002). The increased quality of the *H. erectus* diet might have allowed a reduction of the metabolically expensive tissues of the gut, thus freeing up extra calories for the larger metabolically expensive brain (Aiello & Wells, 2002; Aiello & Wheeler, 1995). Increased energy requirements of *H. erectus* were predicted to be especially high for females, as the energetic cost of lactation in *H. erectus* was estimated to be up to 45% higher than in the australopithecines (Aiello & Key, 2002). However, if *H. erectus* is modeled as having a more human-like reproductive schedule (with reduced interbirth interval, resulting in reduced period of lactation), then the energetic cost of each offspring is reduced, and each mother could potentially produce more offspring (Aiello & Key, 2002). It has also been proposed that changes in body size and diet quality in *H. erectus* led to an increased home-range size in *H. erectus*, and promoted the dispersal out of Africa (Antón *et al.*, 2002).

Additionally, the life-history stages in *H. erectus* were predicted to be different from *H. habilis*. The brain size in *H. erectus* was large enough to seemingly "justify" the insertion (or expansion) of a childhood phase and a reduced duration of the infancy stage (Bogin, 1997, 1999b, this volume; Bogin & Smith, 1996). A reduction in the infancy stage would have given *H. erectus* (and later hominids) a reproductive advantage by reducing the period of nursing and allowing a mother to have more offspring (Aiello & Key, 2002; Bogin, 1997, 1999b, this volume; Bogin & Smith, 1996). See Leigh & Park (1998) for an evaluation and discussion of the evolution of human growth prolongation that considers Bogin's idea of the "insertion" of a childhood phase. Issues relating to the life-history stage of adolescence are discussed above in reference to the adolescent growth spurt.

Another modern human life-history stage that was suggested to appear in *H. erectus* was the female postreproductive stage (Aiello & Key, 2002; Alvarez, 2000; Bogin, 1997, 1999b, this volume; Bogin & Smith, 1996; Hawkes *et al.*, 1998; O'Connell *et al.*, 1999). Reducing the period of lactation means that someone other than the mother can provision the child, and it has been suggested that postmenopausal women filled this role. If postmenopausal women contribute to the parenting of their daughter's offspring, then they increase their daughter's fertility and their own inclusive fitness, and increase selection for extended lifespan (Aiello & Key, 2002). A postreproductive stage of "significant duration" could only evolve after life expectancy increased beyond 50 years (Bogin & Smith, 1996). Predicting life expectancy on the basis of body and brain weight suggests that later *H. erectus* had a life expectancy in the 60s, making it the first hominid to have the potential for a female postreproductive period (Bogin & Smith,1996). However, it is difficult to imagine *H. erectus* individuals living into their 60s. Demographic analysis of *H. heidelbergensis* remains from the Sima de los Huesos site at Atapuerca, Spain, suggests that longevity was no greater than 40 years for this sample, although most females in that sample died at between 16–20 years of age, and no females over 30 years of age are represented (Bermúdez de Castro & Nicolas, 1997). Neandertal mortality patterns give a similar picture, as 80% of individuals died before the age of 40 (Trinkaus, 1995; Trinkaus & Tompkins, 1990). Although it is likely that these samples do not represent the true mortality profile of populations (e.g., Bermúdez de Castro & Nicolas, 1997), it is difficult to imagine a significant proportion of females living into their 60s in these groups, let alone in *H. erectus*.

Homo antecessor and Homo heidelbergensis

Information on growth and development in non-*H. erectus* populations from the Lower and Middle Pleistocene comes primarily from two sites in the Sierra de Atapuerca, Spain: Gran Dolina and Sima de los Huesos. Material from the Lower Pleistocene TD6 level (Aurora stratum) of Gran Dolina, dates to more than 780 000 years, and has been used to define a new species in the genus *Homo, H. antecessor* (see papers in Bermúdez de Castro *et al.*, 1999a). This species is proposed to be the last common ancestor of Neandertals and modern humans (Arsuaga *et al.*, 1999b; Bermúdez de Castro *et al.*, 1997). Although *H. antecessor* slightly predates the evolutionary divergence between Neandertals and modern humans suggested on the basis of mitochondrial DNA comparisons (317–741 ka from Krings *et al.*, 1999; 550–690 ka from Krings *et al.*, 1997; and 365–853 ka from Ovchinnikov *et al.*, 2000), the upper end of the confidence interval for two studies lies close to the dates for this species. Ultimately, further work will be necessary to confirm the specific status of *H. antecessor* and its ancestral relationship to modern humans and Neandertals (Gibbons, 1997; Rightmire, 1998). Material from the Middle Pleistocene Sima de los Huesos site, dating close to 300 ka, has been assigned to the species *H. heidelbergensis* (see papers in Arsuaga *et al.*, 1997a). The Sima de los Huesos hominids, and *H. heidelbergensis* in general, show many Neandertal-like features and are presumed to be ancestral to Neandertals (Arsuaga *et al.*, 1997b; Rightmire, 1998).

Dental remains

Patterns of dental development were studied in Hominids 1–3 from the TD6 level of the Gran Dolina site, and were compared against modern human and chimpanzee dental development standards (Bermúdez de Castro *et al.*, 1999b). When comparing relative dental development of the I2 and M1, and the C and M2, the TD6 Atapuerca specimens clustered closest with modern humans, Neandertals, and late *H. erectus* specimens (although the *H. erectus* specimens were more variable, further supporting the idea of an intermediate dental development pattern in that group, see above). M3 development in the TD6 Atapuerca specimens was within the range of modern human variability, although it was most similar to modern populations of non-European origin (similar to results of Tompkins, 1996). These results were interpreted to mean that *H. antecessor*, a species living at least 800 ka, had a pattern of dental development like that of modern humans, although it did differ in some aspects (like in M3 development) from recent modern humans of European origin. This suggested that modern human dental developmental patterns evolved sometime between the Late Pliocene and late Lower Pleistocene.

Patterns of dental development were also examined in Hominid 18 (XVIII) from the Sima de los Huesos site (Bermúdez de Castro & Rosas, 2001). The dental development patterns for this individual were within the modern range of variation, although Hominid 18 showed a relative delay in canine formation, acceleration of the M2 and M3 relative to other teeth, and advanced M3 calcification. The advanced M2 and M3 development might result from having more space in the mandible (such as the retromolar space) for successively developing teeth, or to genetic differences controlling the onset of dental and skeletal formation (the authors suggested that both mechanisms were possible). When specimens from both Atapuerca sites were considered in the same analysis of relative dental development, development of the I2/M1 and C/M2 was closest to modern humans, Neandertals, and late *H. erectus* (Bermúdez de Castro *et al.*, this volume).

A study of dental development measuring incremental enamel microstructure (surface perikymata counts) has also been carried out on anterior teeth from the Sima de los Huesos site (Bermúdez de Castro *et al.*, this volume). Crown formation times for the upper I2 and C, and lower I1 and I2 were similar to modern humans, while crown formation times in the lower C and upper I1 were shorter than in modern humans. The authors suggest that the crown formation times for the lower C and upper I1 might be a more recent acquisition in human evolution. The results of that study fit very well with the recent findings that *H. erectus* had more ape-like timing and patterns of dental development, while Neandertals had modern rates of dental development (Dean *et al.*, 2001). Although Dean *et al.* note that the "shift in enamel growth rates in the hominin fossil record seems to be with the origin of larger-brained Neanderthals" (2001: 630), they did not include fossils spanning the temporal gap between *H. erectus* and Neandertals. Thus, the tooth crown formation times presented in Bermúdez de Castro *et al.* (this volume) fill that important temporal gap between *H. erectus* and Neandertals, and show that modern rates of enamel growth were present long before the Neandertals. Since *H. antecessor* is the proposed common ancestor of both Neandertals and modern humans, it is possible that both these later groups inherited this extended growth pattern from that species, rather than independently acquiring it (Bermúdez de Castro *et al.*, this volume).

Cranial and mandibular remains

The partial face of juvenile ATD6-69 (assigned to Hominid 3) from the TD6 level of Gran Dolina was first described in 1997, and belongs to the new species, *H. antecessor* (Bermúdez de Castro *et al.*, 1997). This juvenile specimen had a modern pattern of midfacial topography (such as a canine fossa), but showed some more primitive features as well (such as reduced nasal projection). Arsuaga *et al.* (1999b) described the cranial remains from the Gran Dolina site, including a more thorough description of ATD6-69; however, comparisons with other hominids were from a descriptive and phylogenetic point of view, and the facial ontogeny of *H. antecessor* was not considered. The juvenile status of ATD6-69 has caused some speculation about the validity of using its morphology in the description of *H. antecessor* (Gibbons, 1997; Rightmire, 1998). While it is true that certain population- or species-specific aspects of craniofacial morphology are present at a young age in great apes, *A. africanus*, Neandertals, and modern humans (e.g., Ackermann & Krovitz, 2002; Krovitz, 2000, this volume; Ponce de León & Zollikofer, 2001; Strand Viðarsdóttir *et al.*, 2002; Strand Viðarsdóttir & O'Higgins, this volume), future research will be necessary to determine the ontogenetic basis of the distinctive features (particularly of the maxilla) in *H. antecessor*.

Arsuaga *et al.* (1997b) described the *H. heidelbergensis* cranial remains from the Sima de los Huesos site. Cranium 3 is an adolescent (post-M2, pre-M3 eruption), and Crania 7 and 9 probably also represent adolescent individuals. There are other miscellaneous juvenile cranial bones from this site. Although these bones have been described, they have not yet been analyzed in an ontogenetic context.

Postcranial remains

The postcranial remains from the TD6 level of Gran Dolina were described by Carretero *et al.* (1999) and Lorenzo *et al.* (1999) and compared with other hominid samples. Of the two juvenile individuals represented on the basis of appendicular and axial skeletons, one was a child (probably Hominid 2 or Hominid 6), and one an adolescent (probably Hominid 1 or 3). Of the two juvenile individuals represented on the basis of hand and foot remains, one was a juvenile (probably Hominid 3, 10–11 years of age), and one an adolescent (probably Hominid 1, 13–15 years of age). Clavicles, scapulae, and humeri from the Sima de los Huesos site, including many from juvenile individuals, have been described and analyzed by Carretero *et al.* (1997).

A complete adult male pelvis recovered from the Sima de los Huesos site at Atapuerca (Arsuaga *et al.*, 1999a) allows for a discussion of hip and birth canal morphology in *H. heidelbergensis*. This pelvis was mediolaterally broad, with a large biacetabular breadth (possibly for thermoregulatory reasons), although the femora of the Sima de los Huesos sample still had long femoral necks, like that of earlier *Homo* (Arsuaga *et al.*, 1999a). The reconstructed birth canal dimensions would have been large enough for a modern-sized neonate, and the anteroposterior dimension of the midplane was larger than the medio-lateral dimension, showing that rotational birth was possible in the Sima de los Huesos hominid group (Arsuaga *et al.*, 1999a). Ruff (1995) predicted that the pelvic complex of a wide pelvis and long femoral neck that were present in *H. erectus*, and were tied to non-rotational birth and resulting smaller adult cranial capacity, would not be present in "archaic" humans. By the time adult brain size reached modern levels (i.e., around 200 ka), the pelvic-femoral complex Ruff identified disappeared, suggesting that larger brain size was only possible once the shape of the true pelvis had been modified to allow rotational birth (although see Rosenberg & Trevathan (1996) for another view on the evolution of human birth). However in *H. heidelbergensis*, the pelvis shows modern human dimensions, although the femoral neck is still long (Arsuaga *et al.*, 1999a). It is possible that the Sima de los Huesos hominids preserved transitional or mosaic-like aspects of the pelvic–femoral complex, or that Ruff's (1995) model needs to be revised based on this recent fossil find.

Life-history issues

Studies of dental development that have been carried out on the Atapuerca hominids (discussed above) have revealed that *H. antecessor* and *H. heidelbergensis* both had a pattern of dental development like that of modern humans. Because of Smith's predictions that an essentially modern human life-history schedule would appear once cranial capacity reached over 1000 cm^3 (e.g., Smith, 1991), the Atapuerca hominids were reasoned to have had a life-history pattern like modern humans (Bermúdez de Castro *et al.*, 1999b, this volume). Additionally, according to another model of life-history evolution based on cranial capacity, the Atapuerca hominids would also have had an adolescent life-history phase (Bogin & Smith, 1996). It is therefore possible that the Atapuerca hominids could have had a relatively short interbirth interval, high rate of postnatal brain growth, a childhood phase, an extended period of off-spring dependency, a marked adolescent growth spurt, and delayed reproductive cycles (Bermúdez de Castro *et al.*, this volume).

Additionally, the demographics of the Sima de los Huesos Atapuerca sample have been considered (Bermúdez de Castro, 1996; Bermúdez de Castro & Nicolas, 1997). In the Sima de los Huesos fossil assemblage there are no neonates or infants under 2 years in this sample, 53% of individuals died during adolescence with maximum mortality between 15–18 years, only 1/3 of individuals lived into their 30s, and only two individuals over 40 years of age were represented. This mortality profile could not represent a biologically viable population, so it is likely that the true mortality of the Middle Pleistocene Atapuerca population was significantly different from the mortality distribution of the Sima de los Huesos sample.

Summary

In *H. habilis*, aspects of dental growth (Dean *et al.*, 2001; Dean & Reid, 2001; Moggi-Cecchi, 2001; Smith, 1986, 1989, 1991, 1994) and patterns of facial remodeling (Bromage, 1989) were very similar to those of the australopithecines (see Kuykendall, this volume), and suggest a more ape-like period of growth and development for *H. habilis*. In contrast, other aspects of dental development are roughly similar between *H. habilis* and modern humans (Beynon & Wood, 1987), and life-history parameters are predicted to be intermediate between australopithecines and *H. erectus* (Smith *et al.*, 1995), although Bogin (1997, 1999b, this volume; Bogin & Smith, 1996) is equivocal about the presence of a short childhood phase in *H. habilis*. The similarity between maturational patterns in *H. habilis* and the australopithecines is a persuasive piece of evidence used to suggest that *H. habilis* (and *H. rudolfensis*) does not belong in the genus *Homo*, and instead belongs in *Australopithecus* (Wood & Collard, 1999).

In *H. erectus*, several studies suggest a dental developmental pattern more intermediate between modern humans and apes (Moggi-Cecchi, 2001; Smith, 1986, 1989, 1991, 1994), and conceptual models of growth and development also predict a more extended period of development in *H. erectus* than in earlier hominids (Smith, 1989b, 1991, 1994; Smith *et al.*, 1995). However, a recent study has shown that the overall timing and rate of *H. erectus*'s dental development (and thus, overall development) was in the range of fossil and modern apes (Dean *et al.*, 2001). Adolescent facial growth patterns in *H. erectus* are similar to those of chimpanzees and modern humans (Richtsmeier & Walker, 1993), while aspects of neurocranial growth differ between *H. erectus* and modern humans (Antón, 1997, 2002). It is unclear if secondary altriciality was present in *H. erectus*, and it is likely that *H. erectus* had a non-rotational birth process (Ruff, 1995; although see Rosenberg & Trevathan, 1996). There is theoretical evidence for the presence of a childhood life-history phase (Bogin, 1997, 1999b, this volume; Bogin & Smith, 1996), and there is theoretical and material evidence both for (Antón, 2002; Antón & Leigh, this volume; Clegg & Aiello, 1999; Tardieu, 1998) and against (Bogin, 1999a, 1999b, this volume; Dean *et al.*, 2001; Smith, 1993; Thompson & Nelson, 2000) an adolescent life-history stage (and growth spurt) in *H. erectus*. A female postreproductive phase is predicted to be theoretically possible for *H. erectus* (Aiello & Key,

2002; Alvarez, 2000; Bogin, 1997, 1999b, this volume; Bogin & Smith, 1996; Hawkes *et al.*, 1998; O'Connell *et al.*, 1999), although it is unlikely due to practical concerns of hominid life expectancy. It has been suggested that "a major shift in the process of growth and development seems to have occurred in the transition from *Homo habilis* to *Homo erectus*, in which *H. habilis* may be the last representative of an australopithecine-like mode of growth, whereas *Homo erectus* marks the appearance of a basically new mode of growth in hominid evolution" (Moggi-Cecchi, 2001: 132). This is somewhat true, as *H. erectus* does have many aspects of a modern human life-history pattern, although certain aspects of the pattern are still surprisingly ape-like.

The hominid remains from Atapuerca, from both the TD6 level of the Gran Dolina site, *H. antecessor*, and the Sima de los Huesos site, *H. heidelbergensis*, show modern human dental developmental patterns (Bermúdez de Castro & Rosas, 2001; Bermúdez de Castro *et al.*, 1999b, this volume). Predictions of life-history stages based on brain size and dental development suggest that both *H. antecessor* and *H. heidelbergensis* would have had life-history patterns like modern humans, including an adolescent life-history period (Bogin & Smith, 1996; Bermudez de Castro *et al.*, this volume). Although these two Atapuerca sites differ widely in time (i.e., 780 ka versus 300 ka), there are, at present, few grounds for ontogenetic comparison between the species. The juvenile cranial and postcranial remains from both sites have yet to be analyzed in a broad ontogenetic context, and it is clear that much future research will be carried out with this fossil material. Additional juvenile fossil remains from other sites will also be necessary to further elaborate on the growth patterns of these species, and to compare the *H. heidelbergensis* material from the Sima de los Huesos site with juvenile remains from elsewhere in that species range (particularly Africa).

Thus, this summary has reviewed the evolution of patterns of growth and development from the very ape-like patterns documented for australopithecines, to the very modern patterns of dental development and predicted life-history stages in *H. antecessor* and *H. heidelbergensis*. It is clear that the modern human pattern of growth and development evolved in a mosaic manner, and that aspects of dental, cranial, and postcranial growth all evolved at different times, probably following different evolutionary pressures. It is also clear that different studies give different, and often conflicting, results. This might be a result of the different information gained from studies of data collected from actual juvenile hominid fossils versus life-history predictions based on more theoretical or conceptual models of life history evolution, such as the relationship between brain size and dental maturation (e.g., Smith, 1989b; Smith *et al.*, 1995), or the presence of different life-history stages (e.g., Bogin, 1999). Studies of juvenile fossil remains reveal details about the actual patterns of growth and development seen in the dentition, skull, and postcranial skeleton, while conceptual models predict what life-history stages were present and the overall length of the life-history schedule. However, both research designs provide valuable information about the evolution of patterns of growth and development in the genus *Homo*. Continued research on the subject, as well as the slow but continual increase in the juvenile fossil sample for the species discussed here (or others yet named), will also increase our knowledge about this important subject.

References

Ackermann, R. R., & Krovitz, G. E. (2002). Common patterns of facial ontogeny in the hominid lineage. *Anatomical Record,* **269**, 142–147.

Aiello, L. C., & Key, C. (2002). Energetic consequences of being a *Homo erectus* female. *American Journal of Human Biology,* **14**, 551–565.

Aiello, L. C., & Wells, J. C. K. (2002). Energetics and the evolution of the genus *Homo. Annual Review of Anthropology,* **31**, 323–338.

Aiello, L. C., & Wheeler, P. (1995). The expensive tissue hypothesis: The brain and the digestive system in human and primate evolution. *Current Anthropology,* **36**, 199–221.

Alvarez, H. P. (2000). Grandmother hypothesis and primate life histories. *American Journal of Physical Anthropology,* **113**, 435–450.

Antón, S. C. (1997). Developmental age and taxonomic affinity of the Mojokerto Child, Java, Indonesia. *American Journal of Physical Anthropology,* **102**, 497–514.

Antón, S. C. (1999). Cranial growth in *Homo erectus*: How credible are the Ngandong juveniles? *American Journal of Physical Anthropology,* **108**, 223–236.

Antón, S. C. (2002). Cranial growth in *Homo erectus*. In *Human Evolution through Developmental Change*, eds. N. Minugh-Purvis & K. J. McNamara, pp. 349–380. Baltimore: Johns Hopkins University Press.

Antón, S. C., & Leigh, S. (1998). Paedomorphosis and neoteny in human evolution. *Journal of Human Evolution,* **34**, A2.

Antón, S. C., Leonard, W. R., & Robertson, M. L. (2002). An ecomorphological model of the initial hominid dispersal from Africa. *Journal of Human Evolution,* **43**, 773–785.

Arsuaga, J. L., Bermúdez de Castro, J. M., & Carbonell, E. (1997a). The Sima de los Huesos hominid site. *Journal of Human Evolution,* **33**.

Arsuaga, J. L., Martinez, I., Gracia, A., & Lorenzo, C. (1997b). The Sima de los Huesos crania (Sierra de Atapuerca, Spain): A comparative study. *Journal of Human Evolution,* **33**, 219–281.

Arsuaga, J. L., Lorenzo, C., Carretero, J.-M., Gracia, A., Martinez, I., Garcia, N., Bermúdez de Castro, J. M., & Carbonell, E. (1999a). A complete human pelvis from the Middle Pleistocene of Spain. *Nature,* **399**, 255–258.

Arsuaga, J. L., Martinez, I., Lorenzo, C., & Gracia, A. (1999b). The human cranial remains from Gran Dolina Lower Pleistocene site (Sierra de Atapuerca, Spain). *Journal of Human Evolution,* **37**, 431–457.

Asfaw, B., White, T., Lovejoy, O., Latimer, B., Simpson, S., & Suwa, G. (1999). *Australopithecus garhi*: A new species of early hominid from Ethiopia. *Science,* **284,** 629–635.

Bermúdez de Castro, J. M. (1996). European Middle Pleistocene human mortality patterns: The case of the Atapuerca–SH hominids. In *The Last Neandertals, the First Anatomically Modern Humans: A Tale About the Human Diversity – Cultural Change and Human Evolution, the Crisis at 40 KA B.P.,* eds. E. Carbonell & M. Vaquero, pp. 21–38. Tarragona: Universitat Rovira i Virgili.

Bermúdez de Castro, J. M., & Nicolas, M. E. (1997). Palaeodemography of the Atapuerca–SH Middle Pleistocene hominid samples. *Journal of Human Evolution,* **33,** 333–355.

Bermúdez de Castro, J. M., & Rosas, A. (2001). Pattern of dental development in Hominid XVIII from the Middle Pleistocene Atapuerca-Sima de los Huesos site (Spain). *American Journal of Physical Anthropology,* **114,** 325–330.

Bermúdez de Castro, J. M., Arsuaga, J. L., Carbonell, E., Rosas, A., Martínez, I., & Mosquera, M. (1997). A hominid from the Lower Pleistocene of Atapuerca, Spain: Possible ancestor to Neandertals and modern humans. *Science,* **276,** 1392–1395.

Bermúdez de Castro, J. M., Carbonell, E., & Arsuaga, J. L. (1999a). Gran Dolina Site: TD6 Aurora Stratum (Burgos, Spain). *Journal of Human Evolution,* **37**.

Bermúdez de Castro, J. M., Rosas, A., Carbonell, E., Nicolas, M. E., Rodriguez, J., & Arsuaga, J. L. (1999b). A modern human pattern of dental development in Lower Pleistocene hominids from Atapuerca-TD6 (Spain). *Proceedings of the National Academy of Sciences of the USA,* **96,** 4210–4213.

Beynon, A. D., & Wood, B. A. (1987). Patterns and rates of enamel growth in the molar teeth of early hominids. *Nature,* **326,** 493–496.

Bogin, B. (1997). Evolutionary hypotheses for human childhood. *Yearbook of Physical Anthropology,* **40,** 63–89.

Bogin, B. (1999a). Evolutionary perspective on human growth. *Annual Review of Anthropology,* **28,** 109–153.

Bogin, B. (1999b). *Patterns of Human Growth,* 2nd edn. Cambridge: Cambridge University Press.

Bogin, B., & Smith, B. H. (1996). Evolution of the human life cycle. *American Journal of Human Biology,* **8,** 703–716.

Bromage, T. G. (1989). Ontogeny of the early hominid face. *Journal of Human Evolution,* **18,** 751–773.

Bromage, T. G., & Dean, M. C. (1985). Re-evaluation of the age at death of immature fossil hominids. *Nature,* **317,** 525–527.

Carretero, J. M., Arsuaga, J. L., & Lorenzo, C. (1997). Clavicles, scapulae and humeri from the Sima de los Huesos site (Sierra de Atapuerca, Spain). *Journal of Human Evolution,* **33,** 357–408.

Carretero, J. M., Lorenzo, C., & Arsuaga, J. L. (1999). Axial and appendicular skeleton of *Homo antecessor. Journal of Human Evolution,* **37,** 459–499.

Clegg, M., & Aiello, L. C. (1999). A comparison of the Nariokotome *Homo erectus* with juveniles from a modern human population. *American Journal of Physical Anthropology,* **110,** 81–93.

Day, M. H. (1986). *Guide to Fossil Man*, 4th edn. London: Cassell.

Dean, M. C. (1985). The eruption pattern of the permanent incisors and first permanent molars in *Australopithecus (Paranthropus) robustus*. *American Journal of Physical Anthropology*, **67**, 251–257.

Dean, M. C. (1987). The dental developmental status of six East African juvenile fossil hominids. *Journal of Human Evolution*, **16**, 197–213.

Dean, M. C., & Reid, D. J. (2001). Perikymata spacing and distribution on hominid anterior teeth. *American Journal of Physical Anthropology*, **116**, 209–215.

Dean, C., Leakey, M. G., Reid, D., Schrenk, F., Schwartz, G. T., Stringer, C., & Walker, A. (2001). Growth processes in teeth distinguish modern humans from *Homo erectus* and earlier hominins. *Nature*, **414**, 628–631.

Dunsworth, H., & Walker, A. (2002). Early genus *Homo*. In *The Primate Fossil Record*, ed. W. C. Hartwig, pp. 419–435. Cambridge: Cambridge University Press.

Feibel, C. S., Brown, F. H., & McDougall, I. (1989). Stratigraphic context of fossil hominids from the Omo Group deposits: Northern Turkana Basin, Kenya and Ethiopia. *American Journal of Physical Anthropology*, **78**, 595–622.

Gibbons, A. (1997). A new face for human ancestors. *Science*, **276**, 1331–1333.

Hawkes, K., O'Connell, J. F., Blurton-Jones, N. G., Alvarez, H., & Charnov, E. L. (1998). Grandmothering, menopause, and the evolution of human life histories. *Proceedings of the National Academy of Sciences of the USA*, **95**, 1336–1339.

Kennedy, G. E. (1983). A morphometric and taxonomic assessment of a hominine femur from the lower member, Koobi Fora, Lake Turkana. *American Journal of Physical Anthropology*, **61**, 429–436.

Krings, M., Stone, A., Schmitz, R. W., Krainitzki, H., Stoneking, M., & Pääbo, S. (1997). Neandertal DNA sequences and the origin of modern humans. *Cell*, **90**, 19–30.

Krings, M., Geisert, H., Schmitz, R. W., Krainitzki, H., & Pääbo, S. (1999). DNA sequence of the mitochondrial hypervariable region II from the Neandertal type specimen. *Proceedings of the National Academy of Sciences of the USA*, **96**, 5581–5585.

Krovitz, G. E. (1999). Three-dimensional analysis of modern human and Neandertal craniofacial growth patterns. *American Journal of Physical Anthropology*, Suppl. 28, 175–176.

Krovitz, G. E. (2000). Three-dimensional comparisons of craniofacial morphology and growth patterns in Neandertals and modern humans. PhD dissertation, The Johns Hopkins University.

Leigh, S. R. (1996). Evolution of human growth spurts. *American Journal of Physical Anthropology*, **101**, 455–474.

Leigh, S. R., & Park, P. B. (1998). Evolution of human growth prolongation. *American Journal of Physical Anthropology*, **107**, 331–350.

Lorenzo, C., Arsuaga, J. L., & Carretero, J. M. (1999). Hand and foot remains from the Gran Dolina Early Pleistocene site (Sierra de Atapuerca, Spain). *Journal of Human Evolution*, **37**, 501–522.

Moggi-Cecchi, J. (2001). Patterns of dental development of *Australopithecus africanus*, with some inferences on their evolution with the origin of the genus *Homo*. In *Humanity from African Naissance to Coming Millennia: Colloquia in Human Biology and Palaeoanthropology*, eds. P. V. Tobias, M. A. Raath, J. Moggi-Cecchi, & G. A. Doyle, pp. 125–133. Firenze: Firenze University Press.

Montagu, M. F. A. (1955). Time, morphology, and neoteny in the evolution of Man. *American Anthropologist,* **57**, 13–27.

O'Connell, J. F., Hawkes, K., & Blurton Jones, N. G. (1999). Grandmothering and the evolution of *Homo erectus. Journal of Human Evolution,* **36**, 461–485.

Ovchinnikov, I. V., Götherström, A., Romanova, G. P., Kharitonov, V. M., Lidén, K., & Goodwin, W. (2000). Molecular analysis of Neanderthal DNA from the northern Caucasus. *Nature,* **404**, 490–493.

Ponce de León, M. S., & Zollikofer, C. P. E. (2001). Neanderthal cranial ontogeny and its implications for late hominid diversity. *Nature,* **412**, 534–538.

Richmond, B. G., Aiello, L. C., & Wood, B. A. (2002). Early hominin limb proportions. *Journal of Human Evolution,* **43**, 529–548.

Richtsmeier, J. T., & Walker, A. (1993). A morphometric study of facial growth. In *The Nariokotome* Homo erectus *Skeleton*, eds. A. Walker & R. Leakey, pp. 391–410. Cambridge: Harvard University Press.

Rightmire, G. P. (1998). Human evolution in the Middle Pleistocene: The role of *Homo heidelbergensis. Evolutionary Anthropology,* **6**, 218–227.

Rosenberg, K., & Trevathan, W. (1996). Bipedalism and human birth: The obstetrical dilemma revisited. *Evolutionary Anthropology,* **4**, 161–168.

Ruff, C. B. (1995). Biomechanics of the hip and birth in early *Homo. American Journal of Physical Anthropology,* **98**, 527–574.

Shipman, P. (1986). Scavenging or hunting in early hominids. *American Anthropologist,* **88**, 27–43.

Shipman, P., & Walker, A. (1989). The costs of becoming a predator. *Journal of Human Evolution,* **18**, 373–392.

Smith, B. H. (1986). Dental development in *Australopithecus* and early *Homo. Nature,* **323**, 327–330.

Smith, B. H. (1989a). Dental development as a measure of life history in primates. *Evolution,* **43**, 683–688.

Smith, B. H. (1989b). Growth and development and its significance for early hominid behaviour. *Ossa,* **14**, 63–96.

Smith, B. H. (1991). Dental development and the evolution of life history in the Hominidae. *American Journal of Physical Anthropology,* **86**, 157–174.

Smith, B. H. (1993). The physiological age of KNM-WT 15000. In *The Nariokotome* Homo erectus *Skeleton*, eds. A. Walker & R. Leakey, pp. 196–220. Cambridge: Harvard University Press.

Smith, B. H. (1994). Patterns of dental development in *Homo, Australopithecus, Pan* and *Gorilla. American Journal of Physical Anthropology,* **94**, 307–325.

Smith, B. H., & Tompkins, R. L. (1995). Toward a life history of the Hominidae. *Annual Review of Anthropology,* **24**, 257–279.

Smith, R. J., Gannon, P. J., & Smith, B. H. (1995). Ontogeny of australopithecines and early *Homo*: Evidence from cranial capacity and dental eruption. *Journal of Human Evolution,* **29**, 155–168.

Strand Viðarsdóttir, U., O'Higgins, P., & Stringer, C. (2002). A geometric morphometric study of regional differences in the ontogeny of the modern human facial skeleton. *Journal of Anatomy,* **201**, 211–229.

Tardieu, C. (1998). Short adolescence in early hominids: Infantile and adolescent growth of the human femur. *American Journal of Physical Anthropology,* **107**, 163–178.

G. E. Krovitz, J. L. Thompson, & A. J. Nelson

Tardieu, C. (1999). Ontogeny and phylogeny of femoro-tibial characters in humans and hominid fossils: Functional influence and genetic determinism. *American Journal of Physical Anthropology,* **110**, 365–377.

Tardieu, C., & Trinkaus, E. (1994). Early ontogeny of the human femoral bicondylar angle. *American Journal of Physical Anthropology,* **95**, 183–195.

Thompson, J. L., & Nelson, A. J. (2000). The place of Neandertals in the evolution of hominid patterns of growth and development. *Journal of Human Evolution,* **38**, 475–495.

Tobias, P. V. (1991). *Olduvai Gorge,* vol. 4, *The Skulls, Endocasts and Teeth of* Homo habilis. Cambridge: Cambridge University Press.

Tompkins, R. L. (1996). Relative dental development of Upper Pleistocene hominids compared to human population variation. *American Journal of Physical Anthropology,* **99**, 103–118.

Trinkaus, E. (1984). Does KNM-ER 1481A establish *Homo erectus* at 2.0 myr BP? *American Journal of Physical Anthropology,* **64**, 137–139.

Trinkaus, E. (1995). Neanderthal mortality patterns. *Journal of Archaeological Science,* **22**, 121–142.

Trinkaus, E., & Tompkins, R. L. (1990). The Neandertal life cycle: The possibility, probability, and perceptability of contrasts with recent humans. In *Primate Life History and Evolution,* ed. C. J. de Rousseau, pp. 153–180. New York: Wiley-Liss.

Walker, A. (1993). Perspectives on the Nariokotome discovery. In *The Nariokotome* Homo erectus *Skeleton,* eds. A. Walker & R. Leakey, pp. 411–430. Cambridge: Harvard University Press.

Walker, A., & Leakey, R. (eds.) (1993). *The Nariokotome* Homo erectus *Skeleton.* Cambridge: Harvard University Press.

Walker, A., & Ruff, C. B. (1993). The reconstruction of the pelvis. In *The Nariokotome* Homo erectus *Skeleton,* eds. A. Walker & R. Leakey, pp. 221–233. Cambridge: Harvard University Press.

Wood, B. (1992). Origin and evolution of the genus *Homo. Nature,* **355**, 783–790.

Wood, B., & Collard, M. (1999). The human genus. *Science,* **284**, 65–71.

Wood, B., & Richmond, B. G. (2000). Human evolution: Taxonomy and paleobiology. *Journal of Anatomy,* **196**, 16–60.

Part III
The last steps: The approach to modern humans

12 Diagnosing heterochronic perturbations in the craniofacial evolution of Homo (Neandertals and modern humans) and Pan (P. troglodytes and P. paniscus)

F. L. WILLIAMS
Georgia State University

L. R. GODFREY
University of Massachusetts

M. R. SUTHERLAND
University of Massachusetts

Introduction

A number of researchers have suggested that neoteny played an important role in the craniofacial evolution of the genus *Homo* and its close relatives (Abbie, 1947; Bolk, 1926; Brothwell, 1975; de Beer, 1958; Gould, 1977, 2000; Montagu, 1989; Privratsky, 1981; Verhulst, 1999). The same has also been suggested, at least in part, for the evolution of *Pan paniscus* (the "pygmy chimpanzee" or bonobo) (Rice, 1997; Shea, 1983, 1989, 2000, 2002; see also Coolidge, 1933). Comparisons of fossil as well as extant taxa have been brought to bear on the problem. Rozzi (2000) suggested that changes in molar morphology, from *Australopithecus afarensis* to the robust australopithecines, may be attributed to neoteny. Antón & Leigh (1998) held that neoteny helps to explain the evolution of craniofacial form from *Homo erectus* to *Homo sapiens*. Czarnetzki *et al.* (2001) invoked neoteny to explain the development of adult Neandertal traits. Finally, Alba *et al.* (2001) and Alba (2002) characterized the reduced canine size and facial prognathism in both bonobos and *Oreopithecus* as "paedomorphic," and Shea (1984, 2000) characterized the "paedomorphic" facial

Patterns of Growth and Development in the Genus Homo, ed. J. L. Thompson, G. E. Krovitz and A. J. Nelson. Published by Cambridge University Press. © Cambridge University Press 2003.

skeletons of bonobos as "neotenic," relative to a hypothetical *Pan troglodytes*-like ancestor.

The notion that modern humans and bonobos are neotenic has also come under severe criticism. Shea (1989) contested the neoteny hypothesis for human evolution (as did McKinney & McNamara, 1991 and McNamara, 2002) while upholding neoteny for bonobos. Godfrey & Sutherland (1995, 1996) challenged the evidence brought to bear on the arguments for and against neoteny in both humans and bonobos. Ponce de León & Zollikofer (2001) suggested that the postnatal ontogenies of modern humans and Neandertals differ solely in their (coupled) rate of change in size-and-shape. They characterized Neandertals as "rate hypermorphs" (and, by implication, modern humans as "rate hypomorphs") with respect to the other. Williams *et al.* (2002) argued that modern humans make poor "neotenic" Neandertals, and that bonobos make poor "neotenic" chimpanzees.

Evaluations of heterochronic perturbations have been plagued by differences in the meanings accorded heterochronic terms. These differences are sometimes made explicit, but more often they are only implicit in the methodologies used to draw heterochronic inferences. Disagreement over appropriate tools for heterochronic analysis, as well as their implications, has led to a certain degree of dissonance in the field (Godfrey & Sutherland, 1995, 1996; Gould, 2000; McKinney, 1998; McKinney & McNamara, 1991; Shea, 1989, 2000). We embrace here the definitions put forth by Gould (1977).

Gould's (1977) main argument that humans are neotenic was predicated on the notion that growth (change in size) and development (change in shape) are inherently and easily dissociable. By this he meant that descendants might follow their ancestor's shape-change pathway, but at different sizes. Whereas the assumption of inherent dissociability of growth and development has been challenged (Shea, 2000), Gould's model cannot be appropriately tested unless that assumption is embraced in the development of a diagnostic tool. In other words, Gould may have been wrong in his insistence that size and shape can be perturbed independently, but predictions generated from that model cannot be tested without granting the possibility that they can. Furthermore, because heterochronic perturbations are only a subset of possible evolutionary developmental changes, the fit of a model based solely on heterochrony to the real data must be tested.

Here, we present a tool for analyzing heterochrony that incorporates whole suites of traits simultaneously into a multivariate framework. It treats change in size and change in shape as inherently dissociable, *sensu* Gould (1977, 2000). In doing so, it assumes that the products of heterochrony, such as paedomorphosis (descendant adult exhibiting the morphology of the ancestral juvenile) and peramorphosis (descendant adult resembling an "overdeveloped" or "adultified" ancestor) can vary greatly in size.

The program we use, HETPAD (Heterochronic Prediction and Diagnosis), is currently being developed by two of us (MRS and LRG) and others (see Williams *et al.*, 2001). HETPAD's exhaustive search explores all possible combinations of heterochronic processes affecting distinct evolutionary "modules" (or subsets of traits), and it finds the solution that best fits the observations. That is, it finds the modeled descendant (fashioned from an observed ancestor using heterochronic "verbs" or processes) that comes closest to the observed descendant. This allows the investigator to test the efficacy of heterochrony in explaining the evolutionary differences between two taxa. HETPAD also allows the investigator to select any particular heterochronic transformation (for example, global neoteny) and measure how well that transformation, by itself, accounts for observed evolutionary differences. We recognize that no computer model can reflect the full complexity of the natural world, but HETPAD offers a particularly focused way to test how heterochrony accounts for the differences between two taxa.

Although comparisons of modern human and modern ape ontogenies provide a bounty of information upon which to build heterochronic predictions regarding the evolution of both lineages, the fossil record is the ultimate test for these hypotheses. In the absence of ontogenetic series of Plio-Pleistocene apes, we are forced to rely on comparisons of living apes to test heterochronic hypotheses. Thus, Shea's (1983, 1984) hypothesis that *Pan paniscus* is a neotenic *Pan troglodytes* can be tested by selecting *Pan troglodytes* as the hypothetical ancestor. This means selecting either *Pan troglodytes* or *Pan paniscus* to represent the hypothetical "ancestor" in order to test the hypothesis that the other can be derived from that "ancestor" via one or several simple heterochronic perturbations. Effectively, this is equivalent to testing the hypothesis that one species could be derived via simple heterochrony from an ancestor *similar to* the other one.

Plio-Pleistocene fossils belonging to the human lineage are far more numerous than are those belonging to any ape (Pilbeam, 2002), although well-preserved infant and juvenile early hominin specimens (e.g., Taung, Mojokerto (Perning I), Nariokotome (KNM-WT 15000), Ngandong (2, 8)) are rare. The Upper Pleistocene preserves a much richer assemblage of immature fossil humans than does the Pliocene or Lower/Middle Pleistocene. Neandertals unequivocally provide the most complete ontogenetic sequence of any fossil hominid. There are some important though fragmentary immature early modern human fossils (see Minugh-Purvis, 1988, 2002), but the ontogenetic series of Neandertals is far superior. Whereas the actual phylogenetic relationship of Neandertals to modern humans (i.e., the question of whether or not Neandertals are directly ancestral to modern humans) remains unresolved, it is clear that many ancestral (Middle Pleistocene) *Homo* characteristics are manifested in all Upper Pleistocene humans including Neandertals (Stringer *et al.*, 1990). Our purpose in this paper is not to address the question of direct ancestry, but to test the hypothesis that simple heterochronic perturbations can turn a Neandertal or Neandertal-*like* ancestor into a modern human, or vice versa (Williams *et al.*, 2002). Brothwell (1975) and de Beer (1958) described Neandertals as over-developed modern humans, suggesting that modern humans simply stop developing at a pre-adult stage of Neandertal postnatal ontogeny. We ask whether or not this is true. More specifically, we examine the hypothesis of "global craniofacial neoteny" for *Homo* and *Pan*. We ask, is global neoteny sufficient to transform a Neandertal-like ancestor into a modern human-like descendant, or a chimpanzee-like ancestor into a bonobo-like descendant?

Materials and methods

Samples

Linear craniofacial measurements were collected for 643 modern human, Neandertal, and *Pan* specimens of mixed ages currently housed in American, European, and Israeli museums. The modern human sample comprised 294 individuals originating from Europe ($n = 63$), sub-Saharan Africa ($n = 14$), the Middle East ($n = 11$), South-east Asia ($n = 18$), the Americas ($n = 20$), medieval Belgium ($n = 54$), Papua New Guinea ($n = 24$), and from an unknown location ($n = 90$). Specimens were chosen on the basis of completeness, and whether both juveniles and adults could be obtained from a given population. The Neandertal sample comprised 39 original fossils and three casts from a broad geographic range, including Italy, Israel, Croatia, France, Hungary, Uzbekistan, Belgium, Gibraltar, and the Czech Republic. The chimpanzee sample ($n = 156$) comprised individuals largely from central West Africa (subspecies *P. t. troglodytes*), while the bonobo sample ($n = 151$) was drawn almost exclusively from the Republic of the Congo. Details of sample locations are presented in Table 12.1.

Aging methods

All hominin individuals were assigned ages in years based on one or a combination of methods. For adults, suture closure and dental wear were used to separate older from younger individuals (see Williams, 2001). Immature individuals were assigned age estimates on the basis of their stage of dental eruption, using Ubelaker's (1978) dental-aging chart for modern humans and Neandertals. Daily incremental markings for a Tabun C1 first molar indicate that Neandertals and modern humans have similar rates of enamel formation (Dean

Table 12.1 *List of institutions visited and samples measured*

Country	Institution and sample
Belgium	Institut Royal des Sciences Naturelles de Belgique (Spy 1, Spy 2, La Naulette; modern humans, $n = 54$)
	Direction de l'Archéologie, Ministère de la Région Wallonne (Sclayn)
	Université de Liège (Engis 2)
	Koninklijk Museum voor Midden-Afrika (*Pan paniscus*, $n = 146$; *Pan troglodytes*, $n = 7$)
Croatia	Croatian Natural History Museum (Krapina C3, maxillae B, C, D, E, mandibles C, D, E, F, G, H, J, and rami 63 and 66)
Czech Republic	Moravské Museum (Šipka and Ochoz)
United Kingdom	Natural History Museum (Forbes' Quarry, Tabun C1, Devil's Tower)
	Powell-Cotton Museum (*Pan troglodytes*, $n = 86$)
France	Musée de L'Homme (La Chapelle-aux-Saints, La Ferrassie, La Quina 5, Pech de l'Azé, Malarnaud)
	Musée des Antiquités Nationales–Saint Germain-en-Laye (La Quina 18)
	Muséum National de Préhistoire–Les Eyzies-de-Tayac (Roc de Marsal)
	Université de Poitiers (Châteauneuf-sur-Charente)
	Muséum National d'Histoire Naturelle–Paris (*Pan troglodytes*, $n = 22$; *Pan paniscus*, $n = 1$)
Hungary	Természettudományi Muzeum (Subalyuk 1 and 2)
Israel	Tel Aviv University (Amud 1 and 7, Kebara 2, cast of Teshik Tash)
Italy	Museo Preistorico Ethnografico, 'Luigi Pigorini' (Guattari 1)
	Istituto di Paleontologia Umana (Circeo 2 and 3, Archi 1)
The Netherlands	Rijksuniversiteit Groningen (modern humans, $n = 46$)
	Rijksuniversiteit Leiden (modern humans, $n = 86$; *Pan troglodytes*, $n = 1$)
	Nationaal Natuurhistorisch Museum (modern humans, $n = 36$; *Pan troglodytes*, $n = 32$; *Pan paniscus*, $n = 1$)
	Rijksdienst voor Oudheidkundig Bodemonderzoek (modern humans, $n = 5$)
United States of America	American Museum of Natural History (modern humans, $n = 47$)
	Johns Hopkins University School of Medicine (modern humans, $n = 24$)
	Museum of Comparative Zoology, Harvard University (*Pan troglodytes*, $n = 8$; *Pan paniscus*, $n = 3$)

et al., 2001). Smith (1991) infers a similar age at M1 eruption for Neandertals and modern humans on the basis of their great overlap in adult brain sizes and a strong correlation across primates between adult brain size and age at M1 eruption. Indeed, the larger-brained Neandertals may have erupted their first molars slightly *later* than modern humans (Smith, 1991). Wolpoff (1979) argues on the basis of dental wear that the third molar may have erupted *earlier* in

Neandertals than in modern humans. Estimated differences in timing are minor, however, and we accept the schedule for modern human dental eruption as the best available timetable for Neandertals as well. Epiphyseal plate fusion and tympanic ring development were also considered in determining age estimates for humans, following Weaver (1986).

Age estimates for both species of *Pan* were determined using dental eruption schedules developed by Dean & Wood (1981) for wild-caught *Pan troglodytes*. We applied the same standards to both species because all known dental eruption schedules for great apes are similar (Dean & Wood, 1981; Winkler *et al.*, 1996) and because bonobo dental eruption schedules are poorly known (see Smith *et al.*, 1994). The specimens of *Pan troglodytes* in our wild-caught sample displayed an eruption sequence more like that reported by Dean & Wood (1981) than like those reported by other authors (e.g., Anemone *et al.*, 1996; Kuykendall & Conroy, 1996) on the basis of captive specimens. We therefore adopted Dean & Wood's dental aging scheme for *Pan*. We caution, however, that bonobo dental eruption sequences may be more variable than those of chimpanzees (FLW, unpublished data; see also Williams *et al.*, 2002). Schultz's (1933) deciduous-tooth aging chart for *Pan troglodytes* was used for all very young chimpanzees and bonobos (from birth to 1.3 years).

Modeling growth

Twenty-four standard linear measurements were taken on each specimen (Table 12.2), and these measurements were divided into three groups that represented the calotte, face, and mandible. The traits were selected on the basis of the completeness of the Neandertal fossils, and whether the traits followed a monotonic increase in growth. Using piecewise regression, we regressed each trait against age to produce 24 growth curves for each of the four samples.

Modeling growth trajectories on the basis of relatively small cross-sectional samples is necessarily a process of approximation (German & Myers, 1989a, 1989b). We chose to use a version of piecewise regression as our approximating, smoothed curve. We wanted to capture the curvature of early growth but also the "flatline" behavior of "growth" after maturation. An important advantage of piecewise regression is that it succeeds in generating good-fit growth curves with limited ontogenetic samples (e.g., Neandertals).

Neandertal, modern human, chimpanzee and bonobo growth was modeled using a second-order version of piecewise regression. Second-order piecewise regression is one of a number of analytic tools used to model growth over time from cross-sectional studies (de Bruin, 1993). It comprises two segments: a quadratic, and a constant, between which there is a breakpoint age and value

Table 12.2 *Linear distances used in modeling growth*

Calotte	Face	Mandible
Maximum cranial length (glabella–opisthocranion)	Biectomolare breadth (breadth across the most lateral points on the alveolar margin of maxilla)	Mandibular symphysis height (infradentale–gnathion)
Maximum cranial breadth (biparietal)	Palatal length (prosthion–staphylion)	Mandibular corpus height (at mental foramen)
Minimum frontal breadth (maximum postorbital constriction)	Maximum nasal aperture breadth (bialare)	Mandibular corpus thickness (at mental foramen)
		Mandibular length (gonion–gnathion)
Biorbital breadth (across the most lateral extremes of the frontomaxillary suture)	Bizygomatic breadth (bizygion)	Bicondylar breadth (bicondylion breadth)
	Upper facial height (alveolare–nasion)	
Orbital breadth (maxillofrontale to ectoconchion)	Palatal breadth (biendomolare at widest point)	Bigonial breadth
Interbiorbital breadth (biectoconchion)	Basion–prosthion	Ascending ramus height (gonion to the most superior aspect of the mandibular condyle)
Upper cranial height (porion–bregma)		Minimum ascending ramus breadth (shortest distance between posterior and anterior margins of the ascending ramus)
Cranial base height (porion–basion)		Maximum ascending ramus breadth (between the most anterior edge of ascending ramus and the most posterior aspect of the mandibular condyle)

which indicates the age and trait size at which growth ceases. The difference between this model and others, such as von Bertalanffy's, Gompertz's, logistic, and simple power curves, is that piecewise regression can produce a flat line indicative of maturation whereas most other growth models continue to increase asymptotically after growth ceases (Leigh & Terranova, 1998). As opposed to more complex growth models which use third-, fourth-, and even higher-order polynomials, this version of piecewise regression makes fewer assumptions about the growth of traits.

Using Systat's non-linear function, coefficients describing growth rate and duration were obtained from piecewise regression. From a maximum of 50 iterations using 20 half-steps of Gauss–Newton least squares, four coefficients were obtained for each taxon, for each of our 24 craniofacial traits. The four beta (β) coefficients have approximate interpretations as the size at birth ($\beta 0$), the initial growth rate ($\beta 1$), the rate of decline in the growth rate ($\beta 2$), and the age at which growth levels off for a given trait ($\beta 3$). Starting values for the four coefficients provided initial estimates to locate global versus local least squares minima, and thus the best fit of the model to the data. The equation for the piecewise regression used here is:

$$\text{Trait} = [(\beta 0 + \beta 1^* \text{age} + \beta 2^* \text{age}^2) \text{ if age} < \beta 3,$$
$$\text{or } (\beta 0 + \beta 1^* \beta 3 + \beta 2^* \beta 3^2) \text{ if age} \geq \beta 3]$$

The least-squares solution used to approximate the ontogenetic trajectories was constrained by imposing minimum and maximum values for each of the four coefficients. The same constraints were applied to all taxa to make estimates of growth rates and ages at cessation of growth comparable. These constraints are consistent with empirical evidence and theoretical expectations of nonhuman and human growth (Hauspie, 1989, 1998; Leigh & Terranova, 1998). Growth curves for each variable were generated using the estimated coefficients from the piecewise regressions, evaluated at half-year intervals up to 25 years for Neandertals and modern humans, and for quarter-year intervals up to 14 years for *Pan troglodytes* and *P. paniscus*. Comparisons between closely related taxa can be made at each age interval (in this case, 50 for *Homo* and 56 for *Pan*).

HETPAD analysis

The program, HETPAD, has three goals. The first is *prediction*: given a particular ancestor, HETPAD can model descendants generated from that ancestor following particular heterochronic transformations. One can model the effects of different heterochronic processes affecting different modular or morphologically integrated suites of traits, or one can apply a single heterochronic transformation globally to the entire ancestor, and build the specified descendant. The second goal is *diagnosis*: given a particular ancestor and descendant, HETPAD can test a particular hypothesis of heterochronic transformation, measuring the fit of the actual descendant to the descendant modeled from the given ancestor following the specified heterochronic rules. One can test a simple hypothesis (e.g., global neoteny), or a complex hypothesis involving different transformations affecting different modular sets of traits. HETPAD can also search the universe of heterochronic transformations and possible modular suites of traits for the set of transformations that comes closest to capturing an observed descendant, given a particular ancestor. The third goal is *testing the efficacy of heterochrony as an explanatory tool for evolutionary change*. Treating the differences between the observed ancestor and the observed descendant as the "total evolutionary change," and the differences between the modeled and observed descendant as the "residual error of the model," HETPAD can measure the percentage of evolutionary change explained by any particular heterochronic transformation or set of transformations. The "percent evolutionary change explained" is the difference between the "total evolutionary change" and the "residual error of the model." Table 12.3 supplies the mathematical formulas HETPAD uses to measure "total evolutionary change," the "residual error of the model," and the "percent evolutionary change explained."

Heterochrony, as defined by Gould (1977: 2), comprises a subset of evolutionary developmental perturbations – specifically *"changes in developmental timing* that produce *parallels* between the stages of ontogeny and phylogeny" (emphasis in the original). 'Paedomorphosis' and 'recapitulation' (now called 'peramorphosis;' see Alberch *et al.*, 1979) both require the conservation in the descendant of ancestral shape (or developmental) pathways. Gould asked to what extent simple shifts in developmental timing account for differences in size, shape, and age at maturation of closely related taxa: to what extent does ontogeny actually parallel phylogeny? Heterochrony has nothing to do with the appearance of novel developmental pathways. For example, Zelditch (2001) described the difference between spatial versus temporal perturbations, and how the former, via heterotopy, can lead to novel developmental pathways that do not fall under the rubric of heterochrony. However, new relationships among body parts whose development is differentially retarded or accelerated do fall within the rubric of heterochrony, as they preserve ancestral shape pathways within developmental modules. *To preserve ancestral shape pathways, an important constraint built into HETPAD is that individual ancestral growth curves must be modified in groups of two or more.* Ancestral shape pathways can only be conserved if the relationships (or ratios) among traits within modular units remain unperturbed, and all heterochronic shifts, by definition, conserve ancestral shape pathways.

Table 12.4 lists all of the heterochronic transformations or "verbs" that exist in HETPAD's toolkit. Paedomorphosis and peramorphosis are the domains of heterochrony – the shape changes allowable under heterochronic transformation. Between them lies what might be called 'isomorphosis' or 'geometric similarity' – i.e., no change in shape at all. A dissociation of size (Gould's growth) and shape (Gould's development) might lead to a descendant whose size differs from that of its ancestor but whose shape remains the same. These were labeled 'proportioned giants' or 'proportioned dwarfs' by Gould who

Table 12.3 *Mathematical formulas used in HETPAD*

HETPAD metric	Definition	Formula
Total evolutionary change	Sum of the squared differences between observed ancestor and observed descendant measured over k age intervals for i traits	$\sum_{k=1}^{n} \sum_{i=1}^{m} (Y_{ik,\mathrm{OBS.ANC}} - Y_{ik,\mathrm{OBS.DES}})^2$
Average evolutionary change	Total evolutionary change benchmarked against the number of age intervals over which the differences were summed	Total evolutionary change $\div k$
Residual error of the model	Sum of the squared differences between modeled descendant and observed descendant measured over k age intervals for i traits	$\sum_{k=1}^{n} \sum_{i=1}^{m} (Y_{ik,\mathrm{MOD.DES}} - Y_{ik,\mathrm{OBS.DES}})^2$
Average residual error of the model	Residual error of the model benchmarked against the number of age intervals over which the residual error was calculated	Residual error of the model $\div k$
Percent evolutionary change explained	Difference between total evolutionary change and the residual error of the model as a percentage of the total evolutionary change	(Total evolutionary change − residual error of the model) $\div$ (total evolutionary change)

k = number of age intervals over which growth trajectories are measured, from 1 to n.
i = number of traits measured from 1 to m.
Y_{ik} = value of the ith trait at the kth age.
OBS.ANC = observed ancestor.
OBS.DES = observed descendant.
MOD.DES = modeled descendant.

dismissed them as outside the strict realm of heterochrony, because they involve no shape change (Gould, 1977: 250). They do, however, involve a simple change in size/shape linkages, and as such they are allowable outcomes of the heterochronic perturbations in HETPAD's toolkit.

We use the term 'neomorphosis' to describe shapes that cannot be generated via simple heterochronic transformations (Godfrey *et al.*, 1998), and are

Table 12.4 *List of heterochronic processes (verbs) used in HETPAD*

Process	Description of changes from ancestor to descendant	Product	Size/shape dissociation?[a]
Neoteny	Decrease rate of shape change (without concomitant decrease in the rate of growth) with respect to age at maturation	Paedomorph	Yes
Acceleration	Increase rate of shape change (without concomitant increase in the rate of growth) with respect to age at maturation	Peramorph	Yes
Post-displacement	Late onset of size-and-shape change within module relative to age at maturation	Paedomorph	No
Pre-displacement	Early onset of size-and-shape change within module relative to age at maturation	Peramorph	No
Progenesis or time hypomorphosis	Early maturation truncates size-and-shape change	Paedomorph	No
Hypermorphosis or time hypermorphosis	Late maturation extends size-and-shape change	Peramorph	No
Rate hypomorphosis	Decrease rate of change in size-and-shape with respect to age at maturation	Paedomorph	No
Rate hypermorphosis	Increase rate of change in size-and-shape with respect to age at maturation	Peramorph	No
Proportioned dwarfism	Decrease rate of change in size with respect to change in shape and age at maturation	Isomorph	Yes
Proportioned giantism	Increase rate of change in size with respect to change in shape and age at maturation	Isomorph	Yes

[a] Size/shape dissociation occurs when the ancestral linkages between size and shape *within the affected modular set of traits* are not manifested in the descendant.
Source: Modified after Godfrey & Sutherland (1995) and Shea (1988).

therefore neither paedomorphic, peramorphic, nor isomorphic. In HETPAD, these cannot be modeled. However, HETPAD can measure the *degree to which* heterochronic transformations fail to capture observed descendant growth trajectories, and that is one way to capture neomorphosis mathematically. In effect, neomorphosis can be treated as equivalent to the residual error of the heterochronic model. Biologically, it is any evolutionary perturbation *not* produced via a simple heterochronic transformation (see many examples in Zelditch, 2001).

Some heterochronic processes involve 'size/shape dissociation' while others do not (Table 12.4). 'Size/shape dissociation' is said to occur when ancestral linkages between size and shape are broken in the descendant. That is, the descendant follows its ancestor's shape pathway, but at different sizes. Neoteny, as defined by Gould (1977), is one such process. Thus, for example, a descendant adult might be most similar in shape (or trait proportions) to a *10-year old*, immature ancestor, *scaled up by a factor of 2*. The *degree of paedomorphosis* is captured by the ancestral age at maximum shape similarity (e.g., 10 years), and the *overall size* (measured, following Gould, (1977: 253) as the sum of trait sizes in the affected module) is captured by the scaling factor (e.g., 2). A descendant that matures at the same age as its ancestor, is similar in shape at maturity to an ancestral juvenile at age 10, but is twice the size of that juvenile, is 'neotenic.' HETPAD searches for the best scaling factor for all verbs that require size/shape dissociation. It identifies the process (e.g., 'neoteny'), but it also specifies the degree of paedomorphosis and the scaling adjustment.

HETPAD's *exhaustive search* looks among all possible heterochronically derived models for the one that most closely replicates the observed differences between ancestor and descendant. In a multivariate context, HETPAD allows subsets of traits to be perturbed independently of one another. It finds modular sets by testing single transformations first on pairs of variables, and then on larger variable groupings. All traits within modular subsets must be identically perturbed in a manner commensurate with one of the heterochronic processual 'verbs.' They must also exhibit the same amount of paedomorphosis or peramorphosis, and be scaled up or down to the same degree, if at all. If the residual error of the best heterochronic model is large – particularly if the modular units it finds do not make biological sense – then it behooves the researcher to seek an evolutionary explanation that falls outside the realm of heterochrony.

Gould (1977: 483) defined neoteny as 'paedomorphosis (retention of formerly juvenile characters by adult descendants) produced by retardation of somatic development.' He graphically depicted 'pure' neoteny as retardation of shape change with respect to both growth and age at maturation (p. 257). A 'pure' neotene would exhibit the shape of a juvenile ancestor *but the size and age at maturation of the ancestral adult.* Of course, Gould recognized that 'pure' neotenes thus defined must rarely exist, and he applied the term 'neoteny' more broadly. HETPAD uses neoteny in its broadest sense, to encompass any descendant that can be generated from a given ancestor by any degree of retardation of shape change and any scaling factor. This means that we accept *downward* as well as upward rescaling as possible outcomes of neoteny, despite the fact that Gould envisioned upward scaling as its normal corollary. We use Shea's (1984) term 'rate hypomorphosis' to describe shape retardation without rescaling, and we apply the term 'neoteny' to describe any amount of shape retardation with rescaling (or size/shape dissociation). (We also caution that, under shape retardation with downward scaling, there can be no expectation of a weakening of growth allometries; see Godfrey *et al.*, 1998.)

When HETPAD is asked to test a particular solution, such as global neoteny, the universe of heterochronic options is severely limited. For example, to find the 'global neotene' that best matches a given descendant and that can be derived through a single perturbation of a given ancestor, HETPAD must (1) treat all traits as belonging to a single module; (2) scale all traits up or down by a single multiplier; and (3) select the degree to which the rate of shape change should be retarded. HETPAD's search space, under this test condition, is within the realm of paedomorphosis. Global neoteny has prespecified edge or boundary conditions. One is 'rate hypomorphosis' which might be inferred if the descendant is best modeled as having been derived through a retardation of the ancestral rate of shape change *without rescaling.* Under 'rate hypomorphosis,' the descendant adult is the size and shape of an ancestral juvenile. A second edge condition is 'proportioned dwarfism' or 'proportioned giantism,' which may be inferred if the descendant is best modeled as having been derived through rescaling *without any retardation of the ancestral rate of shape change.* Under proportioned dwarfism or proportioned giantism, the descendant adult is the shape of the ancestral adult (i.e., isomorphic), but smaller or larger. Finally, if the descendant is best modeled as having been derived through neither rescaling nor retardation, then the best solution entails no evolutionary change. These edge conditions describe the logical limits of global neoteny. It goes without saying that if the model of global neoteny is a poor one, then the residual error of the 'best' solution under global neoteny will be large – perhaps considerably larger than the residual error of the best solution derived through HETPAD's exhaustive search.

We applied two of the tests to the data we collected for *Homo* and for *Pan*: the global neoteny test and HETPAD's exhaustive search for the best heterochronic solution. As described above, Neandertals were selected as the 'observed ancestor' for modern humans (the 'observed descendant') and chimpanzees were selected as the 'observed ancestor' for bonobos (this pair's 'observed descendant'). Modeled descendants were generated using the constraints of HETPAD's exhaustive search, as well as global neoteny, for each of the above pairs, and then compared to our observed descendants. Our observed ancestors and descendants were described by the 24 trait growth curves generated for each of two taxa of *Homo* and for each of two species of *Pan*, derived from piecewise regression.

Results

The total evolutionary change is virtually identical for both *Homo* and *Pan*, but the residual errors of the global neoteny and the best-fit models are considerably lower for *Pan* than for *Homo* (Table 12.5). Much more of the total evolutionary change is explained by HETPAD's exhaustive search than by neoteny in both comparisons; however the fit for either solution is not terribly good.

Global neoteny completely failed to account for the differences between Neandertals and modern humans. When HETPAD's global neoteny test was applied to the comparison of these two taxa, modern human adults were found to be no more similar in shape to any immature Neandertal than they are to Neandertal adults. Because most of the measured traits are smaller in absolute size in modern humans than in Neandertals, the 'best' solution under the constraint of global neoteny is that modern *Homo* can be derived from a Neandertal ancestor via downward scaling (by a factor of 0.9). This is, of course, one of HETPAD's edge conditions for global neoteny – i.e., no retardation at all, but some rescaling (and thus size/shape dissociation). We do not take this to mean that modern humans *were* derived from Neandertals via downward scaling (or proportioned dwarfism). Modern humans and Neandertals are not the same shape. Rather, we infer that global craniofacial neoteny can be rejected as the mechanism by which modern human craniofacial morphology could be derived from a Neandertal-like ancestor. Global neoteny is unable to explain *any* of the craniofacial *shape* differences between Neandertals and modern humans. Simple downward scaling can 'account for' up to 58% of the observed size differences along the ontogenetic trajectories of the two taxa but *none* of the shape differences.

HETPAD's exhaustive search does little better in explaining the size and shape differences between Neandertals and modern humans. Even with the entire toolkit of heterochronic processes, HETPAD cannot transform Neandertal growth trajectories into those of modern humans. HETPAD's optimal solution has eight modules that together can account for 70% of the sum of squared differences between the 'observed' ancestral and 'observed' descendant growth curves. Some aspects of craniofacial shape in modern human adults appear to

Table 12.5 HETPAD residuals and the percentage of total evolutionary change "explained" by heterochrony in Homo and Pan[a]

Genus	Total evolutionary change	Average evolutionary change	Total residual error for global neoteny	Average residual error for global neoteny	Total residual error for best-fit model	Average residual error for best-fit model	Evolutionary change explained by global neoteny	Evolutionary change explained by best-fit model
Homo	8704	8704/50 = 174.1	3634	3634/50 = 72.7	2558	2558/50 = 51.2	58%	70%
Pan	9830	9830/56 = 175.5	2958	2958/56 = 52.8	1140	1140/56 = 20.4	70%	88%

[a] See formulas, Table 12.3. Note that sums of squared differences were measured over 50 age intervals for Homo and 56 age intervals for Pan.

310   *F. L. Williams* et al.

be 'peramorphic' and others 'paedomorphic' with respect to Neandertal adults. Because the fit of the generated model to the observed descendant is still poor, one cannot take paedomorphosis or peramorphosis as good descriptors of modular parts. One can assert, however, that they better describe the shape differences of isolated parts than does geometric similarity.

For all modular units identified by HETPAD, modern human adults are smaller than Neandertal adults. For four of the eight modules (and about half the traits in our database), modern humans are modeled as 'paedomorphs' the size and shape of Neandertal juveniles or near-adults (ranging from 11.5 to 22 years of age); the heterochronic process that best fits these data is 'rate hypomorphosis.' The affected modules include a portion of the face and most of the mandible, as well as the breadth across the skull in the temporal region (bizygomatic and bicondylar breadths). Within this set of modules, the midfacial region (i.e., the relationships between nasal breadth, palatal breadth, basion–prosthion, and minimum ascending ramus breadth) exhibits the greatest degree of paedomorphosis. A fifth module comprising a hodgepodge of cranial and mandibular variables (e.g., upper cranial height, mandibular symphysis height, maximum breadth of the ascending ramus) similarly exhibits apparent paedomorphosis, but through apparent progenesis. A separate module largely comprising the upper face (upper facial height, minimum frontal breadth, cranial base height, and palatal length) is modeled as similar in shape but smaller than Neandertal adults by a factor of 0.89 (thus geometrically similar).

For two additional modules, modern humans are described as 'peramorphic' with respect to Neandertals. Some aspects of the shape of the calotte and supraorbital region (e.g., calotte length vs. calotte breadth and the breadth across the orbits) of Neandertal adults are closer to 18–21-year-old than to full adult modern humans. From infancy to adulthood, the Neandertal calotte changes less in shape than does that of modern humans (Williams *et al.*, 2002). Both Neandertal and modern human infants have small faces and orbital superstructures. Later in postnatal ontogeny, modern human faces become larger relative to their brow ridges. In Neandertals, the face becomes larger, but so does the torus. Thus, Neandertals maintain their infant proportions more than do modern humans for some aspects of upper versus lower facial shape (Figure 12.1).

Global craniofacial neoteny better describes the differences between *Pan troglodytes* and *P. paniscus* than between Neandertals and modern humans, but the fit of the model is again poor (Table 12.5). Using the constraints of global neoteny, HETPAD finds bonobos to be closest in craniofacial shape to common chimpanzees of age 11.5 years, but *smaller* – i.e., scaled down by a factor of 0.92. Bonobo crania are certainly smaller than those of full adult common chimpanzees. Figure 12.2 compares a bonobo adult to a chimpanzee

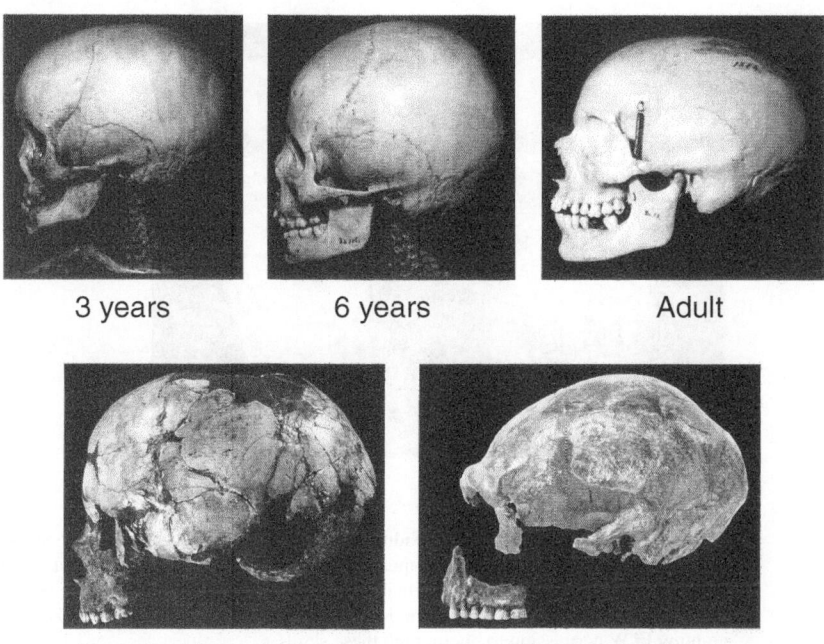

3 years   6 years   Adult

Roc de Marsal, 3 years   Spy 1, Adult

Figure 12.1   A comparison of modern human (top row) and Neandertal (bottom row) postnatal craniofacial ontogeny reveals that infants of both taxa exhibit small splanchnocrania with respect to their neurocrania. In Neandertals, both the face and supraorbital torus greatly increase in size, whereas in modern humans, the face increases considerably more than does the supraorbital torus. Postnatal neurocranial shape change is fundamentally different in Neandertals and modern humans. (We are grateful to G. Maat at the Anatomie Museum, Rijksuniversiteit Leiden for granting FLW permission to photograph the modern human skulls; photograph of Roc de Marsal courtesy of J.-J. Cleyet Merle, photograph of Spy 1 courtesy of Patrick Semal and the family of Prof. Max Lohest.)

subadult: similarity in size, but *not* shape, is readily apparent. 'Global neoteny' succeeds in accounting for 70% of evolutionary change only because the program accepts downward scaling as partial evidence of 'neoteny' (paedomorphosis with size/shape dissociation).

HETPAD's exhaustive search does considerably better than global neoteny in accounting for the size and shape differences between the two species of *Pan*. The optimal solution finds six modules that together 'explain' 88% of the observed evolutionary change in *Pan*. Five of the six modules show a strong conservation of ancestral adult or near-adult shapes (under this solution, the bonobo adult shapes are most similar to those of 12.75–13.75-year-old chimpanzees). However, each of the five modules exhibits downward scaling to varying degrees (with scaling factors ranging from 0.94 to 0.80). Downward

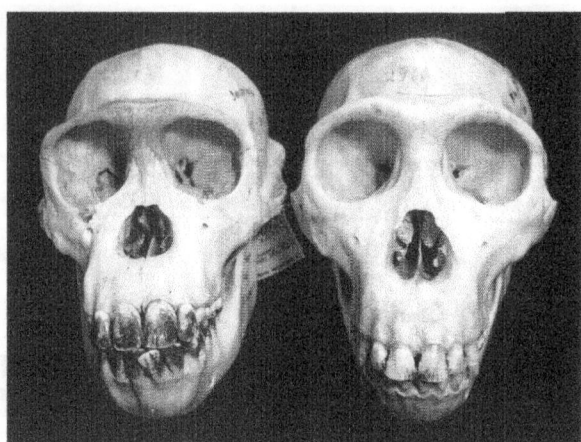

Chimpanzee subadult   Bonobo adult

Figure 12.2   A comparison of a subadult common chimpanzee and an adult bonobo
shows similarities in size, but not shape. Adult bonobos exhibit a narrower, but longer
nasal aperture, a more slender lower maxilla, a taller nasal clivus, and smaller incisors.
(We are grateful to W. Van Neer at the Koninklijk Museum voor Midden-Afrika for
granting FLW permission to photograph these skulls.)

scaling is most marked in the module comprising most of the middle face,
palate, and mandible (palatal length, nasal breadth, bigonial breadth, mandibular
length, mandibular symphysis height, mandibular corpus height, and mandibular corpus thickness). A module encompassing the ascending ramus of the
mandible as well as a portion of the palate and cranial base (ascending ramus height, minimum ascending ramus breadth, palatal breadth, and basion–
prosthion) shows slightly greater paedomorphosis but less downward scaling,
and a module encompassing biorbital breadth, interbiorbital breadth, cranial
base height, and upper facial height shows yet less downward scaling. The
smallest amount of downward scaling characterizes two modules describing the
calotte, temporal, and part of the orbital regions (maximum cranial length and
breadth, bizygomatic breadth, bicondylar breadth; minimum frontal breadth, orbital breadth, upper cranial height). Finally, a sixth module highlights an unusual
relationship between biectomolare breadth and the maximum breadth of the ascending ramus: here bonobo adults are shown to be closest in size and shape to
7.5-year-old chimpanzees, and rate hypomorphosis comes closest to capturing
the difference in their ontogenetic trajectories. The fact that HETPAD's exhaustive search performs better than global neoteny in both *Pan* and *Homo*
underscores the failure of global neoteny to explain differences between
species.

Discussion

We have shown that: (1) 'neoteny' (especially global neoteny) is an inappropriate descriptor of the differences between both Neandertals and modern humans,
and chimpanzees and bonobos; and (2) even complex models that depend solely
on Gould's simple model of heterochronic transformation cannot transform a
Neandertal into a modern human, or a chimpanzee into a bonobo. Differences in
growth trajectories cannot be understood through heterochrony alone if shape
paths differ from infancy onward. This is the case for both species of *Pan*, and for
Neandertals and modern *Homo*. Nevertheless, simple heterochronic shifts can
account for *much more* of the differences between chimpanzees and bonobos
than between Neandertals and modern humans. The residuals for each solution
are lower for *Pan* than for *Homo* and the best-fit heterochronic model is much
more complex in *Homo* than in *Pan*.

The extent to which heterochrony is a significant driver of evolutionary
change must be demonstrated and not merely assumed. If heterochrony provides a useful explanatory scheme, then HETPAD's exhaustive search might
be expected to account for a large portion of the differences between closely
related taxa. It should also identify modules that are biologically meaningful in
embryological, developmental, and/or functional terms. If HETPAD cannot do
so, then we must conclude that heterochrony, as a model, is insufficient singly
to account for the observed ontogenetic differences.

Heterochrony (or changes in developmental timing) conserves the spatial relationships among parts, at least within separately perturbed or modular units.
Heterotopy (or changes in the relative position or location of parts) does not.
It was the conviction that heterotopy explains more of evolutionary change
than generally acknowledged that motivated Zelditch (2001) to title her recent volume on evolutionary developmental change *'Beyond Heterochrony:
The Evolution of Development.'* That same conviction led Alberch (1985) to
warn against the indiscriminate application of heterochronic terms, and Hall
(2001: vii) to ask 'What lies beyond the hegemony of heterochrony?' Some
researchers describe developmental differences between hominid taxa without
invoking the vocabulary of heterochrony (e.g., Krovitz, 2000; Lieberman *et al.*,
2002), and that may well be appropriate.

Nevertheless, heterochronic descriptors continue to be applied widely
and inconsistently in paleoanthropology and elsewhere (Klingenberg, 1998;
Minugh-Purvis & McNamara, 2002). In our view, the inconsistent use of heterochronic terms is problematic, in that it hampers hypothesis testing and stymies
communication. For example, Ponce de León & Zollikofer (2001) describe Neandertals as 'rate hypermorphs' relative to modern humans (which corresponds
to rate hypomorphosis of modern humans relative to a Neandertal-like ancestor),

despite the fact that they acknowledge (indeed demonstrate) that the shape pathways of the two diverge prior to birth. They apply this term apparently because Neandertals grow faster than modern humans, and certain shape differences present at birth are conserved throughout ontogeny. Neandertals and modern humans, according to Ponce de León & Zollikofer (2001), follow roughly parallel *postnatal* shape-change trajectories. Rate hypermorphosis, thus defined, cannot lead to peramorphosis, and does not conserve ancestral size/shape linkages.

What do we know about the ontogenetic differences between Neandertals and modern humans, or between bonobos and chimpanzees? First, relative to dental eruption, Neandertal craniofacial traits generally grow faster than those of modern humans, and craniofacial growth in chimpanzees is faster than it is in bonobos (Williams *et al.*, 2002). Second, the *shape paths* of Neandertals and modern humans, as well as of bonobos and chimpanzees, differ from infancy onward and become increasingly divergent throughout ontogeny (Krovitz, 2000; Ponce de León & Zollikofer, 2001; Tillier, 1989; Williams, 2001). This alone means that a simple heterochronic model for either cannot succeed (Williams *et al.*, 2002). Third, we know that size/shape dissociation is a factor for both. Differential downscaling does more to explain the differences in the skulls of adult Neandertals and modern humans, or of adult chimpanzees and bonobos, than does rate hypomorphosis or neoteny. Adult modern human skulls are not uniformly paeodomorphic with respect to those of Neandertals, but they are smaller. Rate hypomorphosis does better than any other heterochronic process at accounting for the size and shape differences between Neandertal and modern human mandibles and part of the face, but strong modularity (different degrees of size and shape retardation) is required to make rate hypomorphosis 'work' even for this limited portion of the skull. Rate hypomorphosis cannot account for differences in the upper face or calotte. Adult bonobo skulls might be considered very slightly paedomorphic relative to those of chimpanzees, but they are also considerably smaller than those of chimpanzees, and parts of the skull are scaled down more than others. Even the face is not uniformly neotenic, as has been suggested (Shea, 1984, 2000, 2002). It is more appropriately characterized as small.

Conclusion

The lack of a coherent and internally consistent methodology to test for differences in rates of growth and rates of development has plagued interpretations of competing heterochronic models for human and ape evolution. Consensus on what theoretical conditions satisfy neoteny and how to separate heterochronic transitions from other evolutionary perturbations has not yet been attained. There is some agreement that heterochronic analyses would benefit from a multivariate treatment (Klingenberg, 1998), but little consensus beyond that.

We maintain that terms such as 'paedomorphosis' and 'peramorphosis' are only meaningful if they are applied to shape and not size (see also Gould, 1977, 2000), and we therefore embrace Gould's definitions of heterochronic terms. We take seriously the notion that heterochrony describes a *subset* of evolutionary ontogenetic changes, and we argue that its efficacy as an explanatory model must be tested and not simply assumed.

In this chapter, we present a diagnostic tool for identifying heterochronic products and processes, and we examine the efficacy of global neoteny and more complex heterochronic perturbations in explaining the ontogenetic differences between modern humans with respect to Neandertals, and bonobos with respect to common chimpanzees. Contra Brothwell (1975) and de Beer (1958), we show that Neandertals are not overdeveloped modern humans, or conversely, modern humans do not stop developing at a pre-adult stage of Neandertal postnatal ontogeny. We show that global craniofacial neoteny explains neither the craniofacial differences between the two species of *Pan* nor the differences between Neandertals and modern humans. If indeed modern human skulls are neotenic, they do not derive from a Neandertal-like ancestor. For modern humans to be neotenic Neandertals, they would have to resemble Neandertal juveniles. Furthermore, the extent to which skulls of *Pan paniscus* are paedomorphic, or more importantly, smaller than those of *Pan troglodytes*, depends on the craniofacial region considered. Our most important take-home message is that no heterochronic 'process' taken alone, and no set of heterochronic processes in simple combination, succeeds in accounting for the adult shape differences between Neandertals and modern humans, or between chimpanzees and bonobos.

Acknowledgments

We would like to express our thanks to the following individuals for allowing one of us (FLW) to examine the materials in their care: C. Stringer, J.-J. Hublin, I. Pap, E. Potty, R. Orban, J.-J. Cleyet-Merle, D. Buisson, Y. Rak, B. Arensburg, H. de Lumley, P. Vignaud, R. Macchiarelli, L. Bondioli, A. G. Segre, J. Radovčić, J. Jelínek, I. Tattersall, G. Maat, G. N. van Vark, F. Laarman, J. de Vos, C. Smeenk, D. Howlett, W. Van Neer, J. Cuisin, M. Rotzmoser, J. Richtsmeier, and M. Toussaint. We are grateful to G. Maat at the Anatomie Museum, Rijksuniversiteit Leiden, and W. Van Neer at the Koninklijk Museum voor Midden-Afrika, for granting FLW permission to photograph the modern human skulls (Figure 12.1), and those of *Pan* (Figure 12.2) respectively. We thank J.-J. Cleyet-Merle at the Muséum National Préhistoire–Les Eyzies-de-Tayac for the photograph of Roc de Marsal, and Rosine Orban and Patrick Semal (and the family of Prof. Max Lohest) at the Institut Royal des Sciences Naturelles de Belgique for the photograph of Spy 1 (Figure 12.1).

References

Abbie, A. A. (1947). Head form and human evolution. *Journal of Anatomy*, **81**, 233–258.
Alba, D. M. (2002). Shape and stage in heterochronic models. In *Human Evolution through Developmental Change*, eds. N. Minugh-Purvis and K. McNamara, pp. 28–50. Baltimore: Johns Hopkins University Press.
Alba, D. M., Moya-Solá, S., & Kohler, M. (2001). Canine reduction in the Miocene hominoid *Oreopithecus bambolii*: Behavioural and evolutionary implications. *Journal of Human Evolution*, **40**, 1–16.
Alberch, P. (1985). Problems with the interpretation of heterochronic processes. *Systematic Zoology*, **34**, 46–58.
Alberch, P., Gould, S. J., Oster, G. F., & Wake, D. B. (1979). Size and shape in ontogeny and phylogeny. *Paleobiology*, **5**, 296–317.
Antón, S., & and Leigh, S. (1998). Paedomorphosis and neoteny in human evolution. *Journal of Human Evolution*, **34**, A2.
Anemone, R. L., Mooney, M. P., & Siegel, M. I. (1996). Longitudinal study of dental development in chimpanzees of known chronological age: Implications for understanding the age at death of Plio-Pleistocene hominids. *American Journal of Physical Anthropology*, **99**, 119–133.
Bolk, L. (1926). On the problem of anthropogenesis. *Proceedings of the Koninklijke Akademie van Wetenschappen te Amsterdam*, **29**, 465–475.
Brothwell, D. R. (1975). Adaptive growth rate changes as a possible explanation for the distinctiveness of Neanderthalers. *Journal of Archaeological Science*, **2**, 161–163.
Coolidge, H. J. (1933). *Pan paniscus*: Pigmy chimpanzee from south of the Congo River. *American Journal of Physical Anthropology*, **18**, 1–59.
Czarnetzki, A., Gaudzinski, S., & Pusch, C. M. (2001). Hominid skull fragments from Late Pleistocene layers in Leine Valley (Sarstedt, District of Hildesheim, Germany). *Journal of Human Evolution*, **41**, 133–140.
Dean, M. C., & Wood, B. A. (1981). Developing pongid dentition and its use for ageing individual crania in comparative cross-sectional growth studies. *Folia Primatologica*, **36**, 111–127.
Dean, M. C., Leakey, M. G., Reid, D., Schrenk, F., Schwartz, G. T., Stringer, C., & Walker, A. (2001). Growth processes in teeth distinguish modern humans from *Homo erectus* and earlier hominids. *Nature*, **414**, 628–631.
de Beer, G. R. (1958). *Embryos and Ancestors*. Oxford: Clarendon Press.
de Bruin, R. (1993). A mathematical model applied to craniofacial growth. PhD dissertation, Rijksuniversiteit Groningen, Groningen, The Netherlands.
German, R. Z., & Myers, L. L. (1989a). The role of time and size in ontogenetic allometry. I: Review. *Growth, Development and Aging*, **53**, 101–106.

German, R. Z., & Myers, L. L. (1989b). The role of time and size in ontogenetic allometry. II: An empirical study of human growth. *Growth, Development and Aging*, **53**, 107–115.

Godfrey, L. R., & Sutherland, M. R. (1995). What's growth got to do with it? Process and product in the evolution of ontogeny. *Journal of Human Evolution*, **29**, 405–431.

Godfrey, L. R., & Sutherland, M. R. (1996). Paradox of peramorphic paedomorphosis: Heterochrony and human evolution. *American Journal of Physical Anthropology*, **99**, 17–42.

Godfrey, L. R., King, S. J., & Sutherland, M. R. (1998). Heterochronic approaches to the study of locomotion. In *Primate Locomotion*, eds. E. Strasser, J. Fleagle, A. Rosenberger, & H. M. McHenry, pp. 277–307. New York: Plenum Press.

Gould, S. J. (1977). *Ontogeny and Phylogeny*. Cambridge: Belknap Press.

Gould, S. J. (2000). Of coiled oysters and big brains: How to rescue the terminology of heterochrony now gone astray. *Evolution and Development*, **2**, 241–248.

Hall, B. K. (2001). Foreword. In *Beyond Heterochrony: The Evolution of Development*, ed. M. L. Zelditch, pp. vii–ix. New York: John Wiley & Sons.

Hauspie, R. C. (1989). Mathematical models for the study of individual growth patterns. *Revue d'Epidémiologie et Santé*, **37**, 461–476.

Hauspie, R. C. (1998). The human growth curve. In *Cambridge Encyclopedia of Human Growth and Development*, eds. S. J. Ulijaszek, F. E. Johnston, & M. A. Preece, pp. 108–115. Cambridge: Cambridge University Press.

Klingenberg, C. P. (1998). Heterochrony and allometry: The analysis of evolutionary change in ontogeny. *Biological Reviews*, **72**, 79–123.

Krovitz, G. E. (2000). Three-dimensional comparisons of craniofacial morphology and growth patterns in Neandertals and modern humans. PhD dissertation, Johns Hopkins University, Baltimore.

Kuykendall, K. L., & Conroy, G. C. (1996). Permanent tooth calcification in chimpanzees (*Pan troglodytes*): Patterns and polymorphisms. *American Journal of Physical Anthropology*, **99**, 159–174.

Leigh, S. R., & Terranova, C. J. (1998). Comparative perspectives on bimaturism, ontogeny, and dimorphism in lemurid primates. *International Journal of Primatology*, **19**, 723–749.

Lieberman, D. E., McBratney, B. M., & Krovitz, G. E. (2002). The evolution and development of cranial form in *Homo sapiens*. *Proceedings of the National Academy of Sciences of the USA*, **99**, 1134–1139.

McKinney, M. L. (1998). The juvenilized ape myth: Our "overdeveloped' brain. *BioScience*, **48**, 109–123.

McKinney, M. L., & McNamara, K. J. (1991). *Heterochrony: The Evolution of Ontogeny*. New York: Plenum Press.

McNamara, K. J. (2002). Sequential hypermorphosis: Stretching ontogeny to the limit. In *Human Evolution through Developmental Change*, eds. N. Minugh-Purvis & K. McNamara, pp. 102–121. Baltimore: Johns Hopkins University Press.

Minugh-Purvis, N. (1988). Patterns of craniofacial growth and development in Upper Pleistocene hominids. PhD dissertation, University of Pennsylvania, Philadelphia.

Minugh-Purvis, N. (2002). Heterochronic change in the neurocranium and the emergence of modern humans. In *Human Evolution through Developmental Change*, eds.

318   F. L. Williams et al.

N. Minugh-Purvis & K. McNamara, pp. 479–498. Baltimore: Johns Hopkins University Press.
Minugh-Purvis, N. & McNamara, K. J. (eds.) (2002). *Human Evolution through Developmental Change*. Baltimore: Johns Hopkins University Press.
Montagu, M. F. A. (1989). *Growing Young*, 2nd edn. Granby: Bergin & Garvey.
Pilbeam, D. R. (2002). Perspectives on the Miocene Hominoidea. In *The Primate Fossil Record*, ed. W. C. Hartwig, pp. 303–310. Cambridge: Cambridge University Press.
Ponce de León, M. S., & Zollikofer, C. P. E. (2001). Neanderthal cranial ontogeny and its implications for late hominid diversity. *Nature*, **412**, 534–538.
Privratsky, V. (1981). Neoteny and its role in the process of hominization. *Anthropologie*, **19**, 219–230.
Rice, S. H. (1997). The analysis of ontogenetic trajectories: When a change in size or shape is not heterochrony. *Proceedings of the National Academy of Sciences of the USA*, **94**, 907–912.
Rozzi, F. V. R. (2000). Heterochronic process in hominid evolution: The dental development in "robust' australopithecines. *Comptes Rendus de l'Académie des Sciences de Paris*, **331**, 571–577.
Schultz, A. H. (1933). Chimpanzee fetuses. *American Journal of Physical Anthropology*, **18**, 61–80.
Shea, B. T. (1983). Paedomorphosis and neoteny in the pygmy chimpanzee. *Science*, **222**, 521–522.
Shea, B. T. (1984). An allometric perspective on the morphological and evolutionary relationships between pygmy (*Pan paniscus*) and common (*Pan troglodytes*) chimpanzees. In *The Pygmy Chimpanzee: Evolutionary Behavior and Biology*, ed. R. Susman, pp. 89–130. New York: Plenum Press.
Shea, B. T. (1989). Heterochrony in human evolution: The case for neoteny reconsidered. *Yearbook of Physical Anthropology*, **32**, 69–101.
Shea, B. T. (2000). Current issues in the investigation of evolution by heterochrony, with emphasis on the debate over human neoteny. In *Biology, Brains and Behavior*, eds. S. T. Parker, J. Langer, & M. L. McKinney, pp. 181–214. Sante Fe: School of American Research Press.
Shea, B. T. (2002). Are some heterochronic transformations likelier than others? In *Human Evolution through Developmental Change*, eds. N. Minugh-Purvis & K. McNamara, pp. 79–101. Baltimore: Johns Hopkins University Press.
Smith, B. H. (1991). Dental development and the evolution of life history in Hominidae. *American Journal of Physical Anthropology*, **86**, 157–174.
Smith, B. H., Crummett, T. L., & Brandt, K. L. (1994). Ages of eruption of primate teeth: A compendium for aging individuals and comparing life histories. *Yearbook of Physical Anthropology*, **37**, 177–232.
Stringer, C. B., Dean, M. C., & Martin, R. (1990). A comparative study of cranial and facial development in a recent British population and Neandertals. In *Primate Life History and Evolution*, ed. C. J. DeRousseau, pp. 115–152. New York: Wiley-Liss.
Tillier, A-m. (1989). The evolution of modern humans: Evidence from young Mousterian individuals. In *The Human Revolution: Behavioural and Biological Perspectives on the Origins of Modern Humans*, eds. P. Mellers & C. B. Stringer, pp. 286–297. Princeton: Princeton University Press.

Ubelaker, D. H. (1978). *Human Skeletal Remains: Excavation, Analysis and Interpretation.* Chicago: Aldine.

Verhulst, J. (1999). Bolkian and bokian retardation in *Homo sapiens. Acta Biotheoretica,* **47**, 7–28.

Weaver, D. S. (1986). Forensic aspects of fetal and neonatal specimens. In *Forensic Osteology: Advances in the Identification of Human Remains*, ed. K. J. Riechs, pp. 90–100. Springfield: Charles C. Thomas.

Williams, F. L. (2001). Heterochronic perturbations in the craniofacial evolution of *Homo* (Neandertals and modern humans) and *Pan (Pan troglodytes* and *P. paniscus).* PhD dissertation, University of Massachusetts, Amherst.

Williams, F. L., Godfrey, L. R., & Sutherland, M. R. (2001). Diagnosing heterochronic perturbations in the craniofacial evolution of *Homo* and *Pan. American Journal of Physical Anthropology*, Suppl. **32**, 165.

Williams, F. L., Godfrey, L. R., & Sutherland, M. R. (2002). Heterochrony and the evolution of Neandertal and modern human craniofacial form. In *Human Evolution through Developmental Change*, eds. N. Minugh-Purvis & K. McNamara, pp. 405–441. Baltimore: Johns Hopkins University Press.

Winkler, L. A., Schwartz, J. H., & Swindler, D. R. (1996). Development of the orangutan permanent dentition: Assessing patterns and variation in tooth development. *American Journal of Physical Anthropology*, **99**, 205–220.

Wolpoff, M. H. (1979). The Krapina dental remains. *American Journal of Physical Anthropology*, **50**, 67–114.

Zelditch, M. L. (2001). *Beyond Heterochrony: The Evolution of Development.* New York: John Wiley & Sons.

13 Shape and growth differences between Neandertals and modern humans: Grounds for a species-level distinction?

G. E. KROVITZ
Pennsylvania State University

Introduction

There is a great deal of interest in the phylogenetic relationship between Neandertals and anatomically modern humans, and the taxonomic classification of Neandertals is of central importance in the discussion of the origins of anatomically modern humans. There are two prominent hypotheses about modern human origins that offer contrasting views of Neandertal taxonomic status: the recent African origin model posits that Neandertals (*Homo neanderthalensis*) and modern humans (*Homo sapiens*) were separate species (e.g., Stringer, 1989, 1992; Stringer & Andrews, 1988; Stringer *et al.*, 1984); while the multiregional model proposes genetic continuity between Neandertals (*Homo sapiens neanderthalensis*) and early modern humans (*Homo sapiens sapiens*) in Eurasia (e.g., Wolpoff, 1989; Wolpoff *et al.*, 1984); although less extreme models have also been presented (e.g., Bräuer, 1984; Relethford & Harpending, 1994; Smith, 1992; Smith *et al.*, 1989). Historically, morphological comparisons of Neandertals and modern humans (such as those cited above) have focused almost entirely on *adult* morphology. However, the study of adult remains alone has failed to answer the question of whether or not Neandertals are a subspecies of *Homo sapiens*, or a separate species, *Homo neanderthalensis*.

There is growing awareness that developmental shifts are an important component of evolutionary change (for recent examples, see papers in Minugh-Purvis & McNamara, 2002; O'Higgins & Cohn, 2000). Therefore, the identification of growth processes that differentiate Neandertal and modern human

Patterns of Growth and Development in the Genus Homo, ed. J. L. Thompson, G. E. Krovitz and A. J. Nelson. Published by Cambridge University Press. © Cambridge University Press 2003.

craniofacial morphology is potentially informative about whether or not Neandertals and modern humans belong to the same or to different species, and several recent studies have investigated this idea. For example, recent studies by Ponce de León & Zollikofer (2001; see also Zollikofer *et al.*, 1995) and Lieberman *et al.* (2002) suggest that early ontogenetic shape differences between Neandertals and modern humans and "taxon-specific" ontogenetic patterns are consistent with a species-level distinction between Neandertals and modern humans. A study by Schillaci & Froehlich (2001) showed that pair-wise minimum genetic distances calculated from craniometric data differed more between Upper Paleolithic modern humans and Neandertals than between naturally hybridizing and non-hybridizing species of macaques, thus suggesting genetic isolation between Neandertals and Upper Paleolithic modern humans. Another study by Williams *et al.* (2002) found that Neandertals and modern humans are morphologically different from an early age, and that those differences are at least as great as differences between *Pan troglodytes* and *P. paniscus*, which might also support species differences between Neandertals and modern humans. However, developmental studies do not uniformly favor species-level differences, as Minugh-Purvis continues to use developmental data to assert evolutionary continuity between Neandertals and modern humans (Minugh-Purvis, 1988, 2002).

Taxonomic hypothesis

This study uses developmental data to explore whether or not Neandertals and modern humans are the same species. It must be recognized that "there is relatively little precedent for the interpretation of growth pattern data . . . as phylogenetic characters" (Richtsmeier *et al.*, 1993b: 323). Even with the increased awareness of the important link between developmental and evolutionary change, there has been little discussion on the kinds of developmental differences that we would *expect to see* between closely related species. Researchers are currently exploring ways to use morphometric shape data for phylogenetic analysis (see papers in MacLeod & Forey, 2002), and hopefully research will continue to explore phylogenetic differences reflected in growth patterns as well. Most studies that have used developmental data to examine phylogenetic questions base their analysis on a previously determined (usually molecular) primate phylogeny (such as Collard & O'Higgins, 2001; Gomez, 1992). However in this case, there is no previously determined phylogeny of Neandertals and modern humans, making it impossible to use these studies as a model for using developmental data in a taxonomic context.

322   *G. E. Krovitz*

Table 13.1 *Samples used in this study. The modern human sample is a mixture of eight geographically distinct populations. Neandertal individuals marked with an asterisk (*) were analyzed using data collected from casts. Age estimates are based on tooth calcification, and age groups are given in years. See text for further information on age estimation and age groups*

Age Group 1: 0–3 years	Age Group 2: 3.1–6 years	Age Group 3: 6.1–9 years	Age Group 4: 9.1–13.5 years	Age Group 5: Adult
Modern Human				
44	57	55	74	142
Neandertal				
Subalyuk 2	Engis 2	La Quina 18*	Teshik Tash*	Amud 1*
Pech de l'Aze*				La Chapelle 1*
				Circeo 1
				La Ferrassie 1*
				Forbes Quarry
				La Quina 5*
				Saccopastore 1
				Saccopastore 2
				Shanidar 1*
				Shanidar 5*
				Spy 1
				Spy 2
				Tabun 1

This study uses a three-dimensional morphometric method, Euclidean distance matrix analysis, to quantify and compare patterns of craniofacial shape and growth between ontogenetic series of Neandertals and eight geographically distinctive modern human samples. It is assumed that morphological growth trajectories have a genetic basis, and that similarities in morphological patterns are suggestive of genetic relationships (after Richtsmeier & Walker, 1993). Therefore, if Neandertals have a close genetic relationship (i.e., they are conspecific) with anatomically modern *H. sapiens*, they should share similar patterns of craniofacial shape and growth. Conversely, if they do not share similar patterns of craniofacial shape and growth, particularly if the patterns diverge at an early age, then they are not genetically close, and may represent different species.

Materials

The samples used in this study are listed in Table 13.1. Five non-adult and 13 adult Neandertal individuals were compared with modern human samples from eight geographically distinctive populations. The modern human samples included populations from: England (Christ Church Spitalfields; $n_{\text{non-adult}} = 36$, $n_{\text{adult}} = 20$), medieval Denmark ($n_{\text{non-adult}} = 32$, $n_{\text{adult}} = 20$), West Africa ($n_{\text{non-adult}} = 11$, $n_{\text{adult}} = 6$), Nubia ($n_{\text{non-adult}} = 41$, $n_{\text{adult}} = 20$), Edo Period Japan ($n_{\text{non-adult}} = 29$, $n_{\text{adult}} = 16$), Hawaii/Oahu ($n_{\text{non-adult}} = 11$, $n_{\text{adult}} = 20$), St Lawrence Island Yupik Eskimo ($n_{\text{non-adult}} = 27$, $n_{\text{adult}} = 18$), and Indian Knoll ($n_{\text{non-adult}} = 43$, $n_{\text{adult}} = 22$). Modern human specimens were only included if they were relatively complete, undistorted and non-pathological.

The non-adult samples were broken up into four developmental age groups based on tooth formation and eruption sequences: 0–3.0 years (Age Group 1), 3.1–6.0 years (Age Group 2), 6.1–9.0 years (Age Group 3), and 9.1–13.5 years (Age Group 4). Tooth formation was the primary method for dental age estimation (using data from Gorlin & Goldman, 1960; Logan & Kronfeld, 1933; Moorrees *et al.*, 1963; Smith, 1991), although tooth eruption was also used when necessary (see discussion and citations in Krovitz, 2000). These developmental age groups *roughly* coincide with the following developmental criteria (after Minugh-Purvis (1988), although the age groups are defined on the basis of dental age, not these criteria): (1) infancy (birth to completion of deciduous tooth eruption and development), (2) early childhood (period between deciduous tooth development and permanent tooth eruption), (3) mid childhood (eruption of the first permanent teeth), and (4) late childhood (completion of permanent tooth eruption and development, except for the third molar). The fifth developmental age group (Age Group 5) consisted of adult individuals; an individual was considered an adult if it had a closed spheno-occipital synchondrosis, and fused postcranial epiphyses, if available. The adult modern samples contained equal numbers of males and females. Sex of the juvenile individuals was unknown. See Krovitz (2000) for further information on samples and age estimation techniques.

Methods

Landmarks

Three-dimensional landmark coordinate data were collected from each individual skull for 24 landmarks on the face, neurocranium, and basicranium (see Table 13.2 and Figure 13.1). All landmarks listed in Table 13.2 with the description L/R were collected from *either* the right or left side, depending on which side was better preserved in that individual. The method of analysis used in this paper (Euclidean distance matrix analysis, EDMA) is reflection invariant, making it possible to compare individuals with data collected from the right

Table 13.2 *Description of landmarks used in this analysis (see Figure 13.1 for location of landmarks). All landmarks described with L/R were collected from* either *the right or left side, depending on which side was better preserved in that individual*

Landmark Description
1. NAS (Nasion, intersection of internasal suture with nasofrontal sutures)
2. ANS (Anterior nasal spine)
3. IDS (Intradentale superior, between central incisors)
4. L/R PMM (Premaxilla–maxilla junction at alveolar border, between I2 and C)
5. L/R ALA (Alare, widest point on nasal aperture, perpendicular to nasal height)
6. L/R ORB (Top of orbit, halfway between NAS and FZJ)
7. L/R FZJ (Frontal–zygomatic junction at orbital rim)
8. L/R ZYS (Top zygomatic–maxillary suture, at orbital rim)
9. L/R PTN (Pterion, intersection of frontal, parietal, and sphenoid bones)
10. L/R SPH (Intersection of squamous temporal, parietal, and greater wing of sphenoid)
11. L/R EAM (External auditory meatus, uppermost lateral point)
12. L/R AST (Asterion, intersection of parietal, temporal, and occipital bones)
13. N-B (Halfway between nasion and bregma in midline)
14. BRG (Bregma, coronal–sagittal suture intersection in midline)
15. B-L (Halfway between bregma and lambda in midline)
16. LAM (Lambda, sagittal–lambdoid suture intersection, in midline)
17. L-O (Halfway between lambda and opisthion, in midline)
18. ICF (Incisive foramen, marked posteriorly)
19. L/R PAL (Junction on palatine suture with edges/curve of palate)
20. L/R CAR (Carotid canal, posterolateral point)
21. L/R STY (Stylomastoid foramen)
22. BAS (Basion, midpoint of anterior margin of foramen magnum)
23. OPI (Opisthion, midpoint of posterior margin of foramen magnum)
24. L/R CFM (Posterior border of the occipital condyle with foramen magnum)

side of the skull to individuals with data collected from the left side of the skull. It is assumed that the skull is symmetric, and that linear distances between midline and right-sided landmarks are comparable to the same distances between midline and left-sided landmarks. These 24 landmarks were chosen to represent anatomical regions that are generally present in the available juvenile Neandertal individuals. However, all the fossil individuals were broken to some extent, so these landmarks were divided into anatomical subsets to allow different regions of the skull to be analyzed separately. Two subsets were chosen from the face and palate, and four were chosen from the neurocranium and basicranium (see Krovitz, 2000 for definition and further discussion of landmark subsets). Neandertal and modern individuals missing landmarks from a particular subset were not included in analysis of that subset, so not all individuals were used

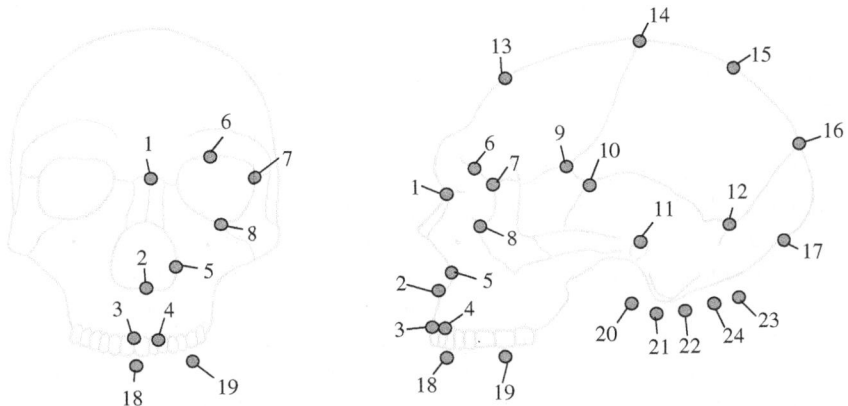

Figure 13.1   Landmarks used in this analysis, illustrated on a Neandertal skull. Numbers and definitions as in Table 13.2.

in every analysis. To give an integrated picture of Neandertal and modern human craniofacial shape and growth patterns, results from the different subset analyses were combined in this paper.

Euclidean distance matrix analysis

Euclidean distance matrix analysis (EDMA) was used in this study to test for differences in craniofacial shape and growth patterns between Neandertals and modern humans. EDMA is a coordinate-system invariant method for form, shape, or growth difference comparisons between two samples (Lele, 1991, 1993, 1999; Lele & Cole, 1996; Lele & Richtsmeier, 1991, 2001; Lele & McCulloch, 2002; Richtsmeier & Lele, 1993; Richtsmeier *et al.*, 1993a). Using landmark coordinates as raw data, a three-dimensional object can be described by the matrix of Euclidean distances between all possible landmark pairs. This matrix of distances is called the form matrix (or FM(A) for object A), and is an equivalent representation of the landmark coordinate data that is invariant to the nuisance parameters of translation, rotation, and reflection (Lele, 1991, 1993, 1999; Lele & McCulloch, 2002; Lele & Richtsmeier, 2001; Richtsmeier *et al.*, 2002). This analysis relies heavily on the parameter of reflection invariance, which permits pooling of data from right-side landmarks for some individuals and left-side landmarks for other individuals. Two different types of EDMA analyses were used in this study: shape difference matrix analysis and growth difference matrix analysis.

EDMA shape difference comparisons

In EDMA shape difference analyses, the form matrix (FM) for each individual was scaled in an effort to minimize the effects of size, because Neandertals were generally larger than modern humans in most craniofacial measurements (e.g., Dodo *et al.*, 1998; Minugh-Purvis, 1988; Stringer *et al.*, 1990; Zollikofer *et al.*, 1995). The geometric mean of all interlandmark distances was chosen as the scaling factor for these analyses because it is a measure of overall size and there was no a priori reason to choose an individual linear distance as a scaling measure (Jungers *et al.*, 1995; Lele & Cole, 1996). Each individual was scaled separately by first calculating the geometric mean of all interlandmark distances from the individual's FM, and then scaling the elements within that individual's FM by dividing each linear distance by the geometric mean calculated for that individual. Once all individuals were individually scaled according to their geometric mean, a mean FM was computed for each sample (Lele, 1993).

A comparison of craniofacial shape between two samples (such as modern humans, A, and Neandertals, B) involves the calculation of a form difference matrix, or FDM(B,A) (following the methodology of Lele & Richtsmeier, 1991, 2001). The FDM(B,A) is a matrix of ratios that was calculated by dividing each element in FM(B) (the scaled mean form matrix for the numerator sample) by the corresponding element in FM(A) (the scaled mean form matrix for the denominator sample), i.e., FDM(B,A) = FM(B)/FM(A). The ratios were viewed in vector format, and sorted from minimum to maximum to identify the specific linear distances that were most similar (i.e., ratios closest to 1.0) or different between samples (i.e., ratios farthest from 1.0). Because sample sizes in this study were too small for statistical testing (e.g., with methods described in Lele & Richtsmeier, 1995), the results illustrated in this paper show individual linear distances that were more than 5% different between samples (i.e., distances with ratios over 1.05 or under 0.95, for the numerator and denominator samples, respectively). Although statistical significance cannot be addressed, identifying extreme differences between mean forms provides descriptive and relevant information about localized differences in morphology between Neandertals and modern humans (see Richtsmeier & Lele, 1990, Richtsmeier & Walker, 1993, and Richtsmeier *et al.*, 1998 for examples of other studies in which small sample size precluded statistical testing). In this study, FDM pair-wise comparisons in each developmental age group were carried out between Neandertals and all modern humans combined, and also between Neandertals and each separate modern human population. All these analyses were carried out separately for each of the six landmark subsets, and results were combined for purposes of illustration. Note that shape difference comparisons between the eight modern human samples were not included in this study; see Krovitz (2000) for detailed descriptions of form differences between these modern human populations.

EDMA growth difference comparisons

Differences in craniofacial growth patterns between Neandertals and modern humans were also evaluated using EDMA (following the methodology of Richtsmeier & Lele, 1993; Richtsmeier *et al.*, 1993a). The process of growth difference analysis is described here for growth between Age Groups 1 and 2, although the same procedure would be followed for analysis of growth between any younger and older samples. The growth pattern between Age Groups 1 and 2 within modern humans, sample A, was quantified by calculating a growth matrix for sample A, written GM(A2,A1). This GM was computed exactly like a FDM in that each element in FM(A2) (the older sample, in the numerator) was divided by the corresponding element in FM(A1) (the younger sample, in the denominator), i.e., GM(A2,A1) = FM(A2)/FM(A1). This determined which distances were growing most and which distances were growing least within the modern human sample. A growth matrix was also calculated for Neandertals, sample B: GM(B2,B1) = FM(B2)/FM(B1). To compare growth patterns between modern humans and Neandertals, samples A and B, the like elements of the growth matrices GM(A2,A1) and GM(B2, B1) were individually compared as ratios. This resulted in a growth difference matrix, written as GDM(A2,A1:B2,B1), i.e., GM(A2,A1)/GM(B2,B1). The process of interpreting growth difference matrix (GDM) results was very similar to that for FDMs, in that the resulting ratios were viewed in vector format and sorted in ascending order to identify linear distances that exhibited growth differences between samples. In growth difference matrix analyses, linear distances with ratios farthest from 1.0 showed the largest differences in growth between the two samples. The results illustrated in this paper show individual linear distances that experienced more than a 5% difference in growth (i.e., distances with ratios over 1.05 or under 0.95, for the numerator and denominator samples, respectively). In this study, GDM pair-wise comparisons were carried out between Neandertals and modern humans in successive age groups (i.e., $1 \rightarrow 2$, $2 \rightarrow 3$, $3 \rightarrow 4$, and $4 \rightarrow 5$). For each of these growth intervals, growth differences were calculated between Neandertals and all modern humans combined, and also between Neandertals and each separate modern human population. All of these growth difference analyses were carried out separately for each of the six landmark subsets, and results were combined for purposes of illustration. Note that growth difference comparisons between the eight modern human samples were not included in this study; see Krovitz (2000) for detailed descriptions of growth differences between these modern human populations.

Results

EDMA shape difference comparisons

Figure 13.2 summarizes the results of shape difference comparisons between Neandertals and modern humans for all five developmental age groups. This figure illustrates individual linear distances that are relatively larger in Neandertals (i.e., distances with ratios over 1.05) compared with modern humans (see Krovitz, 2000 for illustration and discussion of distances relatively larger in modern humans). The landmarks illustrated with an X are missing on the Neandertal individual/s for that age group. For example, the Age Group 2 Neandertal individual (Engis 2) does not preserve the face, and Age Group 3 Neandertal (La Quina 18) is missing the occipital.

Shape differences involving distances that are relatively larger in Neandertals mostly reflect the more anterior location of the Neandertal face relative to the neurocranium and basicranium. This is evident in the anteroposteriorly elongated distances between landmarks on the palate and those on the inferior neurocranium and basicranium, and between upper facial landmarks and those on the inferior neurocranium and basicranium. Within the Neandertal face, the largest shape differences are generally localized to the area between the top of the zygomatic–maxillary suture, the nasal aperture, and the anterior maxillary alveolar region. This reflects the more anterior location of the Neandertal mid and lower facial regions relative to more laterally placed facial landmarks, otherwise known as midfacial and alveolar prognathism. It is important to note that most of these morphological features are not only found in the adult age group (Figure 13.2A), but are also present, at least to some degree, in the non-adult age groups as well (Figure 13.2B–E). These results clearly indicate that patterns of craniofacial shape difference between Neandertals and modern humans are present at a very early age.

Figure 13.3 shows the separate pair-wise comparisons between the non-adult Age Group 4 Neandertal and the eight geographically distinctive Age Group 4 modern human samples; the Age Group 4 sample was randomly chosen for illustration (see Krovitz (2000) for illustration of other age groups). As with the previous figure, the illustrated linear distances are larger in Neandertals relative to modern humans. These comparisons further illustrate Neandertal craniofacial shape (as described above in Figure 13.2), especially the more anterior location of the Neandertal face relative to the neurocranium and basicranium, and the anterior location of the anterior maxillary alveolar region relative to the top of the zygomatic–maxillary suture. Importantly, these results clearly show that Neandertal craniofacial shape differs from each of the modern human samples in almost exactly the same way (note that the Eskimo sample only looks different

A. Adult (Age Group 5)

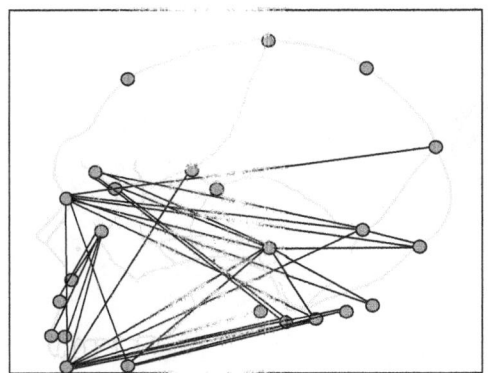

B. Age Group 4

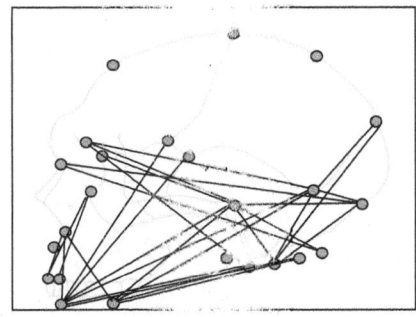

C. Age Group 3

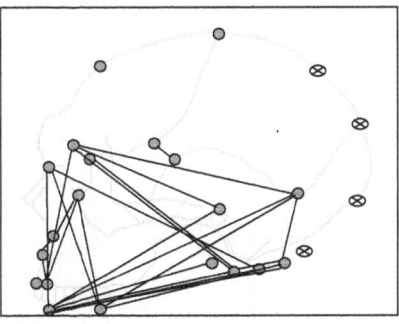

D. Age Group 2

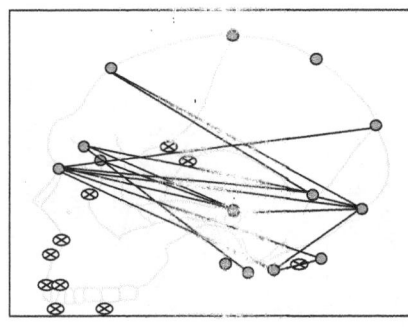

E. Age Group 1

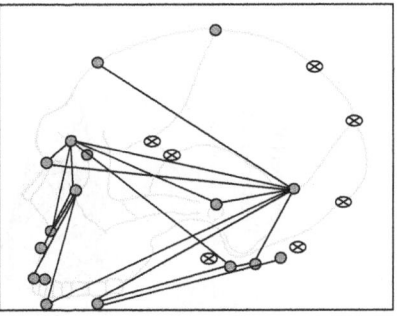

Figure 13.2 Summary of shape differences between Neandertals and modern humans for all five developmental age groups. The linear distances illustrated on this figure are relatively larger in Neandertal individuals. Results of all six landmark subsets are combined in this figure. Missing landmarks are indicated with an encircled X.

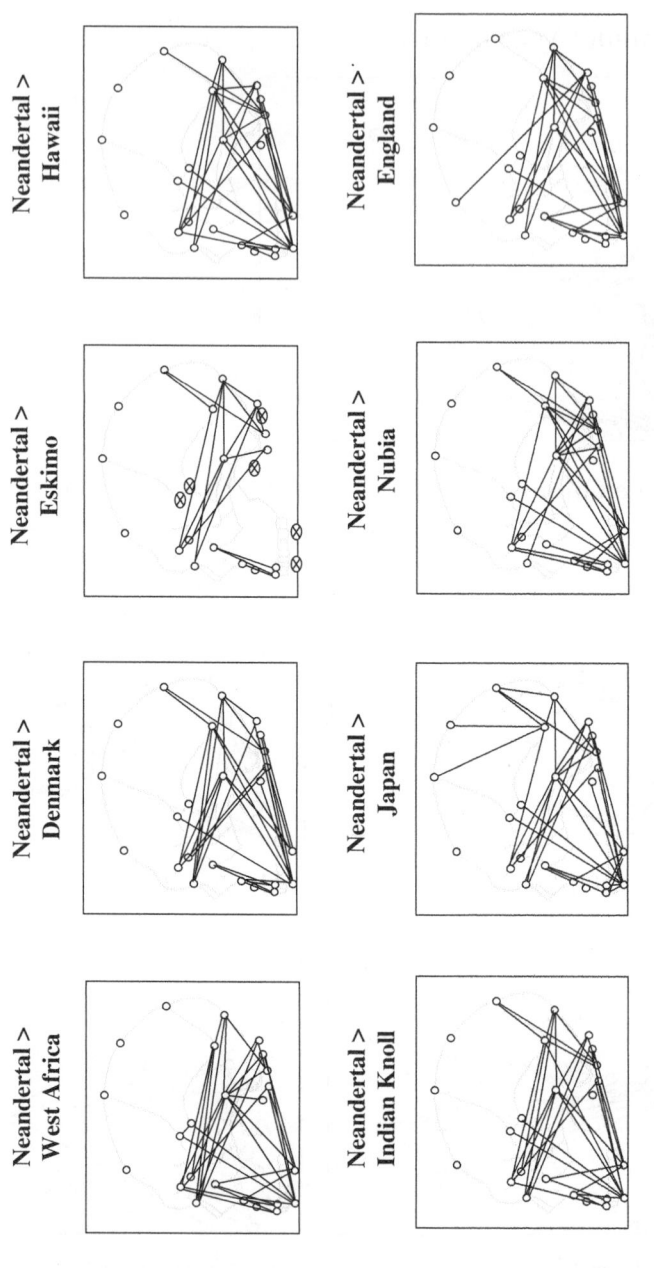

Figure 13.3   Shape difference comparisons between non-adult (Age Group 4) Neandertals and individual modern human populations. The linear distances illustrated on this figure are relatively larger in Neandertal individuals. Results of all six landmark subsets are combined in this figure. Missing landmarks are indicated with an encircled X.

because it is missing several influential landmarks, illustrated with an X). These results are to be expected in analyses of adult individuals, as it has been shown that Neandertal adults differ consistently from all recent modern human adult crania (Bräuer & Rimbach, 1990; Howells, 1989, 1995; Lahr, 1996; Turbón *et al.*, 1997; van Vark *et al.*, 1992). However, these results show that this same pattern also holds true in analyses of non-adult samples. All non-adult Neandertal age groups showed differences from modern human samples that were consistent with those of the Age Group 4 Neandertal shown here; see Krovitz (2000) for individual pair-wise comparisons for all developmental age groups.

EDMA growth difference comparisons

This paper focuses on differences in facial growth between Neandertals and modern humans. Due to space constraints, patterns of neurocranial growth are not addressed here (but see Krovitz (2000) for a full discussion of neurocranial growth in Neandertals and modern humans). Figure 13.4 shows growth

A. Age Groups 1 ➲ 3   **B. Age Groups 3 ➲ 4**   **C. Age Groups 4 ➲ 5**

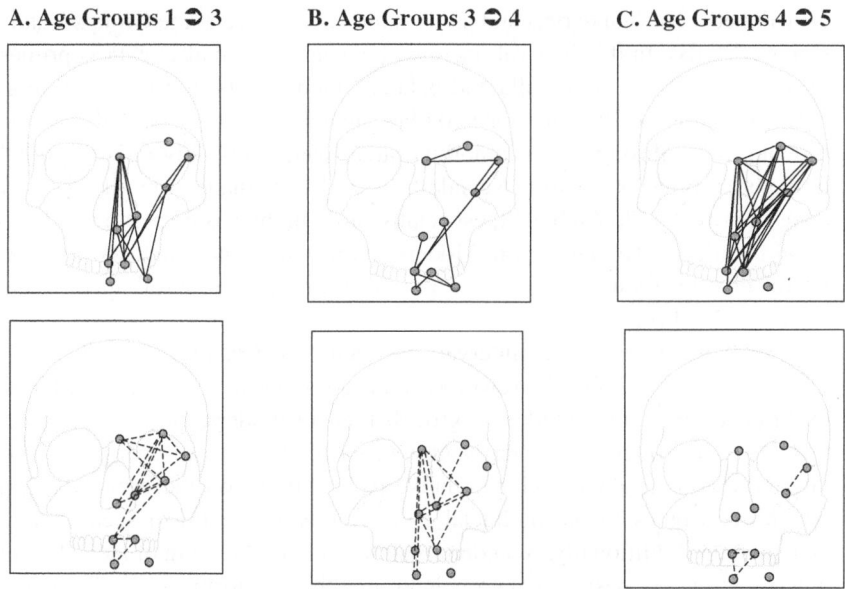

Figure 13.4   Summary of differences in facial growth patterns between Neandertals and modern humans for growth between (A) Age Groups 1 → 3, (B) Age Groups 3 → 4, and (C) Age Groups 4 → 5. The linear distances shown with solid lines are growing relatively more in Neandertals, and distances shown with dashed lines are growing relatively more in modern humans. This figure shows the combined results of two facial landmark subsets.

differences between Neandertals and modern humans in the face (a combination of the two facial landmark subsets). Linear distances shown with solid lines, illustrated on the Neandertal skulls, are growing more in Neandertals than in modern humans, while linear distances shown with dotted lines, illustrated on the modern human skulls, are growing more in the modern humans. There are three growth intervals shown here, growth between Age Group 1 → Age Group 3, Age Group 3 → Age Group 4, and Age Group 4 → Age Group 5. The Age Group 2 Neandertal (Engis 2) does not preserve the face and could not be included in this analysis of facial growth.

At the interval Age Groups 1 → 3 (Figure 13.4A), Neandertals show increased growth in facial height, particularly through the nasal aperture, and in the subnasal maxillary dimensions that lead to alveolar prognathism. While Neandertal growth is primarily oriented along the superoinferior axis, the modern humans are primarily growing in dimensions along the mediolateral axis. The modern humans show increased growth of orbital height and width, through the frontal process of the maxilla, and between the lower face and inferior orbital margin.

As with the previous growth interval, there are localized differences in growth between Neandertals and modern humans between Age Groups 3 and 4 (Figure 13.4B). In this interval, increased Neandertal facial growth is primarily localized to the premaxilla and palate, which results in a relatively more anterior location of the anterior maxillary alveolus. The Neandertals are also growing more through the supraorbital area, which probably reflects growth of the brow ridge. Relative to Neandertals, modern humans primarily show increased growth in facial height, especially through the nasal aperture and subnasal maxillary alveolar region. Modern humans also show increased mediolateral growth through the maxilla, primarily through the nasal aperture and inferior orbital rim.

During the final growth interval, between Age Groups 4 and 5 (Figure 13.4C), Neandertal facial growth is very pronounced, and most distances experience larger magnitudes of growth than in modern humans. Neandertals experience increased anterior growth through the nasal aperture, alveolar, and anterior palatal regions relative to landmarks on the orbital margin, which contributes to the midfacial and alveolar prognathism seen in adult Neandertals. Additionally, Neandertals show increased growth of facial height, anterior growth of nasion, and growth through the supraorbital area. Compared to Neandertals, modern humans show reduced magnitudes of facial growth between Age Groups 4 and 5, and modern human growth is primarily localized to the premaxilla. Modern human populations also show increased growth of the inferolateral portion of the orbit.

Thus, there are localized growth differences between Neandertals and modern humans, and differences in the relative timing of growth events. For example,

**Neandertal >
West Africa**

**Neandertal >
Hawaii**

**Neandertal >
Indian Knoll**

**Neandertal >
Japan**

**Neandertal >
Nubia**

**Neandertal >
Spitalfields**

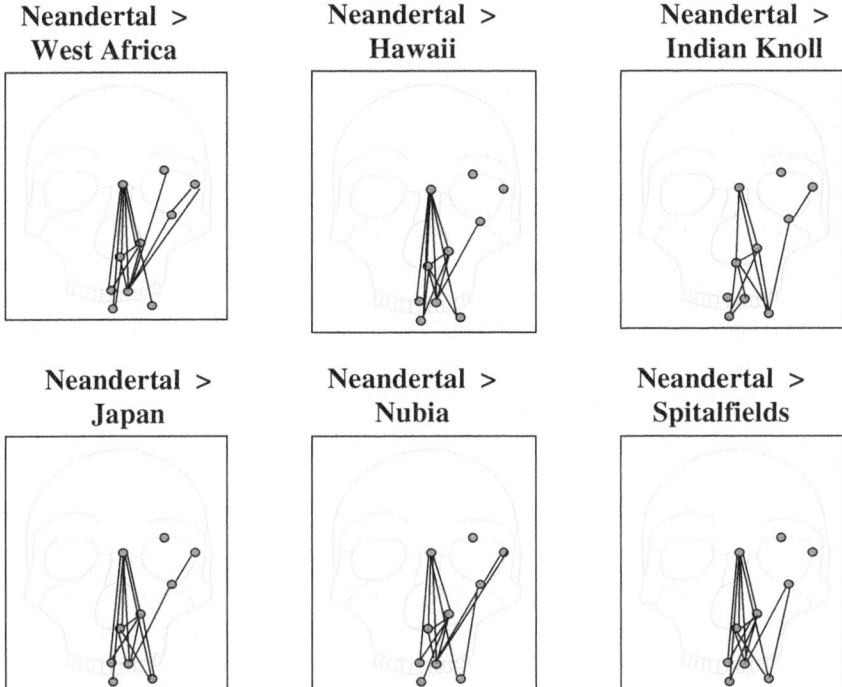

Figure 13.5   Growth difference comparisons between Neandertals and individual modern human populations for facial growth in the interval Age Groups 1 → 3. The linear distances shown in this figure are growing relatively more in Neandertals than in modern humans. Danish and Eskimo modern human populations could not be included in this analysis as there were no Age Group 1 individuals from these populations. This figure shows the combined results of two facial landmark subsets.

Neandertals experience increased growth in facial height and more anterior growth at nasion at an earlier age (Age Groups 1 → 3) relative to modern humans (Age Groups 3 → 4). In contrast, the modern humans grow more in mediolateral dimensions through the orbit at an earlier age (Age Groups 1 → 3) than do the Neandertals (Age Groups 3 → 4). Interestingly, modern humans also grow more through the midfacial region early on, particularly between the nasal aperture and the inferior orbital rim (Age Groups 1 → 3, and 3 → 4). Neandertals grow more than modern humans in this region, which contributes to midfacial prognathism in adult Neandertals, only during the last growth interval (Age Groups 4 → 5).

Figure 13.5 shows the separate pair-wise growth difference comparisons between Neandertals and individual modern human populations for growth taking place in the interval Age Groups 1 → 3; this growth interval was randomly chosen for illustration (see Krovitz (2000) for illustration of other growth intervals).

The illustrated linear distances are experiencing more growth in Neandertals than in modern humans. The Danish and Eskimo modern human populations could not be included in this analysis as there were no Age Group 1 individuals from these populations. Overall, these results are very similar to those shown in Figure 13.4A; the Neandertals are experiencing more superoinferior growth of the face and through the subnasal maxillary region than are modern humans. There are slight differences in the comparisons with different modern human populations, as, for example, the Neandertals are not growing much more than the Indian Knoll sample in facial height. However, the overall picture is very much the same. Neandertals show clear, localized differences in growth when compared to any modern human population considered here. This clear difference between Neandertals and all modern human samples is also seen in other growth intervals, illustrated in Krovitz (2000).

Discussion

The results of this study show that: (1) the distinctive Neandertal craniofacial shape is present by the age of 3 years at the latest, (2) Neandertal craniofacial shape differs from modern human craniofacial shape at all ages, and (3) there are similar patterns of shape difference between Neandertals and all eight modern human samples. This is consistent with previous descriptive research showing that unique Neandertal morphology is present in neonatal or infant Neandertals (e.g., Akazawa *et al.*, 1995; Dodo *et al.*, 1998; Faerman *et al.*, 1994; Golovanova *et al.*, 1999; Ishida *et al.*, 2000; Mallegni & Trinkaus, 1997; Maureille, 2002; Pap *et al.*, 1996; Rak *et al.*, 1994). Additionally, this research is consistent with other geometric morphometric studies that have shown the early appearance of population-specific or species-specific morphology in humans, Neandertals, *Australopithecus africanus*, and great apes (Ackermann & Krovitz, 2002; Krovitz, 2000; O'Higgins & Strand Viðarsdóttir, 1999; Ponce de León & Zollikofer, 2001; Strand Viðarsdóttir *et al.*, 2002).

However, the current study takes this research one step further by showing that the morphological differences between Neandertals and modern humans are not only present at an early age, but are also accentuated through localized differences in growth patterns, with different growth events taking place at different times in Neandertals and modern humans. Unlike analyses of static adult morphology, this study measured differences in the *growth processes* that contribute to the observable differences in adult Neandertal and modern human morphological types. Not only is Neandertal craniofacial shape different from modern human craniofacial shape at every age, but the growth processes that help to create Neandertal craniofacial shape are also extremely different from all modern human samples included in this study. If it is possible to say that genetic differences between species are phenotypically expressed through unique growth processes (after Richtsmeier & Walker, 1993), then the observed growth differences between Neandertals and modern humans probably reflect species-level distinctions between these groups.

The results of this study are very similar to those of Ponce de León & Zollikofer (2001), who also found shape differences between Neandertals and modern humans at an early postnatal age, and suggested that "taxon-specific" ontogenetic patterns are consistent with a species-level distinction between Neandertals and modern humans. It is possible that minor genetic differences acting on growth centers early in development are responsible for many of the observable differences between Neandertals and modern humans (e.g., Brothwell, 1975; Green, 1990; Green & Smith, 1991; Lieberman *et al.*, 2002; Maureille & Bar, 1999). Recent genetic work has shown that Neandertal and modern human mitochondrial DNA gene pools have been separated for at least 500 000 years, and that these groups were genetically distinct with little or no gene flow between them (as would be expected from different hominid species) (Krings *et al.*, 1997, 1999, 2000; Ovchinnikov *et al.*, 2000).

The findings of these studies differ from the published work of Minugh-Purvis (1988, 1998, 2002), in which she supports "the concept of genetic continuity between Neandertals and the early Upper Paleolithic associated remains from Western Europe" (Minugh-Purvis, 1988: 269). However, Minugh-Purvis's analyses do not control for size differences between Neandertals, early modern *H. sapiens*, and recent *H. sapiens*. It is therefore possible that the noted gradation in growth patterns simply reflects the larger size of Neandertals and early modern *H. sapiens* relative to recent *H. sapiens*, and does not reflect growth pattern (and genetic) continuity between these groups. Additionally, most of Minugh-Purvis's work focuses on the neurocranium, which is strongly influenced by brain growth at an early age. Unfortunately, most neonatal or infant Neandertals are not well preserved, and the youngest specimens that preserve a relatively complete neurocranium are Subalyuk 2 and Pech de l'Aze 1, which, at 2–3 years of age (Krovitz, 2000; Minugh-Purvis, 1988), were probably past this early period of rapid brain growth. As it is not currently possible to measure early postnatal brain growth (as reflected in neurocranial remains) in Neandertals, it is likely that we are not detecting distinctive and important early aspects of Neandertal neurocranial growth.

Future work

To help place these results in a broader comparative and evolutionary framework, future work should include a larger sample size of fossil individuals. For example, there are several other young Neandertals preserving relatively complete cranial remains that could be added to the analysis, such as Roc de Marsal 1 (Tillier, 1983), and Dederiyeh 1 or 2 (Dodo *et al.*, 1998; Ishida *et al.*, 2000). Furthermore, inclusion of the adolescent Neandertal Le Moustier 1 (Ponce de León & Zollikofer, 1999; Thompson & Bilsborough, 1997; Thompson & Illerhaus, 1998) would help reduce the age gap between the Age Group 4 (9.1–13.5 years) and Age Group 5 (adult) specimens in this study, and would potentially reveal important information about Neandertal craniofacial growth during puberty.

More importantly, future studies should include ontogenetic samples of early modern *H. sapiens* (including available specimens from Skhul, Qafzeh, and Upper Paleolithic sites), as it will be essential to compare early modern human growth patterns with those observed for Neandertals and recent modern humans. If Neandertals are indeed a separate species, then the recent and early modern humans should exhibit very similar craniofacial growth patterns and the Neandertal growth patterns should differ markedly and consistently from both groups. If Neandertals are not a distinct species and there was considerable admixture between Neandertal and early modern human populations, then the early modern human growth patterns should be intermediate between those observed for recent modern humans and Neandertals, or should differ from both Neandertals and recent modern humans.

Ideally, the comparative context should be extended beyond Neandertals, early modern *H. sapiens*, and recent *H. sapiens* to include patterns of growth and development in other, less well-known, species within the genus *Homo* (such as *H. antecessor*: see Bermúdez de Castro *et al.*, 1997). Such analysis would identify primitive and derived features of craniofacial growth patterns. For example, although Neandertal growth patterns are presented in this study as being "distinctive" relative to modern humans, it is likely that certain aspects of Neandertal growth patterns are actually primitive for *Homo* in general, while other aspects are derived. Likewise, aspects of modern human growth patterns (such as those relating to facial projection and neurocranial globularity: Krovitz, 2000; Lieberman *et al.*, 2002) are probably derived. Unfortunately, the paucity of pre-Neandertal juvenile fossils with relatively complete cranial remains currently makes it difficult to explore the growth processes responsible for the development of adult morphology throughout much of the history of our genus.

Ultimately, the present study provides an important baseline for further studies of juvenile hominids by documenting morphology and growth patterns in Neandertals and a large, geographically variable sample of recent modern humans. Data such as these will be necessary for future studies in which important biological and phylogenetic questions rest on the analysis of individual juvenile fossil specimens. For example, adequate modern and fossil juvenile comparative samples are necessary to ensure that the morphology of juvenile specimens of *H. antecessor* (Bermúdez de Castro *et al.*, 1997) truly reflects species specific ontogenetic patterning and not simply primitive features shared with other juvenile hominids. Additionally, knowledge about skeletal variability in recent modern human, Neandertal, and Upper Paleolithic juvenile samples will be essential for interpreting the morphology of the potential Neandertal–modern human hybrid from the Upper Paleolithic site of Lagar Vello (Duarte *et al.*, 1999). In the future, the continuing discovery of new juvenile hominids and the increasing appreciation of the need for careful comparative analyses contribute to our growing knowledge about the evolution of modern human maturational patterns and the diversity of growth patterns within the genus *Homo*.

Acknowledgments

I thank all the people who gave me access to the skeletal collections in their care, or who helped me obtain or interpret the dental X-rays for the recent human samples: Hisao Baba, Pia Bennike, Jodie Blodgett, Luca Bondioli, Jennifer Clark, Kevin Conley, Jean-Marie Cordy, Kate Hesseldenz, Louise Humphrey, David Hunt, Robert Kruszynski, Helen Liversidge, Niels Lynnerup, Roberto Macchiarelli, Giorgio Manzi, Yuji Mizoguchi, Theya Molleson, Søren Nørby, Nancy O'Malley, Rosine Orban, Doug Owsley, Ildiko Pap, Rick Potts, Edouard Poty, Mary Powell, Patrick Semal, Ib Sewerin, Gabriella Spedini, Chris Stringer, and Erik Trinkaus. Additionally, thanks to Valerie Burke DeLeon, Andrew Nelson and Jennifer Thompson for their comments on this manuscript. This research was supported by grants from the L.S.B. Leakey Society, the National Science Foundation, and the Japanese Society for the Promotion of Science.

References

Ackermann, R. R., & Krovitz, G. E. (2002). Common patterns of facial ontogeny in the hominid lineage. *Anatomical Record*, **269**, 142–147.

Akazawa, T., Muhesen, S., Dodo, Y., Kondo, O., Mizoguchi, Y., Abe, Y., Nishiaki, Y., Ohta, S., Oguchi, T., & Haydal, J. (1995). Neanderthal infant burial from the Dederiyeh Cave in Syria. *Paléorient*, **21**, 77–86.

Bermúdez de Castro, J. M., Arsuaga, J. L., Carbonell, E., Rosas, A., Martínez, I., & Mosquera, M. (1997). A hominid from the Lower Pleistocene of Atapurerca, Spain: Possible ancestor to Neandertals and modern humans. *Science*, **276**, 1392–1395.

Bräuer, G. (1984). A craniological approach to the origin of anatomically modern *Homo sapiens* in Africa and implications for the appearance of modern Europeans. In *The Origins of Modern Humans: A World Survey of the Fossil Evidence*, eds. F. H. Smith & F. Spencer, pp. 327–410. New York: Alan R. Liss.

Bräuer, G., & Rimbach, K. W. (1990). Late archaic and modern *Homo sapiens* from Europe, Africa, and Southwest Asia: Craniometric comparisons and phylogenetic implications. *Journal of Human Evolution*, **19**, 789–807.

Brothwell, D. (1975). Adaptive growth rate changes as a possible explanation for the distinctiveness of the Neanderthalers. *Journal of Archaeological Science*, **2**, 161–163.

Collard, M., & O'Higgins, P. (2001). Ontogeny and homoplasy in the papionin monkey face. *Evolution and Development*, **3**, 322–331.

Dodo, Y., Kondo, O., Muhesen, S., & Akazawa, T. (1998). Anatomy of the Neandertal infant skeleton from Dederiyeh Cave, Syria. In *Neandertals and Modern Humans in Western Asia*, eds. T. Akazawa, K. Aoki, & O. Bar-Yosef, pp. 323–338. New York: Plenum Press.

Duarte, C., Maurício, J., Pettitt, P. B., Souto, P., Trinkaus, E., van der Plicht, H., & Zilhão, J. (1999). The early Upper Paleolithic human skeleton from the Abrigo do Lagar Velho (Portugal) and modern human emergence in Iberia. *Proceedings of the National Academy of Sciences of the USA*, **96**, 7604–7609.

Faerman, M., Zilberman, U., Smith, P., Kharitonov, V., & Batsevitz, V. (1994). A Neanderthal infant from the Barakai Cave, Western Caucasus. *Journal of Human Evolution*, **27**, 405–415.

Golovanova, L. V., Hoffecker, J. F., Kharitonov, V. M., & Romanova, G. P. (1999). Mezmaiskaya Cave: A Neanderthal occupation in the Northern Caucasus. *Current Anthropology*, **40**, 77–86.

Gomez, A. M. (1992). Primitive and derived patterns of relative growth among species of Lorisidae. *Journal of Human Evolution*, **23**, 219–233.

Gorlin, R. J., & Goldman, H. M. (1960). *Oral Pathology*, 5th edn. St Louis: C. V. Mosby.

Green, M. D. (1990). Neandertal craniofacial growth: An ontogenetic model. MA thesis, University of Tennessee, Knoxville.

Green, M. D., & Smith, F. H. (1991). Neandertal craniofacial growth. *American Journal of Physical Anthropology* Suppl., **12**, 164.

Howells, W. W. (1989). *Skull Shapes and the Map: Craniometric Analyses in the Dispersion of Modern Homo*, Papers of the Peabody Museum of Archaeology and Ethnology, vol. 79. Cambridge: Harvard University.

Howells, W. W. (1995). *Who's Who in Skulls: Ethnic Identification of Crania from Measurements*, Papers of the Peabody Museum of Archaeology and Ethnology, vol. 82. Cambridge: Harvard University.

Ishida, H., Kondo, O., Muhesen, S., & Akazawa, T. (2000). A new Neanderthal child recovered at Dederiyeh Cave, Syria in 1997–1998. *American Journal of Physical Anthropology* Suppl., **30**, 186–197.

Jungers, W. L., Falsetti, A. B., & Wall, C. E. (1995). Shape, relative size, and size-adjustments in morphometrics. *Yearbook of Physical Anthropology*, **38**, 137–161.

Krings, M., Stone, A., Schmitz, R. W., Krainitzki, H., Stoneking, M., & Pääbo, S. (1997). Neandertal DNA sequences and the origin of modern humans. *Cell*, **90**, 19–30.

Krings, M., Geisert, H., Schmitz, R. W., Krainitzki, H., & Pääbo, S. (1999). DNA sequence of the mitochondrial hypervariable region II from the Neandertal type specimen. *Proceedings of the National Academy of Sciences of the USA*, **96**, 5581–5585.

Krings, M., Capelli, C., Tschentscher, F., Geisert, H., Meyer, S., von Haeseler, A., Grochmidt, K., Possnert, G., Paunovic, M., & Pääbo, S. (2000). A view of Neandertal genetic diversity. *Nature Genetics*, **26**, 144–146.

Krovitz, G. E. (2000). Three-dimensional comparisons of craniofacial morphology and growth patterns in Neandertals and modern humans. PhD dissertation, Johns Hopkins University, Baltimore.

Lahr, M. M. (1996). *The Evolution of Modern Human Diversity: A Study of Cranial Variation*. Cambridge: Cambridge University Press.

Lele, S. (1991). Some comments on coordinate-free and scale-invariant methods in morphometrics. *American Journal of Physical Anthropology*, **85**, 407–417.

Lele, S. (1993). Euclidean distance matrix analysis (EDMA): Estimation of mean form and mean form difference. *Mathematical Geology*, **25**, 573–602.

Lele, S. (1999). Invariance and morphometrics: A critical appraisal of statistical techniques for landmark data. In *On Growth and Form: Spatio-Temporal Patterning in Biology*, eds. M. A. J. Chaplain, G. D. Singh, & J. McLachlan, pp. 325–336. New York: John Wiley & Sons.

Lele, S., & Cole, T. M. III. (1996). A new test for shape differences when variance–covariance matrices are unequal. *Journal of Human Evolution*, **31**, 193–212.

Lele, S., & Richtsmeier, J. T. (1991). Euclidean distance matrix analysis: A coordinate-free approach for comparing biological shapes using landmark data. *American Journal of Physical Anthropology*, **86**, 415–427.

Lele, S., & Richtsmeier, J. T. (1995). Euclidean distance matrix analysis: Confidence intervals for form and growth differences. *American Journal of Physical Anthropology*, **98**, 73–86.

Lele, S. R., & McCulloch, C. E. (2002). Invariance, identifiability and morphometrics. *Journal of the American Statistical Association*, **97**, 1–11.

Lele, S. R., & Richtsmeier, J. T. (2001). *An Invariant Approach to Statistical Analysis of Shapes*. London: Chapman & Hall.

Lieberman, D. E., McBratney, B. M., & Krovitz, G. (2002). The evolution and development of cranial form in *Homo sapiens*. *Proceedings of the National Academy of Sciences of the USA*, **99**, 1134–1139.

Logan, W., & Kronfeld, R. (1933). Development of the human jaws and surrounding structures from birth to the age of 15 years. *Journal of the American Dental Association*, **20**, 379–427.

MacLeod, N., & Forey, P. L. (eds.) (2002). *Morphology, Shape and Phylogeny*. London: Taylor & Francis.

Mallegni, F., & Trinkaus, E. (1997). A reconsideration of the Archi 1 Neandertal mandible. *Journal of Human Evolution*, **33**, 651–668.

Maureille, B. (2002). A lost Neanderthal neonate found. *Nature*, **419**, 33–34.

Maureille, B., & Bar, D. (1999). The premaxilla in Neandertal and early modern children: Ontogeny and morphology. *Journal of Human Evolution*, **37**, 137–152.

Minugh-Purvis, N. (1988). Patterns of craniofacial growth and development in Upper Pleistocene hominids. PhD dissertation, University of Pennsylvania, Philadelphia.

Minugh-Purvis, N. (1998). The search for the earliest modern Europeans. In *Neandertals and Modern Humans in Western Asia*, eds. T. Akazawa, K. Aoki, & O. Bar-Yosef, pp. 339–352. New York: Plenum Press.

Minugh-Purvis, N. (2002). Heterochronic change in the neurocranium and the emergence of modern humans. In *Human Evolution through Developmental Change*, eds. N. Minugh-Purvis & K. J. McNamara, pp. 479–498. Baltimore: Johns Hopkins University Press.

Minugh-Purvis, N., & McNamara, K. J. (eds.) (2002). *Human Evolution through Developmental Change*. Baltimore: Johns Hopkins University Press.

Moorrees, C. F. A., Fanning, E. A., & Hunt, E. E. (1963). Age variation of formation stages for ten permanent teeth. *Journal of Dental Research*, **42**, 1490–1502.

O'Higgins, P., & Cohn, M. J. (eds.) (2000). *Development, Growth and Evolution: Implications for the Study of the Hominid Skeleton*. London: Academic Press.

O'Higgins, P., & Strand Viðarsdóttir, U. (1999). New approaches to the quantitative analysis of craniofacial growth and variation. In *Human Growth in the Past: Studies from Bones and Teeth*, eds. R. D. Hoppa & C. M. Fitzgerald, pp. 129–160. Cambridge: Cambridge University Press.

Ovchinnikov, I. V., Götherström, A., Romanova, G. P., Kharitonov, V. M., Lidén, K., & Goodwin, W. (2000). Molecular analysis of Neanderthal DNA from the northern Caucasus. *Nature*, **404**, 490–493.

Pap, I., Tillier, A.-M., Arensburg, B., & Chech, M. (1996). The Subalyuk Neanderthal remains (Hungary): A re-examination. *Annales Historico-Naturales Musei Nationalis Hungarici*, **88**, 233–270.

Ponce de León, M. S., & Zollikofer, C. P. E. (1999). New evidence from Le Moustier 1: Computer-assisted reconstruction and morphometry of the skull. *Anatomical Record*, **254**, 474–489.

Ponce de León, M. S., & Zollikofer, C. P. E. (2001). Neanderthal cranial ontogeny and its implications for late hominid diversity. *Nature*, **412**, 534–538.

Rak, Y., Kimbel, W. H., & Hovers, E. (1994). A Neandertal infant from Amud Cave, Israel. *Journal of Human Evolution*, **26**, 313–324.

Relethford, J. H., & Harpending, H.C. (1994). Craniometric variation, genetic theory, and modern human origins. *American Journal of Physical Anthropology*, **95**, 249–270.

Richtsmeier, J. T., & Lele, S. (1990). Analysis of craniofacial growth in Crouzon syndrome using landmark data. *Journal of Craniofacial Genetics and Developmental Biology*, **10**, 39–62.

Richtsmeier, J. T., & Lele, S. (1993). A coordinate-free approach to the analysis of growth patterns: Models and theoretical considerations. *Biological Reviews*, **68**, 381–411.

Richtsmeier, J. T., & Walker, A. (1993). A morphometric study of facial growth. In *The Nariokotome* Homo erectus *Skeleton*, eds. A. Walker & R. Leakey, pp. 391–410. Cambridge: Harvard University Press.

Richtsmeier, J. T., Cheverud, J. M., Danahey, S. E., Corner, B. D., & Lele, S. (1993a). Sexual dimorphism of ontogeny in the crab-eating macaque *(Macaca fasicularis)*. *Journal of Human Evolution*, **25**, 1–30.

Richtsmeier, J. T., Corner, B. D., Grausz, H. M., Cheverud, J. M., & Danahey, S. E. (1993b). The role of postnatal growth pattern in the production of facial morphology. *Systematic Biology*, **42**, 307–330.

Richtsmeier, J. T., Cole, T. M. III, Krovitz, G. E., Valeri, C. J., & Lele, S. (1998). Preoperative morphology and development in sagittal synostosis. *Journal of Craniofacial Genetics and Developmental Biology*, **18**, 64–78.

Richtsmeier, J., DeLeon, V., & Lele, S. (2002). The promise of geometric morphometrics. *Yearbook of Physical Anthropology*, **45**, 63–91.

Schillaci, M. A., & Froehlich, J. W. (2001). Nonhuman primate hybridization and the taxonomic status of Neanderthals. *American Journal of Physical Anthropology*, **115**, 157–166.

Smith, B. H. (1991). Standards of human tooth formation and dental age assessment. In *Advances in Dental Anthropology*, eds. M. A. Kelley & C. S. Larsen, pp. 143–168. New York: Wiley-Liss.

Smith, F. H. (1992). The role of continuity in modern humans. In *Continuity or Replacement: Controversies in* Homo sapiens *Evolution*, eds. G. Bräuer & F. H. Smith, pp. 145–156. Rotterdam: A. A. Balkema.

Smith, F. H., Falsetti, A. B., & Donnelly, S. M. (1989). Modern human origins. *Yearbook of Physical Anthropology*, **32**, 35–68.

Strand Viðarsdóttir, U., O'Higgins, P., & Stringer, C. (2002). A geometric morphometric study of regional differences in the ontogeny of the modern human facial skeleton. *Journal of Anatomy*, **201**, 211–229.

Stringer, C. B. (1989). The origin of early modern humans: A comparison of the European and non-European evidence. In *The Human Revolution: Behavioural and Biological Perspectives on the Origins of Modern Humans*, ed. P. Mellars & C. Stringer, pp. 232–244. Princeton: Princeton University Press.

Stringer, C. B. (1992). Replacement, continuity and the origin of *Homo sapiens*. In *Continuity or Replacement: Controversies in* Homo sapiens *Evolution*, eds. G. Bräuer & F. H. Smith, pp. 9–24. Rotterdam: A. A. Balkema.

Stringer, C. B., & Andrews, P. (1988). Genetic and fossil evidence for the origin of modern humans. *Science*, **239**, 1263–1268.

Stringer, C. B., Hublin, J.-J., & Vandermeersch, B. (1984). The origin of anatomically modern humans in Western Europe. In *The Origins of Modern Humans: A World Survey of the Fossil Evidence*, eds. F. H. Smith & F. Spencer, pp. 51–135. New York: Alan R. Liss.

Stringer, C. B., Dean, M. C., & Martin, R. D. (1990). A comparative study of cranial and dental development within a recent British sample and among Neandertals. In *Primate Life History and Evolution*, ed. C. J. de Rousseau, pp. 115–152. New York: Wiley-Liss.

Thompson, J. L., & Bilsborough, A. (1997). The current state of the Le Moustier 1 skull. *Acta Praehistorica et Archaeologica*, **29**, 17–38.

Thompson, J. L., & Illerhaus, B. (1998). A new reconstruction of the Le Moustier 1 skull and investigation of internal structures using 3-D-uCT data. *Journal of Human Evolution*, **35**, 647–665.

Tillier, A.-m. (1983). L'Enfant néanderthalien du Roc de Marsal (Campagne du Bugue, Dordogne): Le squelette facial. *Annales de Paléontologie*, **69**, 137–149.

Turbón, D., Pérez-Pérez, A., & Stringer, C.B. (1997). A multivariate analysis of Pleistocene hominids: Testing hypotheses of European origins. *Journal of Human Evolution*, **32**, 449–468.

van Vark, G. N., Bilsborough, A., & Henke, W. (1992). Affinities of European Upper Palaeolithic *Homo sapiens* and later human evolution. *Journal of Human Evolution*, **23**, 401–417.

Williams, F. L., Godfrey, L. R., & Sutherland, M. R. (2002). Heterochrony and the evolution of Neandertal and modern human craniofacial form. In *Human Evolution through Developmental Change*, eds. N. Minugh-Purvis & K. J. McNamara, pp. 405–441. Baltimore: Johns Hopkins University Press.

Wolpoff, M. H. (1989). Multiregional evolution: The fossil alternative to Eden. In *The Human Revolution*, eds. P. Mellars & C. B. Stringer, pp. 62–108. Edinburgh: Edinburgh University Press.

Wolpoff, M. H., Wu, X., & Thorne, A. G. (1984). Modern *Homo sapiens* origins: A general theory of hominid evolution involving the fossil evidence from East Asia. In *The Origins of Modern Humans*, eds. F. H. Smith & F. Spencer, pp. 411–483. New York: Alan R. Liss.

Zollikofer, C. P. E., Ponce de León, M. S., Martin, R. D., & Stucki, P. (1995). Neanderthal computer skulls. *Nature*, **375**, 283–285.

14 Ontogenetic patterning and phylogenetic significance of mental foramen number and position in the evolution of Upper Pleistocene Homo sapiens

H. COQUEUGNIOT
Université Bordeaux 1

N. MINUGH-PURVIS
University of Pennsylvania

Introduction

In the last decade, a number of new Pleistocene subadults have been discovered in Europe, Asia and the Levant, from sites including Moula-Guercy, France (Defleur *et al.*, 1999), Sima de los Huesos, Spain (Arsuaga *et al.*, 1997), Lagar Velho, Portugal (Duarte *et al.*, 1999), Mezmaiskaya in the northern Caucasus (Golovanova *et al.*, 1999), and Dederiyeh in Syria (Akazawa *et al.*, 1995, 1999). As the available sample of immature remains dating to the later phases of hominid evolution continues to grow, our opportunities to better understand the developmental patterning underlying the range of morphological variability demonstrated by adults of Upper Pleistocene populations improve.

Neandertal mandibles differ from those of modern humans in a number of features including the usual lack of a mental eminence (Le Gros Clark, 1964), possession of a large retromolar space (de Lumley, 1973), a hook-shaped coronoid process with a vertical height often exceeding that of the condylar process (Minugh-Purvis & Lewandowski, 1992), and unique configuration of the superior ramus (Rak *et al.*, 2002). The anterior mandibular region of *Homo* is particularly enigmatic in that late in hominid evolution its profile changes dramatically with the appearance of the chin, a phenomenon which has long been of interest to paleoanthropologists (Schwartz & Tattersall, 2000). Although

Patterns of Growth and Development in the Genus Homo, ed. J. L. Thompson, G. E. Krovitz and A. J. Nelson. Published by Cambridge University Press. © Cambridge University Press 2003.

some Neandertals possess a fairly vertical labial aspect of the mandibular symphysis or even a rudimentary mental eminence, close scrutiny of this region reveals several differences in the morphological details contributing to the formation of a modern human, as opposed to a Neandertal, chin (Arensburg *et al.*, 1989; Coqueugniot, 1999; Mallegni & Trinkaus, 1997; Minugh-Purvis, 2000; Tillier, 1981; and others). For example, modern humans typically exhibit marked development of the mental fossae and tuber symphoses labially, inferiorly Neandertals possess larger and more posteriorly oriented digastric fossae, and lingually Neandertals tend to possess a genioglossal fossa whereas modern humans have genial tubercles (mental spines), although a few Neandertals possess both (Vlček, 1969).

Among the interesting aspects of Neandertal anterior mandibular morphology that might shed light on the gradual appearance of the chin during Upper Pleistocene hominid evolution are the position and number of apertures associated with the mental foramen. This foramen is a bilateral opening on the anterolateral mandibular corpus where the mental nerve, artery, and vein are transmitted through the bone. Although located laterally, beyond the immediate area of the mental symphysis itself, the vessels and nerves conveyed by the mental foramen supply the mental region and, as such, are part of the developmental complex related to the ontogeny of the chin.

In modern humans the mental foramen is most commonly located below the P_2 tooth (Tebo & Telford, 1950) while in Neandertal adults this feature is typically located below the M_1 (Figure 14.1). This more posterior location in Neandertals compared with the usual modern human condition has led some

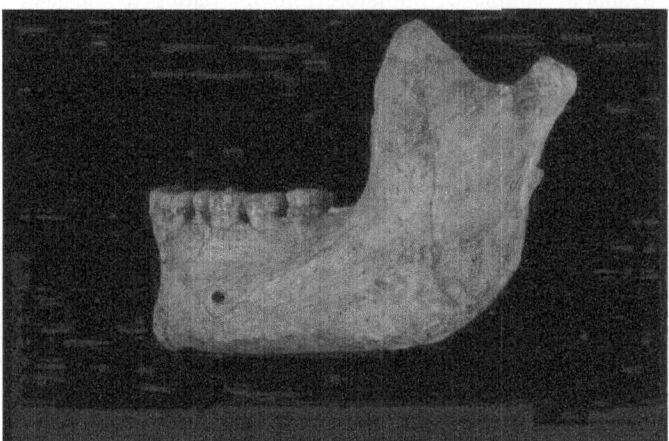

Figure 14.1   Posterior position of the mental foramen on the adult mandible of Amud 1. (Photograph by M. Barazani.)

to consider it a Neandertal autapomorph (e.g. Condemi, 1991; Stringer *et al.*, 1984). Others disagree, including Trinkaus, who asserts (1993) that the positioning of the mental foramen under M_1 is seen in high frequencies in other fossil hominids such as *Homo erectus* and should not, therefore, be considered a diagnostic feature of Neandertals. However, this does not rule out the possibility that a more anterior position of this feature is a modern human autapomorph.

In terms of the ontogeny of this feature, most researchers argue that the Neandertal mental foramen does not assume this relatively posterior position until mandibular growth is nearly completed and the entire permanent dentition erupted (Nara, 1994; Tillier, 1986). In fact, prior to adulthood, the Neandertal mental foramen can appear as far anteriorly as the deciduous canine level of the corpus, indicating that this feature undergoes a posterior migration relative to the tooth row during the course of postnatal growth and development (Coqueugniot, 1999). Apparently, at least some ancient members of the genus *Homo* (e.g., *H. erectus*) shared this pattern of mandibular growth with Neandertals, further strengthening the argument that this may be a shared ancestral trait and clearly not a Neandertal specialization (Coqueugniot, 2000).

In addition to its generally more posterior position in relation to modern humans, the Neandertal mental foramen is often expressed as multiple apertures, a trait noted by some of the earliest students of the hominid fossil record (Gorjanović, 1906; Hamy, 1889; Hrdlička, 1930; Martin, 1926; and others). This situation can and does occur in modern humans (Azaz & Lustman, 1973) but a single opening is by far the most common modern adult morphology (Hauser & de Stefano, 1989). However, modern infants, particularly newborns, are quite another matter. At birth, the mental foramen, often multiple, opens below the dental lamina of the unerupted deciduous canine or first premolar (Williams *et al.*, 1995: 578). Between the first and second postnatal years, the opening(s) of the aperture(s) change(s) direction. Initially these are directed anteriorly, then gradually become posterosuperior in orientation until finally they coalesce into a single opening that assumes its adult position as the aperture turns to face directly posteriorly. These changes must result from remodeling that accompanies mandibular elongation in postincisal length as the neurovascular structures are dragged relatively posteriorly during the first few years of life (Warwick, 1950). In addition, we speculate that remodeling of this region in modern humans during growth and development results in a relative medial displacement of the mental foramen, so that it would eventually transmit only the main trunk of the mental nerve, artery, and vein rather than their peripheral branches, thus reducing the number of bony apertures necessary to convey the neurovascular bundle.

These examples illustrate how anterolateral corpus growth and development could influence the number, position, and orientation of the mental foramen.

Table 14.1 *Sources of recent comparative data*

I. Osteological remains of children of known age at death, eighteenth to twentieth
     centuries AD[a]
   Musée de l'Homme, Paris
   Musée Anatomique Delmas-Orfila-Rouvière, Paris
   Department of Anthropology, Coimbra University, Coimbra
   Christ Church Crypt, Spitalfields, London
II. Archaeological and laboratory collections: 2000 BC to twentieth century AD
   Rajhrad,[a] Narodni Museum, Praha
   Libben Collection,[b] Dept. Sociology and Anthropology, Kent State University, Ohio
   Morton Collection,[a] Tepe Hissar[a, b] and Hasanlu,[a] Iran, and miscellaneous laboratory
   specimens,[a, b] Dept. Anthropology, University of Pennsylvania, Philadelphia
   Saint-Chéron,[a] Chartres
   Saint-Martin de Cognac,[a] Charente
   Saint-Etienne,[a] Toulouse

[a] Non-metric data collected by HC.
[b] Metric data collected by NM-P.

In this paper, we expand upon Coqueugniot's (2000) examination of the non-metric frequencies of mandibular foramen distribution in earlier hominids compared with modern humans and investigate the possibility that the differences in mental foramen number and position between Neandertals and modern humans might be related to distinct patterns of ontogenetic metric change in the anterior mandible of these groups. Ultimately, such information might assist attempts to understand the evolution of this region of the face, particularly as it relates to the appearance of the chin.

Materials

Recent specimens

Data collected separately by both authors from several large skeletal samples of recent, immature *Homo sapiens*, free of developmental pathology, were used in this study. Recent human data utilized in the non-metric analyses were taken by HC on recent European material comprised of osteological collections of subadults of known age ($n = 221$) and from several archaeological samples ($n = 540$). Recent human metric data were collected by NM-P from several sources for a combined sample of comparative metric data ($n = 84$). In several instances both non-metric and metric data were collected from the same sources, all of which are summarized in Table 14.1. For more details regarding these collections, see Coqueugniot (1999) and Minugh-Purvis (1988).

Table 14.2 *Neandertal subadults used in this study*

Specimen	Age group[a]	Approximate age at death (years)[b]	Primary reference(s)
Amud 7[c]	2	<1	Schwartz & Tattersall, 2000
Archi[c, d]	2	3–4[b]	Ascenzi & Segre, 1971; Mallegni & Trinkaus, 1997
Barakai[c]	2	3	Faerman *et al.*, 1994
Châteauneuf I[c]	2	3.5[b]	Patte, 1957
Dederiyeh 1[c]	2	2	Akazawa *et al.*, 1995
Dederiyeh 2[c]	2	2	Akazawa *et al.*, 1999
Le Molare I[c]	2	3–4	Mallegni & Ronchitelli, 1987
Pech de l'Azé[c, d]	2	2.5–3[b]	Patte, 1957
Roc de Marsal[d, e]	2	3[b]	Tillier, 1983; Madre-Dupouy, 1992
Combe Grenal I[c, d]	3	7.5[b]	Genet-Varcin, 1982
Devil's Tower[d, e]	3	4.5–5[b]	Tillier, 1982
Le Fate II[c, f]	4	8[b]	Giacobini & de Lumley, 1983
L'Hortus II[c, d]	4	9[b]	de Lumley, 1973
Malarnaud 1[c, d]	4	11–13	Heim & Granat, 1995
Montgaudier[c]	4	12.5–14.5	Duport & Vandermeersch, 1976; Mann & Vandermeersch, 1997
Le Moustier 1[c, g]	4	15.5–16[b]	Thompson & Bilsborough, 1997
Sclayn 4 A-I/9[e]	4	10–11	Toussaint, 1996; Toussaint *et al.*, 1998
Šipka 1	4	10	Minugh-Purvis, 2000
Teshik Tash[c, d]	4	10–10.5[b]	Gremyatskij, 1949
Zaskalnaya VI[c, g]	4	9–10[b]	Kolossov *et al.*, 1975

[a] For details of age group boundaries see text.
[b] Ages marked with [b] from Minugh-Purvis (1988); otherwise, approximate age in years taken from primary reference(s).
[c] Non-metric data taken from primary reference(s).
[d] Metric data collected directly from original by NM-P.
[e] Non-metric data collected directly from original by HC.
[f] Metric data taken from primary reference(s).
[g] Metric data collected from a primary cast by NM-P.

Fossil specimens

Forty-nine Upper Pleistocene and late Middle Pleistocene subadults from Europe, Asia, and the Levant provided the immature fossil data used in this study. This sample includes Neandertals ($n = 20$, Table 14.2), early anatomically modern (Skhul/Qafzeh) ($n = 4$, Table 14.3), and Upper Paleolithic associated Europeans ($n = 25$, Table 14.4). For the exact position and number of mental foramina in the fossil specimens, see Coqueugniot (2000). The majority of metric data were collected by NM-P (see Minugh-Purvis, 1988).

Table 14.3 *Skhul/Qafzeh early anatomically modern subadults used in this study*

Specimen	Age group[a]	Approximate age at death (years)[b]	Primary reference(s)
Skhul 1[c, d]	2	4.5[b]	McCown & Keith, 1939
Qafzeh 10[d]	3	6	Tillier, 1999
Qafzeh 4[c, d]	3	6.5[b]	Tillier, 1979
Qafzeh 11[c, d]	4	12[b]	Tillier, 1984

[a] For age group boundaries see text.
[b] Ages marked with [b] from Minugh-Purvis (1988); otherwise, approximate age in years taken from primary reference(s).
[c] Metric data collected directly from original by NM-P.
[d] Non-metric data taken from primary reference(s).

Methods

For the non-metric portion of this study, all subadults, fossil and modern, were classified into four age groups based on visual assessment of the dentition: Age Group 1 (newborn without dentition at any phase of eruption); Age Group 2 (subadults with deciduous dentition erupting or in place); Age Group 3 (subadults with first permanent molar erupted); and Age Group 4 (subadults with permanent teeth in addition to the M1 erupted). Unfortunately, no fossil specimens from the first age group were found suitable for this study. For the metric portion of this study, fossil and modern subadults were aged by combining radiographic assessments of dental calcification status and visual assessment of tooth formation and eruption as outlined in Minugh-Purvis (1988).

For all specimens, the number and position of mental foramina relative to the tooth row were recorded on each hemi-mandible with the result that complete mandibles were scored twice as outlined in Coqueugniot (2000). To compare the distribution of mental foramen position, a non-metric feature, between fossil and modern specimens, the bilateral Fisher test was used. This statistical approach provides an exact probability of the correlation between the tested groups. If the probability $P < 0.05$, then we can reject the null hypothesis that the distribution of mental foramen position is independent of classification.

In attempting to examine whether any of the variations in mental foramen number and position coincide with growth of the anterior mandible, we also studied metric aspects of the mandibular anterior alveolar arch for many of the fossil and comparative specimens. The anterior alveolar arch is that portion of

Table 14.4 *Upper Paleolithic associated subadults used in this study*

Specimen	Age group[a]	Approximate age at death (years)[b]	Primary reference(s)
Baoussé Roussé GE1[c]	2	3	Henry-Gambier, 2001
Baoussé Roussé GE2[c]	2	1.5	Henry-Gambier, 2001
Le Figuier[c]	2	3	Billy, 1979
Fontéchevade[c]	2	5–6	Gambier, 1989
Isturitz 65[c]	2	1–1.5	Gambier, 1990/1
Isturitz 67[c]	2	3	Gambier, 1990/1
Isturitz 118[d]	2	2–3	Gambier, 1990/1
Isturitz 119[d]	2	4–5	Gambier, 1990/1
Lagar Velho[c]	2	4	Duarte *et al.*, 1999
La Madeleine 4[d]	2	3	Heim, 1991
Isturitz 64[c, e]	3	6	Gambier, 1990/1
Isturitz 116[d]	3	6–7	Gambier, 1990/1
Isturitz 68[c, e]	3	7	Gambier, 1990/1
Předmostí 2[c]	3	7–8[b]	Matiegka, 1934
Kostenki 3[e]	–	6–7[b]	Yakimov, 1957
Isturitz 66[c]	4	9	Gambier, 1990/1
Kostenki 4[e]	–	10[b]	Debetz, 1961
Ksar' Akil 1[c]	4	8	Bergman & Stringer, 1989
Miesslingtal[e]	–	9–10[b]	Szombathy, 1950
Le Morin 1[c]	4	11–12	Bouvier, 1971
Předmostí 25[c]	4	10[b]	Matiegka, 1934
Les Rois A[c, e]	4	10.5[b]	Gambier, 1989
Romanelli 7[c]	4	14	Fabbri, 1987
Sungir' 2[c, e]	4	11.5[b]	Minugh-Purvis, pers. obs.; Tillier, unpublished data
Sungir' 3[c, e]	4	10.5[b]	Minugh-Purvis, pers. obs.; Tillier, unpublished data

[a] For age group boundaries see text.
[b] Ages marked with[b] from Minugh-Purvis (1988); otherwise, approximate age in years taken from primary reference(s).
[c] Non-metric data taken from primary reference(s).
[d] Non-metric data collected directly from original by HC.
[e] Metric data collected directly from original by NM-P.

the jaw housing the deciduous then successional dentition, i.e., the incisors, canines and premolars. Specifically, we examined two metric dimensions: anterior alveolar arch breadth, defined as the external transverse breadth of the mandibular alveolus at the distal edge of the dm_2/P_2; and anterior alveolar arch length, defined as the anterior–posterior length from the labial surface of the symphyseal alveolus to its intersection with the line defining anterior alveolar arch breadth.

Table 14.5 *Distribution of mental foramen position in modern children*

	dc–dm$_1$ C–P$_1$	dm$_1$ P$_1$	dm$_1$–dm$_2$ P$_1$–P$_2$	dm$_2$ P$_2$	dm$_2$–M$_1$ P$_2$–M$_1$
Age Group 1	150	72	0	0	0
Age Group 2	74	76	12	0	0
Age Group 3	20	144	60	0	0
Age Group 4	3	56	269	30	2

Table 14.6 *Distribution of mental foramen position in Neandertal children*

	dc–dm$_1$ C–P$_1$	dm$_1$ P$_1$	dm$_1$–dm$_2$ P$_1$–P$_2$	dm$_2$ P$_2$	dm$_2$–M$_1$ P$_2$–M$_1$
Age Group 2	1	9	6	0	0
Age Group 3	1	3	3	1	0
Age Group 4	1	1	7	6	4

Table 14.7 *Distribution of mental foramen position in children of the Skhul/Qafzeh early anatomically modern group*

	dc–dm$_1$ C–P$_1$	dm$_1$ P$_1$	dm$_1$–dm$_2$ P$_1$–P$_2$	dm$_2$ P$_2$	dm$_2$–M$_1$ P$_2$–M$_1$
Age Group 2	0	0	1	0	0
Age Group 3	0	0	3	0	0
Age Group 4	0	0	2	0	0

Results

Mental foramen position relative to the tooth row

The distribution of mental foramen position is presented for the modern (Table 14.5), Neandertal (Table 14.6), early anatomically modern Skhul/Qafzeh group (Table 14.7), and Upper Palaeolithic samples (Table 14.8). Coqueugniot (1999, 2000) found that in modern humans, the mental foramen acquires a more and more posterior position relative to the tooth row during growth. At

Table 14.8 *Distribution of mental foramen position in Upper Paleolithic children*

	dc–dm_1 C–P_1	dm_1 P_1	dm_1–dm_2 P_1–P_2	dm_2 P_2	dm_2–M_1 P_2–M_1
Age Group 2	3	10	0	0	0
Age Group 3	0	1	4	0	0
Age Group 4	0	4	6	1	0

birth, the mental foramina most frequently open at a level below the dc to dm_1 (67.6%) or below the dm_1 root sockets (32.4%). When the permanent dentition is complete, the mental foramen is observed below the P_1–P_2 (74.7%), P_2 (8.3%), or sometimes below the P_2–M_1 (0.6%) alveolus. Apparently, growth of the mandibular corpus contributes to this increasingly posterior positioning of the mental foramen during ontogeny.

Compared to modern humans, Age Group 2 Neandertals have mental foramina in a significantly more posterior location (Table 14.6). In that group, the foramen usually occurs under the dm_1 (56.25%) or below the dm_1–dm_2 level of the tooth row (37.5%) while in the Age Group 2 modern subadults it is positioned under the dc–dm_1 (45.7%) or dm_1 (46.9%). In Age Group 3, the Neandertal mental foramen is again more posterior than in modern humans, occurring underneath the alveolus between the dm_1–dm_2 interspace in 37.5% of the sample and under the dm_2 in 12.5% of hemi-mandibles. This frequency of posteriorly located foramina far exceeds the pattern found in modern specimens for whom 64.3% occur underneath the dm_1, 26.8% underneath the dm_1–dm_2 interspace, and 0% underneath the dm_2 alveolus. Finally, in Age Group 4, the mental foramen is usually below the P_1–P_2 (36.8%), P_2 (31.6%), or P_2–M_1 (21%) in the Neandertals. In contrast, in modern adolescents 74.7% were below the P_1–P_2, but only 8.3% and 0.6% below the P_2 and P_2–M_1 alveoli respectively.

The difference between Neandertals and modern humans in frequencies for all three age categories was found to be significant at the 0.05 level (Coqueugniot, 2000). Thus, posterior localization of the mental foramen in Neandertals appears earlier in ontogeny than has been previously assumed.

An examination of the Upper Paleolithic associated European material reveals a somewhat different pattern (Table 14.8). In Age Group 2, 23% of mental foramina were inferior to the dc–dm_1 alveolus, while the majority (77%) were located inferior to the dm_1. This latter is the one statistically significant example of the mental foramen being more posterior in Upper Paleolithic than in the modern samples. In Upper Paleolithic Age Group 3, 20% of mental foramina

Table 14.9 *Distribution of the number of mental foramina per hemi-mandible by age and phylogeny*

Age group	Modern				Neandertal			Skhul/Qafzeh (early anatomically modern)			Upper Paleolithic		
	1	2	3	4	2	3	4	2	3	4	2	3	4
One foramen	176	141	214	328	10	0	5	1	3	2	13	5	11
Two foramina	36	17	10	30	4	1	6	0	0	0	0	0	0
Three or more foramina	2	0	0	0	0	2	1	0	0	0	0	0	0

were located under the dm_1 position while 80% lie underneath the dm_1–dm_2 alveolar interface – a slightly more posterior position than in the moderns, but not a statistically significant one. For Age Group 4, the Upper Paleolithic specimens include 36.3% of all mental foramina underneath the P_1 position; 54.5% underneath the P_1–P_2, and 9.1% underneath the P_2, illustrating a continued posterior movement of the foramen. However, without any statistically significant differences from the modern human distribution this cannot be regarded as the same pattern of increasing posterior migration of the mental foramen through growth as seen in the Neandertals (Coqueugniot, 2000). Finally, although it must be noted that the number of subadults from the Skhul/Qafzeh group is insufficient to statistically compare the distribution of their mental foramen position with those of the modern sample, 100% of all their mental foramina, for Age Groups 2, 3, and 4, occur underneath the dm_1–dm_2 (P_1–P_2) alveolus (Table 14.7). This contrasts markedly with the Neandertal pattern in which 12.5% of Age Group 3 individuals and 52.6% of Age Group 4 individuals exhibit mental foramina at or more posterior than underneath the P_2 position.

Number of mental foramina

In comparing the number of mental foramina per hemi-mandible (Table 14.9), Coqueugniot (1999) found that Neandertals consistently exhibit a higher incidence of multiple mental foramina than modern humans. Among 214 recent human infants, 18% exhibited multiple mental foramina prior to the eruption of the deciduous teeth. However, by the time the deciduous dentition was fully erupted, only 11% possessed more than one aperture, which is essentially the same frequency as in the adult members of these samples.

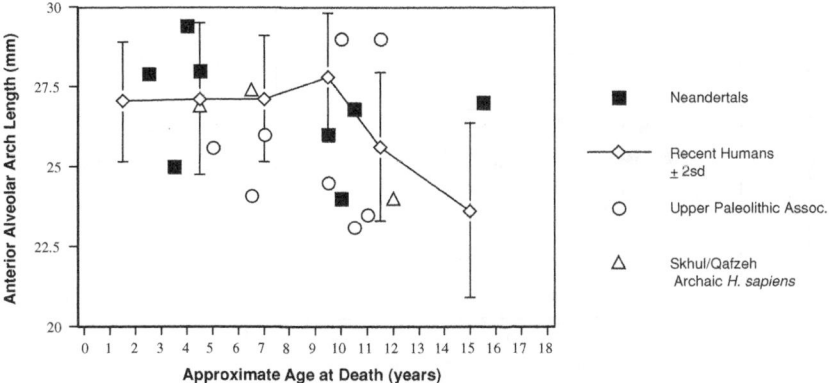

Figure 14.2   Anterior alveolar arch length through growth.

Although no neonatal Neandertal mandibles are sufficiently well preserved for comparison with the youngest modern age group, 29% of Neandertals in Age Group 2 exhibit more than one aperture. In Age Group 3, 100% had more than one opening at the site of the mental foramen, while multiple mental foramina were present in 58% of the Neandertals from Age Group 4. In contrast, in a sample of 22 Upper Paleolithic associated subadults representing the same three developmental intervals, 100% possessed only a single mental foramen.

Growth in anterior alveolar arch length

Following the eruption of all the deciduous teeth, anterior alveolar arch length changes very little through mid childhood in modern individuals (Figure 14.2). However, upon eruption of the permanent premolars, which occupy less mesiodistal (M–D) alveolar space than their deciduous predecessors, the M–D length of the tooth row, from incisors through the premolars, and the corpus housing it, actually shrinks (Weidenreich, 1936). This effect is seen as a decrease in anterior alveolar arch length in late childhood and adolescence among recent modern individuals.

Looking at Neandertal anterior alveolar arch length, we found a decline in anterior alveolar arch length at the same developmental intervals (Figure 14.2) as in the modern comparative sample. This pattern is also seen in the Upper Paleolithic associated specimens as well as in the Skhul/Qafzeh subadults (Figure 14.2). We acknowledge, however, that the extremely small sample sizes available preclude statistical analysis of this pattern at meaningful developmental intervals so that it is not possible to test whether these differences are significant.

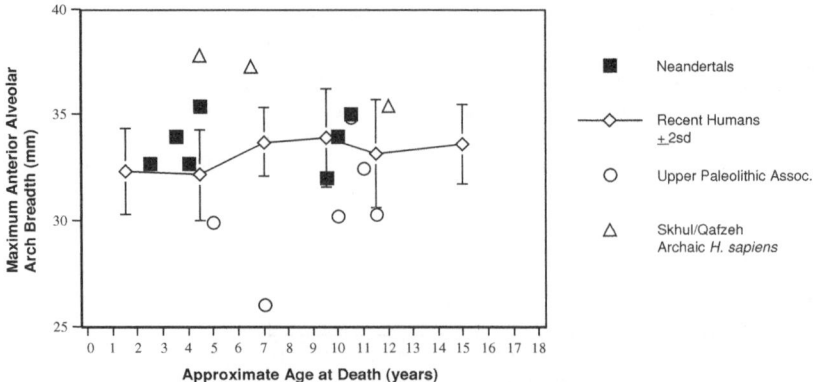

Figure 14.3   Anterior alveolar arch breadth through growth.

Growth in anterior alveolar arch breadth

Anterior alveolar arch breadth in modern humans appears to be rather dynamic prior to mandibular maturity. Initially, this dimension increases gradually followed by a dramatic increase from early to mid childhood as the developing permanent canines and premolars enlarge within their bony crypts.

However, as these teeth begin to erupt, anterior alveolar arch breadth shows a decreasing trend in the recent modern humans. In Neandertals, despite the wide scatter (Figure 14.3), there is no evidence of similar change. Instead, anterior alveolar arch breadth remains rather consistent following early childhood, leading us to suspect that less remodeling characterized the Neandertal as compared to the modern lateral mandibular corpus following early childhood. We note that the Upper Paleolithic associated specimens more closely follow the modern pattern with a decrease in anterior alveolar arch breadth in late childhood, suggesting that they experienced lateral corpus remodeling more similar to that of modern individuals (Figure 14.3).

Discussion

Metric growth of the anterior portions of the mandible is highly dynamic due, in part, to the close relationship between the developing dentition and the alveolar bone housing it (van der Linden & Duterloo, 1976). Initially the alveolus must accommodate encrypted deciduous dental germs, which, upon eruption, create a temporary vacancy until this space is again filled by the growing germs of the successional teeth. Presumably, the larger the size of the encrypted developing crowns the more metric dimensions of the local alveolar bone will enlarge to accommodate them. Following exfoliation of the deciduous dentition, the alveolus is again vacant except for the permanent dental roots.

Both growing Neandertals and modern humans experience a lengthening of the anterior mandibular corpus, which our results suggest causes a posterior displacement of the mental foramen. However, it appears that, generally speaking, differences in dental volume alone do not entirely explain the differences characterizing Neandertals and modern humans in both number and position of mental foramina (Coqueugniot, 1999; Coqueugniot & Minugh-Purvis, this study). In modern humans, posterior mental foramen displacement is somewhat offset by the loss of anteroposterior length in the anterior alveolar arch as the large deciduous molars are replaced by their smaller permanent successors in late childhood. However, Neandertal subadults experience a similar loss in anteroposterior anterior alveolar arch length. Yet despite the same initial position relative to the tooth row as in modern infants, the Neandertal mental foramen eventually remodels to a location even more posterior relative to the tooth row. This suggests that a greater amount of *relative as well as absolute* growth in mandibular corpus length occurred during Neandertal, relative to modern human, ontogeny.

This more posterior position of the mental foramen throughout ontogeny cannot be used as a diagnostic trait for Neandertals, since it is already present in earlier groups such as Asian and African *Homo erectus* subadults (Coqueugniot, 2000; Trinkaus, 1993). Among archaic members of *H. sapiens*, the Irhoud 3 subadult from North Africa possesses a single mental foramen equivalent to the position observed in modern subadults of the same dental age (Hublin & Tillier, 1981). In contrast, the North African archaic *H. sapiens* adolescent from Rabat possesses two mental foramina located under the P_2 (Hublin, 1991), a more posterior position compared with modern subadults. Thus, these two specimens add little clarification to our knowledge of mental foramen position and number among archaic members of our species.

The changes in anterior mandibular alveolar breadth during ontogeny, resulting in a narrower transverse dimension in modern humans as opposed to Neandertals, undoubtedly involve resorptive remodeling of the anterolateral mandibular corpus. In modern humans, the resorptive remodeling of this region through postnatal ontogeny obviously displaces and keeps the mental foramen in a more medial location so that it transmits the main trunk of the mental nerve, artery, and vein prior to their peripheral branching. This seems to have eliminated high frequencies of multiple mental foramina in modern humans. Most likely, Neandertal newborns possessed multiple mental foramina as do most modern babies (Warwick, 1950), but our data strongly suggest this region did not narrow metrically in Neandertals as in modern subadults with the result that multiple apertures would have been more frequently retained in Neandertals.

Of all the features of the modern human mandible, the chin is indisputably its most distinctive autapomorph. As such, it is a trait whose presence or absence is often utilized in phylogenetic judgments and taxonomic assignments. However, the relative development of this feature is somewhat obscured at some developmental ages by the bulging of the developing dental germs within the alveolus so that it is often difficult to interpret when attempting to assess the taxonomic affinities of subadult material (Minugh-Purvis, 1988, 2000). As shown in this study, for those developmental intervals where chin development is difficult to assess (Age Group 3 in this study), a single, fairly anteriorly positioned mental foramen is already in place in the majority of subadult *H. s. sapiens*, so that this autapomorph should be regarded as a valuable diagnostic criterion for the identification of modern individuals in the absence of clear expression of the mental eminence.

Summary and conclusions

Numerous evolutionary changes accompanied the transformation of the anterior mandible as the modern human chin emerged during the Upper Pleistocene. This work suggests that a shift in anterior mandibular ontogeny, both in absolute dimensional changes in alveolar arch length and breadth and in the relative changes reflected in the number and position of the mental foramen relative to the tooth row, comprises an alteration in developmental patterning which contributed to the appearance of modern human morphology. Based on our findings here and those of Coqueugniot (2000), we consider the anterior position of a single mental foramen, as seen in our modern, Upper Paleolithic, and Skhul/Qafzeh sample, to be a modern autapomorph in both adults as well as subadults, unlike the chin which is difficult to fully evaluate in subadults at certain stages of the mixed dentition (Minugh-Purvis, 1988, 2000). On the other hand, as noted by Trinkaus (1993) and Coqueugniot (2000), it appears that multiple mental foramina, and a more posterior positioning of those foramina, should be regarded as plesiomorphic as they are found in Neandertals and other, more archaic members of *Homo*. Finally, these plesiomorphs apply equally to immature as well as to adult members of our genus (Coqueugniot, 2000).

Acknowledgments

We thank all the persons who so generously permitted us to examine the specimens in their care, including the late H. Bach, O. Bar-Yosef, B. Boissavit-Camus, D. Castex, J.-J. Cleyet Merle, É. Crubézy, E. Cunha, M. Dobisikova, L. Duport, J.-L. Heim, J.-J. Hublin, J. Jelínek, D. Joly, A. Kozinov, A. Langaney, J.-P. Lassau, C.O. Lovejoy, H. de Lumley, M.-A. de Lumley, A.E. Mann,

T. Molleson, J.M. Monge, M.M. Phillipe, D. Pillbeam, M. Sansilbano-Collilieux, P. Smith, M. Stloukal, C.B. Stringer, J. Szilvássy, M.-H. Thiault, A.-m. Tillier, B. Vandermeersch, and A. Zubov.

References

Akazawa, T., Muhesen, S., Dodo, Y., Kondo, O., Mizoguchi, Y., Abe, Y., Nishiaki, Y., Ohta, S., Oguchi, T., & Haydal, J. (1995). Neanderthal infant burial from the Dederiyeh Cave in Syria. *Paléorient,* **21**, 77–86.

Akazawa, T., Muhesen, S., Ishida, H., Kondo, O., & Griggo, C. (1999). New discovery of a Neanderthal child burial from the Dederiyeh Cave in Syria, *Paléorient,* **25**, 129–142.

Arensburg, B., Kaffe, I., & Littner, M. (1989). The anterior buccal mandibular depressions: Ontogeny and phylogeny. *American Journal of Physical Anthropology,* **78**, 431–437.

Arsuaga, J. L., Martinez, I., Gracia, A., & Lorenzo, C. (1997). The Sima de los Huesos crania (Sierra de Atapuerca, Spain): A comparative study. *Journal of Human Evolution,* **33**, 219–281.

Ascenzi, A., & Segre, A. G. (1971). A new Neandertal child mandible from an Upper Pleistocene site in southern Italy. *Nature,* **233**, 280–283.

Azaz, B., & Lustman, J. (1973). Anatomical configurations in dry mandibles. *British Journal of Oral Surgery,* **11**, 1–9.

Bergman, C. A., & Stringer, C. B. (1989). Fifty years after Egbert an early Upper Palaeolithic juvenile from Ksar Akil, Lebanon. *Paléorient,* **15**, 99–111.

Billy, G. (1979). L'Enfant magdalénien de la Grotte du Figuier (Ardèche). *L'Anthropologie (Paris),* **83** (2), 223–252.

Bouvier, J.-M. (1971). Les Mandibules humaines du Magdalénien français. Doctoral thesis, Université Paris VII, Paris.

Condemi, S. (1991). Some considerations concerning Neandertal features and the presence of Neandertals in the near east. *Rivista di Antropologia,* **69**, 27–38.

Coqueugniot, H. (1999). *Le Crâne d' Homo sapiens en Eurasie: Croissance et Variation Depuis 100 000 Ans.* British Archaeological Reports International Series no. 822. Oxford: British Archaeological Reports.

Coqueugniot, H. (2000). La Position du foramen mentonnier chez l'enfant: Révision ontogénétique et phylogénétique. *Bulletins et Mémoires de la Société d'Anthropologie (Paris),* n.s., **12** (3–4), 227–246.

Debetz, G. F. (1961). The skull from the Upper Paleolithic burial in Pokrovskij Ravine (Kostenki XVIII). *Kratkie Soobshcheniya Instituta Arkheologgi,* **82**, 120–127.

Defleur, A., White, T., Valensi, P., Slimak, L., & Crégut-Bonnoure, E. (1999). Neanderthal cannibalism at Moula-Guercy, Ardèche, France. *Science,* **286**, 128–131.

de Lumley, M. A. (1973) *Anténéandertaliens et Néandertaliens du Bassin Méditerranéen Occidental Européen,* Etudes quaternaires no. 2. Paris: Editions du CNRS.

Duarte, C., Mauricio, J., Pettitt, P. B., Souto, P., Trinkaus, E., van der Plicht, H., & Zilhao, J. (1999). The early Upper Paleolithic human skeleton from the Abrigo do Lagar Velho (Portugal) and modern human emergence in Iberia. *Proceedings of the National Academy of Sciences of the USA,* **96**, 7604–7609.

Duport, L., & Vandermeersch, B. (1976). La Mandibule moustérienne de Montgaudier (Montbron, Charente). *Comptes Rendus de l'Académie des Sciences de Paris,* **283**, 1161–1164.

Fabbri, P. F. (1987). Restes humains retrouvés dans la grotte Romanelli (Lecce, Italie): Etude anthropologique. *Bulletins et Mémoires de la Société d'Anthropologie (Paris),* **4**, série **XIV** (4), 219–248.

Faerman, M., Zilberman, U., Smith, P., Kharitonov, V., & Batsevitz, V. (1994). A Neanderthal infant from the Barakai cave, Western Caucasus. *Journal of Human Evolution,* **27**, 405–415.

Gambier, D. (1989). Fossil hominids from the early Upper Palaeolithic (Aurignacian) of France. In *The Human Revolution: Behavioural and Biological Perspectives on the Origin of Modern Human,* eds. P. Mellars & C. Stringer, pp. 194–211. Edinburgh: Edinburgh University Press.

Gambier, D. (1990/1). Les Vestiges humains du gisement d'Isturitz (Pyrénées-Atlantiques): Etude anthropologique et analyse des traces d'action humaine intentionnelle. *Antiquités Nationales,* **22/3**, 9–26.

Genet-Varcin, E. (1982). Vestiges humains du Würmien inférieur de Combe Grenal, commune de Domme (Dordogne). *Annales de Paléontologie,* **68** (2), 133–169.

Giacobini, G., & de Lumley, M.-A. (1983). Restes humains néandertaliens de la Caverna delle Fate (Finale, Ligurie Italienne). *L'Anthropologie (Paris),* **87** (1), 142–144.

Golovanova, L. V., Hoffecker, J. F., Kharitonov, V. M., & Romanova, G. P. (1999). Mezmaiskaya Cave: A Neanderthal occupation in the Northern Caucasus. *Current Anthropology,* **40** (1), 77–86.

Gorjanović, D. (1906). *Der Diluviale Mensch von Krapina in Kroatien: Ein Beitrag zur Paleoanthropologie.* Wiesbaden: Kreidels.

Gremyatskij, M. M. (1949). Cerep rehenka neandertalsa iz grota teshik-tash, ioujnil ouzbekistan. In *Teshik-Tash, Paleoliticeskij celovek* ed M. M. Gremyatskij & M. F. Nesturkh, pp. 137–181. Moscow: Trudy naucno-issledovatel'skogo instituta antropologii, Moskva, Izdatel'stvo Moskovskogo gosudarstvennogo universiteta. (In Russian.)

Hamy, E. T. (1889). *Nouveaux Matériaux pour Servir à l'Etude de la Paléontologie Humaine. Paris:* Congrès International d'Anthropologie et d'Archaéologie Préhistoriques.

Hauser, G., & de Stefano, G. F. (1989). *Epigenetic Variants of the Human Skull.* Stuttgart: Schweizerbart.

Heim, J.-L. (1991). L'Enfant magdalénien de La Madeleine. *L'Anthropologie (Paris),* **95**, 611–638.

Heim, J.-L., & Granat, J. (1995). La Mandibule de l'enfant néandertalien de Malarnaud (Ariège): Une nouvelle approche anthropologique par la radiographie et la tomodensitométrie. *Anthropologie et Préhistoire,* **106**, 75–96.

Henry-Gambier, D. (2001). *La Sépulture des Enfants de Grimaldi (Baoussé Roussé, Italie): Anthropologie et Palethnologie Funéraire des Populations de la Fin du Paléolithique Supérieur.* Paris: Editions du Comité des Travaux Historiques et Scientifiques, Réunion des Musées Nationaux.

Hrdlička, A. (1930). *The Skeletal Remains of Early Man,* Smithsonian Miscellaneous Collections no., **83**. Washington, DC: Smithsonian Institution.

Hublin, J.-J. (1991). L'Emergence des *Homo sapiens* archaïques: Afrique du Nord-Ouest et Europe occidentale. Doctoral thesis, Université Bordeaux 1, Bordeaux.

Hublin, J.-J., & Tillier, A.-m. (1981). The Mousterian juvenile mandible from Irhoud (Morocco): A phylogenetic interpretation. In *Aspects of Human Evolution*, Symposia of the Society for the Study of Human Biology no. **21**, ed. C. B. Stringer, pp. 167–185. London: Taylor & Francis.

Kolossov, Y. G., Kharitonov, V. M., & Yakimov, V. P. (1975). Palaeoanthropic specimens from the site Zaskalnaya VI in the Crimea. In *Palaeoanthrology, Morphology and Palaeoecology*, ed. R. H. Tuttle, pp. 419–428. The Hague: Mouton.

Le Gros Clark, W. E. (1964). *The Fossil Evidence for Human Evolution*, 2nd edn. Chicago: University of Chicago Press.

van der Linden, F. P. G. M., & Duterloo, H. S. (1976). *Development of the Human Dentition: An Atlas*. Hagerstown: Harper & Row.

Madre-Dupouy, M. (1992). *L'Enfant du Roc de Marsal: Etude Analytique et Comparative*. Paris: Cahiers de Paléoanthropologie, Editions du CNRS.

Mallegni, F., & Ronchitelli, A. (1987). Découverte d'une mandibule néandertalienne à l'abri du Molare près de Scario (Salerno-Italie): Observations stratigraphiques et palethnologiques. I: Étude anthropologique. *L'Anthropologie (Paris)*, **91** (1), 163–174.

Mallegni, F., & Trinkaus, E. (1997). A reconsideration of the Archi 1 Neandertal mandible. *Journal of Human Evolution*, **33**, 651–668.

Mann, A. E., & Vandermeersch, B. (1997). An adolescent female Neandertal mandible from Montgaudier Cave, Charente, France. *American Journal of Physical Anthropology*, **103**, 507–527.

Martin, H. (1926). Machoire humaine moustèrienne trouvée dans la station de La Quina. *L'Homme Préhistorique*, **22**, 3–21.

Matiegka, J. (1934). *Homo Predmostensis: Fosilny človek z Předmostí na Moravě. I: Lebky. (L'Homme fossile de Predmosti en Moravie. I: Les Crânes)*. Prague: Česka Akademie Věd a Umeni. (In Czech and French.)

McCown, T., & Keith, A. (1939). *The Stone Age of Mount Carmel*, vol. 2, *The Fossil Human Remains from the Levalloiso-Mousterian*. Oxford: Clarendon Press.

Minugh-Purvis, N. (1988). Patterns of craniofacial growth and development in Upper Pleistocene hominids. PhD dissertation, University of Pennsylvania, Philadelphia.

Minugh-Purvis, N., (2000). Ontogeny and morphology of the child's mandible from Šipka – Moravia, Czech Republic. *Anthropologie*, **38**, 71–82.

Minugh-Purvis, N., & Lewandowski, J. (1992). Functional anatomy, ontogeny, and behavioral implications of coronoid process morphology in Upper Pleistocene hominines. *American Journal of Physical Anthropology*, Suppl. **14**, 124–125.

Nara, T. (1994). Etude de la variabilité de certains caractères métriques et morphologiques des Néandertaliens. Doctoral thesis, Université Bordeaux 1, Bordeaux.

Patte, E. (1957). *L'Enfant néandertalien du Pech de l'Azé*. Paris: Masson.

Rak, Y., Ginzburg, A., & Geffen, E. (2002). Does *Homo neanderthalensis* play a role in modern human ancestry? The mandibular evidence. *American Journal of Physical Anthropology*, **119**, 199–204.

Schwartz, J. H., & Tattersall, I. (2000). The human chin revisited: What is it and who has it? *Journal of Human Evolution*, **38**, 367–409.

Stringer, C. B., Hublin, J.-J., & Vandermeersch, B. (1984). The origin of anatomically modern humans in Western Europe. In *The Origins of Modern Humans,* eds. F.H. Smith & F. Spencer, pp. 51–135. New York: Alan R. Liss.

Szombathy, J. (1950). Der menschliche Unterkiefer aus dem Miesslingtal bei Spitz, N-Ö. *Archaeologia Austriaca,* **5,** 1–5.

Tebo, H. G., & Telford, I. R. (1950). An analysis of the relative positions of the mental foramina. *Anatomical Record,* **106,** 254.

Thompson, J. L., & Bilsborough, A. (1997). The current state of the Le Moustier 1 skull. *Acta Praehistorica et Archaeologica,* **29,** 17–38.

Tillier, A.-m. (1979). Restes crâniens de l'enfant moustérien Homo 4 de Qafzeh (Israël): La mandibule et les maxillaires. *Paléorient,* **5,** 67–85.

Tillier, A.-m. (1981). Evolution de la région symphysaire chez les *Homo sapiens* juvéniles du Paléolithique moyen: Pech de l'Azé, Roc de Marsal et La Chaise 13. *Comptes Rendus de l'Académie des Sciences de Paris,* **293,** 724–727.

Tillier, A.-m. (1982). Les Enfants néanderthaliens de Devil's Tower (Gibraltar). *Zeitschrift für Morphologie und Anthropologie,* **73,** 125–148.

Tillier, A.-m. (1983). L'Enfant néanderthalien du Roc de Marsal (Campagne du Bugue, Dordogne): Le squelette facial. *Annales de Paléontologie,* **69,** 137–149.

Tillier, A.-m. (1984). L'Enfant Homo 11 de Qafzeh (Israël) et son apport à la compréhension des modalités de la croissance des squelettes moustériens. *Paléorient,* **10,** 7–48.

Tillier, A-m. (1986). Quelques aspects de l'ontogenèse du squelette cranien des Néanderthaliens. In *Fossil Man: New Facts, New Ideas,* eds. V. V. Novotny & A. Mizerova, *Anthropos (Brno),* **23,** 207–216.

Tillier, A.-m. (1999). *Les Enfants moustériens de Qafzeh: Interprétations Phylogénétique et Paléoauxologique,* Cahiers de Paléoanthropologie. Paris: Editions du CNRS.

Toussaint, M. (1996). D'Engis à Sclayn, les Néandertaliens mosans. In, *Neandertal: Catalogue d'Exposition,* ed. D. Bonjean, pp. 48–70. Sclayn: Editions ASBL, Archéologie Anennaise.

Toussaint, M., Otte, M., Bonjean, D., Bocherens, H., Falgueres, C., & Yokoyama, Y. (1998). Les Restes humains néandertaliens immatures de la couche 4A de la grotte Scladina (Andenne, Belgique). *Comptes Rendus de l'Académie des Sciences de Paris,* **326,** 737–742.

Trinkaus, E. (1993). Variability in the position of the mandibular mental foramen and the identification of Neandertal apomorphies. *Rivista di Antropologia,* **71,** 259–274.

Vlček, E. (1969). *Neanderthaler der Tschechoslowakei.* Prague: Böhlau.

Warwick, R. (1950). The relation of the direction of the mental foramen to the growth of the human mandible. *Journal of Anatomy,* **84,** 116–120.

Weidenreich, F. (1936). The mandible *of Sinanthropus pekinensis*: A comparative study. *Paleontologia Sinica,* Series D, **7**(3), 1–150.

Williams, P. L., Bannister, L. H., Berry, M. M., Collins, P., Dyson, M., Dussek, J. E., & Ferguson, M. W. J. (1995). *Gray's Anatomy: The Anatomical Basis of Medicine and Surgery,* 38th edn. New York: Churchill Livingstone.

Yakimov, V. P. (1957). The Upper Paleolithic child from the burial at Gorodtsof's site at Kostenki. *Sbornik Muzeya Antropologiiy Ethnografii,* **17,** 500–529. (In Russian.)

15 A new approach to the quantitative analysis of postcranial growth in Neandertals and modern humans: Evidence from the hipbone

T. MAJÓ
Université Bordeaux 1

A-M. TILLIER
Université Bordeaux 1

Introduction

Growth and development studies in Paleolithic populations have, for a long time, focused on craniodental remains as there was simply less documentation available for similar studies of the postcranial skeleton. Yet the appendicular skeleton and the hipbone are thought to provide an indication of differences in robusticity and morphology between human groups within the Late Pleistocene hominid sample (e.g., Churchill, 1994; Trinkaus, 1992; Trinkaus *et al.*, 1998). The morphology of the hipbone (note: we follow the *Nomina Anatomica* for the use of appropriate nomenclature, in agreement with Tuttle's note published in 1988) has been used to distinguish Neandertals from their close relatives, anatomically modern humans (e.g., McCown & Keith, 1939; Rak, 1990; Rosenberg, 1988; Stewart, 1960; Trinkaus, 1976). However, none of the Neandertal specimens in Europe is sufficiently preserved to estimate relative sizes of the iliac, ischial, and pubic elements in order to produce accurate analysis of the hipbone morphocomplex. By far the most complete and least distorted specimen within the Middle Paleolithic hominid sample is represented by Kebara 2 from the southern Levant. While the identification of several Near Eastern specimens as parts of the Neandertal sample is still debated (e.g., Arensburg, 1989; Arensburg & Belfer-Cohen, 1998; Mann, 1995; Trinkaus, 1992), the Kebara specimen is usually described as a presumed Neandertal (see Aiello & Dean, 1990: 455).

Patterns of Growth and Development in the Genus Homo, ed. J. L. Thompson, G. E. Krovitz and A. J. Nelson. Published by Cambridge University Press. © Cambridge University Press, 2003.

On the basis of the European fossil record, the Neandertal adult hipbone is primarily described as exhibiting an unusually elongated and slender superior pubic ramus. In addition, it demonstrates a prominent iliac pillar, an extreme protrusion of the anterior iliac spines, a very shallow greater sciatic notch and a large acetabulum. However, evidence from other fossil hominids suggests that a similar pattern of pubic elongation could be observed among early anatomically modern humans from the Levant (i.e., Skhul and Qafzeh hominids). Indeed, according to McCown & Keith (1939: 80) and Arensburg & Belfer-Cohen (1998: 317), one adult hominid from Skhul, Skhul 9, manifests an elongated superior pubic ramus.

Although infant and juvenile skeletons reflect only a specific point in the time trajectory of an individual's development, they have much to contribute in the search for anatomical differences between Neandertals and other fossil hominids. Little work has been done to understand the development of hipbone morphology in Neandertals and their contemporaries, primarily because there is little relevant fossil material (e.g., Heim, 1982; Majó, 1995, 2000; Tillier, 1999; Tompkins & Trinkaus, 1987). The aim of the present contribution is to discuss some aspects of developmental changes in the hipbone of Neandertal and early modern subadults using documented skeletal collections as reference samples. We will examine the growth of the pubis, but will also include other measurements of the hipbone in order to examine the growth of the whole anatomical element.

Materials

Analyses of growth-related changes in the postcranial skeleton of Late Pleistocene hominid samples have relied upon specimens that have been recovered from sites that are geographically and chronologically dispersed. During the last decade discoveries in Syria (Akazawa *et al.*, 1995, 1999), northern Caucasus (Golanova *et al.*, 1999), and Portugal (Duarte *et al.*, 1999) have substantially increased the sample for further investigation in skeletal growth within the Late Pleistocene sample. However, all of these fossil juvenile remains have only been described in a preliminary fashion.

Few sites have provided documentation of hipbone morphology in Neandertal children and early anatomically modern juveniles (Table 15.1). Unfortunately, in most cases the remains are fragmentary, thus limiting the morphometric analysis. Of the sites listed in Table 15.1, only two sites, La Ferrassie in France and Qafzeh in Israel, have delivered individual bones complete enough to be studied metrically and to permit an assessment of hipbone growth in Middle Paleolithic hominids. Furthermore, in both sites, adult

Table 15.1 *Hipbone remains of Middle Paleolithic immature specimens (n = 17) and their age class distribution. Many of these specimens are too fragmentary to produce the measurements used in this study*

	n	Age group (years)	Original reference
European hominids			
Roc de Marsal (France)	1	1–4	Madre-Dupouy, 1992
La Ferrassie (France)	3	0 and 1–4	Heim, 1982
René Simard (France)	1	Unknown	Duport, 1958
Kiik Koba (Russia)	1	1–4	Tillier, unpublished data; Vlček, 1973
Krapina (Croatia)	3	Unknown	Radovčić *et al.*, 1988; Smith, 1976
Mezmaiskaya (Caucasus)	1	0	Golanova *et al.*, 1999
Le Moustier 1 (France)	1	15–19	Herrmann, 1977
African hominids			
Jebel Irhoud (Morocco)	1	Unknown (10–14?)	Tixier *et al.*, 2001
Near Eastern hominids			
Dederiyeh (Syria)	2	1–4	Akazwa *et al.*, 1999; Dodo *et al.*, 1998
Skhul (Israel)	2	1–4	McCown & Keith, 1939
Qafzeh (Israel)	3	0, 1–4, and 5–9	Tillier, 1999

specimens retain isolated portions of the hipbone (La Ferrassie 1) or a complete but seriously distorted pelvic girdle (Qafzeh 9). Such a situation permits us to examine the proportion of adult pubic length achieved by the children. This paper will focus on the analysis of the subadults from La Ferrassie and Qafzeh.

From the La Ferrassie rockshelter located near the city of Le Bugue in Dordogne, several Neandertal human remains were recovered at the beginning of the twentieth century. Among these remains were several infants and children (Heim, 1982) found in the same layer. While this layer has not been dated radiometrically, the geological age and the associated artefacts and fauna are consistent with its assignment to oxygen isotope stage 3. Three of the immature individuals provided information on the development of juvenile hipbones (Table 15.2). From La Ferrassie 4b nearly complete right and left ilia were preserved and attributed by Heim (1982) to a neonate (on the basis of their size and ossification pattern these two iliac bones might belong to two separate individuals). La Ferrassie 8, a child ca. 2 years old, preserves the two nearly complete ilia, a large portion of the two superior pubic rami, and one fragment of the left ischium. The hipbone from the third individual, La Ferrassie 6

Table 15.2 *Number of specimens studied within each age group*

Fossil children	Age at death[a, b]	Hipbone remains[c]
La Ferrassie 4b	Neonate	ilium (f)
La Ferrassie 8	ca. 2 yrs	ilium, ischium (ff), pubis (f)
La Ferrassie 6	ca. 3 years[a] 4–5 yrs[b]	ilium (f), ischium (f), pubis
Qafzeh 13	Neonate	ilium (f)
Qafzeh 21	ca. 3 yrs	ilium (f), ischium, pubis
Qafzeh 10	ca. 6 yrs	ilium, ischium, pubis
Documented samples	Age group	Number of individuals
Spitalfields	under 1 year	22
	1–4 yrs	30
	5–9 yrs	2
	10–14 yrs	3
Coimbra	5–9 yrs	6
	10–14 yrs	12

[a] After Heim (1982).
[b] After Tompkins & Trinkaus (1987).
[c] (f) incomplete bone, (ff) fragmentary bone.

(Figure 15.1), consists of a right pubis and ischium (minus the superior part of the acetabulum), two-thirds of the left ilium, as well as fragments of the left ischium and right ilium. As no teeth are available for this specimen, the age at death has been estimated on the basis of bone maturation and long-bone lengths to be in the range of 3 years (Heim, 1982) to 4–5 years (Tompkins & Trinkaus, 1987).

In the Qafzeh Cave, situated near the city of Nazareth in Israel, a total of 15 individuals were found. Originating from the Mousterian layers XV–XVII, they have been dated by means of thermoluminescence and electron spin resonance to ca. 95 ± 5 ka (Schwarcz *et al.*, 1988; Valladas *et al.*, 1988). This hominid sample consists of anatomically modern-looking people (Tillier, 1999; Vandermeersch, 1981) and exhibits a prevalence of non-adult specimens. Three of the children were examined for the purposes of this study (Table 15.2) and their age at death was estimated on the basis of dental development. The Qafzeh 13 hipbone is represented by a left incomplete ilium of a stillborn or neonate. From Qafzeh 21, a child ca. 3 years old at death, the hipbone is represented by the right ilium, ischium, and pubis, and the upper part of the left iliac wing (Figure 15.2). The last specimen, Qafzeh 10, is attributed to a child ca. 6 years old at death. It is complete, although some portions of the left pubis and ischium are missing (see Figure 15.2).

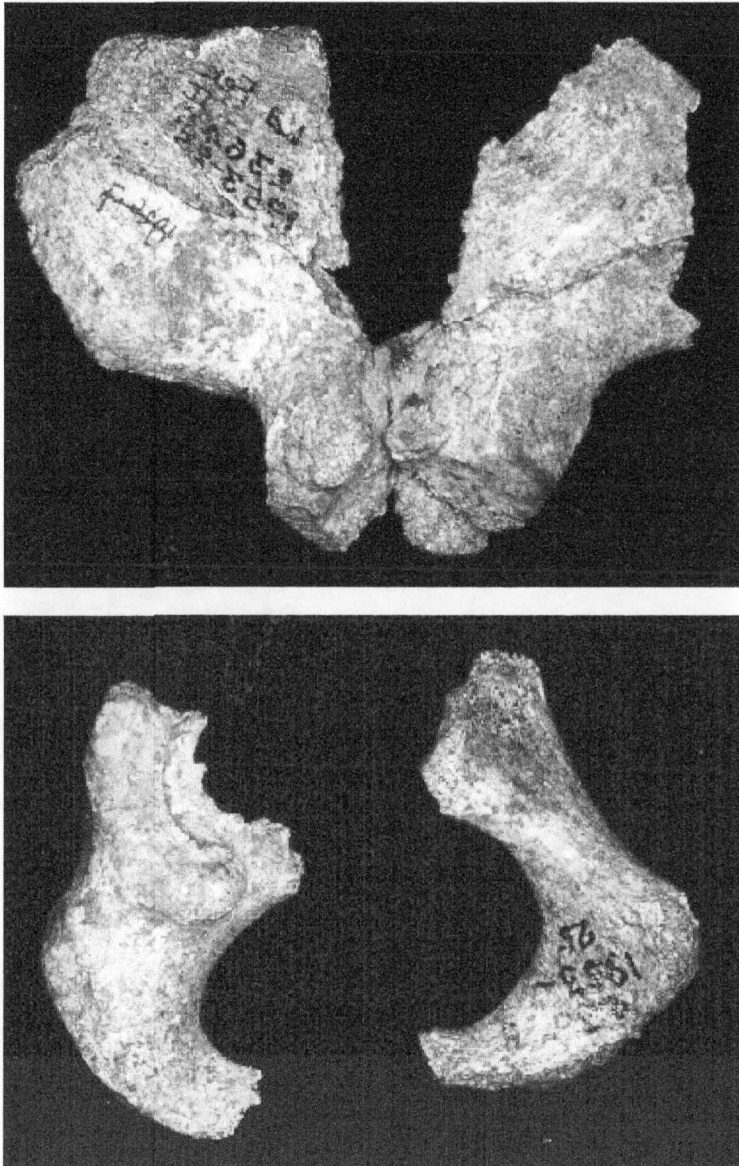

Figure 15.1   The left ilium (ventral view), right ischium and pubis (lateral view) of the La Ferrassie 6 child.

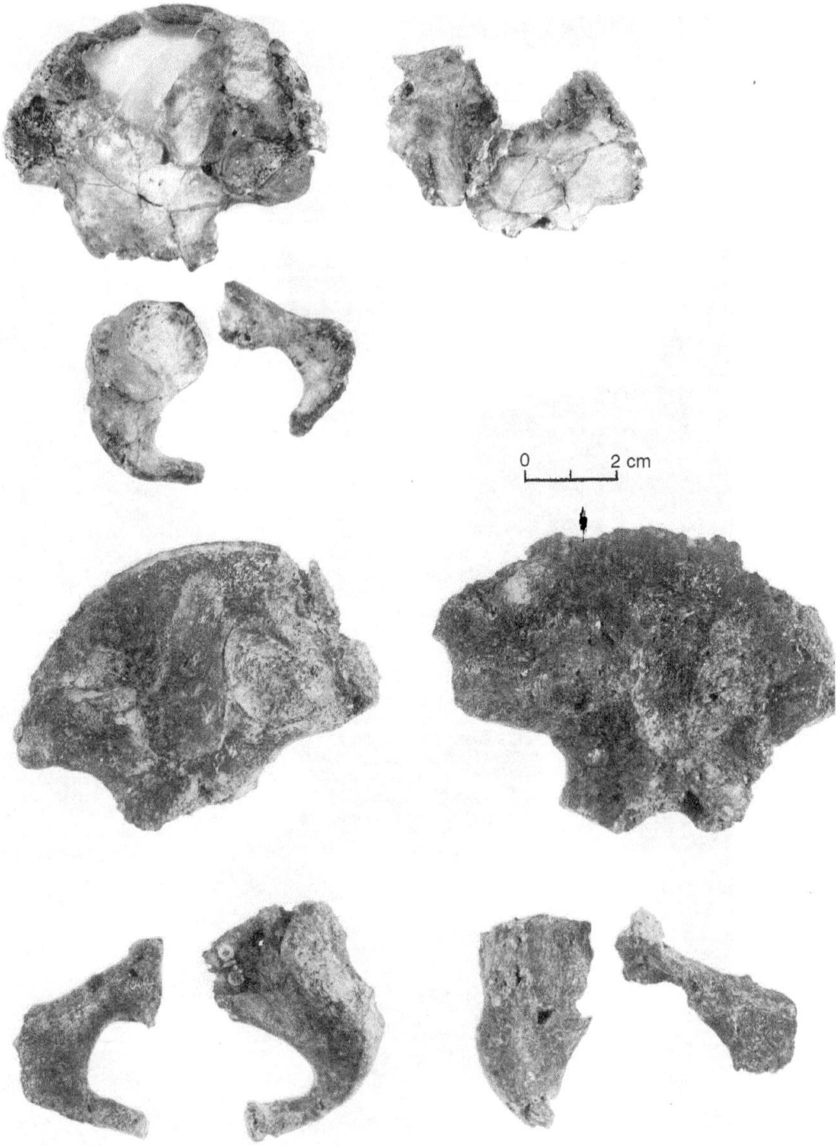

Figure 15.2   The Qafzeh 21 (top) and Qafzeh 10 (bottom) hipbones.

Information on hipbone development that derives from the study of documented dry bones is rare. This study is based on a personal investigation conducted by one of us (TM) of two European samples of known age and sex. The largest series is derived from the Spitalfields collection of eighteenth- and nineteenth-century skeletons housed in the Department of Palaeontology of the Natural History Museum in London (Molleson *et al.*, 1993). A total of 58 hipbones (24 females and 34 males) was available for study with ages at death that ranged from birth to 10 years (Table 15.2). Although we were aware of possible deficiencies and growth disturbances in some of the children originating from the Spitalfields reference sample (Majó, 2000; Molleson *et al.*, 1993), we have not included any juvenile bones manifesting severe alterations.

The second sample originated from the Portuguese skeletal collection housed at the Department of Anthropology of Coimbra University that dates from the beginning of the present century. This series consisted of 23 individuals (14 girls and 9 boys) with ages at death from 7 to 16 years (Anonymous, 1985). While the Portuguese sample consists of older individuals than those of La Ferrassie and Qafzeh, the metric data from the two first age classes (see Table 15.2) are employed in order to have a better age spectrum in the analysis of growth profiles.

Methods

Several factors complicate the analytical framework for growth studies of prehistoric populations. Regarding the specificity of immature bones and the lack of osteometric points currently defined on adult bones, a direct application of standard anthropological measurements (Bräuer, 1988) cannot be made. Other methodological difficulties are inherent to sampling and aging of the skeletal juveniles in archaeological sites (Saunders, 1992; Saunders & Hoppa, 1993; Tillier, 2000).

In their study of fetal skeletons housed in Szeged University, Fazekas & Kósa (1978) recorded five measurement variables of the hipbones: length and width of ischium and ilium, pubic length. More recently Humphrey (1998), investigating growth patterns of 59 skeletal variables measured in two samples of known age and sex, also included iliac width.

Other data were published in studies of archaeological populations originating in Europe and North America (Duday *et al.*, 1995; Hoppa, 1992; Merchant & Ubelaker, 1977; Sundick, 1978; Saunders & Hoppa, 1993). Interestingly most analyses have used variables measured on the ilium and there has been very little information on age-related growth in individual bones of the complete pelvic girdle.

One of us (TM) has defined in her study of infant and juvenile hipbones a total of 36 linear dimensions and other measurement variables (see Majó, 2000 for details). To this end, the developmental stage of the three bones (ischium, ilium, and pubis) and their preservation state in archaeological skeletal samples were taken into consideration. Here, only eight of the measurements were selected to permit a comparative metric study of the pubis ramus and to estimate its relative development to the two other hipbone components.

Maximum pubic length (PU1) was taken, as the distance between the plane of the symphyseal surface and the acetabulum, in a parallel direction with the superior ramus. Minimum pubic length (PU8) was defined by us as the direct distance from the medial point of the ramus (inflexion point between the ramus and the symphyseal surface) to the superior margin of the acetabular subchondral bone. This measurement differs slightly from the acetabulo-symphyseal length proposed by Tompkins & Trinkaus (1987). Minimum vertical height (PU4) and the dorsoventral breadth (PU5) of the pubic ramus were measured on the ventral surface at the middle of the sulcus obturatorius.

Among the iliac dimensions, iliac height (IL2) was measured on the ventral surface as the maximum distance between the lower limit of the arcuate line and the iliac crest. Maximum iliac breadth (IL3) is the direct distance between the anterior superior and posterior superior spines. These two dimensions are respectively equivalent to the width and length measurements defined by Fazekas & Kósa (1978). The minimum iliac breadth (IL6) is the distance between the ossification point of the posterior inferior iliac spine and that of the anterior inferior iliac spine. This last measurement is comparable to the iliac breadth previously employed by Tompkins & Trinkaus (1987).

The maximum height of the ischium (IS1) is the greatest distance between the inferior margin of the ramus and the superior end of the iliac articular surface. The position of the superior articular point depends on the developmental stage of the ischium and differs slightly between infants and older children.

For the present analysis we have used the diaphyseal length of the femur (intermetaphyseal maximum length, FE1) (Duday *et al.*, 1995) as a reasonable proxy for stature. Indeed two of the specimens, i.e., La Ferrassie 6 and Qafzeh 10, were sufficiently complete to evaluate the correlation between hipbone metrics and stature, based on this femoral estimate.

Metric comparative analyses between fossil and recent children were performed. To compare growth and age in the hipbone of the Middle Paleolithic and modern specimens, cross-sectional growth "curves" were calculated from the Spitalfields and Coimbra documented collections combined, in order to increase sample sizes over a wider age range. We totally agree with

Saunders (1992: 14) that such curves "are not true growth curves . . . since of course the dimensions represent deceased individuals who never reached maturity" and this is an important aspect to keep in mind in the discussion of the results.

The curves were prepared by use of fourth-degree polynomials (Majó, 2000) and only the Qafzeh and La Ferrassie specimens were incorporated to these skeletal growth profiles. Among other specimens listed in Table 15.1, some are still unpublished (e.g., René Simard, Mezmaiskaya, Dederiyeh). Others (e.g., Roc de Marsal, Skhul 1) are not sufficiently preserved to estimate the osteometric measurements. The La Ferrassie and Qafzeh fossils were plotted on the modern human line according to their estimated ages and using their measurements. In the case of La Ferrassie 6, no dental criteria were available for estimating the age of death of the individual and a mean age (i.e., 4 years) between the Heim and Tompkins & Trinkaus predictions was consequently employed.

Linear regressions between pairs of variables were also introduced into the analysis. With regard to the age at death distribution of La Ferrassie and Qafzeh children which ranged from birth to ca. 6 years, we selected the Spitalfields sample, although we were aware of the presence of deficiencies and growth disturbances in some of the children originating from this reference sample (Majó, 2000; Molleson *et al.*, 1993). We have therefore excluded the juvenile bones manifesting severe alterations from our statistical analysis. Simple linear regressions between pairs of variables were employed and the probability of the fossil specimens being outside the modern range of variation was represented by z-score values that were computed for each of the fossil residuals. The probability associated with each z-score represents the probability of the fossil to belong to the recent sample.

Results

In Table 15.3 are reported linear dimensions of the hipbones and diaphyseal length of femur in the La Ferrassie and Qafzeh children. Our investigation of the relationship between chronological age and hipbone development is illustrated here by the presentation of cross-sectional growth curves built for five dimensions (IL2, IL3, IL6, PU8, and IS1). Special attention is given to pubic morphology, as this is distinctive in Neandertals, and we present the results of linear regressions between the minimum pubic ramus length (PU8) and four different variables (IL2, IL3, IL6, and FE1). Later we will attempt a comparative study between non-adult and adult pubic configurations in fossils and we will refer to three other measurements (PU1, PU4, PU5).

370   *T. Majó & A.-m. Tillier*

Table 15.3 *Measurements (in millimeters) of the hipbone elements[a] in La Ferrassie and Qafzeh subadults*

	PU1	PU8	PU4	PU5	IL2	IL3	IL6	IS1	FE1
La Ferrassie 4b					33.4				
La Ferrassie 8		14.9			46.0	50.0	34.6		
La Ferrassie 6	37.0	28.7	8.8	6.4	63.7		47.7		165.0
Qafzeh 13					22.0	25.0			
Qafzeh 21	32.1	22.5	8.9	7.1	56.6	68.0	48.4	39.5	
Qafzeh 10	40.2	30.0	10.6	9.0	70.3	86.9	64.5	46.9	233.0

[a]Maximum (PU1) and minimum (PU8) public lengths, minimum vertical height (PU4) and dorsoventral breadth (PU5) of the pubic ramus. Height (IL2), maximum (IL3) and minimum (IL6) iliac breadths. Maximum height of the ischium (IS1). Length of the femoral diaphysis (FE1). (Definitions of the measurements are in the text.)

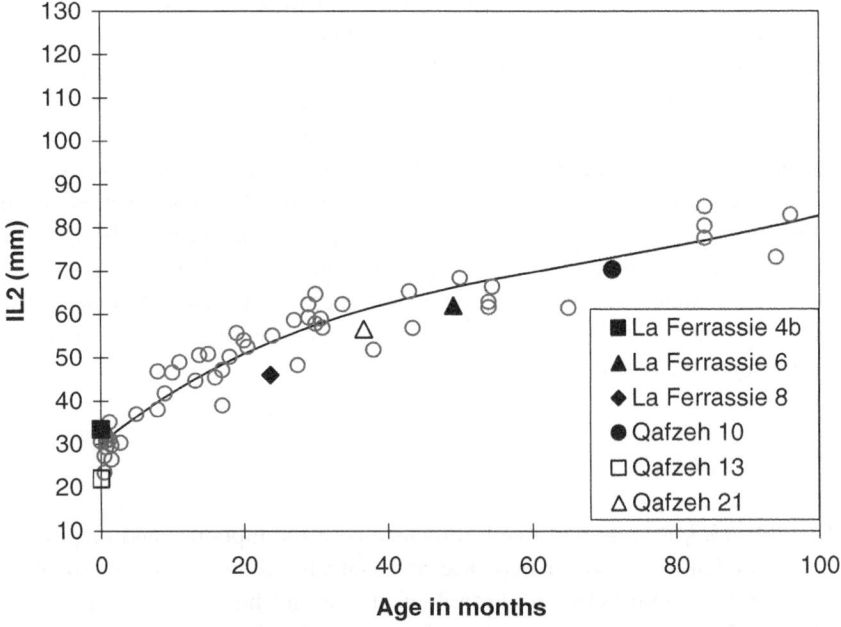

Figure 15.3   Fourth-degree polynomial regression showing the relation of age to the iliac height in the documented skeletal sample (correlation: 96.6%). Note the position of the La Ferrassie and the three Qafzeh children; open circles represent recent children.

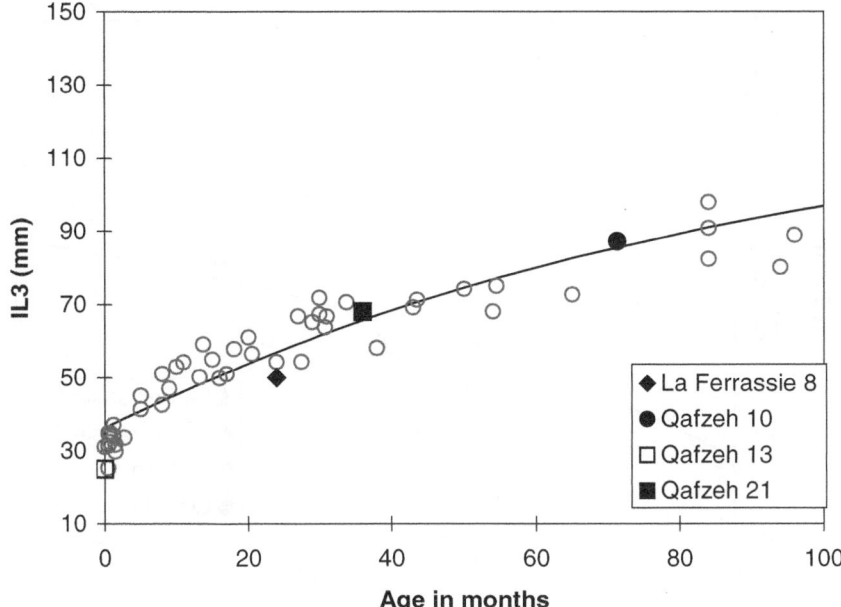

Figure 15.4   Fourth-degree polynomial regression curve showing the relation of age to the maximum iliac breadth in the documented skeletal sample (correlation: 95.4%). Note the position of the Neanderthal child La Ferrassie 8 and of the three Qafzeh specimens; open circles represent recent children.

Ilium

The "growth curve" for iliac height was the only one that permitted a comparison between the overall fossil sample ($n = 6$) and the modern children (Figure 15.3). Compared to modern neonate bones, the ilium of La Ferrassie 4b appeared quite large (Figure 15.3), while that Qafzeh 13 was characterized by small size. However, an increase in height cannot be generalized to all La Ferrassie children. Indeed La Ferrassie 8, ca. 2 years old at death, possessed a short ilium relative to its dental age (Figure 15.3). In fact with regard to its maximum and minimum iliac breadths (Figures 15.4 and 15.5), La Ferrassie 8 showed the same tendency for small size.

The positions of La Ferrassie 6 on the "growth profiles" showing the relation of age to iliac height (Figure 15.3) and to minimum iliac breadth (Figure 15.5), do not contradict a developmental age for this specimen around 4 years.

Interestingly Qafzeh 21, ca. 3 years old at death, possessed a minimum iliac breadth close to that of La Ferrassie 6, while its ilium is slightly shorter than that of the Neandertal child (Table 15.3 and Figure 15.3). Such relative

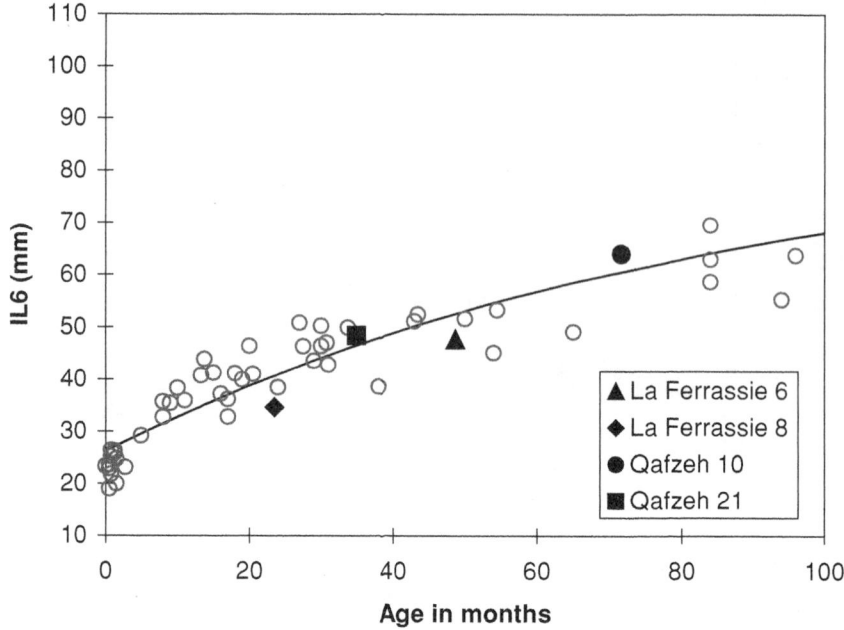

Figure 15.5   Fourth-degree polynomial regression curve showing the relation of age to the minimum iliac breadth in the documented skeletal sample (correlation %: 95.3). Position of the two La Ferrassie and two Qafzeh specimens.

enlargement of iliac breadth in the Qafzeh specimens was confirmed by Qafzeh 10, in relationship between age and minimum iliac breadth (Figure 15.5).

Pubis

The "growth curve" for minimum pubic length proves to be very enlightening (Figure 15.6), although Qafzeh 13 and La Ferrassie 4b could not be included in the analysis. Within the Middle Paleolithic sample, only La Ferrassie 8, exhibiting a relatively short ramus, had a "pubic age" that conformed the chronological one following the modern human line. By contrast, older specimens such as La Ferrassie 6, Qafzeh 21, and Qafzeh 10, exhibited rather long pubic rami when the bones were metrically compared to those of similarly aged modern children (Figure 15.6). Therefore, a discrepancy exists between the developmental age suggested by minimum pubic length and the previous age estimates for the three specimens. If the fossil subadults are thought to follow the same growth trajectory demonstrated by recent children for growth of the pubis, then

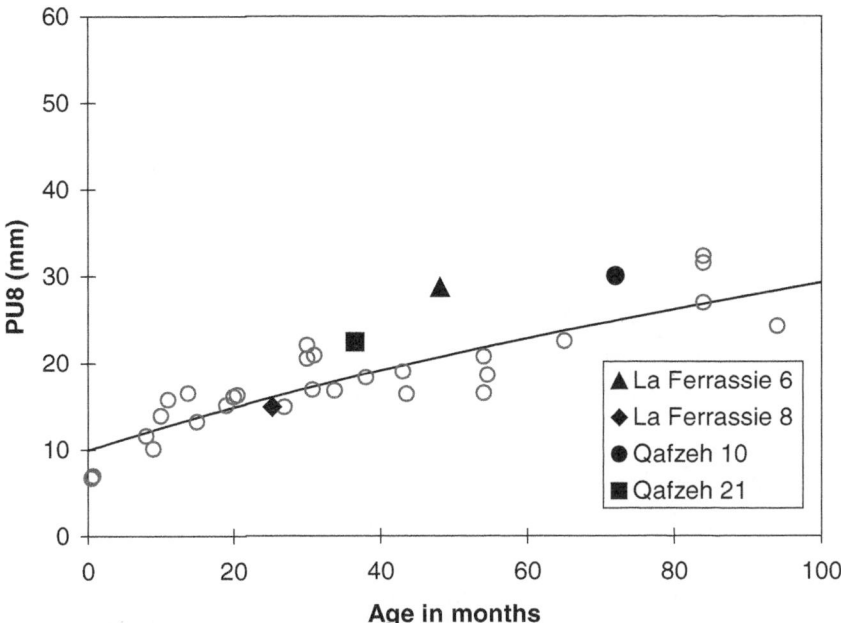

Figure 15.6   Fourth-degree polynomial regression curve showing the relation of age to the minimum pubic length in the documented skeletal sample (correlation: 94.2%). Three of the fossil children, La Ferrassie 6, Qafzeh 21 and Qafzeh 10, exhibit a superior pubic ramus that is rather long compared to those of recent children of similar ages (open circles).

the age at death estimates of the Neandertal La Ferrassie 6 child, as well as that of Qafzeh 10, might need to be reconsidered.

In order to substantiate this difference in the pattern of pubic development between fossil and recent children, we have employed simple linear regressions between pairs of variables and the probability of the fossil specimens being outside the modern range of variation ($P < 0.05$) was represented by z-score values. When minimum pubic ramus length is plotted against maximum iliac breadth (Figure 15.7), La Ferrassie 8 fell within the distribution of the Spitalfields sample (z-score 0.94, $P = 0.1736$). By contrast Qafzeh 21 and Qafzeh 10 (near the upper limit of variation) exhibited significantly higher z-scores using the linear regression formula (respectively $+2.44$, $P = 0.0073$, and $+1.8$, $P = 0.0359$). The use of minimum iliac breadth (Figure 15.8) allowed the inclusion of La Ferrassie 6 in the comparative analysis, and demonstrated that this child clearly fell above the distribution of the modern historical sample while the z-scores results put Qafzeh 21 and Qafzeh 10 at the limit of the modern variation ($+1.54$, $P = 0.0618$).

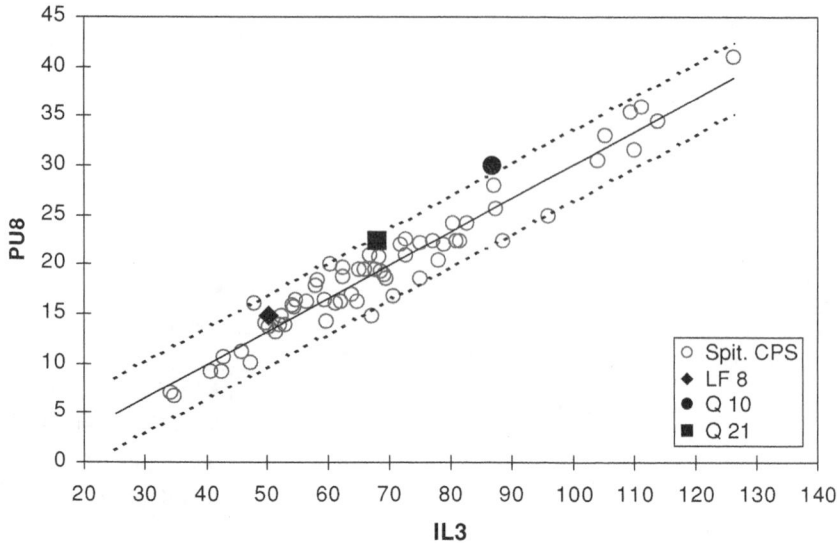

Figure 15.7   Minimum pubic length (PU8) versus maximum iliac breadth (IL3).
La Ferrassie 8, Qafzeh 21, and Qafzeh 10 are plotted on the linear regression based on
the Spitalfields reference sample ($n = 62$) (correlation: 95%). Measurements in
millimeters.

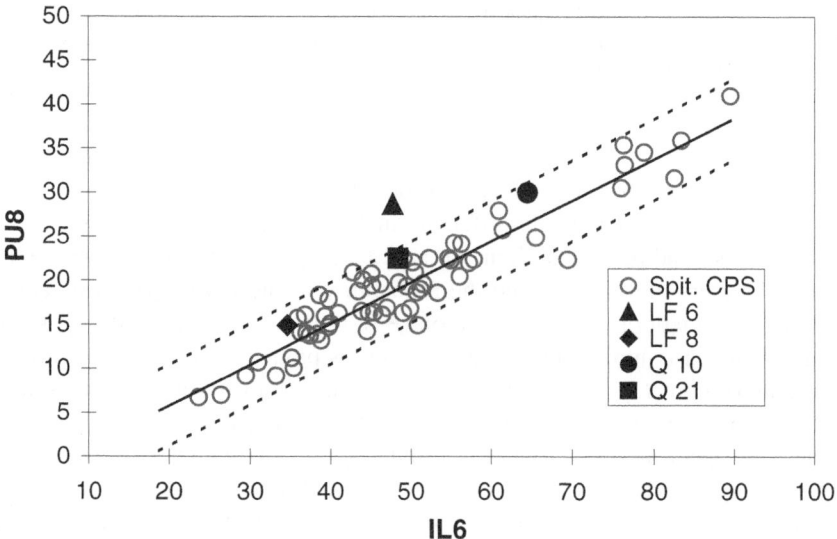

Figure 15.8   Minimum pubic length (PU8) versus minimum iliac breadth (IL6). On
this linear regression using the Spitalfields children ($n = 63$), the two La Ferrassie
children can be compared to the two Qafzeh ones (correlation: 95%). Measurements
in millimeters.

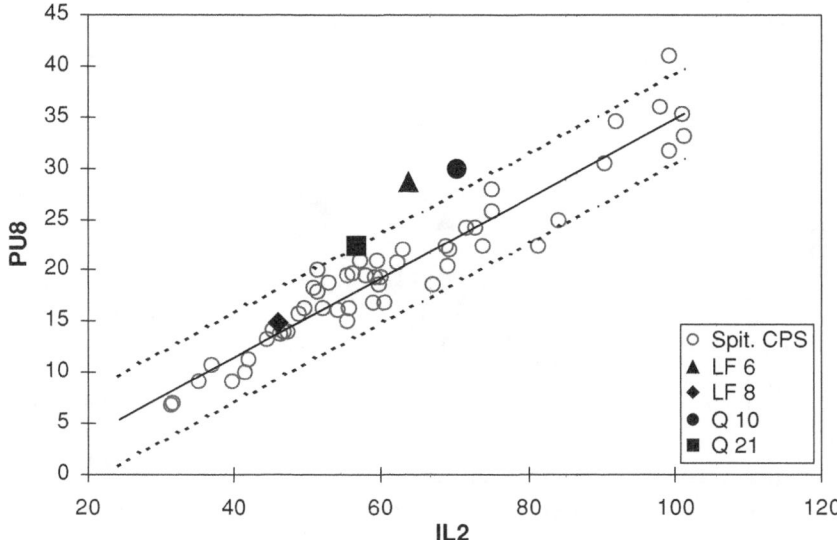

Figure 15.9   Minimum pubic length (PU8) versus iliac height (IL2). Graph shows the four fossil children on the linear regression ($n = 52$) (correlation: 95%). Measurements in millimeters.

The elongation of the superior pubic ramus was identified only on the oldest child in La Ferrassie and on the Qafzeh juveniles. Evidence of age-related change in the elongation of the pubic ramus between the two La Ferrassie children was further shown when minimum pubic length was plotted against iliac height. On this diagram (Figure 15.9), only La Ferrassie 8, the youngest in developmental age, appeared to follow the modern "growth curve." By contrast La Ferrassie 6 *and* Qafzeh 10 fell above the distribution of the modern sample. In addition the z-score results ($+2.12$, $P = 0.0170$) showed that Qafzeh 21 was positioned at the upper limit of the modern distribution.

Similarities between La Ferrassie 6 and Qafzeh 10 were also seen in the linear regression on minimum pubic ramus length on diaphyseal length of the femur (Figure 15.10). While La Ferrassie 6 stood outside the distribution of the modern sample, the plotted position of Qafzeh 10 appeared quite informative regarding the discriminated z-score value ($+1.98$, $P = 0.0239$). Again La Ferrassie 6 *and* Qafzeh 10 exhibit a minimum pubic length that is longer, relative to femur length, than is the case with the modern Spitalfields children.

Ischium

Among the La Ferrassie bones, the most complete specimen, La Ferrassie 6 (Figure 15.1) lacks a large part of the acetabular region. However, following

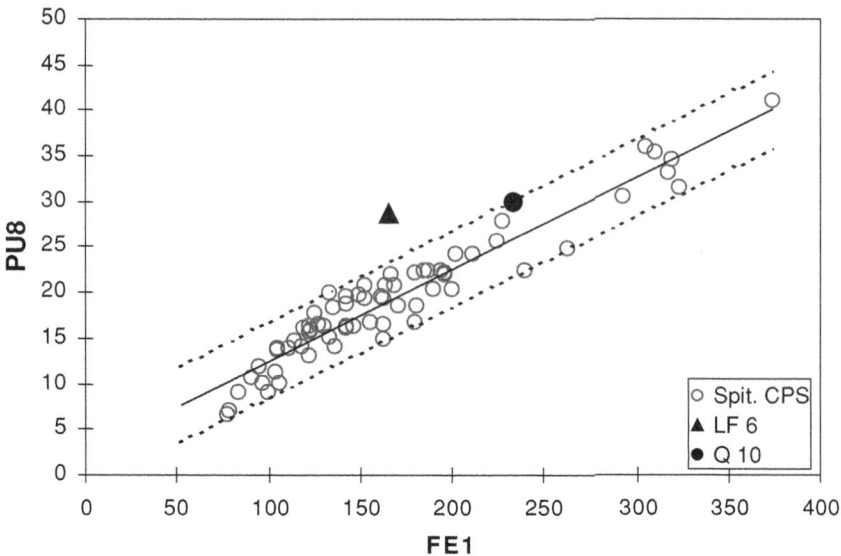

Figure 15.10   Minimum pubic length (PU8) versus diaphyseal femur length (FE1).
La Ferrassie 6 and Qafzeh 10 are plotted on the linear regression based upon the
Spitalfields reference sample ($n = 66$) (correlation: 95%). Measurements in
millimeters.

Heim (1982: 84) an enlargement of the obturator foramen size was detected
for this specimen, while La Ferrassie 8 was described as similar to modern
children in terms of ischial configuration (Heim 1982: 71). With respect to the
relationship between age and ischium size, reference was made in the present
study only to complete bones. As illustrated by the Figure 15.11, Qafzeh 21
and Qafzeh 10 appeared similar to recent children in terms of ischium length.

Discussion

Our results bring new insights in the ongoing debate over the presence or
absence of differences in developmental patterns between Neandertals and early
modern humans. A difference in size of the ilium can be observed between La
Ferrassie 4b and Qafzeh 13, with the former being larger that the latter. This
distinction in iliac size between Neandertals and modern humans is contradicted
by the examination of La Ferrassie 8, which is older in chronological age, but
exhibits a small ilium when metrically compared to modern immature hipbones.
Minimum pubic length for La Ferrassie 8 also clearly falls within the range of
similarly aged recent bones.

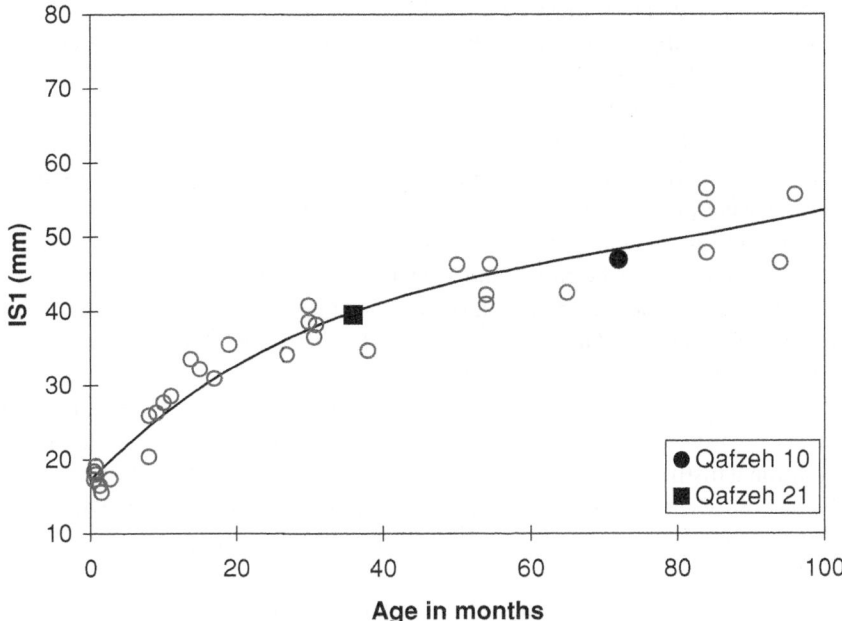

Figure 15.11 Fourth-degree polynomial regression curve showing the relation of age to the ischium length in the documented skeletal sample (correlation: 97.2%). The sizes of Qafzeh 21 and Qafzeh 10 bones are similar to those of recent bones (open circles) of similar chronological ages.

By contrast La Ferrassie 6, obviously older in developmental age than La Ferrassie 8, falls above the modern range for most variables considered here. Indeed, there is enough evidence to suggest that the published age estimate for La Ferrassie 6 might be incorrect, and this individual may well be older than 3 years old (long-bone morphology, following Tompkins & Trinkaus, 1987: 235; this study).

The data collected from La Ferrassie 8 and La Ferrassie 6 seem to suggest individual variation and age related change in pubic development among young European Neandertals. In their preliminary report of the Dederiyeh 1 child from Northern Syria, Dodo *et al.* (1998: 336, Figure 14) showed that the acetabulo-symphyseal index (acetabulo-symphyseal length/iliac breadth × 100) of the Syrian specimen stood above the range of variation of a modern Japanese sample. The Dederiyeh 1 pubis is significantly longer than that of modern Japanese children. As the ages at death of Dederiyeh 1 and of La Ferrassie 8 are comparable (ca. 2 years), individual variation among young Neandertals might be considered (see also Kondo & Ishida, this volume). Regarding the geographical distance between the La Ferrassie 8 and Dederiyeh 1 individuals, such an interpretation must be considered as equally plausible. In the literature, indeed, various lines of evidence suggest that there are marked differences between the European Neandertals (so called "classic" Neandertals) and the Near Eastern samples from Tabun, Amud, Shanidar, and Kebara sites (e.g., Arensburg, 1989; Arensburg & Belfer-Cohen, 1998; Mann, 1995; Suzuki & Takai, 1970; Trinkaus, 1992). Apart from this, we also have to resolve the question of what proportion of the differences in the results might be attributed to the biological characteristics of each modern sample selected for comparison in the studies of La Ferrassie and Dederiyeh children. It is evident that further analyses of new immature hipbones are therefore required before accurate conclusions in relation to age related change can be drawn in the case of Neandertals.

Regarding the pubic configuration in adult Neandertals, a superoinferior thinness of the ventral margin of the superior ramus has been described as an additional distinctive feature (e.g., McCown & Keith, 1939; Rak, 1990; Rosenberg, 1988; Stewart, 1960; Trinkaus, 1976). Among adults from the La Ferrassie and Qafzeh sites, this feature is indeed well expressed on the La Ferrassie 1 male specimen, while it is missing on the unique adult hipbone found at Qafzeh, which is attributed to a young woman, Qafzeh 9 (Table 15.4; Figure 15.12). When the proportional characteristics of the superior pubic ramus are compared between subadults and adults (Table 15.4), there are clear age differences in the two Neandertal specimens. In contrast with the adult shape, no variation in height between the ventral and the dorsal margins of the superior ramus can be seen in the immature La Ferrassie 6 pubis. As previously mentioned in studies restricted to subadults (Majó, 2000; Tillier, 1999; Tompkins & Trinkaus, 1987), La Ferrassie 6, like the Qafzeh children, lack the ventral thinning of the superior pubic ramus. Interestingly, observation of the Dederiyeh 1 pubis (Akazawa et al., 1995: 83, Figure 15.11) leads us to suggest that the Syrian child exhibited a similar pattern of a lack of ventral thinness.

The examination of juvenile specimens confirms that there are differences in the timing of the appearance of adult anatomical features in the pubis between Neandertal and modern human populations. This observation corroborates the conclusion reached in previous approaches of Neandertal ontogeny that supported the evidence of heterochrony in the development of Neandertal skull features during growth (Tillier, 1986, 1995).

In comparison of the relationship between ischial length and age in fossil and recent children, it will be important to improve our interpretation of growth patterns with additional data from the Dederiyeh specimen.

Osteological studies using recent skeletal specimens as comparative samples have advocated the possibility of advanced postcranial skeletal maturation in Neandertals (Dodo et al., 1998; Heim, 1982; Tompkins & Trinkaus, 1987).

Table 15.4 *Measurements (in millimeters) of the superior pubic ramus selected for a comparison between children and adults*

	La Ferrassie 6	La Ferrassie 1	Qafzeh 21	Qafzeh 10	Qafzeh 9
Age estimates by authors	>3 years[a]	40–45 years[b]	ca. 3 years[c]	ca. 6 years[c]	17–20 years[d]
Length					
Acetabulo-symphyseal length[e]		93.0			66.6[f]
Maximum pubic length (PU1)	37.0		32.1	40.1	
Percentage of adult size attained	39.8		48.1	60.2	
Height					
Superior pubic height[g]		17.0			14[a]
Ventral thickness[g]		10.2			
Pubic height (PU4)[h]	8.8		8.9	10.6	
Percentage of adult size attained	51.8		63.6	75.7	
Breadth					
Superior pubic breadth[g]		16.0			15[a]
Pubic breadth (PU5)[h]	6.4		7.1	9.9	
Percentage of adult size attained	40.0		47.3	66.0	
Index of pubic robusticity[g]	41.1	35.5	49.8	51.1	43.5

[a] Present study.
[b] Heim (1976).
[c] Tillier (1999).
[d] Vandermeersch (1981).
[e] Defined by McCown & Keith (1939).
[f] After Bruzek & Vandermeersch (1997).
[g] Defined by Trinkaus (1976, 1983) and measured at the middle of the obturator groove.
[h] Measurements taken at the middle of the obturator groove; data from Majó (2000).

Others (Majó, 1995; Nelson & Thompson, 1999; Tillier, 1986, 1995, 2000) have commented on the difficulties in detecting major changes in human growth rate from the analysis of isolated specimens and the use of cross-sectional growth studies. Normally, comparative analysis of human skeletal growth in past populations cannot be carried out without reference to an adult sample. In our analysis, however, the use of measurement variables defined on non-fused bones limits the evaluation of the percentage of adult size attained at different developmental stages, and the juvenile specimens cannot be compared with

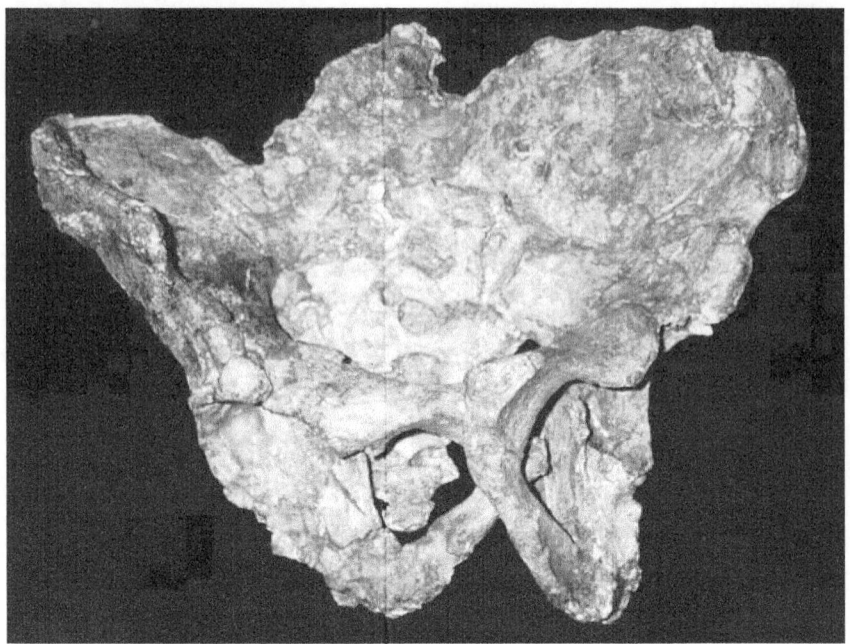

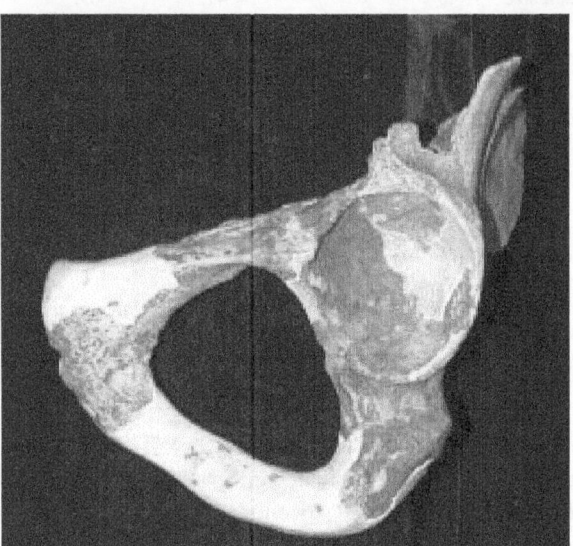

Figure 15.12   The Qafzeh 9 (top) and La Ferrassie 1 (bottom) hipbones (ventral view).

adults. This represents a major methodological problem in conducting growth analyses based on metric data.

Table 15.4 compared measurements of the pubic ramus between children and adults from the La Ferrassie and Qafzeh sites. These measurements were selected to obtain an objective quantification of the anatomical regions by the use of similar anatomical references. Given this, only three pubic linear dimensions were employed for comparison. Interestingly, the differences between La Ferrassie and Qafzeh specimens in the percentage of adult size attained for pubic ramus length do not appear to indicate an acceleration in skeletal maturation in Neandertals for this anatomical region which differed somewhat from that of recent populations. Unfortunately, the skeletal age estimation of the La Ferrassie 6 child is derived from bone maturation patterns that are not the most appropriate standards for estimating chronological age. Moreover there is no way to analyze the relationship between the pattern of hipbone growth and the development of sexual dimorphism.

Conclusions

In the context of the renewed debate regarding the uniqueness of Neandertal hipbone morphology, the conclusions reached by our study of the La Ferrassie and Qafzeh juveniles are meaningful in understanding the appearance of the anatomical differences. Unfortunately, the study of hipbone growth in the two juvenile fossil groups is incomplete because there is little available material (see Table 15.1) and important periods of growth such as adolescence are still totally undocumented. For example the unique Neandertal specimen attributed to this age group, Le Moustier 1, preserves a fragmentary hipbone, but the pubic ramus is not represented (Herrmann, 1977: 134).

The deficiency of well-preserved skeletons is a common problem for growth-related research in prehistoric populations which limits the evaluation of individual variation. In addition, it is difficult to evaluate the importance of sexual dimorphism in the manifestation of morphological variation during the growth period.

Our results reflect similarities in the pattern of pubic development between the La Ferrassie Neandertals and the Qafzeh hominids, but also raise the question of variation in the progress toward the attainment of adult pubic size in Late Pleistocene hominids. From the examination of the La Ferrassie children, the assumption of age-related changes in the elongation of the superior pubic ramus among European Neandertals cannot be rejected.

In our analysis we have used as comparative samples subadults originating from recent populations, for lack of published comparative data from early modern human Europeans at equivalent dental stages. An important question we are left with is how Upper Paleolithic associated subadults differ from Middle Paleolithic children (i.e., Neandertals and Qafzeh specimens) in terms of hipbone proportional characteristics. Future research on Upper Paleolithic specimens and additional data of non-adult individuals that permit comparisons based on both dental and skeletal maturational criteria are required in order to place our results in a broader comparative framework. The reconstruction of skeletal maturational events would benefit from the accumulation of data with new discoveries and further studies.

Acknowledgments

This research was supported by the French Ministry of Foreign Affairs (Mission Archéologique de Qafzeh), the Centre National de la Recherche Scientifique, the UMR 5809 and the Fyssen Fundation (TM). The authors are very thankful to the following institutes and curators of the skeletal collections who made the specimens available to them: the Department of Antiquities in Jerusalem, the Department of Anatomy and Anthropology (Tel Aviv University) and there, especially B. Arensburg and Y. Rak; T. Molleson, L. Humphrey and C. B. Stringer (Department of Palaeontology, Natural History Museum, London), E. Cunha (Departmento de Antropologia, Universidad de Coimbra); A. Langaney and M. Chech (Musée de l'Homme, Paris). Special thanks are due to F. Calafell (Universitat Pompeu Fabra, Barcelona) for useful discussions throughout this research project. The authors are indebted to the editors of this volume, especially to Andrew Nelson, for constructive comments on the manuscript.

References

Aiello, L., & Dean, C. (1990). *An Introduction to Human Evolutionary Anatomy.* London: Academic Press.

Akazawa, T., Muhesen, S., Dodo, Y., Kondo, O., & Mizoguchi, Y. (1995). Neanderthal infant burial from the Dederiyeh Cave in Syria. *Paléorient*, **21**(2), 77–86.

Akazawa, T., Muhesen, S., Ishida, H., Kondo, O., & Griggo, C. (1999). New discovery of a Neanderthal burial from the Dederiyeh cave in Syria. *Paléorient*, **25**(2), 129–142.

Anonymous (1985). *Cem Anos de Antropologia em Coimbra.* Coimbra: Museu e Laboratorio Antropologico.

Arensburg, B. (1989). New skeletal evidence concerning the anatomy of Middle Paleolithic populations in the Middle East: The Kebara skeleton. In *The Human Revolution: Behavioral and Biological Perspectives in the Origins of Modern Humans,* eds. P. Mellars & C. B. Stringer, pp. 165–171. Edinburgh: Edinburgh University Press.

Arensburg, B., & Belfer-Cohen, A. (1998). Sapiens and Neandertals: Rethinking the Levantine Middle Paleolithic hominids. In *Neandertals and Modern Humans in*

Western Asia, eds. T. Akazawa, K. Aoki, & O. Bar Yosef, pp. 311–322. New York: Plenum Press.

Bräuer, G. (1988). Osteometrie. In *Anthropologie: Handbuch der vergleichenden Biologie des Menschen,* Zugleich 4, *Auflage des Lehrbuchs des Anthropologie begründet von martin R.*, Band I, *Wesen und Methoden der Anthropologie,* ed. R. Knussman, pp. 160–232. New York: Gustav Fischer.

Bruzek, J., & Vandermeersch, B. (1997). Reassessment of the sex of the Qafzeh 9 individual based on multivariate statistical analysis. *American Journal of Physical Anthropology,* Suppl. **24**, 84.

Churchill, S. E. (1994). Human upper body evolution in the Eurasian Later Pleistocene. PhD dissertation, University of New Mexico, Albuquerque.

Dodo, Y., Kondo, O., Muhesen, S., & Akazawa, T. (1998). Anatomy of the Neandertal infant skeleton from Dederiyeh Cave, Syria. In *Neandertals and Modern Humans in Western Asia,* eds. T. Akazawa, K. Aoki, & O. Bar Yosef, pp. 323–338. New York: Plenum Press.

Duarte, C., Mauricio, J., Pettitt, P. B., Souto, P., Trinkaus, E., Van Der Plicht, H., & Zilhao, J. (1999). The Early Upper Paleolithic human skeleton from the Abrigo do Lagar Velho (Portugal) and modern human emergence in Iberia. *Proceedings of the National Academy of Sciences of the USA,* **96**, 7604–7609.

Duday, H., Laubenheimer, F., & Tillier, A-m. (1995). *Sallèles-d'Aude: Nouveau-nés et Nourrissons Gallo-romains,* Annales Littéraires de l'Université de Besançon no. 563; Centre de Recherches d'Histoire Ancienne no. 144, série Amphores no. 3. Paris: Les Belles Lettres.

Duport, L. (1958). Le Gisement moustérien de Puymoyen (Charente), Grotte René Simard. *Bulletins et Mémoires de la Société Archéologique de la Charente,* 35–39.

Fazekas, I. G., & Kósa F. (1978). *Forensic Fetal Osteology.* Budapest: Akadémiai Kiadô.

Golanova, L., Hoffecker, J. F., Kharitonov, V., & Romanova, G. (1999). Mezmaiskaya Cave: A Neanderthal occupation in the Northern Caucasus. *Current Anthropology,* **40**, 77–86.

Heim, J-L. (1982). *Les Enfants Néandertaliens de La Ferrassie,* Fondation Singer Polignac. Paris: Masson.

Herrmann, B. (1977). Über die Reste des postcranialen Skelettes des Neanderthalers von Le Moustier. *Zeitschrift für Morphologie und Anthropologie,* **68**, 129–149.

Hoppa, R. D. (1992). Evaluating human skeletal growth: An Anglo-Saxon example. *International Journal of Osteoarchaeology,* **2**, 275–288.

Humphrey, L. (1998). Growth patterns in the modern human skeleton. *American Journal of Physical Anthropology,* **105**, 57–72.

Madre-Dupouy, M. (1992). *L'Enfant Néandertalien du Roc de Marsal: Etude Analytique et Comparative,* Cahiers de Paléoanthropologie. Paris: Editions du CNRS.

Majó, T. (1995). Quelques aspects de la croissance de l'os coxal: Application aux Néandertaliens de la Ferrassie. *Anthropologie et Préhistoire,* **106**, 57–64.

Majó, T. (2000). L'Os coxal non-adulte: Approche méthodologique de la croissance et de la diagnose sexuelle. Application aux enfants du Paléolithique moyen. Thèse en Sciences Biologiques et Médicales, Université Bordeaux 1, Bordeaux.

Mann, A. E. (1995). Modern human origins: Evidence from the Near East. *Paléorient,* **21**(2), 35–46.

McCown, T. D., & Keith, A. (1939). *The Stone Age of Mount Carmel*, vol. 2. Oxford: Clarendon Press.

Merchant, V. L., & Ubelaker, D. H. (1977). Skeletal growth of the protohistoric Arikara. *American Journal of Physical Anthropology*, **46**, 61–72.

Molleson, T. I., Cox, M., Waldron, A. H., & Whitaker, D. K. (1993). *The Spitalfields Project*, vol. 2, Anthropology CBA Research Report no. 86. York: Council for British Archaeology.

Nelson, A. J., & Thompson, J. L. (1999). Growth and development in Neandertals and other fossil hominids: Implications for the evolution of hominid ontogeny. In *Human Growth in the Past: Studies from Bones and Teeth*, eds. R. H. Hoppa & C. M. Fitzgerald, pp. 88–110. Cambridge: Cambridge University Press.

Radovčić, J., Smith, F., Trinkaus, E., & Wolpoff, M. H. (1988). *The Krapina Hominids: An Illustrated Catalog of Skeletal Collection*. Zagreb: Mladost.

Rak, Y. (1990). On the differences between two pelvises of Mousterian context from Qafzeh and Kebara Caves, Israel. *American Journal of Physical Anthropology*, **81**, 323–332.

Rosenberg, K. R. (1988). The functional significance of the Neandertal pubic length. *Current Anthropology*, **29**, 595–617.

Saunders, S. R. (1992). Subadult skeletons and growth related studies. In *Skeletal Biology of Past Peoples: Research Methods*, eds. S. R. Saunders & M. A. Katzenberg, pp. 1–20. New York: Wiley-Liss.

Saunders, S. R., & Hoppa, R. D. (1993). Growth deficit in survivors and non-survivors: Biological mortality bias in subadult skeletal samples. *Yearbook of Physical Anthropology*, **36**, 127–151.

Schwarcz, H. P., Grün, R., Vandermeersch, B., Bar Yosef, O., Valladas, H., & Tchernov, E. (1988). ESR dates for the hominid burial site of Qafzeh in Israel. *Journal of Human Evolution*, **17**, 733–737.

Smith, F. H. (1976). *The Neandertal Remains from Krapina: A Descriptive and Comparative Study*, Department of Anthropology Report of Investigations, no. 15. Knoxville: University of Tennessee.

Stewart, T. D. (1960). Form of the pubic bone in Neanderthal man. *Science*, **131**, 1437–1438.

Sundick, R. I. (1978). Human skeletal growth and age determination. *Homo*, **29**(4), 228–249

Suzuki, H., & Takai, F. (eds.) (1970). *The Amud Man and His Cave Site*. Tokyo: University of Tokyo.

Tillier, A-m. (1986). Quelques aspects de l'ontogenèse du squelette cranien des Néanderthaliens. In *Fossil Man: New Facts, New Ideas*, eds. V. V. Novotny & A. Mizerova, *Anthropos (Brno)*, **23**, 207–216.

Tillier, A-m. (1995). Neanderthal ontogeny: A new source for critical analysis. *Anthropologie (Brno)*, **33**(1/2), 63–68.

Tillier, A-m. (1999). *Les Enfants Moustériens de Qafzeh: Interprétation Phylogénétique et Paléoauxologique*, Cahiers de Paléoanthropologie. Paris: Editions du CNRS.

Tillier, A-m. (2000). Palaeoauxology applied to Neanderthals: Similarities and contrasts between Neanderthal and modern children. In *Children in the Past:*

Paleoauxology, Demographic Anomalies, Taphonomy and Mortuary Practices, ed. A-m. Tillier, *Anthropologie (Brno)*, **38**(1), 109–120.

Tixier, J., Brugal, J-Ph., Tillier, A-m., Bruzek, J., & Hublin, J-J. (2001). Irhoud 5, un fragment d'os coxal non-adulte des niveaux moustériens marocains. *Actes des 1ᵉ Journées Nationales de l'Archéologie et du patrimoine au Maroc*, 1–4 July 1998, vol. 1, *Préhistoire*, pp. 149–53. Rabat: Société Marocaine d'Archéologie et du Patrimoine.

Tompkins, R. L., & Trinkaus, E., (1987). La Ferrassie 6 and the development of Neandertal pubic morphology. *American Journal of Physical Anthropology*, **73**, 223–239.

Trinkaus, E. (1976). The morphology of European and Southwest Asian Neandertal pubic bones. *American Journal of Physical Anthropology*, **44**, 95–104.

Trinkaus, E. (1992). Morphological contrasts between the Near Eastern Qafzeh-Skhul and Late Archaic human samples: Grounds for a behavioral difference? In *The Evolution and Dispersal of Modern Humans in Asia*, eds. T. Akazawa, K. Aoki, & T. Kimura, pp. 277–294. Tokyo: Hokusen-sha.

Trinkaus, E., Ruff, C. B., & Churchill, S. E. (1998). Upper limb versus lower limb loading patterns among Near Eastern Middle Paleolithic Hominids. In *Neandertals and Modern Humans in Western Asia*, eds. T. Akazawa, K. Aoki & O. Bar Yosef, pp. 391–404. New York: Plenum Press.

Tuttle, R. H. (1988). The hipbone and *Nomina Anatomica*. *American Journal of Physical Anthropology*, **77**, 133–134.

Valladas, H., Reyss, J. L., Joron, J. L., Valladas, G., Bar Yosef, O., & Vandermeersch, B. (1988). Thermoluminescence dating of Mousterian "Proto-Cro-Magnon" remains from Israel and the origin of modern man. *Nature,* **331**, 614–616.

Vandermeersch, B. (1981). *Les Hommes Fossiles de Qafzeh (Israël)*, Cahiers de Paléoanthropologie. Paris: Editions du CNRS.

Vlček, E. (1973). Postcranial skeleton of a Neandertal child from Kiik Koba, USSR. *Journal of Human Evolution*, **2**, 537–546.

16 *Ontogenetic variation in the Dederiyeh Neandertal infants: Postcranial evidence*

O. KONDO
University of Tokyo

H. ISHIDA
University of the Ryukyus

Introduction

Neandertal growth and development has been a recent focus in paleoanthropological studies. Although it has long been recognized that the study of the emergence of the modern human pattern of growth is essential and important for human evolution (e.g. Brothwell, 1975), the practical studies have been limited. This was mainly due to the small number of available specimens of both fossils and modern humans. However, recent progress in fieldwork and research methods has been restimulating us to the ontogenetic point of view in paleoanthropological records. Most of these studies have focused on the cranium and dentition, because of abundance of materials and close association of growth assessment with an individual's age estimation. The studies based on postcranial bones have been rare. Thus, several key hypotheses concerning the evolution of human growth have been produced mainly on the dental and cranial evidence.

In this situation, a few scenarios have been proposed for the growth of Neandertals, but the field has yet to arrive at a consensus. We review them here briefly, based on dental, cranial, and postcranial evidence in order.

Studies based on the dentition have dealt with a variety of sources of data: the timing and sequence of dental development (Dean *et al.*, 1986; Smith, 1991b), the rate of enamel formation (Dean *et al.*, 2001), and the relative timing of crown formation (Tompkins, 1996). The dental developmental schedule for Neandertals has been interpreted to be like that of modern humans.

Patterns of Growth and Development in the Genus Homo, ed. J. L. Thompson, G. E. Krovitz and A. J. Nelson. Published by Cambridge University Press. © Cambridge University Press 2003.

For example, Dean *et al.* (2001) inferred that pre-*Homo erectus* hominids possessed an ape-like rapid enamel formation pattern while the modern human-like slow crown formation appeared in Neandertals after their acquisition of large brains. On the other hand, the analysis of relative timing of crown formation has revealed different developmental patterns between Neandertals and modern humans (Tompkins, 1996).

Comparative cranial analyses of Pleistocene immature fossils have indicated that many diagnostic Neandertal characters appeared early during ontogeny (Minugh-Purvis, 1988; Rak *et al.*, 1994; Tillier, 1989) suggesting that Neandertal cranial ontogeny was different from that of modern humans (Ponce de León & Zollikofer, 2001; Williams, 2000). In particular, it has been suggested that the Neandertal cranium grew at a faster rate than that of modern humans (Dean *et al.*, 1986; Stringer *et al.*, 1990; but contra Trinkaus & Tompkins, 1990).

Arguments based on postcranial evidence are rare. Descriptions of individual fossil specimens have inferred possible timings of appearance for several postcranial characters that are thought to be diagnostic for Neandertals. Some are considered to appear early in ontogeny, such as the mediolateral elongation of the superior pubic ramus (Tompkins & Trinkaus, 1987), large distal first phalanx of the hand (Heim, 1982; but contra Tillier, 1999), short distal limb segments (Heim, 1982; Madre-Dupouy, 1992; Vlček, 1973). Others appear later in ontogeny, such as superoinferior thinning of the superior pubic ramus (Tompkins & Trinkaus, 1987), and the dorsal sulcus pattern of the scapular axillary border (Madre-Dupouy, 1992; Vlček, 1973).

Although the postcranial evidence still remains fragmentary, it is important to consider how it can contribute to the understanding of the growth and development of fossil hominids. The postcranial bones are essential to reconstruct an individual's growth profile, which should reflect a variety of growth aspects from the different parts of the skeleton. The postcranial bones can also provide a vital assessment for any hypothetical scenarios based on dental or cranial evidence. In practice, the number of immature fossils that preserve postcranial elements may not be sufficient for the construction of valid statistical tests. However, we can make progress by assessing each fossil specimen within a range of variation of modern humans.

In order to assess the fossil specimens from an ontogenetic point of view, age-related changes of respective traits are needed for comparison. In this study, therefore, we focused our attention on the analysis and assessment of the growth profile, and on individual variation in growth. In order to visualize these features, we preferred the absolute dimensions for measurements or parameters and plotted them against age. Consideration of ratios or any shape characters also discloses one aspect of growth and development, but it is difficult to compare them in terms of age-related changes, or the rate of growth.

In this study, length, circumference, and cross-sectional geometric properties of the lower limb bones were examined in two Neandertal infants, Dederiyeh 1 and 2, which were unearthed during recent excavations (Akazawa *et al.*, 1995a, 1995b, 1999). These fossils expand the chance for us to examine immature postcrania of fossil hominids. They were compared with those of modern humans and several Neandertal, Skhul–Qafzeh and European Upper Paleolithic individuals. Interestingly, the lower limb bones of these two infants are not similar in size and robusticity although their dental developmental stages are almost the same (Akazawa *et al.*, 1999; Ishida *et al.*, 2000). Therefore, they present a good opportunity to consider within-site variation.

Through this analysis, we will propose several results as follows. Considering the Neandertal sample as a whole, limb-bone lengths at any particular age are comparable to those of modern humans as well as other hominid fossils. With respect to shaft circumferences, Neandertals are on average larger for their ages compared with those of modern humans. Dederiyeh 1, in particular, possesses significantly thicker shafts of the lower limb bones and also exhibits significantly greater cross-sectional geometric properties of the tibia, relative to modern humans. Dederiyeh 2, on the other hand, demonstrates both length and robusticity that are well within the range of modern human variation. Although the differences between Dederiyeh 1 and 2 appear to be great, the degree of difference is not beyond interindividual variation within a large sample of modern humans.

These results may indicate that there are no significant differences in postcranial growth patterns between Neandertals and modern humans. However, before we can come to this conclusion, several limitations or prerequisites should be considered.

Materials

Fossil specimens considered here include Dederiyeh 1 and 2, and several European Neandertal, Upper Paleolithic, and Levantine Skhul–Qafzeh subadults (Table 16.1).

Dederiyeh 1 and 2

Two Neandertal specimens, Dederiyeh 1 and 2, were discovered in the Dederiyeh Cave in Syria during excavations from 1993 to 1998. Dederiyeh 1 is quite well preserved, having almost complete postcranial bones (Figure 16.1A) (Akazawa *et al.*, 1995a, 1995b). Dederiyeh 2 is a partial skeleton with a well-preserved cranium and several postcranial bones (Figure 16.1B) (Akazawa *et al.*, 1999). Neither specimen exhibits evidence of trauma or disease.

Table 16.1 *Specimens used in this study*

	Estimated age	Reference (age, postcranial dimension)
Neandertal		
Dederiyeh 1	1.3–2.5	Dodo *et al.*, 2003; Kondo & Dodo, 2003; Mizoguchi, 2003; Sasaki *et al.*, 2003
Dederiyeh 2	1.2–2.5	Akazawa *et al.*, 1999; Ishida & Kondo, 2003; Kondo & Ishida, 2003
Roc de Marsal	ca. 3	Madre-Dupouy, 1992
La Ferrassie 6	3–5	Tompkins & Trinkaus, 1987; personal data
Teshik Tash	ca. 9	Rochlin, 1949; Sinel'nikov & Gremyatskij, 1949
Early modern		
Skhul 1	4–4.5	McCown & Keith, 1939
Skhul 8	ca. 8	McCown & Keith, 1939
Qafzeh 10	ca. 6	Tillier, 1999
Upper Paleolithic modern		
Madeleine 4	ca. 3	Heim, 1991; personal data
Grimaldi 1	3–4	Minugh-Purvis, 1988; personal data
Grimaldi 2	ca. 2	Legoux, 1966 (cited in Minugh-Purvis, 1988); personal data

Comparative specimens

In addition to Dederiyeh 1 and 2, immature fossil hominids examined here include Neandertals (including Roc de Marsal, La Ferrassie 6, and Teshik Tash) and early modern *Homo sapiens* (Skhul–Qafzeh and European Upper Paleolithic specimens) (Table 16.1).

Our modern comparative sample consists of three geographically different populations; modern Japanese (mainly from the University of Tokyo and Kyoto University, $n = 37$), Romano-British and recent English (Poundbury, $n = 30$, Spitalfields, $n = 15$), and modern Black South Africans (Dart collection in the University of the Witwatersrand and K2 site in the University of Pretoria, $n = 44$).

Methods

Aging

Accurate age estimation is crucial for comparison of age-related changes of the postcranial morphology. For subadult individuals, the analysis of tooth formation stages has proven to be the most accurate among many methods used to

(A) 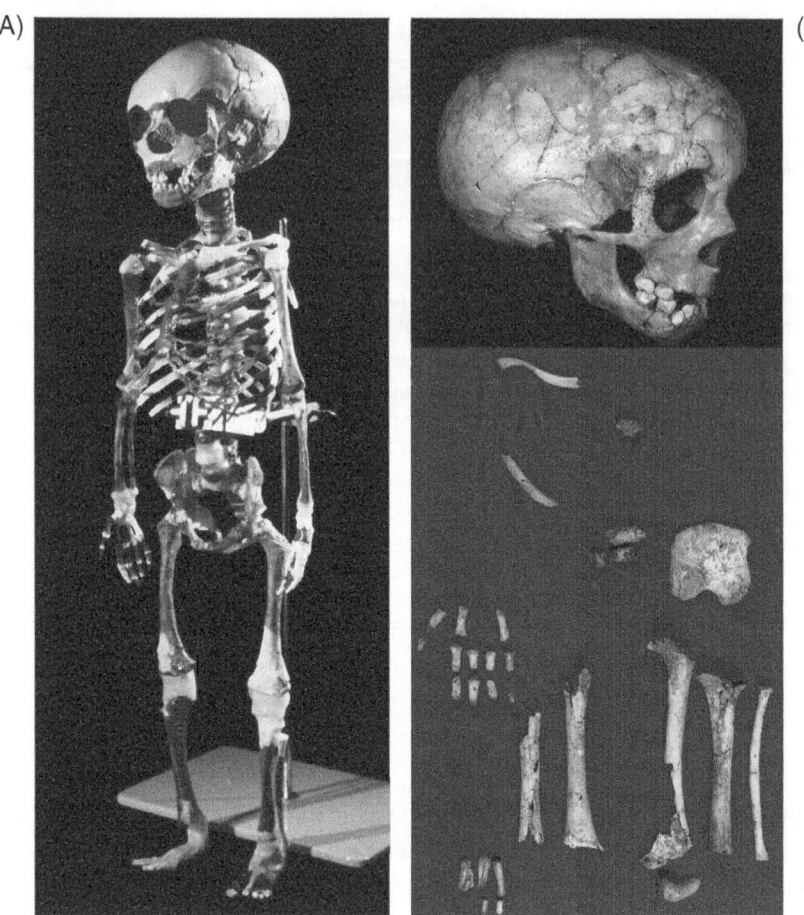 (B)

Figure 16.1   (A) Dederiyeh 1, reconstructed in standing posture, and
(B) Dederiyeh 2, skull reconstruction and postcrania.

assess skeletal maturation (Saunders *et al.*, 1993). In practice, however, several
factors produce uncertainty; e.g., the process of observation (radiography or
naked eye), methods of scoring, types of teeth observed, and so on.

The ages for all the modern individuals were estimated through inspection
of mandibular radiographs. We used Smith's revised scoring system (Smith,
1991a) for age estimation.

For the Dederiyeh individuals, age was estimated in several ways according
to a variety of methods (Dodo *et al.*, 2003; Ishida & Kondo, 2003; Mizoguchi,
2003; Sasaki *et al.*, 2003). Among them, we used those based on Smith's method
(Smith, 1991a) as the representative for ages of Dederiyeh 1 and Dederiyeh 2.

The ages of 2.0 for males and 1.9 for females were reported (Dodo *et al.*, 2003) and the average of 1.95 was used for Dederiyeh 1, while the ages of 2.5/ 1.9 years for males, 2.5/1.8 years for females, and the average of 2.15 for Dederiyeh 2 (Ishida & Kondo, 2003).

Age estimation for comparative fossil specimens follows those published in previous descriptions and papers, which were largely based on dental development but sometimes only on postcranial evidence (e.g., La Ferrassie 6 and Skhul 8). In tables, figures, and calculations, we present the estimated ages for each specimen as a fixed and invariable point, although we should keep in mind that the age estimation always holds a degree of uncertainty, which is usually expressed as a range of the estimation, as in the case for the Dederiyeh infants.

Measurements and analyses

The femur, tibia, and fibula were measured by the authors, or previously published data were used (as indicated in Table 16.1). For all three bones, the maximum length was measured without epiphyses, between the proximal and distal end of the shaft. Midshaft circumference was measured for the femur and fibula, and minimum circumference was measured for the tibia.

For the Dederiyeh infants and modern specimens, cross-sectional geometric properties were also calculated for the assessment of mechanical robusticity. Cross-sectional images were reconstructed using the outline contour of the midshaft and cortical thickness measurements obtained from the biplanar radiographs (Runestad *et al.*, 1993). Polar moment of area (J) and cortical area (CA) were then calculated using a specialized PC program (Sakaue, 1997; modified from Takahashi & Adachi, 1982).

The growth trajectories of the modern sample were presented as plots of each measurement against the estimated dental age. The fossil data were added to the graphs and their relative positions were statistically assessed. First, we fit the line of locally weighted regression smoothing (lowess) to the modern data and calculated the residuals, i.e., the deviations of each data point from the smoothed curve. Then, we calculated z-scores for each fossil specimen. In practice, we checked the normal distribution of the modern residuals by means of quantile–quantile (Q–Q) plots, and calculated the z-scores and the associated probabilities for each fossil specimen using the standard deviation of the modern residuals. As a result, we were able to determine the relative position and degree of difference of the fossil specimen from the modern average.

To measure the difference between Dederiyeh 1 and Dederiyeh 2 and to assess within-site variability, we made a multivariate assessment using the standardized residuals. The use of the residuals, which are supposedly independent

Table 16.2 *Assessment of fossil specimens based on z-scores calculated for femoral length and circumference*

Femur	Z-score for maximum length	Z-score for midshaft circumference
Neandertals		
Dederiyeh 1	1.35	2.63**
Dederiyeh 2	0.05	−0.07
Roc de Marsal	0.14	0.95
La Ferrassie 6	−0.97	0.59
Teshik Tash	–	0.80
Early modern		
Skhul 1	−0.54	0.15
Qafzeh 10	0.14	0.48
Upper Paleolithic modern		
Madeleine 4	1.13	0.83
Grimaldi 1	0.67	–
Grimaldi 2	0.59	–

Significance level: *<5%, **1%.

of age, is expected to eliminate the growth factors from the analysis. We calculated the Mahalanobis' distances based on the residuals from the six linear measurements, length and circumference of the femur, tibia, and fibula, and from the two cross-sectional geometric properties of the femur and tibia. All the measurements and properties are available for both Dederiyeh 1 and Dederiyeh 2. The statistical procedure was carried out using the Windows-based program, S-PLUS (MathSoft, Inc.).

Results

Length and circumference

Z-scores, based on maximum length and midshaft circumference of the femur, are presented in Table 16.2 and age-related changes of absolute measures are plotted in Figures 16.2A and 16.2B. In maximum length of the femur (Figure 16.2A), four Neandertals and five early modern *H. sapiens* specimens fall within the range of variation of the modern sample. Thus, both fossil groups appear to follow the regression curve fitted to the modern sample. For midshaft circumference (Figure 16.2B), Dederiyeh 1 falls above the modern distribution and the associated z-score indicates that it is significantly larger than the moderns, while Dederiyeh 2 is similar to the modern average. As a whole,

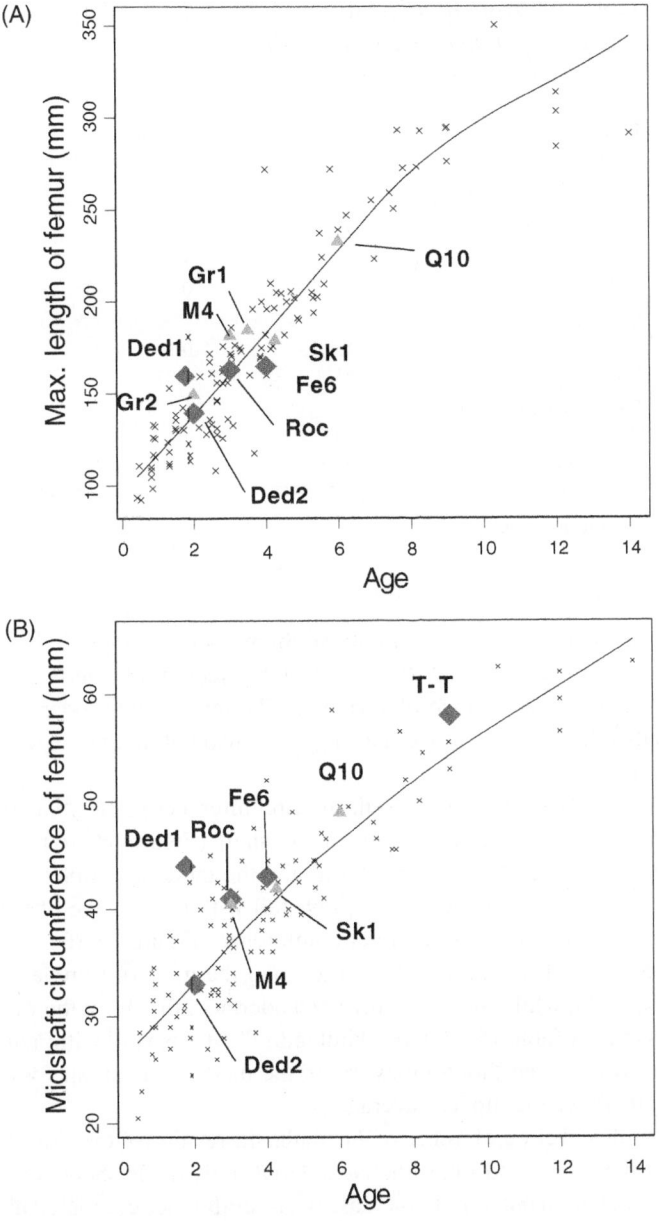

Figure 16.2   Age-related changes of the length (A) and circumference (B) of the femur. Ded1 and Ded2, Dederiyeh 1 and 2; Roc, Roc de Marsal; Fe6, La Ferrassie 6; T-T, Teshik Tash; Gr1 and Gr2, Grimaldi 1 and 2; M4, Madeleine 4; Sk1 and Sk8, Skhul 1 and 8; Q10, Qafzeh 10.

Table 16.3 Z-*scores of fossil specimens calculated for length and circumference of the tibia*

Tibia	Z-score for maximum length	Z-score for minimum circumference
Neandertals		
Dederiyeh 1	1.11	2.81**
Dederiyeh 2	−0.19	0.68
Roc de Marsal	0.03	1.50
La Ferrassie 6	−1.10	0.71
Early modern		
Skhul 1	0.13	−0.22
Skhul 8	0.53	1.07
Upper Paleolithic modern		
Grimaldi 1	0.43	–
Grimaldi 2	0.51	–

Significance level: **<1%.

Neandertal subadults seem to fall above the modern average, indicating their possession of thicker femoral shafts, although, except for Dederiyeh 1, they are still within the modern range of variation. The three early modern *H. sapiens* individuals fall above the modern average but also within the range of modern human variation.

Similar results are given for length and circumference of the tibia (Table 16.3; Figure 16.3). For maximum length of the shaft (Figure 16.3A), there is no significant difference between the fossil and the modern samples, as both the Neandertals and early modern *H. sapiens* fall within the range of variation of the modern sample. For minimum circumference (Figure 16.3B), on the other hand, Dederiyeh 1 is significantly larger than the modern sample (z-score = 2.81, $P < 0.01$), while the other three Neandertals fall within the range of the modern sample (Table 16.3). Two Skhul individuals also fall within the modern range. However, even though they are in the modern range, all fossils except Skhul 1 are above the modern average.

For length and circumference of the fibula, the results follow a similar pattern to that of the femur and tibia; the Neandertal individuals show no significant difference in length but a slight increase in circumference compared to the modern sample (Table 16.4; Figure 16.4). Although all three Neandertals are above the modern average in circumference, only Dederiyeh 1's fibula is significantly larger than that of the moderns (z-score = 3.78, $P < 0.01$ in Table 16.4). Unfortunately, no early modern specimens could be compared in fibular dimensions.

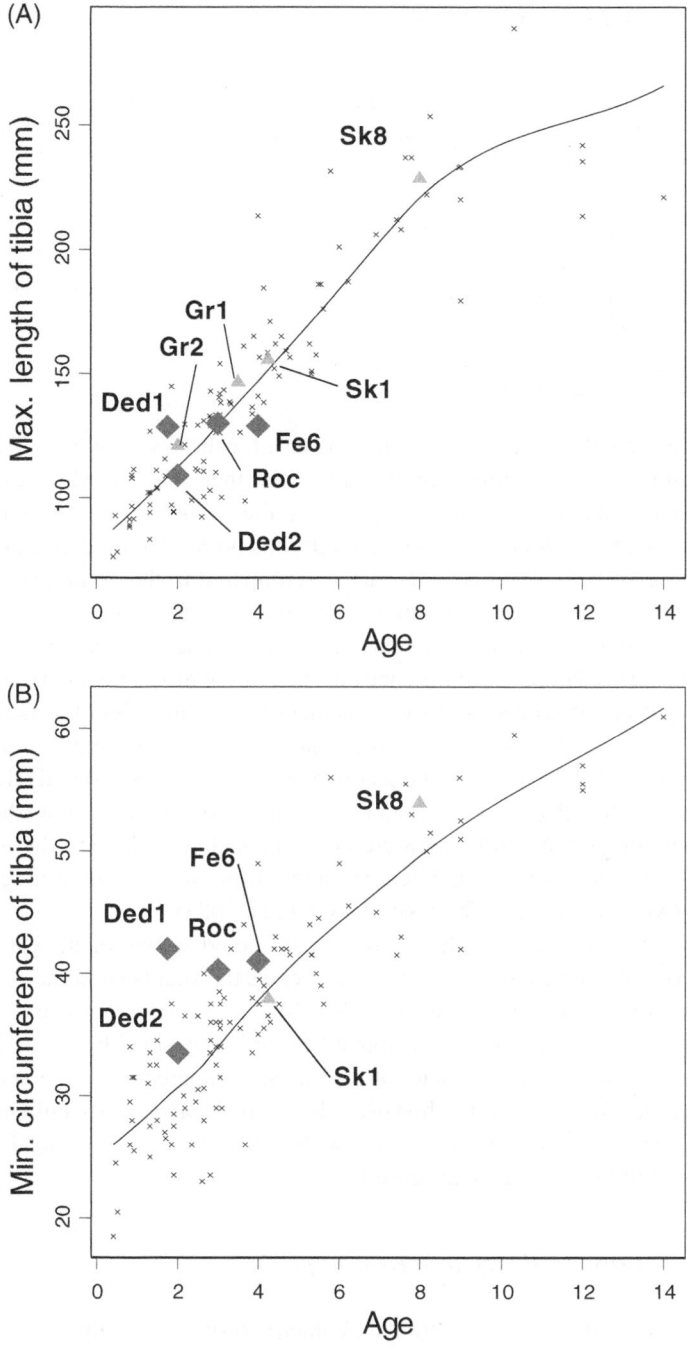

Figure 16.3   Age-related changes of the length (A) and circumference (B) of the tibia. Abbreviations for specimens are the same as in Figure 16.2.

Table 16.4 Z-*scores of fossil specimens calculated for length and circumference of the fibula*

Fibula	Z-score for maximum length	Z-score for midshaft circumference
Dederiyeh 1	1.30	3.78**
Dederiyeh 2	0.03	1.55
La Ferrassie 6	−1.19	0.45

Significance level: **<1%.

These results exhibit a consistent pattern for these young Neandertals; i.e., the length does not differ significantly from that of the modern sample and falls within the modern range of variation, the circumference (or thickness) is greater than the modern average but still falls within the modern range except for Dederiyeh 1. When Neandertals are compared to the young early modern fossils, even though few in number, we do not find meaningful differences. However, concerning the tibia, that of young Neandertals does seem shorter in length and thicker in circumference than that of young early modern *H. sapiens*. This will be visualized when we calculate the crural index (the ratio of tibia length to femur length), and the robustness index (circumference/length). We have reported elsewhere that the crural indices of young Neandertals (Dederiyeh 1, Dederiyeh 2, Roc de Marsal, and La Ferrassie 6) are smaller than the modern average, though the differences are not statistically significant; the robustness indices of tibia generally fall near the upper limit of the modern range of variation (Kondo & Dodo, 2003; Kondo & Ishida, 2003).

In terms of ontogeny, these can be considered as one of the early indications of typical Neandertal characters, such as the small crural index and large robusticity index. It is conceivable that these early indications of shape characters (that is, the ratios) are attributable to probably small but clear difference in the growth of each parameter of the ratios. Compared in a context of growth trajectories (or growth rate), however, the young Neandertals would not exhibit any clear peculiarity in postcranial growth, at least as far as absolute dimensions of lower limb bones are considered.

Cross-sectional geometric properties

Dederiyeh 1 shows cross-sectional geometric properties of the femur and tibia that reach the upper limit of variation of equivalently aged modern individuals (Table 16.5; Figures 16.5 and 16.6). Particularly in the tibia, both the polar moment of area (J) and the cortical area (CA) are significantly greater for

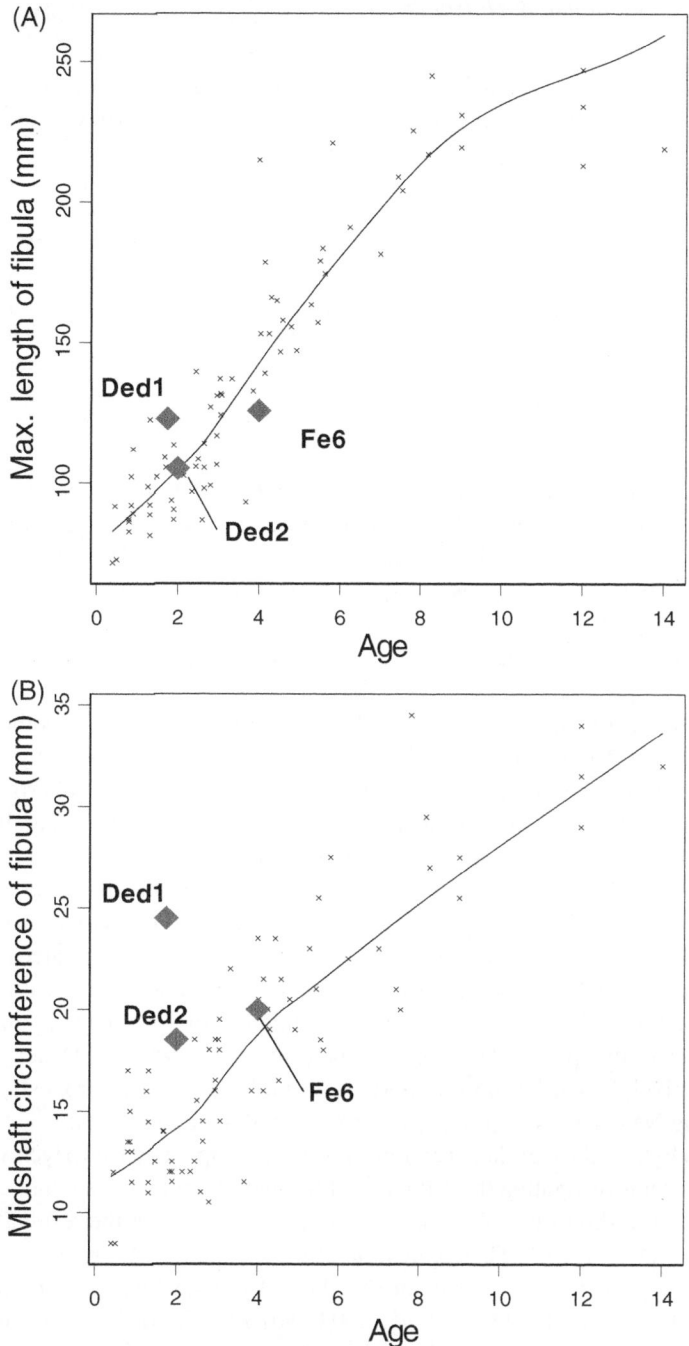

Figure 16.4   Age-related changes of the length (A) and circumference (B) of the fibula. Abbreviations for specimens are the same as in Figure 16.2.

Table 16.5 Z-scores for Dederiyeh 1 and Dederiyeh 2
calculated from cross-sectional geometric properties
(J and CA) of the femur and tibia

	Femur		Tibia	
	J	CA	J	CA
Dederiyeh 1	1.33	1.48	2.44**	2.29*
Dederiyeh 2	−0.93	−1.04	−0.62	−1.03

Significance level: *<5%, **<1%.

this individual (at levels of significance of $P = 0.01$ for J and $P = 0.05$ for CA, respectively; see Table 16.5 and Figures 16.6A and 16.6B). Dederiyeh 2, on the other hand, exhibits negative z-scores (Table 16.5), indicating smaller values than the modern averages. The circumferences of the femur and tibia of Dederiyeh 2 were roughly equivalent to the modern averages. Thus, the difference between Dederiyeh 1 and Dederiyeh 2 seems to be greater in cross-sectional properties than in linear measurements.

In terms of ontogeny, it is difficult to draw any meaningful hypotheses based on the two Dederiyeh infants. However, a previous study of ontogeny in postcranial robusticity has reported that two young Neandertals of La Ferrassie 6 and Teshik Tash 1 have large cortical areas of their femoral shafts, and argued an implication of a modified developmental pattern (Ruff et al., 1994). In their analysis, after adjustment to differences in relative body breadth, relative cortical areas (%CA) of the young Neandertals, fall in the upper end of modern humans (Ruff et al., 1994). Taking this into consideration with the greater robusticity in Dederiyeh 1, we may infer a more general appearance of great robusticity in the young Neandertals, although the number of specimens is quite small. Rather in this paper, the observed great difference between Dederiyeh 1 and Dederiyeh 2 will be more important for understanding the robusticity of the young Neandertals, especially in terms of within-site variation.

In light of the significant extreme position occupied by Dederiyeh 1, it would be worth investigating the influence of the age estimate on its relative position. Table 16.6 shows the effect of different age estimates on the residual analyses for the Dederiyeh 1. The application of the method of Moorrees et al. (1963a, 1963b), provides an age estimate of 1.3 to 2.1 years, while the method of Smith (Smith, 1991a; also Dodo et al., 2003; Mizoguchi, 2003) yields an estimate of from 1.6 to 2.5 years. Taking the maximum range of 1.3 to 2.5 years into consideration, we recalculated the z-scores on circumference of the femur, tibia,

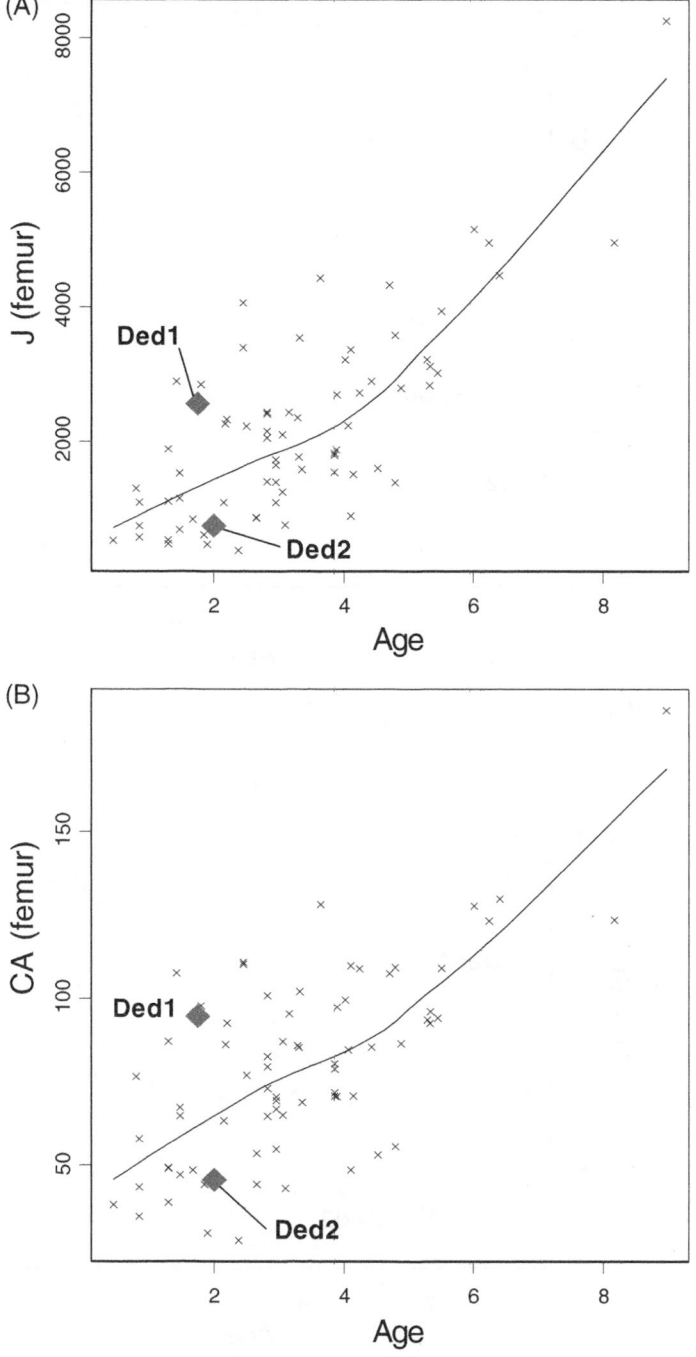

Figure 16.5 Age-related changes of (A) polar moment of area (J), and (B) cortical area (CA) at the midshaft cross-section of the femur.

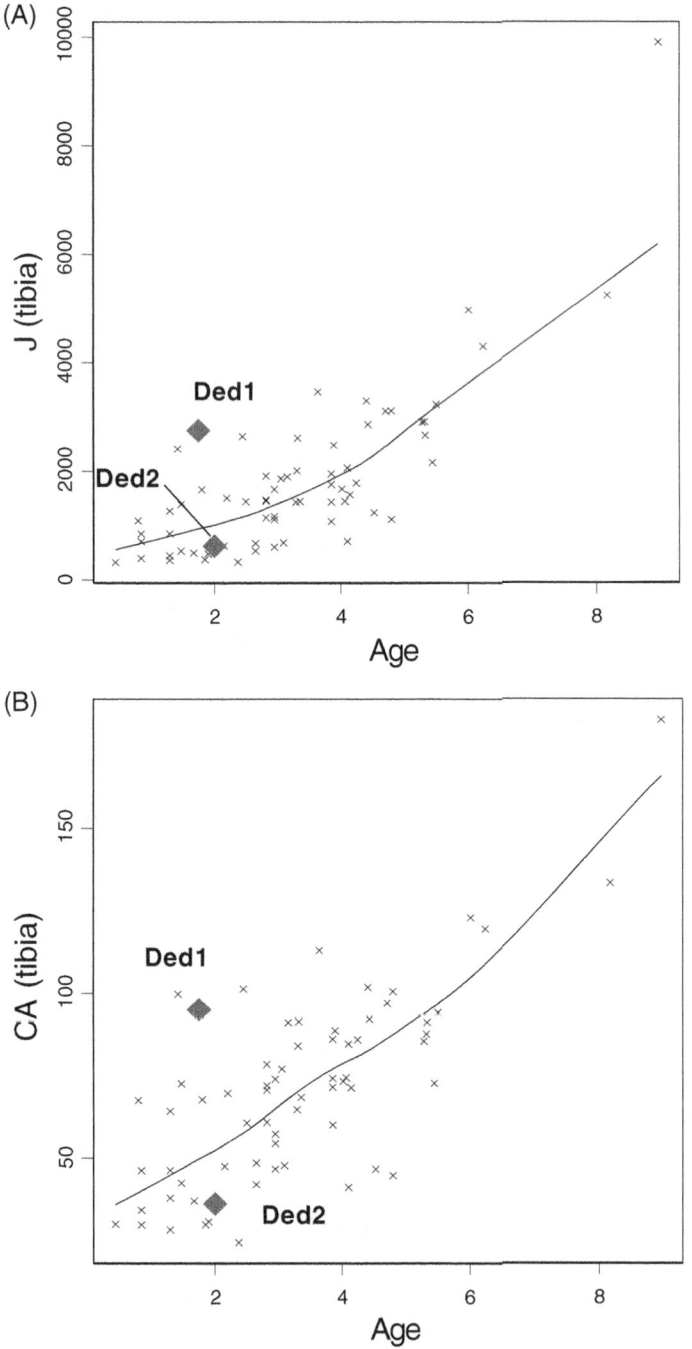

Figure 16.6   Age-related changes of (A) polar moment of area (J), and (B) cortical area (CA) at the midshaft cross-section of the tibia.

Table 16.6 Z-*scores for the estimated maximum*
and minimum ages of Dederiyeh 1 (scores based
on circumferences of the femur, tibia, and fibula)

Estimated age for Dederiyeh 1	Femur (midshaft)	Tibia (minimum)	Fibula (midshaft)
1.3	2.97**	3.03**	4.02**
2.5	2.01*	2.24*	3.40**

Significance level: *<5%, **<1%.

and fibula for the respective ages. In all cases, significant results ($P < 0.05$) for circumference were found when the maximum age estimate (2.5 years) was used (Table 16.6). This means that the Dederiyeh 1 invariably occupies the extreme position relative to the modern sample, no matter what age estimate is used. At the same time, however, the respective z-score decreases by between 0.6 and 1.0 unit with the change of the estimated age from 1.3 to 2.5 years. We should keep this decrease in mind, which corresponds to the 0.6-to-1.0 standard deviation change of the modern sample.

Difference between Dederiyeh 1 and Dederiyeh 2

An assessment of difference in the lower limb bones of Dederiyeh 1 and Dederiyeh 2 was carried out using a multivariate indicator of distance, Mahalanobis' distance based on both linear measurements (length and circumference) and cross-sectional geometric properties. The results are shown in Figures 16.7 and 16.8, respectively. Figures 16.7A and 16.8A show two-dimensional presentations of the distance matrix using multidimensional scaling, and Figures 16.7B and 16.8B show histograms of the interindividual distances of the modern sample. In the latter, the distance between Dederiyeh 1 and Dederiyeh 2 is also indicated.

Based on the linear measurements, the Mahalanobis' distance between Dederiyeh 1 and Dederiyeh 2 (the value is 11.65) is not large compared to the interindividual distances among modern individuals. A probabilistic assessment of the distance between Dederiyeh 1 and Dederiyeh 2 ($P = 0.41$) is not significant (Figure 16.7). This probability is calculated as the proportion of the frequencies of the modern individuals whose interindividual distance exceeds that between Dederiyeh 1 and Dederiyeh 2. The result based on the cross-sectional geometric properties emphasizes the peculiarity of the position of Dederiyeh 1

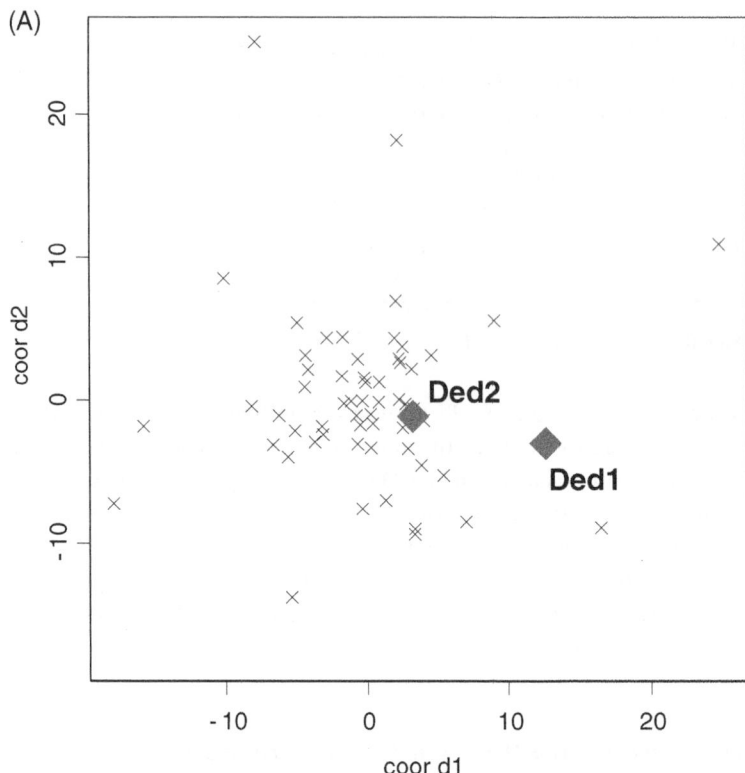

Figure 16.7   Two-dimensional representation (A) and histogram (B) of Mahalanobis'
distances based on the age-independent residuals of the length and circumference of
femur, tibia, and fibula of Dederiyeh 1 and Dederiyeh 2.

(Figure 16.8A). However, the difference between Dederiyeh 1 and Dederiyeh 2
still does not exceed the modern range of variation (the Mahalanobis' distance
is 13.19, $P = 0.16$, in Figure 16.8B). These results reconfirm the peculiarity of Dederiyeh 1, but also indicate that the difference between Dederiyeh 1
and Dederiyeh 2 does not exceed the range of variation of modern immature
individuals.

Discussion

One of the main results from above analyses is the non-peculiarity in length
and thickness of the lower limb bones among Neandertals and early modern
H. sapiens, at least in comparison to the range of modern human variation. In

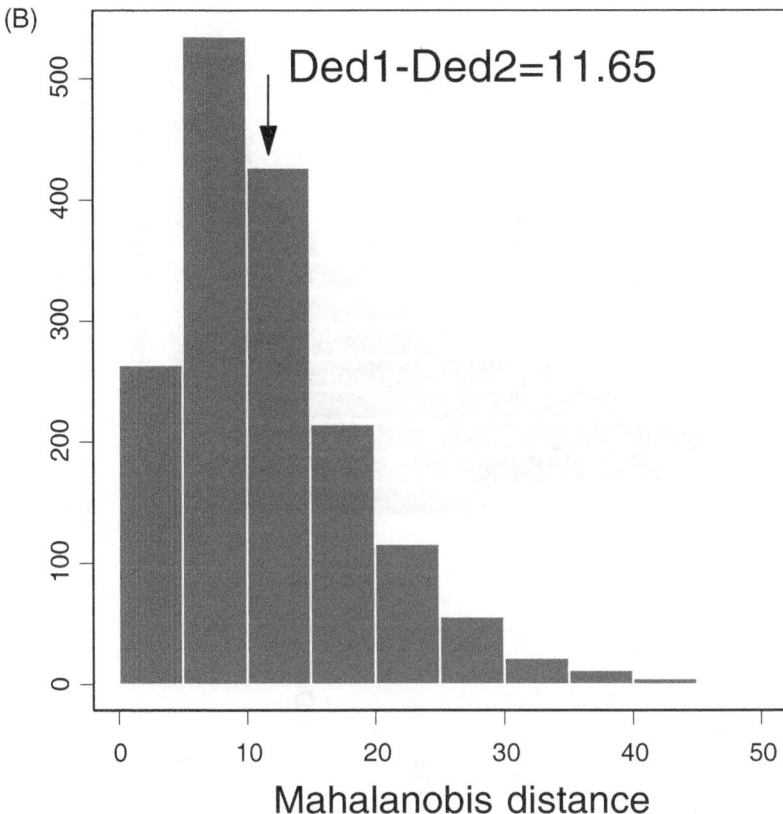

(B)

Ded1-Ded2=11.65

Mahalanobis distance

Figure 16.7 (*cont.*).

more concrete points, one Neandertal individual, Dederiyeh 1, exhibits greater robusticity in some postcranial traits, which significantly exceed the modern range of variation. From a general ontogenetic point of view, this supports the hypothesis that the Neandertals and modern humans did not differ much in postcranial growth patterns. However, there still remain to be solved several prerequisites or related issues, which we discuss here.

Sample size as a comparative unit

First we should address the small sample sizes for Neandertals and early modern *H. sapiens* as population units of comparison. For the comparison of growth patterns, the sample size should ideally be large enough to express a normal range of variation within each growth stage. This cannot be reasonably assumed

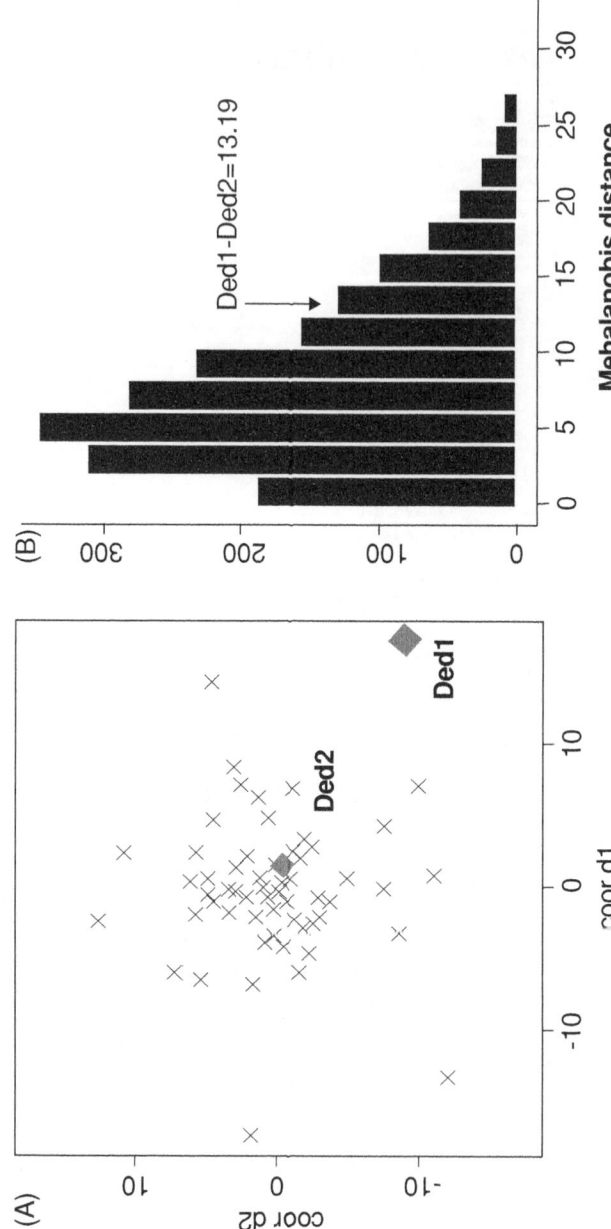

Figure 16.8 Two-dimensional representation (A) and histogram (B) of Mahalanobis' distances based on the age-independent residuals of the cross-sectional geometric properties of femur and tibia of Dederiyeh 1 and Dederiyeh 2.

in the present situation, because of the limited number and the fragmentary preservation of immature fossil hominids, particularly in respect of postcranial bones. Therefore we can at best assess an individual's position relative to modern humans and accumulate comparable data for future analyses.

This small sample size strongly obscures the interpretation of the average or range of variation in fossil hominids. In this analysis, we could detect the significantly strong robusticity for several traits in Dederiyeh 1's postcranium, although a multivariate measure of the distance between Dederiyeh 1 and Dederiyeh 2 did not exceed the modern range of variation. Nevertheless, it is uncertain to what degree this single specimen contributes to the interpretation of Neandertal growth because the small sample size makes it difficult for us to assess the peculiarity of this specimen. In the end, this specimen will expand the range of variation for immature Neandertals but might not change the average. Thus, it can not exhibit any clearly different growth trajectories for Neandertals from those of the modern humans.

Comparison of growth patterns

It has been argued that the analysis of growth patterns requires measurements of growth processes and thus the comparison of absolute values cannot fit the case even within a particular developmental stage. For example, Trinkaus & Tompkins (1990) indicated that the bigger brains of Neandertal infants in the absolute sense does not simply mean faster brain growth, because the cranial volume of adult Neandertals generally exceeds that of recent modern humans. Thompson & Nelson (2000) addressed this issue by considering femoral length as a proportion of the adult average as a relative growth indicator. When they compared *relative* femoral growth to dental development they suggested that Neandertal growth was retarded in terms of body (femur) growth or advanced in terms of dental development.

These arguments are reasonable from the point of view of comparison of growth patterns, but still at issue is the assessment of individual variation. It is difficult to assess individual variation of growth in the fossil specimens because longitudinal data are not available for the fossils. Specifically, we cannot know the actual adult size of any immature fossil skeleton, so we have to use an adult average as a representation of the adult size when calculating the proportional growth pattern of the skeleton. This proportional growth analysis may lead to over- or underestimates of individual variations of growth, depending on how accurately the adult mean is estimated or on how close the individual would have been to the adult mean, had they survived to adulthood. In order to clarify the individual growth variation for fossil specimens, therefore, we feel that we will have to deal with the absolute values of the respective specimens and make an assessment of individual or interindividual differences. Both of the absolute and relative proportional assessment will be useful or should be considered together.

Following the above discussion, we should mention another approach using bivariate relationship. For example, Nelson & Thompson (2002) used bivariate comparisons of postcranial dimensions (length and cross-sectional area) in a broader ontogenetic perspective and argued that distal limb shortening in Neandertals appeared early, while the greater robusticity did not appear until puberty. We should note here, however, that their approach does not directly demonstrate growth trajectories or age-related variation. Thus, although there are several approaches to studying growth patterns, each has respective advantages and disadvantages. Understanding all these aspects and then integrating results from different approaches will be required for interpretation of skeletal growth in fossil hominids.

Uncertainty of age estimation

In addition to the above discussion, age estimation can be one of the most significant issues related to the postcranial growth analyses.

The age estimates used in this paper were based on dental developmental patterns. However, this practice itself can be a focus of discussion. For example, the relative pattern of dental development has been compared between Neandertals and modern humans and a degree of difference has been documented (Tompkins, 1996). Tompkins's study indicated that the relative timing of crown formation in Neandertals was similar to that in early modern *H. sapiens* including Skhul–Qafzeh specimens and European Upper Paleolithic specimens, and all these pre-recent fossil groups differ from the modern sample, although without a consistent pattern of difference. If this is the case, it becomes unwarranted to apply the modern human categories of dental development to Neandertals for estimation of age at death. More detailed studies will be required concerning the degree of difference in dental developmental patterns, and their effect on age estimation and ultimately, on the comparison of the postcranial growth patterns.

Issues in the study of growth of postcranial bones

It is unfortunate that postcranial bones do not show any definitive traits of growth increments such as perikymata or cross-striations in the crown of the dentition. Thus, the postcranial data usually express only an outcome at the time of death and they do not exhibit the process or duration of growth. That is, they can not tell the true history of an individual's growth. The dentition, on the other hand, records the growth increments in the enamel or dentine, and therefore, can directly provide one aspect of the individual's growth history. In this sense, we can recognize the postcranial growth data as complementary to those from the dentition. As such, we should integrate these different aspects of growth data for understanding the growth of fossil hominids.

In addition, it is well known that skeletal growth is greatly affected by environmental factors such as the social or nutritional status of individuals. In practice, the investigation of the effect from such extrinsic factors is one of the main purposes for postcranial growth studies (Saunders, 1992). Growth differences based on postcranial data accordingly include the effect of difference in such environmental factors. We should keep in mind that the study of growth patterns in the skull, dentition, and postcranial bones reflects different levels of the effects from these extraneous factors.

Conclusion

Our approach for investigating the postcranial growth of fossil hominids is to interpolate the absolute values into growth trajectories of the modern sample. This approach clarifies individual growth variations in the modern sample and thus provides the relative position of each fossil both visually and statistically.

The results indicate that there is no significant difference in postcranial growth between Neandertals and modern humans, though a more detailed reviewing is preferable. The analysis of length and circumference of the femur, tibia, and fibula provides the following conclusions:

1. The young Neandertals and early modern *H. sapiens* do not exhibit any clear peculiarity in length and they fall within the modern range of variation in circumference.
2. Among the young fossils, only Dederiyeh 1 occupies an extreme position for shaft circumference.
3. Based on cross-sectional geometric properties of the femur and tibia, the shaft robusticity of Dederiyeh 1 is significantly great and thus the difference between Dederiyeh 1 and Dederiyeh 2 is remarkable.
4. The observed difference between Dederiyeh 1 and Dederiyeh 2 does not exceed the modern range of variation.

These results provide a basis for discussion and interpretation of the growth patterns in Neandertals. At the same time, we should consider several issues that appear in the study of the growth of postcranial bones of fossil hominids, such as the small sample size, imprecise age estimation, and the questionable grounds for the application of modern aging standards to fossils, influence of environmental factors, and so on. In addition, we should take great care in the selection of approaches or methods because each method has different implications. An understanding of these issues and an integration of the results based on dental, cranial, and postcranial evidence will allow us to approach a reasonable consensus on the growth and development of fossil hominids.

Acknowledgments

We are grateful to Drs J. Thompson, G. Krovitz, and A. Nelson for inviting us to contribute to this volume. We thank Drs B. Vandermeersch, A.-m. Tillier, and B. Maureille, Université de Bordeaux I; Drs A. Langeney and J. L. Heim, Musée de l'Homme; Dr Jean-Jacques Cleyet-Merle, Musée National de Préhistoire; Drs C. Stringer and R. Kruszynski, Natural History Museum; Drs G. Suwa and T. Kamiya, University Museum, University of Tokyo; Dr H. Ishida, Kyoto University, for permission to access original fossil specimens or casts and modern human collections. Thanks also to Dr K. Sakaue, Tohoku University School of Medicine for taking radiographs and calculating the cross-sectional geometric properties, and to the editors of this volume, for kindly correcting our English text.

O. Kondo would like to thank Dr T. Molleson for permission to examine the Spitalfields and Poundbury collection at the Natural History Museum, London with the special appreciation to Drs R. Kruszynski and L. Humphrey who helped to access the materials and to take the radiographs. He also thanks Dr K. L. Kuykendall, University of the Witwatersrand and Dr M. Steyn, University of Pretoria for permission to investigate the child specimens of the South African Blacks in their care and for providing the radiographs of the mandibular dentitions.

This study was supported in part by a Grant-in-Aid for Scientific Research and Overseas Scientific Research from the Ministry of Education, Science and Culture and from the Japan Society for the Promotion of Science.

References

Akazawa, T., Muhesen, S., Dodo,Y., Kondo, O., & Mizoguchi, Y. (1995a). Neanderthal infant burial. *Nature*, **377**, 584–585.
Akazawa, T., Muhesen, S., Dodo, Y., Kondo, O., Mizoguchi, Y., Abe, Y., Nishiaki, Y., Ohta, S., Oguchi, T., & Haydal, J. (1995b). Neanderthal infant burial from the Dederiyeh Cave in Syria. *Paléorient*, **21**, 77–86.

Akazawa, T., Muhesen, S., Ishida, H., Kondo, O., & Griggo, C. (1999). New discovery of a Neanderthal child burial from the Dederiyeh Cave in Syria. *Paléorient*, **25**, 127–140.

Brothwell, D. (1975). Adaptive growth rate changes as a possible explanation for the distinctiveness of the Neanderthalers. *Journal of Archaeological Science*, **2**, 161–163.

Dean, M. C., Stringer, C. B., & Bromage, T. G. (1986). Age at death of the Neanderthal child from Devil's Tower, Gibraltar and the implications for studies of general growth and development in Neanderthals. *American Journal of Physical Anthropology*, **70**, 301–309.

Dean, C., Leakey, M. G., Reid, D., Schrenk, F., Schwartz, G. T., Stringer, C., & Walker, A. (2001). Growth processes in teeth distinguish modern humans from *Homo erectus* and earlier hominins. *Nature*, **414**, 628–631.

Dodo, Y., Kondo, O., & Nara, T. (2003). The skull of the Neanderthal child of the burial No.1. In *Neanderthal Burials: Excavation of the Dederiyeh Cave, Afrin, Syria*, eds. T. Akazawa & S. Muhesen, pp. 93–137. Rome: L'Erma di Bretschneider.

Heim, J. L. (1982). *Les Enfants néandertaliens de La Ferrassie*. Paris: Masson.

Heim, J. L. (1991). L'Enfant magdalénien de la Madeleine. *L'Anthropologie (Paris)*, **95**, 611–638.

Ishida, H., Kondo, O., Muhesen, S., & Akazawa, T. (2000). New Neanderthal child remains recovered at Dederiyeh cave, Syria in 1997–1998. *American Journal of Physical Anthropology*, Suppl. **30**, 186–187.

Ishida, H., & Kondo, O. (2003). The skull of the Neanderthal child of burial No.2. In *Neanderthal Burials: Excavation of the Dederiyeh Cave, Afrin, Syria*, eds. T. Akazawa & S. Muhesen, pp. 271–297. Rome: L'Erma di Bretschneider.

Kondo, O., & Dodo, Y. (2003). The postcranial bones of the Neanderthal child of the burial No.1. In *Neanderthal Burials: Excavation of the Dederiyeh Cave, Afrin, Syria*, eds. T. Akazawa & S. Muhesen, pp. 139–214. Rome: L'Erma di Bretschneider.

Kondo, O., & Ishida, H. (2003). The postcranial bones of the Neanderthal child of the burial No.2. In *Neanderthal Burials: Excavation of the Dederiyeh Cave, Afrin, Syria*, eds. T. Akazawa & S. Muhesen, pp. 299–321. Rome: L'Erma di Bretschneider.

Madre-Dupouy, M. (1992). *L'Enfant du Roc de Marsal: Étude Analytique et Comparative*, Cahiers de Paléonthropologie. Paris: Editions du CNRS.

McCown, T. D., & Keith, A. (1939). *The Stone Age of Mount Carmel*, vol. 2. Oxford: Clarendon Press.

Minugh-Purvis, N. (1988). Patterns of craniofacial growth and development in Upper Pleistocene hominids. PhD dissertation, University of Pennsylvania, Philadelphia.

Mizoguchi, Y. (2003). Dental remains excavated at the Dederiyeh Cave during the 1989–1994 seasons. In *Neanderthal Burials: Excavation of the Dederiyeh Cave, Afrin, Syria*, eds. T. Akazawa & S. Muhesen, pp. 221–262. Rome: L'Erma di Bretschneider.

Moorrees, C. F. A., Fanning, E. A., & Hunt, E. E. Jr (1963a). Age variation of formation stages for ten permanent teeth. *Journal of Dental Research*, **42**, 1490–1502.

Moorrees, C. F. A., Fanning, E. A., & Hunt, E. E. Jr (1963b). Formation and resorption of three deciduous teeth in children. *American Journal of Physical Anthropology*, **21**, 205–213.

Nelson, A. J., & Thompson, J. L. (2002). Neandertal adolescent postcranial growth. In *Human Evolution through Developmental Change*, eds. N. Minugh-Purvis & K. McNamara, pp. 442–463. Baltimore: Johns Hopkins University Press.

Ponce de León, M. S., & Zollikofer, C. P. (2001). Neanderthal cranial ontogeny and its implications for late hominid diversity. *Nature*, **412**, 534–538.

Rak, Y., Kimbel, W. H., & Hovers, E. (1994). A Neanderthal infant from Amud Cave, Israel. *Journal of Human Evolution*, **26**, 313–324.

Rochlin, D. G. (1949). A few radiological observations on the skeleton of the Neandertal child from the cave of Teshik-Tash. In *Teshik-Tash*, eds. M. A. Gremyatskij & M. F. Nesturkh, pp. 109–121. Moscow: Trudy Uzbekist. Fil. Akad. Nauk. (In Russian.)

Ruff, C. B., Walker, A., & Trinkaus, E. (1994). Postcranial robusticity in *Homo*. III: Ontogeny. *American Journal of Physical Anthropology*, **93**, 35–54.

Runestad, J. A., Ruff, C. B., Nieh, J. C., Thorington, R. W., & Teaford, M. F. (1993). Radiographic estimation of long bone cross-sectional geometric properties. *American Journal of Physical Anthropology*, **90**, 207–213.

Sakaue, K. (1997). Bilateral asymmetry of the humerus in Jomon people and modern Japanese. *Anthropological Science*, **105**, 231–246.

Sasaki, C., Suzuki, K., Mishima, H., & Kozawa, Y. (2003). Age determination of the Dederiyeh 1 Neanderthal child using enamel cross-striations. In *Neanderthal Burials: Excavation of the Dederiyeh Cave, Afrin, Syria*, eds. T. Akazawa & S. Muhesen, pp. 263–267. Rome: L'Erma di Bretschneider.

Saunders, S. R. (1992). Subadult skeletons and growth related studies. In *Skeletal Biology of Past Peoples: Research Methods*, eds. S. R. Saunders & M. A. Katzenberg, pp. 1–20. New York: Wiley-Liss.

Saunders, S. R., DeVito, C., Herring, A., Southern, R., & Hoppa, R. (1993). Accuracy tests of tooth formation age estimation for human skeletal remains. *American Journal of Physical Anthropology*, **92**, 173–188.

Sinel'nikov, N. A., & Gremyatskij, M. A. (1949). Bones of the skeleton of the Neandertal child from the cave of Teshik-Tash, southern Uzbekistan (in Russian). In *Teshik-Tash*, eds. M. A. Gremyatskij & M. F. Nesturkh, pp. 123–135. Moscow: Trudy Uzbekist. Fil. Akad. Nauk.

Smith, B. H. (1991a). Standards of human tooth formation and dental age assessment. In *Advances in Dental Anthropology*, ed. M. A. Kelly & C. S. Larsen, pp. 143–168. New York: Wiley-Liss.

Smith, B. H. (1991b). Dental development and the evolution of life history in Hominidae. *American Journal of Physical Anthropology*, **86**, 157–174.

Stringer, C. B., Dean, M. C., & Martin, R. D. (1990). A comparative study of cranial and dental development within a recent British sample and among Neandertals. In *Primate Life History and Evolution*, ed. C. J. De Rousseau, pp. 115–152. New York: Wiley-Liss.

Takahashi, H., & Adachi, K. (1982). A computer program calculating geometric properties of two-dimensional shapes by using the digitizer. *Interface*, **4**, 200–214. (In Japanese.)

Thompson, J. L., & Nelson, A. J. (2000). The place of Neandertals in the evolution of hominid patterns of growth and development. *Journal of Human Evolution*, **38**, 475–495.

Tillier, A.-m., (1989). The evolution of modern humans: Evidence from young Mousterian individuals. In *The Human Revolution*, eds. P. Mellars & C. Stringer, pp. 286–297. Edinburgh: Edinburgh University Press.

Tillier, A.-m. (1999). *Les Enfants Moustériens de Qafzeh: Interprétation Phylogénétique et Paléoauxologique*, Cahiers de Paléoanthropologie. Paris: Editions du CNRS.

Tompkins, R. L. (1996). Relative dental development of Upper Pleistocene hominids compared to human population variation. *American Journal of Physical Anthropology*, **99**, 103–118.

Tompkins, R. L., & Trinkaus, E. (1987). La Ferrassie 6 and the development of Neandertal pubic morphology. *American Journal of Physical Anthropology*, **73**, 233–239.

Trinkaus, E., & Tompkins, R. L. (1990). The Neandertal life cycle: The possibility, probability, and perceptibility of contrasts with recent humans. In *Primate Life History and Evolution*, ed. C. J. De Rousseau, pp.153–180. New York: Wiley-Liss.

Vlček, E. (1973). Postcranial skeleton of a Neanderthal child from Kiik-Koba, U.S.S.R. *Journal of Human Evolution*, **2**, 537–544.

Williams, F. L. (2000). Heterochrony and the human fossil record: Comparing Neandertal and modern human craniofacial ontogeny. In *Neanderthals on the Edge*, eds. C. B. Stringer, R. N. E. Barton, & J. C. Finlayson, pp. 257–267. Oxford: Oxbow Books.

17 Hominid growth and development in Upper Pleistocene Homo

A. J. NELSON
University of Western Ontario

G. E. KROVITZ
Pennsylvania State University

J. L. THOMPSON
University of Nevada, Las Vegas

> I have now, as it appears to me, satisfactorily shown that not only in its general, but equally so in its particular characters, has the fossil under consideration the closest affinity to the apes. Only a few points of proximate resemblance have been made out between it and the human skull; and these are strictly peculiar to the latter in the *fœtal state*. The cranium of the human fœtus, however, possesses the lofty dome, the forward position of the frontal respectively to the outer orbital processes, the greatest width at the parietal centers of ossification and the vertical occiput, which are so conspicuous in the adult, but which are remarkably non-characteristic of the Neanderthal skull.
>
> (King, 1994 (1864): 30)

Introduction

Much of the research regarding the evolution of the hominid pattern of growth and development has focused on material from the Upper Pleistocene, particularly on the Neandertals. The Neandertal fossil record is notable not only in terms of absolute numbers of adult and juvenile fossils, but because it preserves skeletons with associated cranial and postcranial remains. William King (in 1864) was responsible for the first publication of the taxonomic name *Homo neanderthalensis* (as discussed in Meikle & Parker, 1994). As the above quote demonstrates, King anticipated the concept of neoteny. Although it does not quite match our current definition of this term, the importance of this observation

Patterns of Growth and Development in the Genus Homo, ed. J. L. Thompson, G. E. Krovitz and A. J. Nelson. Published by Cambridge University Press. © Cambridge University Press 2003.

412

is that even in 1864, King was aware of the importance of the comparison of adult and subadult specimens in order to come to valid scientific conclusions.

We are fortunate to have as contributors to this volume several scholars who have made substantive contributions to this body of research over many years (particularly Tillier and Minugh-Purvis) as well as other scholars who bring new data and new analytical techniques to the table. The papers in this volume that analyze Upper Pleistocene remains fall into two general categories, which reflect the work in this field as a whole. The first category includes one study that uses the relatively abundant fossil record of the Upper Pleistocene to test models of evolutionary change, particularly neoteny. The second category includes detailed examinations of morphological traits that have phylogenetic implications. Typically, such studies attempt to identify when traits used to characterize the adult forms of different fossil samples (autapomorphs) appear in the course of growth and development. These studies seek to differentiate "phylogenetically ingrained" (Vlček, 1973: 537) traits, which reflect fundamental genetic differences, from traits that could arise by other means, such as responses to environmental conditions or behavior.

This chapter will summarize these studies of Upper Pleistocene material within the context of previous research on evolutionary models of morphological change between populations, and phylogenetic relationships among populations of Upper Pleistocene hominids. This latter debate, which focuses on the question of the origin of anatomically modern *Homo sapiens*, is one of the most exciting areas of research in hominid evolution and there is a great deal that the study of growth and development can contribute to this subject. Before we can ask the question "When did modern humans appear?", we need to ask the question "When did the modern human pattern of growth and development appear?", as adult modern humans are the endpoint of a complex sequence of growth and developmental events that is not shared with other hominids. Once we focus on the question of modern human origins in this manner, we are led to other questions such as "Is the modern human pattern of growth and development shared with non-modern Upper Pleistocene hominids?", or "Are there profound enough differences in growth and development between modern human and other Upper Pleistocene hominids to warrant a species-level distinction between those hominids (principally Neandertals) and anatomically modern *H. sapiens?*" It is the fundamental goal of these papers, this summary chapter, and this volume to address these questions about the evolution of human growth patterns.

Models of evolutionary change

Many scholars have noted that Neandertals were bigger than modern humans in almost every dimension: cranial capacity, femoral head diameter, shaft

diameters, and so on. Despite that fact, until recently the only body mass estimate in the literature for Neandertals was 56.8 kg (Oleksiak, 1986) – a figure
that is *less* that the generally accepted mean for modern humans (see Nelson
et al., Conclusion, this volume). However, more recent studies have produced
much larger estimates (Kappelman, 1996; Nelson *et al.*, 1992; Ruff *et al.*,
1997), which exceed the modern human body mass mean. De Beer (1958) and
Brothwell (1975) are among those who first suggested that Neandertals were,
in essence, overgrown humans, while Gould (1977) invoked the process of
neoteny in order to explain the origin of modern *H. sapiens*. Neoteny and other
variants of heterochrony are now frequently discussed in the literature regarding
hominid evolution (e.g., Antón & Leigh, this volume; Godfrey & Sutherland,
1996; Minugh-Purvis & McNamara, 2002; Shea, 1989).

The study by Williams *et al.* (this volume) does not test specific hypotheses
of ancestry, rather it examines whether particular heterochronic perturbations
could turn a hypothetical chimpanzee-like ancestor (*Pan troglodytes*) into a
bonobo (*P. paniscus*), or a hypothetical Neandertal-like ancestor into a modern human. They use a computer program called HETPAD as their analytical
tool, which is designed to (1) predict possible descendants given assumed ancestors and a particular set of heterochronic transformations, and (2) test the
explanatory power of heterochrony by calculating a mathematical treatment of
neomorphosis (a novel form not arising from heterochrony) as the "residual
error" between a modeled and observed descendant. In this study, the specific
test was to use the "rules" of global craniofacial neoteny in order to investigate
the transition from a hypothetical Neandertal-like ancestor for modern humans,
and modern *H. sapiens*.

The real power of an exercise such as this is that it forces the researchers to
be rigorous in terms of laying out how they see heterochrony as working in the
evolutionary process. The Hardy–Weinberg equilibrium model is an important
contribution to evolutionary science, as it forces us to identify clearly the elements that can lead to evolutionary change, and provides us with a means of
quantifying that change. Likewise, the program HETPAD can test evolutionary
models, based on a clear exposition of possible heterochronic processes. However, we must also recognize that any model is, by necessity, a simplification
of the world, and may not include enough complexity to fully understand all
possible aspects of a situation.

Williams *et al.* (this volume) focus on an ontogenetic analysis of craniofacial
measurements, and come to the conclusion that modern humans are very poor
neotenes relative to Neandertals. Furthermore, breaking down the craniofacial
complex into component modules demonstrates that a simple model of heterochronic change is insufficient to explain a transition from a Neandertal to a
modern human pattern of growth and development (also see Williams *et al.*,
2002). Williams *et al.* (this volume) identify several important ontogenetic

differences between Neandertals and modern humans, including: (1) most Neandertal craniofacial traits grew faster, relative to dental eruption, than those of modern humans; (2) the shape paths of Neandertals and modern humans are different from a very early age and continue to diverge throughout ontogeny; and (3) although modern humans, with respect to Neandertals, are slightly juvenilized in certain aspects of the face, and slightly "adultified" in some aspects of calvarial morphology, global neoteny fails to explain these outcomes.

This contribution illustrates the importance of being rigorous in the construction of evolutionary models. Furthermore, it highlights the complexity of changes in growth programs between a hypothetical Neandertal-like ancestor and modern humans. Clearly, modern humans are not simply underdeveloped Neandertals. Does this conclusion negate a transition from Neandertals to modern humans? Could modern humans be derived from a different ancestor through the process of neoteny? Unfortunately, we lack the crucial ante-Neandertal (both European and from elsewhere) material necessary to answer this question. It is worthy of note that Anton & Leigh (this volume) and others (e.g., Montagu, 1955) have suggested that modern humans might be neotenous with respect to *H. erectus*. That suggestion is certainly fodder for future research.

Morphological studies

The remainder of the studies in this section focus on the analysis of morphological traits or trait complexes. Kondo & Ishida (this volume), Coqueugniot & Minugh Purvis (this volume), and Majó & Tillier (this volume) all examine particular traits that have been used in phylogenetic analyses. While they do not take explicit phylogenetic positions, this kind of detailed study is critical to the appropriate interpretation of those traits within a phylogenetic context. Krovitz (this volume) examines craniofacial morphology and growth patterns from a multivariate point of view, with the explicit goal of testing a phylogenetic hypothesis.

One important question for researchers working in the Upper Pleistocene is: When and where did modern humans arise? That question is, of course, closely related to the focus of this volume: When did the modern human pattern of growth and development arise? The two major schools of thought are "Regional Continuity" (also Multiregional Evolution") (e.g., Wolpoff, 1980, 1996; Wolpoff & Caspari, 1997; Wolpoff *et al.*, 1984) and "Out of Africa" (also the "Replacement Hypothesis") (e.g., Stringer, 1994; Stringer & Andrews, 1988; Stringer *et al.*, 1984; Tattersall, 1999). These positions are well known (as are the various permutations of them), so they will not be discussed in detail here. However, the key difference has to do with when and how *H. sapiens* appeared.

As recently elaborated (Wolpoff, 1996), Regional Continuity sees the species *H. sapiens* as arising immediately after *H. habilis*, so the hominids previously ascribed to *H. erectus* and Neandertals are all considered to be conspecific with *H. sapiens*. Under this model, evolution within *H. sapiens* has been an ebb and flow of traits, due to gene flow, between populations of this polytypic species, with regional continuity from earlier to later (modern) populations. By contrast, the Out of Africa model suggests that *H. sapiens* arose as a cladogenetic event in sub-Saharan Africa some 150 000 to 200 000 years ago, with this population moving out to replace the populations of hominids they encountered in Europe and Asia. Hence, these two positions have very different expectations in terms of how Neandertals, early modern humans, and recent *H. sapiens* relate to each other, and how traits should be distributed among these populations.

The postcranium: Limb length and circumference

Linear growth in stature, or in long-bone length, is one of the key features used to characterize growth and development programs. In addition, adult Neandertals and modern humans are well known to have differed in terms of limb robusticity and limb proportions (e.g., Nelson & Thompson, 1999, 2002; Ruff *et al.*, 1993; Stringer & Gamble, 1993; Trinkaus, 1983a, 1997). Thus, an understanding of the growth and development of the postcranium is extremely important in terms of understanding overall body form, and how these well-recognized adult differences in form arise during ontogeny.

Kondo & Ishida (this volume) do not take an explicit position on the phylogenetic status of Neandertals and modern humans, but they do conclude that the pattern of growth and development demonstrated in the Neandertal postcranium cannot be differentiated from that of modern humans. Despite the young age of Dederiyeh 1 and Dederiyeh 2, these hominids have been identified as Neandertals based on cranial, mandibular, dental, and postcranial characters (Akazawa *et al.*,1995; Dodo *et al.*, 1998). Vlček (1973) described similar postcranial traits for the 5–7-month-old Kiik-Koba infant and interpreted the traits as being distinct from modern humans. The early appearance of unique Neandertal postcranial traits has also been noted by Golovanova *et al.* (1999), Heim (1982), and Madre-Dupouy (1992) for the Mezmaiskaya, La Ferrassie, and Roc de Marsal individuals, respectively. Recently, Duarte *et al.* (1999) used limb proportions as a centerpiece of their argument identifying the Lagar Velho subadult as a Neandertal/modern human hybrid. Thus, traits such as robusticity and shortened distal segments have been recognized as "phylogenetically ingrained" (*sensu* Vlček, 1973) diagnostic traits for Neandertals, which appear very early in ontogeny.

However, the discovery of the more gracile, but comparably aged Dederiyeh 2 individual stimulated the detailed analysis presented in this volume. In this study, Kondo & Ishida examine age-related growth in length and circumference of the leg bones and cross-sectional properties of the femur and tibia. They conclude that young Neandertals and modern humans differ very little in terms of linear growth of the long bones. In addition, while the Neandertal long bones are somewhat larger in circumference than modern human subadults of the same age, most Neandertal subadults are encompassed within the range of variation of modern humans. Thus, Kondo & Ishida conclude that the pattern of growth in the length and circumference of young Neandertal long bones is not statistically different from that of modern humans.

Considerations

Kondo & Ishida's findings that, at a very young age, Neandertal long-bone length, circumference, and cross-sectional properties do not differ significantly from modern humans, might seem to contradict the findings of Vlček (1973), Heim (1982), and Madre-Dupouy (1992). There are two potential reasons why these results might not truly be contradictory: (1) Kondo & Ishida are able to examine variability in trait expression rather than relying on a single individual to be representative of Neandertals as a whole; and (2) the age range that is being examined in this particular study is quite narrow.

While the postcranium of Dederiyeh 1 is robust and that of Dederiyeh 2 is gracile, the Mahalanobis' distance analysis found that the distance between these two individuals was not beyond that expected within the modern human sample. When combined with the fact that Neandertals do tend to have shaft diameters that are somewhat larger than those of modern humans, one could suggest that the whole Neandertal range of variation in these dimensions is shifted positively relative to the modern sample. Unfortunately, due to the small fossil sample size we cannot test this proposition, and we are left with the reality that the difference between individual fossil data points and the modern human sample is not statistically significant. The marked morphological difference between Dederiyeh 1 and Dederiyeh 2 is an extremely important cautionary tale for the analysis of single individuals from other sites, as there is no way to know how a single specimen fits into the normal range of variation for that population or species.

It is also possible that the young age of the Dederiyeh specimens is obscuring differences between Neandertals and modern humans. A recent study by Nelson & Thompson (2002), which did not include hominids from the Levant, but did include older Neandertal subadults, showed that it was difficult to see clear differences between Neandertals, Upper Paleolithic individuals, and a sample of climatically diverse modern human individuals at early ages. Nelson &

Thompson (2002) concluded that limb robusticity (length and shaft dimensions considered together) in Neandertals did not become truly distinct from modern humans until adolescence. Kondo & Ishida's work greatly expands our understanding of the early stages of Neandertal postcranial ontogeny and shows that the limb robusticity that characterizes adult Neandertals is variably expressed and not clearly identifiable at early ages. See Ruff *et al.* (1994) and Trinkaus & Ruff (1996) for other discussions on the ontogeny of lower limb robusticity in Upper Pleistocene hominids.

Additionally, multiple specimens from the same site, like Dederiyeh, allow the analysis of within-sample variability that provides an important baseline for any study that attempts to use postcranial characters as phylogenetically significant traits in the identification of young fossil hominids, as has been done with the Lagar Velho specimen (Duarte *et al.*, 1999; for a more detailed critique see Tattersall & Schwartz, 1999).

The Kondo & Ishida study is also important from the perspective of documenting modern human variability, with regard to postcranial growth and development, allowing the calculation of confidence intervals and residuals. It must be noted that Vlček's (1973) conclusions regarding the robusticity of the Kiik-Koba infant were made on the basis of comparison with *one* modern human individual.

The postcranium: Pelvic dimensions

Majó & Tillier (this volume) focus on one of the most distinctive elements of the Neandertal postcranial skeleton: the elongation of the pubis. They note that Neandertal pubic bone morphology (i.e., a thin ventral margin and elongated superior pubic ramus) has been recognized as an important phylogenetic trait since the descriptions of Skhul and Tabun in the 1930s (McCown & Keith, 1939). However, they also note that the pubis is not well represented in the fossil record. Thus, it is vitally important to extract as much information out of available pubic bones as possible, and to place the pubis in the overall context of the innominate (hipbone).

Neandertals are represented in the Majó & Tillier study by La Ferrassie 8, an individual of about 2 years of age, and La Ferrassie 6, an individual of approximately 3 years of age. The development of Neandertal pubic bone morphology has previously been studied by Tompkins & Trinkaus (1987), although that study only included La Ferrassie 6, and did not include La Ferrassie 8. Majó & Tillier find that, in comparison to an ontogenetic sample of modern humans, La Ferrassie 8 does not have a distinctively long pubis, while La Ferrassie 6 does. However, while the pubis of La Ferrassie 6 is long, it does not demonstrate the ventral thinning that is characteristic of adult Neandertals. There are three potential explanations for this pattern: (1) that the Neandertal pubis became elongated at between 2 and 3 years of age; (2) that the aging for La Ferrassie 6 (which does not have associated dental remains) is inaccurate (and the specimen could be older than 3 years); or (3) that La Ferrassie 8 and La Ferrassie 6 show individual variation (in a similar manner to Dederiyeh 1 and Dederiyeh 2), and that pubic length is variably expressed in young individuals. At present, we cannot distinguish between these three potential explanations.

Considerations

It is important to note that both Kondo & Ishida and Majó & Tillier include the Upper Pleistocene hominids from Skhul and Qafzeh in their analyses. The Skhul–Qafzeh hominids are generally believed to be early anatomically modern *H. sapiens* on the basis of their large cranial capacity, high-vaulted crania, short face, and relatively gracile long bones (Vandermeersch, 1969). However, both Kondo & Ishida and Majó & Tillier document metric features (femur and tibia circumference and pubis length, respectively) that cannot differentiate these young early "anatomically modern" humans from similarly aged Neandertals, but do differentiate both samples from recent modern humans. This lack of differentiation has been noted before for other traits, including relative crown formation timing (Tompkins, 1996) and supraorbital torus dimensions (Wolpoff, 1996), which are seen to be similar in Neandertals and the Skhul–Qafzeh specimens, and quite different in recent modern humans. These traits and their distribution in these fossil samples could be interpreted as evidence for regional continuity (Wolpoff, 1996). They could also be evidence that young individuals have not yet developed all the postcranial differences that are commonly seen between adults from their respective populations. Finally, it is possible that these traits are primitive retentions, shared between Neandertals and the Skhul–Qafzeh sample, inherited from an earlier common ancestor. If so, they would not indicate a unique relationship between these two samples. This may well be the case for the elongation of the pubis, which has been reported for the earlier hominids from Atapuerca (Arsuaga *et al.*, 1999).

The skull: The mandible

Coqueugniot & Minugh-Purvis's contribution to this volume focuses on the mandible. A number of mandibular traits have been identified as being characteristic of Neandertals, including: a usual lack of mental eminence (chin) (Schwartz & Tattersall, 2000), the possession of a larger retromolar space, and a hook-shaped coronoid process with vertical height that exceeds that of the condylar process. In addition, Rak & Kimbel (1995) and Rak *et al.* (2002) have identified the form of the mandibular notch as a Neandertal autapomorph. However, Jabbour *et al.* (2002) recently undertook a study of the mandibular condyle region, and concluded that the trait is not in fact autapomorphic, and that the region "cannot be constructively compared without first considering subadult morphology on its own functional and developmental terms" (Jabbour *et al.*, 2002: 114).

Coqueugniot & Minugh-Purvis's careful study does exactly that: it examines the position of the mental foramen and whether it is expressed as a single or multiple opening in an integrated metric and non-metric evaluation of the mandible. The posterior positioning of the mental foramen has been suggested to be a Neandertal autapomorph by Condemi (1991) and Stringer *et al.* (1984). Minugh-Purvis (1988, 2000b) had previously established that the chin is not a reliable taxonomic trait in subadults, as it is obscured by developing tooth germs, making it difficult to evaluate in subadults, and because the modern human chin has a complex ontogenetic sequence of development. In this study Coqueugniot & Minugh-Purvis demonstrate that the anterior position and single opening of the mental foramen is an autapomorph for modern humans, differentiating modern *H. sapiens* from Neandertals and other earlier members of the genus *Homo* who share, with the Neandertals, the primitive pattern of a more posterior position combined with multiple openings. In particular, Coqueugniot & Minugh-Purvis demonstrate that the position and number of mental foramen openings is a function of differences in the pattern of growth in the anterior part of the mandible.

Considerations

The taxonomic importance of Coqueugniot & Minugh-Purvis's findings lies in the interpretation of the autapomorphic status of the position of the mental foramen. As noted above, several scholars (e.g., Condemi, 1991; Stringer *et al.*, 1984) have described the posterior position of the foramen as an autapomorphic trait for Neandertals. Others (e.g., Trinkaus, 1993) have argued that the posterior foramen position is a primitive trait, and so cannot be used to uniquely identify Neandertals. Coqueugniot & Minugh-Purvis's conclusions are in accord with the latter view, but add that the autapomorphic condition is not the *posterior* position of the foramen in Neandertals or other earlier members of the genus *Homo*, rather it is the *anterior* position of the foramen in the anatomically modern *H. sapiens* sample (recent modern humans as well as individuals from the Skhul–Qafzeh and Upper Paleolithic associated sites).

In this contribution, Coqueugniot & Minugh-Purvis do not take an explicit position on the question of how and when the shift took place from the primitive posterior position of the mental foramen to the derived anterior position

that characterizes anatomically modern *H. sapiens*. Their use of nomenclature (*H. sapiens sapiens* and archaic *H. sapiens*) implies a belief in conspecificity between Neandertals and modern humans, and Minugh-Purvis (1988,
2000a) has previously argued for an ontogenetic continuum among all later
(Upper Pleistocene) hominids. In the case of this particular trait, which reflects the ontogenetic growth of the critically important anterior portion of the
mandible, the shift from a currently unidentified ancestor who showed the primitive pattern of growth, must have occurred at least by Skhul–Qafzeh times, with
later Neandertals retaining the primitive condition alongside the derived early
anatomically modern individuals.

The skull: The craniofacial complex

The adult Neandertal craniofacial complex is quite distinctive, and various
scholars have identified traits that they argue are taxonomically unique, i.e.,
autapomorphic for Neandertals. These include the long and low, but volumin-
ous vault with its protruding occipital bun (chignon), projecting midface, a
suprainiac fossa, a juxtamastoid crest, asymmetric attachment of the m. semi-
spinalis capitis and an oval-shaped foramen magnum (Aiello & Dean, 1990;
Hublin, 1978, 1980; Rak *et al.*, 1994, 1996; Santa Luca, 1978).

Krovitz's contribution to this volume is an important addition to the emerging
field of three-dimensional morphometric analysis of patterns of growth and
development, particularly with regard to how such patterns can be used to
test taxonomic hypotheses. Krovitz uses Euclidean distance matrix analysis
(EDMA) to take the distances between all possible pairs of 24 anatomical
landmarks and to collapse them into usable summary matrices, one set used to
examine differences in craniofacial shape and the other to examine differences
in growth patterns across ontogenetic samples for Neandertals and a diverse
sample of modern humans (see also Krovitz, 2000). The key assumption is that
differences in growth and development in a multivariate representation of the
craniofacial complex that occur early in ontogeny are likely to have greater
powers of phylogenetic discrimination than those that occur late in ontogeny.

Krovitz finds that craniofacial shape differences between Neandertals and
modern humans are apparent by the age of 3 years at the latest (see also Krovitz,
2000; Ponce de León & Zollikofer, 2001; Williams *et al.*, this volume). These
differences lie principally in the face, particularly in the lower and midface,
and in the spatial positioning of the face relative to the neurocranium and
basicranium. Localized differences in growth patterns between Neandertals
and modern humans were present throughout ontogeny, but are particularly
accentuated after the age at which M2 erupts. Unfortunately, early anatomically

modern human specimens from Skhul, Qafzeh, and Upper Paleolithic sites were not included in the analysis, and need to be included in future studies.

The conclusion reached in the Krovitz study is that Neandertals and modern humans each have localized, unique aspects of growth and development in the craniofacial complex, and that these unique ontogenetic pathways reflect fundamental phylogenetic differences between these two samples. Hence, she opts for the use of the *nomen H. neanderthalensis* for the Neandertals, highlighting their difference from modern *H. sapiens* by using nomenclature that implies that the two samples were reproductively isolated and were following unique evolutionary trajectories.

Considerations

With reference to the theme of the book, when and/or how did the modern human pattern of growth and development of the craniofacial complex appear, Krovitz's contribution cannot actually tell us when this event happened, but her taxonomic conclusion clearly suggests that the modern human pattern did not evolve from a Neandertal ancestor. These results are similar to those of other recent geometric morphometric studies of craniofacial growth in Neandertals and modern humans (e.g., Ponce de León & Zollikofer, 2001). Krovitz (this volume) argues that elements of the Neandertal pattern of growth and development are possibly primitive for the genus *Homo*, but that others, particularly those of the lower and midface (as highlighted above) are probably uniquely derived for the Neandertals. Unfortunately, the poor fossil record for juveniles in the Lower and Middle Pleistocene make it extremely difficult to explore this matter directly.

Summary

In this section of this volume, we have had five contributions, ranging from modeling heterochronic shifts to detailed analysis of cranial and postcranial traits. With regard to the question at hand: When and how did the modern human pattern of growth and development appear?, we do not appear to have a consensus. Rather, the conclusions can be summarized as follows:

1. Neandertals make poor ancestors for modern humans under a model of global craniofacial neoteny.
2. At a young age, modern humans (including the Skhul–Qafzeh hominids) and Neandertals are not significantly different in limb length and circumference. Growth differences at later ages are implied as a mechanism for producing differing adult morphologies in Neandertals and modern humans.

3. Relative to modern humans, growth in the superior pubic ramus appears to be accelerated in Neandertals after the age of 2, a pattern shared with the Skhul–Qafzeh early modern humans.
4. Different patterns of growth in Neandertal versus modern human (including Skhul–Qafzeh) mandibles leads to differences in position and number of openings for the mental foramen. The anterior position, single-opening condition is autapomorphic for modern humans. The more posterior mental foramen position is not unique to Neandertals. Rather it is plesiomorphic and the Neandertals share this character state with earlier members of the genus *Homo*.
5. Differences in craniofacial shape and localized growth patterns, as expressed in a three-dimensional geometric morphometric analysis, suggest that Neandertals and modern humans are separate species (see also McBratney-Owen & Lieberman, this volume).

In sum, some aspects of the modern human pattern of growth and development are shared with Neandertals, some are not, and some aspects of the modern human growth and development pattern are shared with the Skhul–Qafzeh group, and some are not. If we go to the broader literature we can find even more combinations of features shared or not shared between Upper Pleistocene populations. What factors could account for these differences in results, and is there some way we can harmonize these disparate findings?

Interpretive differences

Methodological factors

There are a variety of methodological factors that could, in part, explain some of the different conclusions offered in this volume and in the literature. First, there is a wide variety of statistical methods used, particularly with regard to curve-fitting techniques. Growth is a complex and often non-linear process that is quite difficult to model. Examples of curve fitting techniques used in this volume (and in paleoanthropological literature) include lowess regression (Kondo & Ishida, this volume), fourth-order polynomial regression (Majó & Tillier, this volume), linear regression (least squares, major axis, and reduced major axis) (Majó & Tillier, this volume), and locally weighted piecewise regression (Williams *et al.*, this volume). Each of these curve-fitting techniques has different underlying assumptions and different capabilities in terms of calculating residuals, confidence intervals, and so on.

A second difference lies in the consideration of single traits or trait complexes. Kondo & Ishida (this volume), Majó & Tillier (this volume), and

Coqueugniot & Minugh-Purvis (this volume) examine one or a few traits in detail in order to better understand their specific ontogenetic pattern. However, Krovitz (this volume) and Williams *et al.* (this volume) use techniques that consider many variables at once. While the methodology of Krovitz can identify individual linear distances that differ between samples, it is typically difficult to assess the behavior of individual dimensions (traits) in traditional multivariate analyses. Nonetheless, multivariate analysis is a popular and reliable method for examining the relationship among suites of traits.

A third difference lies in the treatment of variables. Individual variables can be assessed individually against age (or dental proxy for age) to approximate a chronological growth curve (e.g., Kondo & Ishida, this volume; Majó & Tillier, this volume), or they can be compared to other variables, modeling how the variables change relative to each other through growth (Majó & Tillier, this volume). They can also be treated as raw values (e.g., Kondo & Ishida, this volume), or normalized to some estimate of an individual's body size (e.g., Krovitz, this volume) or to a mean expected adult outcome (e.g., Thompson & Nelson, 2000).

Finally, one of the most important methodological differences between studies lies in the choice of fossils included and the modern comparative samples used to provide the estimate of modern variability. As Kondo & Ishida (this volume) clearly illustrate, there is considerable variability among the few available fossil individuals. Thus, individual data points may not represent the central tendency of the population. It is also unfortunately true that the basic descriptions of several important fossils (such as Teshik Tash) are only presented in publications that are not widely available, or are printed in languages that are not widely understood. In addition, the modern comparative samples chosen are extremely important to the understanding of the relationship between the fossils and modern patterns of growth and development. Many of the modern samples are derived from agricultural or industrial societies, rather than hunter–gatherer societies (see Humphrey, this volume). Finally, many modern comparative samples are small (especially Vlček's (1973) sample of one!), which creates difficulties in calculating confidence intervals. The contributions to this volume all demonstrate a keen appreciation of these issues, and have sought to maximize their fossil samples and to amass large comparative samples from a variety of individual collections.

Conceptual factors

Genetics or behavior?
An important conceptual factor that can determine how one interprets patterns of variability in the fossil record is whether traits are seen to be genetically

determined (or phylogenetically ingrained), as we have assumed to this point, or whether environment and/or behavior can also shape morphology. This distinction is not trivial, as traits that are developmentally plastic can appear in different taxa without indicating any genetic relationship. Examination of these traits from an ontogenetic perspective can shed light on this issue from two perspectives: (1) traits that appear in very young individuals are less likely to represent adult behaviors and are more likely to be phylogenetically meaningful (although the maternal environment can certainly play a role in shaping morphology) (see Ackermann & Krovitz, 2002; Strand Viðarsdóttir & O'Higgins, this volume), and (2) traits that can be tracked throughout growth can potentially demonstrate when the behaviors came into play (assuming that we can clearly identify what behavior causes the trait).

In the case of the Neandertals, two trait complexes have frequently been discussed from the behavioral perspective: the facial–dental complex (e.g., Brace, 1995) and postcranial robusticity (e.g., Ruff *et al.*, 1994; Trinkaus, 1983a, 1983b). Brace (1995) argues that many of the distinctive features of the Neandertal and modern human face and dentition are the product of the use of "teeth as tools" (for biomechanical treatment of this issue, see Demes, 1987; Rak, 1986; Trinkaus, 1987). In this scenario, dental reduction and facial gracilization came about as a result of a technological shift, leading to an increase in the use of fire and more sophisticated tools. Increased postcranial robusticity in Neandertals (as shown on the basis of increased muscle markings and cross-sectional diaphyseal properties) has been suggested to represent increased activity levels in these hominids (Ruff *et al.*, 1994; Trinkaus, 1983a). Trinkaus (1983a) presented evidence that a large crest for the opponens pollicis muscle on Neandertal first metacarpals indicated the increased capability of exerting high levels of force during normal grips. He noted that this trait was present on four Neandertal subadults, including Kiik-Koba (ca. 5–6 months), La Ferrassie 6 (ca. 3 years), La Ferrassie 3 (ca. 10 years), and Zaskalnaya VI (ca. 10–12 years). The presence of this trait in individuals as young as 5–6 months old would seem to contradict the interpretation that this muscle crest is a dynamic response to behavioral patterns, as a 5–6-month-old infant lacks the necessary coordination and duration of activity to produce the osteological response to muscle tension. This trait could be seen as a "preadaptation," but the fact that it exists before the behavior that is supposed to elicit it, actually weakens the causal relationship between trait and behavior.

The behavioral basis for increased cross-sectional properties of the diaphysis in Neandertals and pre-recent *Homo* has been well documented in adults (e.g., Ruff *et al.*, 1993; Trinkaus, 1997), but the ontogeny of these properties has yet to be fully documented (but see Ruff *et al.*, 1994). Kondo & Ishida (this volume) show that the young Neandertal individuals from Dederiyeh varied considerably in their cross-sectional properties. Ruff *et al.* (1994)

documented elevated cortical thickness in two young Neandertals, La Ferrassie 6 (ca. 3 years) and Teshik Tash (ca. 8–10 years), and suggested that Neandertals might have had an "increase in mechanical loading throughout life" (Ruff *et al.*, 1994: 53). While one might well wonder if a 3-year-old child could reasonably be expected to be participating in the level of activity necessary to produce such an osteogenic response, it is clear that some aspects of a powerful musculature were present in at least one adolescent Neandertal (Le Moustier 1: Thompson & Nelson, in press).

Paradigm

As outlined earlier, there are two main perspectives regarding the origin of modern *H. sapiens*: Regional Continuity and Out of Africa/Replacement Hypothesis. These two models differ not only in terms of the particular question of modern human origins but also in terms of how they interpret variability and whether or not traits that appear early in ontogeny are relevant to questions of taxonomy and phylogeny. As they are currently elaborated, these models of the origin of our species are not merely scenarios. Instead, they are paradigms. A paradigm is a world view that influences what scientists think, what they study, how they measure things and how they interpret their data (Kuhn, 1970). There are few places in paleoanthropology where the effects of paradigm are as far-reaching as they are in the field of research into modern human origins (see Clark & Willermet, 1997; Lieberman, 1995; Willermet & Clark, 1995).

With respect to the papers in this volume, the central phylogenetic question is whether continuity can be documented from Neandertals to modern humans. Scholars who adhere to the Replacement Hypothesis would argue that such continuity cannot be demonstrated. However, we must recognize that any evaluation of Regional Continuity must also consider the fossil record outside of Europe. Wolpoff & Caspari (1997) have suggested that even if the Neandertals died without issue in Europe, the model of Regional Continuity would still not be falsified. Rather, it would only demonstrate that replacement was the mode of multiregional evolution in Europe, but that replacement would also need to be demonstrated elsewhere (e.g., south-east Asia: – see Wolpoff *et al.*, 1984) to refute the model. That cannot be done with the current fossil record (especially with respect to young individuals). However, Wolpoff & Caspari (1997) also claim that any proof that Neandertals did evolve into modern humans in Europe would be sufficient to disprove the Out of Africa model.

A key difference between these two paradigms of modern human origins is how they interpret variability. Under the Regional Continuity model, variation in post-1.5-million-year-old hominids is explicitly interpreted within an intraspecific context. Hence, everything after *H. habilis* is included in the species *H. sapiens* (Wolpoff, 1996). Researchers who follow the Out of Africa model tend to see variability within an interspecific context, typified by Tattersall's (1992) study of lemurs. In this study, Tattersall found that species diversity in lemurs, which can be demonstrated on the basis of pelage, behavior, and individual interfertility grounds, tends to be underestimated when looking at osteological evidence. This has led him to the conclusion that species diversity within the fossil hominid record is also underestimated. Hence, (almost) any morphological variability expressed between two fossil samples is possible evidence for species-level differences (Schwartz & Tattersall, 2000; Tattersall, 1986; although also see Conroy, 2002; Kimbel & Martin, 1993).

Paradigms and the origins of the modern human pattern of growth and development

Although we are unlikely to resolve the question of Regional Continuity versus Out of Africa in this volume, the distinction between these two models is important. This is because the paradigm chosen by individual contributors has an effect on their interpretations of their findings.

One way to identify the paradigm of individual researchers is to look carefully at the names they choose to use. Over the years, many names have been used to refer to the Neandertals, including: Neandertals/Neanderthals, *H. sapiens neanderthalensis, H. neanderthalensis,* and European late archaic *H. sapiens.* The choice of a particular name (in taxonomic terms, a *nomen)* is not trivial – in this case "a rose by any other name is *not* as sweet." The choice is important, as different names reflect different perspectives regarding who these hominids were, and how they relate to modern *H. sapiens.* If Neandertals and modern *H. sapiens* are conspecific then *H. sapiens neanderthalensis,* and European late archaic *H. sapiens* would be appropriate *nomina* for Neandertals. If Neandertals represent a separate species, following a unique evolutionary trajectory with genetic isolation from other groups of hominids, then *H. neanderthalensis* would be the appropriate *nomen* for Neandertals.

Among the papers in this volume, Majó & Tillier, Kondo & Ishida, and Williams *et al.* steer a moderate course by using the non-taxonomic terms Middle Paleolithic hominids or Neandertals. Coqueugniot & Minugh-Purvis demonstrate a conspecific perspective by using the terms *H. sapiens sapiens* and archaic *H. sapiens* to refer to modern humans and Neandertals (plus Rabat and Irhoud), respectively. Krovitz demonstrates her rejection of conspecificity by using the term *H. neanderthalensis.* Thus, without even exploring the content of the various papers, one can see that this section incorporates a variety of perspectives on how Neandertals relate to modern humans.

Since perspective (paradigm) is so important, it is probably appropriate to make our phylogenetic perspective explicit. We are all on record (e.g., Krovitz, 2000, this volume; Lieberman *et al.*, 2002; Nelson & Thompson, 2002;

Thompson & Nelson, 2000) as presenting analyses suggesting that patterns of growth and development in Neandertals were different from those demonstrated in modern humans. Krovitz (this volume; Lieberman *et al.*, 2002, McBratney-Owen & Lieberman, this volume) has been most explicit in terms of stating her taxonomic interpretation of these differences, but we are all in agreement that it is unlikely that modern humans evolved from a Neandertal or Neandertal-like ancestor.

With these perspectives in mind, how do the different contributors deal with variability? Kondo & Ishida (this volume) use statistical significance to assess how the young Neandertals compare to modern humans, but they highlight the differences between the two individuals from Dederiyeh. Majó & Tillier are very cautious in their interpretation of variability, emphasizing the fact that Skhul–Qafzeh shared aspects of pelvic morphology with the Neandertals. In the past, Tillier (e.g., 1989) has taken a stronger stance by recognizing the existence of shared traits between Skhul–Qafzeh and Neandertals, but arguing that these particular morphological traits do not support the Regional Continuity model. Coqueugniot & Minugh-Purvis's conclusions could be used to support the Replacement Hypothesis, but previous work by Minugh-Purvis (e.g., 1988, 2000a) has shown that she favors the Regional Continuity model. It is also worth considering Minugh-Purvis's recent analysis of Krapina 1 (Minugh-Purvis, 2000a). Krapina 1 has been described as a "morphologically transitional specimen" by Wolpoff (1980: 314) and others. While that conclusion would certainly aid the cause of continuity, Minugh-Purvis (2000a) objectively weighed the evidence and decided that this individual demonstrated a typical Neandertal morphology, and therefore could not be used as evidence to support Regional Continuity (for a similar story, see also Bräuer & Broeg, 1998). Finally, Krovitz clearly views the observable variability from an interspecific perspective, using her modern sample as the barometer of how much variability to expect within a species.

Discussion

Where does all this leave us in terms of the central questions being addressed in this volume? Can we answer the question: When and how did the modern human pattern of growth and development appear? The honest answer is "Not precisely." Can we decide between the Regional Continuity and the Out of Africa models for the origin of modern humans? The honest answer must be no. What we can do is define several important parameters of the debate, which should serve to shape research in this field for some years to come.

At the very least, it is clear that the modern human pattern of growth and development did not come together as a complete package until fairly recently (at least in geological terms). As the chapters in this volume show, components of the modern human growth pattern have accumulated over the course of human evolution, but it is probably only within the last 100 000 years that all the elements fell into place. The Neandertal pattern of growth and development was clearly different from that of modern humans, including the early appearance of many diagnostic Neandertal characters. However, at the same time, the Neandertals possessed some of the elements of the modern human pattern of growth and development, including the five ontogenetic phases outlined by Bogin (this volume; Nelson & Thompson, 1999, in press; Thompson *et al.*, this volume): the infancy, childhood, juvenile, adolescence, and adulthood stages. However, it is clear that the Neandertal adolescent stage was somewhat different than that of modern humans, with a greater proportion of total growth occurring during this critical time (Nelson & Thompson, in press).

An extremely important issue that arises from the conclusion that the modern human pattern of growth and development did not appear until very recently, and which is addressed by several of the studies included in this volume, is that the Skhul–Qafzeh individuals and some other early anatomically modern *H. sapiens* from Europe present traits that are found in both Neandertals and modern *H. sapiens*. Future research should focus on these fossil samples and on fossils from the Middle Pleistocene in order to examine whether or not these shared traits are indicative of evolutionary continuity between Neandertals and anatomically modern *H. sapiens*, or if these traits are primitive retentions (plesiomorphic characters), which cannot clarify the relationship among these groups (which appears to be the case with the elongated pubis; see above).

Many contributions to this volume have emphasized the importance of appreciating variability, both in the fossil record and among modern comparative samples. With regard to the fossil record, it is critical that researchers maximize the use of these rare fossils. This is something that will require new discoveries (such as Mazmaiskaya Cave: Golovanova *et al.*, 1999), increased access to original fossils, and better availability of rare publications (including translation of original descriptions). Two recent developments in this area are worthy of note: the recovery of the Le Moustier 2 infant (Maureille, 2002) gives us one more individual with a fairly complete skeleton from the early part of the ontogenetic sequence, and the forthcoming publication of a volume dedicated to the Le Moustier 1 adolescent (Ullrich, in press), including detailed descriptions of the cranial reconstruction (Thompson & Bilsborough, in press) and a postcranial description collected from original descriptions, casts and examination of the remaining original fossils (Thompson & Nelson, in press) will provide additional information for the latter part of the ontogenetic sequence. On the other hand, it must also be noted that the Starosele Child must be removed from discussions of Upper Pleistocene hominid evolution, as a reconsideration of its stratigraphic associations have led to the conclusion that it actually dates to late medieval times (Marks *et al.*, 1997).

It is also clear that we will need to broaden our horizons to examine subadult material outside of Europe and the Levant. A small sampling of this material includes the following sites. From South Africa, the site of Border Cave has yielded a 4–6-month-old infant that is associated with Middle Stone Age industries, but is described as early anatomically modern *H. sapiens* (Beaumont *et al.*, 1978). Also from South Africa, the Middle Pleistocene site of Hoedjiespunt has yielded dental remains of a subadult (Stynder *et al.*, 2001, 2002). Late Upper Pleistocene material has been described from the site of Beli Lena in Sri Lanka. Beli Lena No. 2 is a 10–11 year old subadult consisting of a cranium and mandible (Kennedy *et al.*, 1986). Additional subadult material from various time periods has been described from China (Etler, 1996; Xinzhi & Zhenbiao, 1985), Japan (Yamashita-cho 1: Trinkaus & Ruff, 1996) and from Australia (Talgai 1 and various individuals from Kow Swamp: Oakley *et al.*, 1975).

It is only by taking a broad comparative perspective, by undertaking both detailed morphological studies and more synthetic morphometric and modeling studies, by focusing on issues of derived versus primitive traits, and by comparing the well-studied European Upper Pleistocene fossil record to geologically earlier and more geographically diverse samples that we will finally be able to fully address the questions posed in this volume. However, the studies presented here, as well as work being done by other scholars, are laying the groundwork and pointing the way.

References

Ackermann, R. R., & Krovitz, G. E. (2002). Common patterns of facial ontogeny in the hominid lineage. *Anatomical Record*, **269**, 142–147.

Aiello, L., & Dean, C. (1990). *An Introduction to Human Evolutionary Anatomy.* London: Academic Press.

Akazawa, T., Muhesen, S., Dodo, Y., Kondo, O., Mizoguchi, Y., Abe, Y., Nishiaki, Y., Ohta, S., Oguchi, T., & Haydal, J. (1995). Neanderthal infant burial from the Dederiyeh Cave in Syria. *Paléorient*, **21**, 77–86.

Arsuaga, J. L., Lorenzo, C., Carretero, J.-M., Gracia, A., Martinez, I., Garcia, N., Bermúdez de Castro, J. M., & Carbonell, E. (1999). A complete human pelvis from the Middle Pleistocene of Spain. *Nature*, **399**, 255–258.

Beaumont, P. B., de Villiers, H., & Vogel, J. (1978). Modern man in sub-Saharan Africa prior to 49 000 years BP: A review and evaluation with particular reference to Border Cave. *South African Journal of Science*, **74**, 409–419.

Brace, C. L. (1995). *The Stages of Human Evolution,* 5th edn. Englewood Cliffs: Prentice-Hall.

Bräuer, G., & Broeg, H. (1998). On the degree of Neandertal–Modern continuity in the earliest Upper Palaeolithic crania from the Czech Republic: Evidence from non-metrical features. In *The Origins and Past of Modern Humans: Towards Reconciliation*, eds. K. Omoto & P. V. Tobias, pp. 106–125. Singapore: World Scientific Press.

Brothwell, D. R. (1975). Adaptive growth rate changes as a possible explanation for the distinctiveness of Neanderthalers. *Journal of Archaeological Science*, 2, 161–163.

Clark, G. A., & Willermet, C. M. (eds.) (1997). *Conceptual Issues in Modern Human Origins Research*. New York: Aldine de Gruyter.

Condemi, S. (1991). Some considerations concerning Neandertal features and the presence of Neandertals in the near east. *Rivista di Antropologia*, 69, 27–38.

Conroy, G. C. (2002). Speciosity in the early *Homo* lineage: Too many, too few, or just about right? *Journal of Human Evolution*, 43, 759–766.

de Beer, G. R. (1958). *Embryos and Ancestors*. Oxford: Clarendon Press.

Demes, B. (1987). Another look at an old face: Biomechanics of the Neandertal facial skeleton reconsidered. *Journal of Human Evolution*, 16, 297–303.

Dodo, Y., Kondo, O., Muhesen, S., & Akazawa, T. (1998). Anatomy of the Neandertal infant skeleton from Dederiyeh Cave, Syria. In *Neandertals and Modern Humans in Western Asia*, eds. T. Akazawa, K. Aoki, & O. Bar Yosef, pp. 323–338. New York: Plenum Press.

Duarte, C., Mauricio, J., Pettitt, P. B., Souto, P., Trinkaus, E., Van Der Plicht, H., & Zilhao, J. (1999). The Early Upper Paleolithic human skeleton from the Abrigo do Lagar Velho (Portugal) and modern human emergence in Iberia. *Proceedings of the National Academy of Sciences of the USA*, 96, 7604–7609.

Etler, D. (1996). The fossil evidence for human evolution in Asia. *Annual Review of Anthropology*, 25, 275–301.

Godfrey, L. R., & Sutherland, M. R. (1996). Paradox of peramorphic paedomorphosis: Heterochrony and human evolution. *American Journal of Physical Anthropology*, 99, 17–42.

Golovanova, L. V., Hoffecker, J. F., Kharitonov, V. M., & Romanova, G. P. (1999). Mezmaiskaya Cave: A Neanderthal occupation in the Northern Caucasus. *Current Anthropology*, 40, 77–86.

Gould, S. J. (1977). *Ontogeny and Phylogeny*. Cambridge: Harvard University Press.

Heim, J.-L. (1982). *Les Enfants Néandertaliens de La Ferrassie*, Fondation Singer Polignac. Paris: Masson.

Hublin, J-J. (1978). Quelques caractères apomorphes de crâne néandertalien et leur interprétation phylogénique. *Comptes Rendus de l'Académie des Sciences de Paris*, 287, 923–926.

Hublin, J.-J. (1980). A propos de restes inédits de gisement de La Quina (Charente): Un trait méconnu des Néandertaliens et des Prénéandertaliens. *L'Anthropologie*, 84, 81–88.

Jabbour, R. S., Richards, G. D., & Anderson, J. Y. (2002). Mandibular condyle traits in Neanderthals and other *Homo*: A comparative, correlative and ontogenetic study. *American Journal of Physical Anthropology*, 119, 144–155.

Kappelman, J. (1996). The evolution of body mass and relative brain size in fossil hominids. *Journal of Human Evolution*, 30, 243–276.

Kennedy, K. A. R., Disotell, T., Roertgen, W. J., Chiment, J., & Sherry, J. (1986). Biological anthropology of Upper Pleistocene hominids from Sri Lanka: Batadomba Lena and Beli Lena Caves. *Journal of the Archaeological Surveys Department, Sri Lanka,* **6**, 68–168.

Kimbel, W. H., & Martin, L. B. (eds.) (1993). *Species, Species Concepts, and Primate Evolution.* New York: Plenum Press.

King, W. (1994 (1864)). The reputed fossil man of the Neanderthal. In *Naming Our Ancestors: An Anthology of Hominid Taxonomy,* eds. W. E. Meikle & S. T. Parker, pp. 22–35. Prospect Heights: Waveland Press.

Krovitz, G. E. (2000). Three-dimensional comparisons of craniofacial morphology and growth patterns in Neandertals and modern humans. PhD dissertation, Johns Hopkins University, Baltimore.

Kuhn, T. (1970). *The Structure of Scientific Revolutions,* 2nd edn. Chicago: University of Chicago Press.

Lieberman, D. E. (1995). Testing hypotheses about recent human evolution from skulls. *Current Anthropology,* **36**, 159–197.

Lieberman, D. E., McBratney, B. M., & Krovitz, G. E. (2002). The evolution and development of cranial form in *Homo sapiens. Proceedings of the National Academy of Sciences of the USA,* **99**, 1134–1139.

Madre-Dupouy, M. (1992). *L'Enfant Néandertalien du Roc de Marsal: Etude Analytique et Comparative,* Cahiers de Paléoanthropologie. Paris: Editions du CNRS.

Marks, A. E., Demidenko, Yu. E., Monigal, K., Usik, V. I., Ferring, C. R., Burke, A., Rink, J., & McKinney, C. (1997). Starosele and the Starosele child: New excavations, new results. *Current Anthropology,* **38**, 112–213.

Maureille, B. (2002). A lost Neanderthal neonate found. *Nature,* **419**, 33–34.

McCown, T. D., & Keith, A. (1939). *The Stone Age of Mount Carmel,* vol. 2. Oxford: Clarendon Press.

Meikle, W. E, & Parker, S. T. (eds.) (1994). *Naming Our Ancestors: An Anthology of Hominid Taxonomy.* Prospect Heights: Waveland Press.

Minugh-Purvis, N. (1988). Patterns of craniofacial growth and development in Upper Pleistocene hominids. PhD dissertation, University of Pennsylvania, Philadelphia.

Minugh-Purvis, N. (2000a). Krapina 1: A juvenile Neandertal from the Early Late Pleistocene of Croatia. *American Journal of Physical Anthropology,* **111**, 393–424.

Minugh-Purvis, N. (2000b). Ontogeny and morphology of the child's mandible from Sipka – Moravia, Czech Republic. *Anthropologie,* **38**, 71–82.

Minugh-Purvis, N., & McNamara, K. J. (eds.) (2002). *Human Evolution through Developmental Change.* Baltimore: Johns Hopkins University Press.

Montagu, M. F. A. (1955). Time, morphology, and neoteny in the evolution of Man. *American Anthropologist,* **57**, 13–27.

Nelson, A. J., & Thompson, J. L. (1999). Growth and development in Neandertals and other fossil hominids: Implications for hominid phylogeny and the evolution of hominid ontogeny. In *Human Growth in the Past: Studies from Bones and Teeth,* eds. R. D. Hoppa & C. M. Fitzgerald, pp. 88–110. Cambridge: Cambridge University Press.

Nelson, A. J., & Thompson, J. L. (2002). Neandertal adolescent postcranial growth. In *Human Evolution through Developmental Change,* eds. N. Minugh-Purvis & K. McNamara, pp. 442–463. Baltimore: Johns Hopkins University Press.

Nelson, A. J., & Thompson, J. L. (in press). Le Moustier 1 and the interpretation of stages in Neandertal growth and development. In *The Neandertal Adolescent Le Moustier 1: New Aspects, New Results,* ed. H. Ullrich. Berlin: Staatliche Museen zu Berlin/Preussischer Kulturbesitz.

Nelson, A. J., Gauld, S. C., & Austin, J. K (1992). Models of fossil hominid body mass prediction using cranial and post-cranial bone thickness. Paper presented to the 3rd *International Congress on Human Paleontology,* Jerusalem.

Oakley, K. P., Campbell, B. G., & Molleson, T. I. (eds.) (1975). *Catalogue of Fossil Hominids.* London: Trustees of the British Museum (Natural History).

Oleksiak, K. (1986). The estimation of body weights for Neandertals and early anatomically modern humans. *American Journal of Physical Anthropology,* **69**, 248.

Ponce de León, M. S., & Zollikofer, C. P. (2001). Neanderthal cranial ontogeny and its implications for late hominid diversity. *Nature,* **412**, 534–538.

Rak, Y. (1986). The Neanderthal: A new look at an old face. *Journal of Human Evolution,* **15**, 151–164.

Rak, Y., & Kimbel, W. H. (1995). Diagnostic Neandertal characters in the Amud 7 infant. *American Journal of Physical Anthropology,* Suppl. **20**, 117–118.

Rak, Y., Kimbel, W. H., & Hovers, E. (1994). A Neandertal infant from Amud Cave, Israel. *Journal of Human Evolution,* **26**, 313–324.

Rak, Y., Kimbel, W. H., & Hovers, E. (1996). On Neandertal autapomorphies discernible in Neandertal infants: A response to Creed-Miles *et al. Journal of Human Evolution,* **30**, 115–158.

Rak, Y., Ginzburg, A., & Geffen, E. (2002). Does *Homo neanderthalensis* play a role in modern human ancestry? The mandibular evidence. *American Journal of Physical Anthropology,* **119**, 199–204.

Ruff, C. B., Trinkaus, E., Walker, A., & Larsen, C. S. (1993). Postcranial robusticity in *Homo.* I: Temporal trends and mechanical interpretation. *American Journal of Physical Anthropology,* **91**, 21–53.

Ruff, C. B., Walker, A., & Trinkaus, E. (1994). Postcranial robusticity in *Homo.* III: Ontogeny. *American Journal of Physical Anthropology,* **93**, 35–53.

Ruff, C. B., Trinkaus, E., & Holliday, T. W. (1997). Body mass and encephalization in Pleistocene *Homo. Nature,* **387**, 173–176.

Santa Luca, A. P. (1978). A re-examination of presumed Neandertal-like fossils. *Journal of Human Evolution,* **7**, 619–636.

Schwartz, J. H. & Tattersall, I. (2000). The human chin revisited: What is it and who has it? *Journal of Human Evolution,* **38**, 367–409.

Shea, B. T. (1989) Heterochrony in human evolution: The case for neoteny reconsidered. *Yearbook of Physical Anthropology,* **32**, 69–101.

Stringer, C. B. (1994). Out of Africa: A personal history. In *Origins of Anatomically Modern Humans,* eds. M. H. Nitecki & D. V. Nitecki, pp. 149–172. New York: Plenum Press.

Stringer, C. B., & Andrews, P. (1988). Genetic and fossil evidence for the origin of modern humans. *Science,* **239**, 1263–1268.

Stringer, C. B., & Gamble, C. (1993). *In Search of the Neanderthals.* New York: Thames & Hudson.

Stringer, C. B., Hublin, J.-J., & Vandermeersch, B. (1984). The origin of anatomically modern humans in Western Europe. In *The Origins of Modern Humans*, eds. F. H. Smith & F. Spencer, pp. 51–135. New York: Alan R. Liss.

Stynder, D., Moggi-Cecchi, J., Berger, L., & Parkington, J. (2001). Human mandibular incisors from the late Middle Pleistocene locality of Hoedjiespunt 1, South Africa. *Journal of Human Evolution*, **41**, 369–383.

Stynder, D., Moggi-Cecchi, J., Berger, L., & Parkington, J. (2002). Human incisors and molars from the Late Middle Pleistocene locality of Hoedjiespunt 1, South Africa. *American Journal of Physical Anthropology*, Suppl. **34**, 151.

Tattersall, I. (1986). Species recognition in human paleontology. *Journal of Human Evolution*, **15**, 165–175.

Tattersall, I. (1992). Species concepts and species identification in human evolution. *Journal of Human Evolution*, **22**, 341–349.

Tattersall, I. (1999). *The Last Neanderthal: The Rise, Success, and Mysterious Extinction of Our Closest Relatives*, revised edn. New York: Nevraumont.

Tattersall, I., & Schwartz, J. H. (1999). Hominids and hybrids: The place of Neanderthals in human evolution. *Proceedings of the National Academy of Sciences of the USA*, **96**, 7117–7119.

Thompson, J. L., & Bilsborough, A. (in press). The skull of Le Moustier 1. In *The Neandertal Adolescent Le Moustier 1: New Aspects, New Results*, ed. H. Ullrich. Berlin: Staatliche Museen zu Berlin/Preussischer Kulturbesitz.

Thompson, J. L., & Nelson, A. J. (2000). The place of Neandertals in the evolution of hominid patterns of growth and development. *Journal of Human Evolution*, **38**, 475–495.

Thompson, J. L., & Nelson, A. J. (in press). The postcranial skeleton of Le Moustier 1. In *The Neandertal Adolescent Le Moustier 1: New Aspects, New Results,* ed. H. Ullrich. Berlin: Staatliche Museen zu Berlin/Preussischer Kulturbesitz.

Tillier, A.-m. (1989). The evolution of modern humans: Evidence from young Mousterien individuals. In *The Human Revolution*, eds. P. Mellars & C. Stringer, pp. 288–297. Princeton: Princeton University Press.

Tompkins, R. L. (1996). Relative dental development of Upper Pleistocene hominids compared to human population variation. *American Journal of Physical Anthropology*, **99**, 103–118.

Tompkins, R. L., & Trinkaus, E. (1987). La Ferrassie 6 and the development of Neandertal pubic morphology. *American Journal of Physical Anthropology*, **73**, 233–239.

Trinkaus, E. (1983a). Neandertal postcrania and the adaptive shift to modern humans. In *The Mousterian Legacy*, British Archaeological Reports International Series no. 164, ed. E. Trinkaus, pp. 165–200. Oxford: British Archaeological Reports.

Trinkaus, E. (1983b). *The Shanidar Neandertals*. New York: Academic Press.

Trinkaus, E. (1987). The Neandertal face: Evolutionary and functional perspectives on a recent hominid face. *Journal of Human Evolution*, **16**, 429–443.

Trinkaus, E. (1993). Variability in the position of the mandibular mental foramen and the identification of Neandertal apomorphies. *Rivista di Antropologia*, **71**, 259–274.

Trinkaus, E. (1997). Appendicular robusticity and the paleobiology of modern human emergence. *Proceedings of the National Academy of Sciences of the USA*, **94**, 13367–13373.

Trinkaus, E., & Ruff, C. B. (1996). Early modern human remains from eastern Asia: The Yamashita-cho 1 immature postcrania. *Journal of Human Evolution*, **30**, 299–314.

Ullrich, H. (ed.) (in press). *The Neandertal Adolescent Le Moustier 1: New Aspects, New Results*. Berlin: Staatliche Museen zu Berlin/Preussischer Kulturbesitz.

Vandermeersch, B. (1969). Les nouveaux squelettes moustériens découverts à Qafzeh (Israel) et leur signification. *Comptes Rendus de l'Académiè des Sciences de Paris*, **268**, 2562–2565.

Vlček, E. (1973). Postcranial skeleton of a Neanderthal child from Kiik-Koba, U.S.S.R. *Journal of Human Evolution*, **2**, 537–544.

Willermet, C. M., & Clark, G. A. (1995). Paradigm crisis in modern human origins research. *Journal of Human Evolution*, **29**, 487–490.

Williams, F. L., Godfrey, L. R., & Sutherland, M. R. (2002). Heterochrony and the evolution of Neandertal and modern human craniofacial form. In *Human Evolution through Developmental Change*, eds. N. Minugh-Purvis & K. McNamara, pp. 405–441. Baltimore: Johns Hopkins University Press.

Wolpoff, M. H. (1980). *Paleoanthropology*. New York: Alfred A. Knopf.

Wolpoff, M. H. (1996). *Human Evolution*, 1996/7 edn. New York: McGraw-Hill.

Wolpoff, M. H., & Caspari, R. (1997). *Race and Human Evolution*. New York: Simon & Schuster.

Wolpoff, M. H., Xin Zhi, W., & Thorne, A. G. (1984). Modern *Homo sapiens*: A general theory of hominid evolution involving the fossil evidence from East Asia. In *The Origins of Modern Humans: A World Survey of the Fossil Evidence*, eds. F. Smith & F. Spencer, pp. 411–483. New York: Alan R. Liss.

Xinzhi, W., & Zhenbiao, Z. (1985). *Homo sapiens* remains from Late Palaeolithic and Neolithic China. In *Palaeoanthropology and Palaeolithic Archaeology in the People's Republic of China*, eds. W. Rukang & J. W. Olsen, pp. 107–133. Orlando: Academic Press.

18 Conclusions: Putting it all together

A. J. NELSON
University of Western Ontario

J. L. THOMPSON
University of Nevada, Las Vegas

G. E. KROVITZ
Pennsylvania State University

Introduction

The primary goal of this volume has been to address the question: "When and how did the modern human pattern of growth and development appear?" This is a similar question to: "When did modern humans appear?" – but the emphasis on growth and development stresses that the adult morph, which is the usual focus of analysis, is only the end product of a long and complicated ontogenetic sequence. Furthermore, for much of the time that the genus *Homo* has been in existence, almost half of any individual's lifetime was spent growing up (see Krovitz *et al.*, Introduction, this volume). Thus, a growth and development perspective leads to a much broader inquiry than one simply focused on adult individuals. Questions such as "When do distinctive traits appear?" and "How do trait complexes work together throughout growth to produce the final outcome?" become relevant, and the whole of an individual's life history becomes the research focus.

In this volume, the parts were designed to (1) provide an understanding of what the modern human pattern of growth and development is, and how it compares to our primate relatives, (2) examine the Lower and Middle Pleistocene fossil evidence for the genus *Homo*, and (3) examine the fossil evidence from the Upper Pleistocene. The summary chapters in this volume (Thompson *et al.*, Krovitz *et al.*, and Nelson *et al.*) provide discussions on the modern context and evolutionary origins of modern human developmental patterns. Early research in this field suggested that modern human developmental patterns and life-history characteristics had great antiquity (Mann, 1968, 1975). However, the research presented here suggests that the full program of modern human

Patterns of Growth and Development in the Genus Homo, ed. J. L. Thompson, G. E. Krovitz and A. J. Nelson. Published by Cambridge University Press. © Cambridge University Press 2003.

growth and development is a very recent phenomenon. Australopithecines had a unique mosaic of developmental traits that sets them apart from both modern apes and modern humans (Kuykendall, this volume). Thus, it is to the genus *Homo* that we must look for an answer to our central question. However, as we see in the papers presented here, the complete growth and development program does not appear to have been fully expressed in Neandertals, nor even in early anatomically modern *H. sapiens* (i.e., specimens from Skhul and Qafzeh). This conclusion might seem at odds with other conclusions presented in this volume and elsewhere in the literature. For example, Bermúdez de Castro *et al.* (this volume) suggest that modern patterns of dental development were present in *H. antecessor* and *H. heidelbergensis*, and Bogin (this volume) suggests that the childhood stage may have appeared as early as the beginning of the Pleistocene. The answer to the paradox is "All of the above." The modern human pattern of growth and development has gradually evolved in a mosaic manner over the course of several million years, becoming increasingly different from that of our non-human ancestors. Components of the modern pattern do indeed appear in earlier hominids, and over time the elements of the modern human pattern gradually fell into place. However, the complete modern human pattern probably did not come together until some time in the last 100 000 years. Thus, claims that a particular trait or ontogenetic trajectory is shared between an earlier hominid and modern *H. sapiens* must be tempered with the qualification that that trait or ontogenetic trajectory only reflects one aspect of the total modern human pattern of growth and development.

Finally, it should be emphasized that the modern human pattern of growth and development is not a static entity. The first part of this volume contains a series of papers that are designed to document the modern pattern of growth and development, and to emphasize the importance of understanding normal ontogenetic variation in the modern human skeleton. Several papers that focus on fossil remains also touch on the same theme. Thus, we must be very careful to avoid the trap of typological thinking when we create an overall picture of human growth and development. However, with that caveat in mind, there are several features of the human ontogenetic program that clearly differentiate humans from non-human primates and from earlier hominids.

What is the modern human pattern of growth and development and when did it appear?

Growth in brain size/body size

As used here, the term *growth* refers to an increase in total size, body proportions, and body segment size (Thompson *et al.*, this volume). From the perspective of hominid evolution, perhaps the most important parameters from the growth and size perspective are body size and brain size. This subject was not dealt with directly in this volume, so it will be touched on here. It is quite difficult to define these parameters precisely for modern humans as a whole, as both vary considerably across all modern human populations. One study that has considered this topic is Pakkenberg & Voigt (1964), who presented sample means for modern Danes of 60 kg and 1350 g for body weight and brain weight respectively. These figures are comparable to the "living worldwide" figures presented by Ruff *et al.* (1997). Kappelman (1996) presented figures compiled from a variety of sources suggesting mean values of 71.9 kg for males and 57.2 kg for females for body weight and 1424.5 g and 1285.2 g respectively for brain size. While these individual figures for brain size and body size may vary, it is the relationship between them, generally expressed as an encephalization quotient (see Jerrison, 1973; Martin, 1983) that distinguishes modern humans from earlier hominids and other primates.

Compared to modern humans, the australopithecines and the habilines were all fairly small bodied (McHenry, 1992). However, with the appearance of *H. erectus* we see a hominid with a body size (ca. 58 kg) and body proportions that are similar to modern humans, but with a smaller brain size (ca. 856 g) (Ruff *et al.*, 1997). By the time of the Neandertals, both brain size and body size (ca. 1442 g and 76 kg, respectively) exceeded these measures in modern humans (Ruff *et al.*, 1997). While there is some debate regarding the estimates used in these studies (see Nelson, 1995) and the actual body weights used vary widely, there does seem to be a consensus that the final transition to modern humans included a reduction in both brain size and body size from a Middle Pleistocene ancestor (the identity of which depends on your perspective on the debate on the origin of modern humans) (Henneberg, 1998; Kappelman,1996; Mai *et al.*, 1992; Ruff *et al.*, 1997). However, there are few studies that take an ontogenetic perspective on brain/body size relationships (Pakkenberg & Voigt, 1964, do include some ontogenetic data), and of the studies considered to this point, only the Mai *et al.* (1992) study has taken an ontogenetic perspective on brain/body evolution. Mai *et al.* (1992) argued that a significant change in encephalization quotient could be brought about by a reduction in body size while maintaining brain size, since the major proportion of somatic growth takes place once brain growth is complete. Thus, the final push to the highly encephalized modern human condition may have involved a reduction in body size by halting somatic growth after the completion of brain growth rather than an increase in brain size (see also Henneberg, 1998; Kappelman, 1996). This dynamic view of brain size and body size evolution is very different from the positive allometry model of Pilbeam & Gould (1974), and has important implications for heterochronic models of human evolution (Williams *et al.*, this volume) and for the reconstruction of life-history parameters (Antón & Leigh, this volume:

Bogin, this volume). Suffice to say, the present relationship between brain and body growth was not established until sometime within the last 35 000 years (Henneberg, 1998; Kappelman, 1996; Mai *et al.*, 1992; Ruff *et al.*, 1997). The development of the modern brain/body size relationship is an area that definitely deserves further research, particularly with an ontogenetic perspective.

Development

Bogin (2001, this volume) has defined five major stages in the development of modern humans: *infancy, childhood, juvenile (puberty), adolescence,* and *adulthood* (see also Thompson *et al.*, this volume). An important component of infancy in modern humans is the secondary altriciality that allows a helpless infant to continue rapid brain growth outside of the womb. As discussed above, it is a large brain relative to body size that is the hallmark of modern humans. Secondary altriciality is crucial in the trade-off between infant growth (allowing fetal growth rates to continue after birth) and the mother's capacity to bear and successfully deliver a neonate (see Thompson *et al.*, this volume). The childhood and adolescent stages are proposed to distinguish life-history patterns in modern humans from non-human primates. These stages extend the total growth period, allowing for, among other things, increased opportunities for social learning.

Kuykendall's study (this volume) documents the fascinating and unique mosaic of human-like and ape-like developmental patterns demonstrated by the australopithecines. However, on balance, the australopithecines clearly do not follow a modern human pattern of growth and development. Rather they share many primitive ontogenetic elements with modern apes. Thus, they had taken on the "modern" characteristic of upright bipedalism, but it is extremely unlikely that they possessed childhood or adolescent growth stages or any sort of secondary altriciality (Bogin, this volume; Bogin & Smith, 1996; Smith & Tompkins, 1995).

It is not clear when secondary altriciality developed in our lineage. In 1993, Walker (1993) and Walker & Ruff (1993) suggested that the form of the *H. erectus* pelvis of KNM-WT 15000 indicated that secondary altriciality existed at least by 1.8 million years ago. However, there are difficulties in any attempt to draw conclusions about the obstetrical pelvis on the basis of an adolescent male and the presence of this stage in *H. erectus* has yet to be corroborated (see A. Walker pers. comm. in Krovitz *et al.*, this volume). Indeed, Ruff (1995) later suggested that early *H. erectus* demonstrated non-rotational birth in a manner similar to the australopithecines.

The childhood stage is defined as an extended period of dependence, when somatic growth continues slowly. Brain growth ceases by the end of this stage (Bogin, this volume). Mann *et al.* (1996) argued that this stage was present in

Neandertals on the basis of stasis between the eruption of the lateral incisors and second molars, while Bogin (1997, 1999; Bogin & Smith, 1996) has used brain size to suggest that *H. habilis* might have had a childhood stage.

The adolescent stage with its associated growth spurt may have been present in *H. erectus*, but the evidence is equivocal (see discussion in Krovitz *et al.*, this volume). As noted above, *H. erectus* overall body size was comparable to that of modern humans – a profound change from the australopithecines and habilines. Smith (1993) has argued that the *H. erectus* subadult KNM-WT 15000 does not demonstrate a growth spurt, as he had completed a large proportion of his expected adult outcome in body size at an early age (see also Thompson & Nelson, 2000). Furthermore, Dean *et al.* (2001) suggested that KNM-WT 15000's dental development was more ape-like than modern human-like. However, Antón (2002) and Antón & Leigh (this volume) argue that he still had a lot of facial growth to complete, and caution against dismissing the possibility of a growth spurt. Perhaps *H. erectus* had an adolescent phase that was somewhat different from ours, as KNM-WT 15000 had already completed much of his linear growth, but had yet to experience a growth spurt in facial dimensions.

Bermúdez de Castro *et al.* (this volume) have proposed that *H. antecessor* and *H. heidelbergensis* had an adolescent stage on the basis of patterns of enamel formation, that also indicate a prolonged pattern of development. The adolescent stage in Neandertals was somewhat different from that of modern humans (Nelson & Thompson, in press), as Neandertals experienced a greater proportion of their total postcranial growth during this stage. The existence and nature of a Neandertal growth spurt is as yet unclear (Nelson & Thompson, in press; Thompson & Nelson, 2000), but it was during this time that Neandertal limb robusticity became distinct (Nelson & Thompson, 2002), and when the adult degree of craniofacial shape was established (Krovitz, 2000, this volume).

Thus, while the childhood stage may have first appeared in *H. habilis* and the adolescent stage in *H. erectus*, there were important differences between the expressions of these stages in earlier hominids and their expression in modern humans. Furthermore, while the presence of these stages implies similarity to modern humans, these hominids still retained many primitive aspects of their growth and development program (see Krovitz *et al.*, this volume, for further discussion). In other words, the presence of these stages does not, in and of itself, define the modern human pattern of growth and development.

The origin of modern humans

As discussed in the previous chapter (Nelson *et al.* this volume), we cannot make a definitive argument here for either the Out of Africa (Replacement) nor the Regional Continuity (Multiregional Evolution) models of the origin of modern humans. However, many traits that are characteristic of adult Neandertals appear very early in ontogeny, suggesting that they have a large genetic component and are of phylogenetic importance. The paper by McBratney-Owen and Lieberman (this volume) contends that the modern human pattern of facial retraction is a unique derived feature. Furthermore, it is clear that there are distinct differences in the ontogenetic trajectories of Neandertals and modern humans in the craniofacial complex (Krovitz, 2000, this volume; Ponce de León & Zollikofer, 2001; Thompson & Nelson, 2001; Williams *et al.*, 2002, this volume) and in postcranial development after an early age (Coqueugniot & Minugh-Purvis, this volume; Majó & Tillier, this volume; Nelson & Thompson, 2002; Thompson & Nelson, 2000).

This evidence, interpreted within the context of the Out of Africa paradigm (see Nelson *et al.*, this volume), is suggestive of species-level differences between Neandertals and modern humans (Krovitz, this volume). This would argue that the two populations were on separate evolutionary trajectories, and that there was no evolutionary continuity between Neandertals and modern humans in Europe. However, the traits that are shared between Neandertals, the Skhul–Qafzeh hominids, and individuals from the Early Upper Paleolithic of Europe (Kondo & Ishida, this volume; Majó & Tillier, this volume) do require careful scrutiny. Once the functional and ontogenetic basis of these traits is better understood and documented (e.g., McBratney-Owen & Lieberman, this volume), these traits might turn out to be primitive retentions (thus still supporting separate specific status for Neandertals), or evidence of biological continuity, or behaviorally plastic traits that cannot give phylogenetic information. The elaboration of sophisticated and testable models of heterochronic change (e.g., Williams *et al.*, this volume) that incorporate this material will also help to clarify this question.

This debate will only be solved with a closer examination of the Skhul–Qafzeh, Early Upper Paleolithic, *H. antecessor*, and *H. heidelbergensis* fossil material with these issues in mind. In addition, material from outside of Europe and the Levant needs to be incorporated into these analyses. Coqueugniot & Minugh Purvis (this volume; see also Minugh-Purvis, 1988) have taken an important step in this direction by including the North African specimens of Rabat and Irhoud in their analysis.

Variability

The studies in this volume highlight the importance of documenting variability in both the fossil groups and in the modern comparative samples. As clearly demonstrated by several papers in the first section of this volume, there is a degree of variability in terms of what constitutes the "modern human" pattern of growth and development. Recognizing this variability is crucial to interpreting whether or not individual fossils conform to the modern human pattern. Liversidge's (this volume) study of tooth formation is extremely important in this regard, as dental formation provides the age proxy used in most studies of growth and development. It is clear that dental formation times differ between apes and humans (Kuykendall, this volume), but how, and by how much modern human populations differ has not been systematically explored before. Thus, the conclusion that there is some population variability in dental eruption patterns that is greater than that seen in crown and root formation is a crucial piece of information (Liversidge, this volume). Strand Viðarsdóttir & O'Higgins (this volume) continue with the theme of variability, focusing on interpopulation variability in facial growth within samples of modern humans. These authors document that interpopulation variability in facial ontogeny within modern humans may sometimes be as great as interspecies variability in non-human primates. This has implications not only in terms of characterizing modern humans, but also in terms of assessing confidence intervals around ontogenetic trajectories.

Humphrey's (this volume) study rounds out the coverage of ontogenetic variation in modern humans. Her study of femoral growth in a variety of archaeological populations, in comparison with a modern clinical reference sample, provides an important cautionary tale for the interpretation of longitudinal body growth, something often studied using femoral length as a proxy for stature in both modern and fossil contexts. The conclusion she presents is that archaeological populations are universally retarded in femoral growth relative to age (dental age) in comparison to the modern clinical standard, especially at early ages. Thus, modern clinical reference samples must be used with great caution in modern or fossil comparative studies, as they may not represent the pattern of growth and development expressed by the majority of non-industrial human societies, let alone fossil hominid samples.

Conclusions

The goal of this volume was to explore a fairly specific question: "When and how did the modern human pattern of growth and development appear?" In addressing this question, the contributors to this volume have demonstrated how far this field has come in recent years. Thirty years ago, a single modern infant was sufficient for a comparative sample, but now it is clear that researchers must include a variety of collections in their comparative sample, and that they must embrace both central tendencies as well as variability in both the modern and the fossil record. This represents an important move away from typological analyses.

These papers also clearly illustrate the important contribution that the ontogenetic perspective can bring to the modern study of paleoanthropology. Growth trajectories themselves can be treated as complex characters, and individual traits can be understood in a much more dynamic manner than is possible from the study of adult remains alone (see, for example, McBratney-Owen & Lieberman, this volume). There is no question that there are many difficulties inherent in the analysis of fossil material, but these analyses, which range from detail-oriented studies to sophisticated modeling experiments, are just beginning to tap the potential of the juvenile material that is available. A concerted effort to get data into the public domain, to maximize inclusivity of study samples, and to analyze the data objectively and present the results widely will go a long way toward the continued advancement of this field. As demonstrated by the various contributions to this volume, this work sets the stage for future analyses dealing with patterns of growth and development in the genus *Homo*.

References

Antón, S. C. (2002). Cranial growth in *Homo erectus*. In *Human Evolution through Developmental Change*, eds. N. Minugh-Purvis & K. J. McNamara, pp. 349–380. Baltimore: Johns Hopkins University Press.

Bogin, B. (1997). Evolutionary hypotheses for human childhood. *Yearbook of Physical Anthropology,* **40**, 63–89.

Bogin, B. (1999). *Patterns of Human Growth*, 2nd edn. Cambridge: Cambridge University Press.

Bogin, B. (2001). *The Growth of Humanity*. New York: Wiley-Liss.

Bogin, B., & Smith, B. H. (1996). Evolution of the human life cycle. *American Journal of Human Biology*, **8**, 703–716.

Dean, C., Leakey, M. G., Reid, D., Schrenk, F., Schwartz, G. T., Stringer, C., & Walker, A. (2001). Growth processes in teeth distinguish modern humans from *Homo erectus* and earlier hominins. *Nature*, **414**, 628–631.

Henneberg, M. (1998). Evolution of the human brain: Is bigger better? *Clinical and Experimental Pharmacology and Physiology*, **25**, 745–749.

Jerrison, H. (1973). *Evolution of the Brain and Intelligence*. New York: Academic Press.

Kappelman, J. (1996). The evolution of body mass and relative brain size in fossil hominids. *Journal of Human Evolution*, **30**, 243–276.

Krovitz, G. E. (2000). Three-dimensional comparisons of craniofacial morphology and growth patterns in Neandertals and modern humans. PhD dissertation, Johns Hopkins University, Baltimore.

Mai, L. L., Gauld, S. C., Nelson, A. J., & Austin, J. K. (1992). Evolutionary context of hominid body mass prediction models. Paper presented to the 3rd *International Congress on Human Paleontology*, Jerusalem.

Mann, A. (1968). The paleodemography of *Australopithecus*. PhD dissertation, University of California, Berkeley.

Mann, A. (1975). *Paleodemographic Aspects of the South African Australopithecines*, University of Pennsylvania Publications in Anthropology. Philadelphia: University of Pennsylvania.

Mann, A., Lampl, M., & Monge, J. M. (1996). The evolution of childhood: Dental evidence for the appearance of human maturation patterns. *American Journal of Physical Anthropology*, Suppl. **21**, 156.

Martin, R. D. (1983). *Human Brain Evolution in an Ecological Context*, 52nd James Arthur Lecture on the Evolution of the Human Brain. New York: American Museum of Natural History.

McHenry, H. M. (1992). How big were early hominids? *Evolutionary Anthropology* **1**, 15–19.

Minugh-Purvis, N. (1988). Patterns of craniofacial growth and development in Upper Pleistocene hominids. PhD dissertation, University of Pennsylvania, Philadelphia.

Nelson, A. J. (1995). Cortical thickness in the primate and hominid postcranium: Taxonomy and allometry. PhD dissertation, University of California, Los Angeles.

Nelson, A. J., & Thompson, J. L. (2002). Neandertal adolescent postcranial growth. In *Human Evolution through Developmental Change*, eds. N. Minugh-Purvis & K. McNamara, pp. 442–463. Baltimore: Johns Hopkins University Press.

Nelson, A. J., & Thompson, J. L. (in press). Le Moustier 1 and the interpretation of stages in Neandertal growth and development. In *The Neandertal Adolescent Le Moustier 1: New Aspects, New Results*, ed. H. Ullrich. Berlin: Staatliche Museen zu Berlin/Preussischer Kulturbesitz.

Pakkenberg, H., & Voigt, J. (1964). Brain weight of the Danes. *Acta Anatomica*, **56**, 297–307.

Pilbeam, D., & Gould, S. J. (1974). Size and scaling in human evolution. *Science*, **186**, 892–901.

Ponce de León, M. S., & Zollikofer, C. P. (2001). Neanderthal cranial ontogeny and its implications for late hominid diversity. *Nature*, **412**, 534–538.

Ruff, C. B. (1995). Biomechanics of the hip and birth in Early *Homo*. *American Journal of Physical Anthropology*, **98**, 527–574.

Ruff, C. B., Trinkaus, E., & Holliday, T. W. (1997). Body mass and encephalization in Pleistocene *Homo*. *Nature*, **387**, 173–176.

Smith, B. H. (1993). The physiological age of KNM-WT 15000. In *The Nariokotome* Homo erectus *Skeleton*, eds. A. Walker & R. Leakey, pp. 196–220. Cambridge: Harvard University Press.

Smith, B. H., & Tompkins, R. L. (1995). Toward a life history of the Hominidae. *Annual Review of Anthropology*, **24**, 257–279.

Thompson, J. L., & Nelson, A. J. (2000). The place of Neandertals in the evolution of hominid patterns of growth and development. *Journal Human Evolution*, **38**, 475–495.

Thompson, J. L., & Nelson, A. J. (2001). Relative postcranial and cranial growth in Neandertals and modern humans. *American Journal of Physical Anthropology*, Suppl. **32**, 149.

Walker, A. (1993). Perspectives on the Nariokotome discovery. In *The Nariokotome Homo erectus Skeleton*, eds. A. Walker & R. Leakey, pp. 411–430. Cambridge: Harvard University Press.

Walker, A., & Ruff, C. B. (1993). The reconstruction of the pelvis. In *The Nariokotome Homo erectus Skeleton*, eds. A. Walker & R. Leakey, pp. 221–233. Cambridge: Harvard University Press.

Williams, F. L., Godfrey, L. R., & Sutherland, M. R. (2002). Heterochrony and the evolution of Neandertal and modern human craniofacial form. In *Human Evolution through Developmental Change*, eds. N. Minugh-Purvis & K. J. McNamara, pp. 405–441. Baltimore: Johns Hopkins University Press.

Index